VOLUME ONE HUNDRED AND THREE

Current Topics in DEVELOPMENTAL BIOLOGY

Animal Metamorphosis

CURRENT TOPICS IN DEVELOPMENTAL BIOLOGY

"A meeting-ground for critical review and discussion of developmental processes"
A.A. Moscona and Alberto Monroy (Volume 1, 1966)

VOLUME ONE HUNDRED AND THREE

CURRENT TOPICS IN DEVELOPMENTAL BIOLOGY

Animal Metamorphosis

Edited by

YUN-BO SHI

Section on Molecular Morphogenesis
Laboratory of Gene Regulation and Development
Program in Cellular Regulation and Metabolism
Eunice Kennedy Shriver National Institute Child
Health and Human Development (NICHD)
National Institutes of Health (NIH)
Bethesda, MD, USA

ELSEVIER

AMSTERDAM • BOSTON • HEIDELBERG • LONDON
NEW YORK • OXFORD • PARIS • SAN DIEGO
SAN FRANCISCO • SINGAPORE • SYDNEY • TOKYO
Academic Press is an imprint of Elsevier

Academic Press is an imprint of Elsevier
525 B Street, Suite 1900, San Diego, CA 92101-4495, USA
225 Wyman Street, Waltham, MA 02451, USA
The Boulevard, Langford Lane, Kidlington, Oxford, OX51GB, UK
32, Jamestown Road, London NW1 7BY, UK
Radarweg 29, PO Box 211, 1000 AE Amsterdam, The Netherlands

First edition 2013

ISBN: 978-0-12-385979-2
ISSN: 0070-2153

Printed and bound by CPI Group (UK) Ltd, Croydon, CR0 4YY

Transferred to digital print 2013

13 14 15 16 11 10 9 8 7 6 5 4 3 2 1

CONTENTS

CONTRIBUTORS

Eric H. Baehrecke
Department of Cancer Biology, University of Massachusetts Medical School, Worcester, Massachusetts, USA

Patrice Bilesimo*
Laboratoire d'Evolution des Regulations Endocrinienne, Muséum National d'Histoire Naturelle, Département Régulation, Développement et Diversité Moléculaire, UMR 7221 CNRS, Paris, France

Daniel R. Buchholz
Department of Biological Sciences, University of Cincinnati, Cincinnati, Ohio, USA

Nicolas Buisine
Laboratoire d'Evolution des Regulations Endocrinienne, Muséum National d'Histoire Naturelle, Département Régulation, Développement et Diversité Moléculaire, UMR 7221 CNRS, Paris, France

B.A. Demeneix
Laboratoire d'Evolution des Régulations Endocriniennes, Muséum National d'Histoire Naturelle, UMR 7221 CNRS, Department RDDM, Paris, France

Robert J. Denver
Department of Molecular, Cellular and Developmental Biology, and Department of Ecology and Evolutionary Biology, The University of Michigan, Ann Arbor, Michigan, USA

Richard P. Elinson
Department of Biological Sciences, Duquesne University, Pittsburgh, Pennsylvania, USA

J.B. Fini
Laboratoire d'Evolution des Régulations Endocriniennes, Muséum National d'Histoire Naturelle, UMR 7221 CNRS, Department RDDM, Paris, France

Alexis G. Grimaldi
Laboratoire d'Evolution des Regulations Endocrinienne, Muséum National d'Histoire Naturelle, Département Régulation, Développement et Diversité Moléculaire, UMR 7221 CNRS, Paris, France

Takashi Hasebe
Department of Biology, Nippon Medical School, Kawasaki, Kanagawa, Japan

Kiyoshi Hiruma
Faculty of Agriculture and Life Sciences, Hirosaki University, Hirosaki, and Graduate School of Agricultural Sciences, Iwate University, Morioka, Japan

*Present address: Translational Research Center, Perelman School of Medicine, University of Pennsylvania, Philadelphia, Pennsylvania, USA

Guillaume Holzer
Molecular Zoology Team, Institut de Génomique Fonctionnelle de Lyon, École Normale Supérieure de Lyon, UMR5242 CNRS, Université Lyon, France

Atsuko Ishizuya-Oka
Department of Biology, Nippon Medical School, Kawasaki, Kanagawa, Japan

Carlena K. Johnson
Department of Biology, Spinal Cord and Brain Injury Research Center, University of Kentucky, Lexington, Kentucky, USA

Yu Kaneko
Faculty of Agriculture and Life Sciences, Hirosaki University, Hirosaki, and Graduate School of Agricultural Sciences, Iwate University, Morioka, Japan

Kirst King-Jones
Department of Biological Sciences, University of Alberta, Edmonton, Alberta, Canada

Saurabh S. Kulkarni
Department of Biological Sciences, University of Cincinnati, Cincinnati, Ohio, USA

Vincent Laudet
Molecular Zoology Team, Institut de Génomique Fonctionnelle de Lyon, École Normale Supérieure de Lyon, UMR5242 CNRS, Université Lyon, France

Sarah K. McMenamin
Department of Biology, University of Washington, Seattle, Washington, USA

G. Morvan-Dubois
Laboratoire d'Evolution des Régulations Endocriniennes, Muséum National d'Histoire Naturelle, UMR 7221 CNRS, Department RDDM, Paris, France

Michael B. O'Connor
Department of Genetics, Cell Biology and Development, University of Minnesota, Minneapolis, Minnesota, USA

Qiuxiang Ou
Department of Biological Sciences, University of Alberta, Edmonton, Alberta, Canada

David M. Parichy
Department of Biology, University of Washington, Seattle, Washington, USA

Kim F. Rewitz
Department of Biology, Cell and Neurobiology, University of Copenhagen, Denmark

Laurent M. Sachs
Laboratoire d'Evolution des Regulations Endocrinienne, Muséum National d'Histoire Naturelle, Département Régulation, Développement et Diversité Moléculaire, UMR 7221 CNRS, Paris, France

Alexander M. Schreiber
Laboratory of Vertebrate Metamorphosis, St. Lawrence University, Biology Department, Canton, New York, USA

Kirsten Tracy
Department of Cancer Biology, University of Massachusetts Medical School, Worcester, Massachusetts, USA

S. Randal Voss
Department of Biology, Spinal Cord and Brain Injury Research Center, University of Kentucky, Lexington, Kentucky, USA

Naoki Yamanaka
Department of Genetics, Cell Biology and Development, University of Minnesota, Minneapolis, Minnesota, USA

PREFACE

Metamorphosis occurs throughout the animal kingdom, including dramatic and well-studied examples in insects, fish, and amphibians. Even species that lack classical metamorphosis experience transitions resembling metamorphosis sometime during their development, such as the postembryonic perinatal period in mammals. This volume intends to provide a progress report on our current understanding of selected topics in animal metamorphosis with broad implications for vertebrate development in general.

The first section of the book covers insect metamorphosis. Chapter 1, by Rewitz et al., reviews the signaling transduction pathways and the mechanisms governing developmental checkpoints and feedback circuits that control metamorphic timing and different stages of metamorphosis in Drosophila. The biosynthesis of two key hormones controlling insect metamorphosis is the focus of the next two chapters. Chapter 2, by Ou and King-Jones, describes the roles of transcription factors in regulating ecdysone biosynthesis, with an emphasis on the transcription factors that are themselves regulated by ecdysone in Drosophila, while Chapter 3, by Hiruma and Kaneko, summarizes the regulation of juvenile hormone synthesis at different stages and the function of the hormone in regulating insect metamorphosis, with an emphasis on Bombyx mori. The last chapter on insect metamorphosis by Tracy and Baehrecke discusses autophagy, an important process involved in the transformation of many organs during metamorphosis, and its regulation by steroid hormones and nutrient restriction (Chapter 4).

Teleost metamorphosis is the focus of next two chapters. Chapter 5, by McMenamin and Parichy, provides an overview of metamorphosis in teleosts, where the authors describe the broad spectrum of changes in diverse species and review current understanding of the endocrine controls of teleost metamorphosis. In Chapter 6, Schreiber reviews the morphological and anatomical changes during flatfish metamorphosis, focusing on eye migration from one side of the head to the other, as well as the control of this process by thyroid hormone.

Amphibian metamorphosis occupies the largest section of the book, covering subjects ranging from salamander pedomorphosis, direct development, to anuran metamorphosis, as well as the role of thyroid hormone during embryogenesis. First, Denver, in Chapter 7, reviews the endocrine circuits that regulate metamorphosis in response to both internal and external cues, particularly the synergism between the hypothalamo–pituitary–thyroid and hypothalamo–pituitary–interrenal axes. Chapter 8, by Johnson and Voss, first summarizes salamander life history variations and the hormonal regulation of pedomorphosis, followed by more recent genetic and genomic studies with an emphasis on Mexican axolotl. In Chapter 9, Elinson reviews studies on the cryptic metamorphosis of direct developing frogs that lack a larval/tadpole period, discussing similarities and differences between direct developers and metamorphosing frogs and potential mechanisms underlying the elimination of the larval period. Thyroid hormone receptor function during anuran metamorphosis is the focus of the next chapter, where Grimaldi et al. review recent progress in high-throughput, genome-wide analyses for gene expression, genome-wide receptor binding, and long-range interactions among different chromatin regions (Chapter 10). In Chapter 11, Ishizuya-Oka and Hasebe describe studies on the formation of the adult intestinal epithelium during Xenopus metamorphosis. The authors summarize recent advances and highlight the conservations in the formation of the adult intestinal stem cells between amphibians and mammals. Next, Kulkarni and Buchholz provide a comprehensive summary of microarray analyses of anuran metamorphosis, highlighting the common as well as tissue-dependent gene regulation programs (Chapter 12). They also discuss applications of microarrays to identify biomarkers for endocrine disruptors of thyroid hormone signaling during metamorphosis. Continuing on the subject of endocrine disruption, Morvan-Dubois et al., in Chapter 13, review studies that have led them to conclude that thyroid hormone is also important for vertebrate embryogenesis and that amphibian embryogenesis offers an opportunity to study endocrine disruptors during early development.

The book ends with a chapter by Holzer and Laudet, who review evidence supporting a role of thyroid hormone in regulating postembryonic development across vertebrates, particularly in amniotes (Chapter 14). The authors compare the development of amniotes to amphibians and provide a unifying view for the presence of thyroid hormone-dependent functional remodeling of organs during postembryonic development in amniotes, enabling an ontological shift to a new ecological niche.

A comprehensive coverage of animal metamorphosis is not possible in this volume. I apologize for the omission of many other important aspects of animal metamorphosis, but fortunately for this field of biology, my apology signifies a healthy research community engaged in the study of animal metamorphosis.

Yun-Bo Shi

CHAPTER ONE

Developmental Checkpoints and Feedback Circuits Time Insect Maturation

Kim F. Rewitz[*,1], **Naoki Yamanaka**[†], **Michael B. O'Connor**[†,1]
[*]Department of Biology, Cell and Neurobiology, University of Copenhagen, Denmark
[†]Department of Genetics, Cell Biology and Development, University of Minnesota, Minneapolis, Minnesota, USA
[1]Corresponding authors: e-mail address: Kim.Rewitz@bio.ku.dk; moconnor@umn.edu

Contents

Abstract

The transition from juvenile to adult is a fundamental process that allows animals to allocate resource toward reproduction after completing a certain amount of growth. In insects, growth to a species-specific target size induces pulses of the steroid hormone ecdysone that triggers metamorphosis and reproductive maturation. The past few years have seen significant progress in understanding the interplay of mechanisms that coordinate timing of ecdysone production and release. These studies show that the neuroendocrine system monitors complex size-related and nutritional signals, as well as external cues, to time production and release of ecdysone. Based on results discussed here, we suggest that developmental progression to adulthood is controlled by checkpoints that regulate the genetic timing program enabling it to adapt to different

Current Topics in Developmental Biology, Volume 103
ISSN 0070-2153
http://dx.doi.org/10.1016/B978-0-12-385979-2.00001-0

environmental conditions. These checkpoints utilize a number of signaling pathways to modulate ecdysone production in the prothoracic gland. Release of ecdysone activates an autonomous cascade of both feedforward and feedback signals that determine the duration of the ecdysone pulse at each developmental transitions. Conservation of the genetic mechanisms that coordinate the juvenile-adult transition suggests that insights from the fruit fly *Drosophila* will provide a framework for future investigation of developmental timing in metazoans.

1. INTRODUCTION

The timing of insect metamorphosis is not a simple function of chronological age, but instead adjusts itself depending on environmental conditions. Holometabolous insects develop through a series of larval stages called instars, each punctuated by a molt in which the old cuticle is shed and replaced by a new larger one. After a series of molts, the number of which is species-specific, a terminal larval phase is reached during which extensive growth occurs until a predetermined target size is reached. At this point, the larva enters the pupal stage and undergoes metamorphosis to form the reproductively mature adult (Mirth & Riddiford, 2007). The processes of metamorphosis has fascinated classical entomologist for almost a century in part because it touches upon a fundamental biological question: how are developmental transitions controlled? Answering this question has implications beyond just understanding developmental timing mechanisms in insects. It also will help us understand how growth and maturation processes are coordinated such that reproductively mature organisms are reliably produced despite potentially different environmental conditions encountered by each individual during development.

In insects growth and maturation are distinct processes with growth being restricted primarily to the larval stages (although some growth occurs during metamorphosis: Okamoto et al., 2009; Slaidina, Delanoue, Gronke, Partridge, & Leopold, 2009), which means that the size of the larva at the onset of metamorphosis largely determines the final size of the adult (Edgar, 2006; Mirth & Riddiford, 2007). Final body size is determined by the rate of mass accumulation and the duration of growth. Interestingly, recent studies show that members of the evolutionarily conserved insulin-like growth factor family coordinate both growth rate and duration by impinging on the endocrine system that controls onset of metamorphosis (Caldwell, Walkiewicz, & Stern, 2005; Colombani, Andersen, & Leopold, 2012; Colombani et al., 2005;

Garelli, Gontijo, Miguela, Caparros, & Dominguez, 2012; Mirth, Truman, & Riddiford, 2005; Ou, Magico, & King-Jones, 2011; Walkiewicz & Stern, 2009). At the cellular level, two things determine body dimensions: the number of cells and their size. It is clear that the great size diversity among animals is largely determined by differences in cell number rather than cell volume. However, insects utilize both strategies to grow as many larval tissues sustain their rapid growth and metabolic activity by endoreplication cycles, a process of chromosomal replications without cell division (Edgar & Orr-Weaver, 2001; Lee, Davidson, & Duronio, 2009; Smith & Orr-Weaver, 1991). Increasing the DNA content allows cells to dramatically increase in volume without dividing and is responsible for most of the size increase during larval growth. During metamorphosis, the breakdown of obsolete endoreplicated larval tissues provides nutrients for growth and differentiation of mitotic neuroblasts and imaginal disk tissues that form the adult body.

Unlike the well-characterized genetic control of embryonic development, we are only now beginning to understand the developmental timing system that ensures unidirectional progression of developmental transitions. Recent genetic studies in the fruit fly *Drosophila* have provided new insight into how the developmental timing system evaluates larval growth and energy stores and integrates external cues such as photoperiod to time the onset of metamorphosis (Caldwell et al., 2005; Colombani et al., 2005, 2012; Garelli et al., 2012; Layalle, Arquier, & Leopold, 2008; McBrayer et al., 2007; Mirth et al., 2005). All of these signals are eventually communicated to the endocrine system responsible for producing the pulses of the steroid hormone 20-hydroxyecdysone (for simplicity, hereafter referred to as ecdysone) that triggers molting and metamorphosis (Gilbert, Rybczynski, & Warren, 2002). In *Drosophila*, a single pulse of ecdysone triggers each of the first two larval molts (Warren et al., 2006). During the terminal third instar stage, three low-level pulses followed by a high-level peak of ecdysone initiate the physiological and behavioral changes necessary to transform a food seeking larva into a nonfeeding immobile pupa. Although the roles of the initial low-level ecdysone peaks are not completely understood, the final major high-level peak triggers pupariation and the onset of metamorphosis.

Ecdysone is produced in the prothoracic gland (PG) by a series of reactions mediated by a Rieske oxygenase and enzymes encoded by the Halloween family of genes that include several cytochrome P450s and one dehydrogenase/reductase (Niwa et al., 2010; Ono et al., 2006; Petryk et al., 2003; Warren et al., 2002, 2004; Yoshiyama, Namiki, Mita, Kataoka, & Niwa, 2006;

Yoshiyama-Yanagawa et al., 2011). Once released into circulation, ecdysone is taken up by several peripheral tissues including the gut, fat body and Malphigian tubules where it is converted to the active hormone 20-hydroxyecdysone by the P450 enzyme product of the *shade* gene (Petryk et al., 2003). Circulating 20-hydroxyecdysone then induces a systemic genetic response in multiple tissues by binding to a complex of the ecdysone receptor (EcR) and ultraspiracle, both members of the nuclear receptor family (King-Jones & Thummel, 2005). Peripheral tissues interpret the ecdysone signal in a cell-type specific manner, determined by the combination of specific EcR isoforms and cofactors expressed in the target tissue (Yamanaka, Rewitz, & O'Connor, 2013). The response to ecdysone includes the destruction of obsolete larval tissues by induction of cell death activators and the morphogenesis and differentiation of the imaginal disk tissues to adult structures (Baehrecke, 2000; D'Avino & Thummel, 2000; Siaussat, Porcheron, & Debernard, 2009).

The mechanisms timing the production and release of ecdysone are key to understanding developmental timing of metamorphosis. These are intimately linked to size cues because growth beyond a characteristic size triggers ecdysone release (Edgar, 2006; Mirth & Riddiford, 2007; Mirth & Shingleton, 2012). Here, we discuss progress from *Drosophila* studies that examine how the rates of ecdysone synthesis and inactivation/removal are regulated and illustrate how these studies provide insight into the mechanisms that time ecdysone pulses and their duration (Caceres et al., 2011; Caldwell et al., 2005; Colombani et al., 2005; Gibbens, Warren, Gilbert, & O'Connor, 2011; Layalle et al., 2008; McBrayer et al., 2007; Mirth et al., 2005; Ou et al., 2011; Rewitz, Larsen, et al., 2009; Rewitz, Yamanaka, Gilbert, & O'Connor, 2009; Rewitz, Yamanaka, & O'Connor, 2010).

2. CHECKPOINT CONTROLS FOR PROGRESSION OF DEVELOPMENT

2.1. Increasing complexity for control of ecdysone production and release

Early studies in Lepidoptera led to the development of a key concept known as "critical weight" that appears to switch larvae between a nutrition-dependent growth phase and a nutrition-independent timing program (Beadle, Tatum, & Glancy, 1938; Nijhout & Williams, 1974a,1974b). Starvation before larvae reach critical weight suspends growth and delays developmental progression

(Fig. 1.1). However, in larvae that have attained the correct target size, starvation prevents further growth but does not delay pupariation. Before attaining critical weight, larvae must pass another checkpoint called minimal viable weight, which is the minimal size at which sufficient nutrient storage has occurred to enable the larva to survive through metamorphosis without further

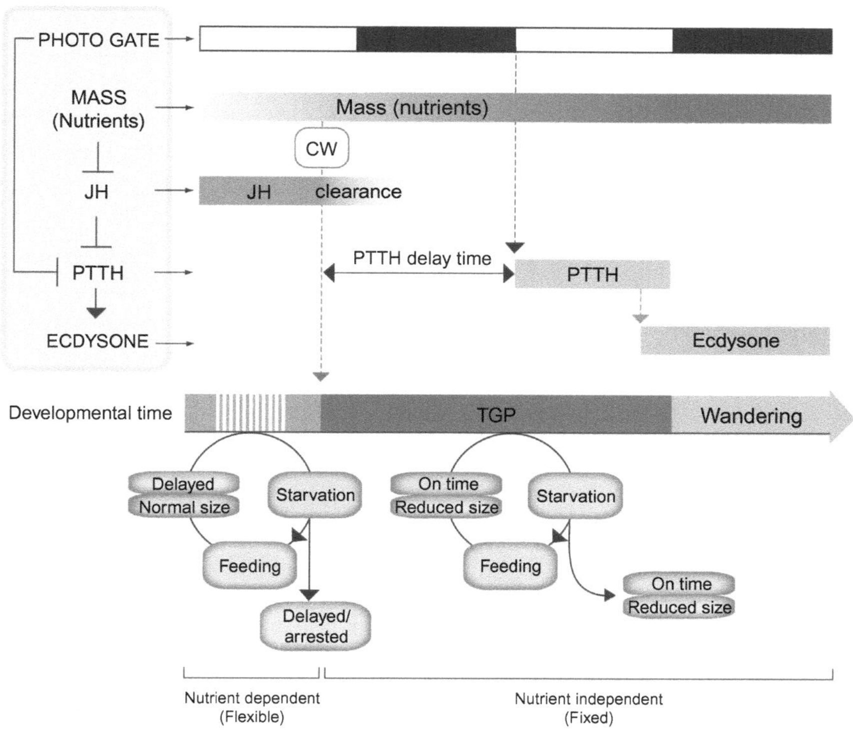

Figure 1.1 Hormonal regulation of body size and timing of metamorphosis. According to classical work the larval accumulation of mass corresponding to the critical weight results in the breakdown of JH that inhibits PTTH release. Above critical weight, PTTH is release after a delay period determined by the time of JH clearance and the photoperiod gating of PTTH. Starvation before critical weight results in a developmental delay of pupariation. If feeding is resumed critical weight is attained and normal body size achieved. In contrast, larvae starved above critical weight, when the metamorphic program is activated, pupariate on time but with a reduced size. As critical weight is independent of nutrition, final size is determined by the amount of growth in the interval between critical weight and cessation of feeding (wandering) called the terminal growth period (TGP). Photo gate, a PTTH gating mechanism imposed by the photoperiod; CW, critical weight. (See Color Insert.)

feeding (Mirth & Riddiford, 2007; Mirth & Shingleton, 2012). Starvation before minimal viable weight results in larval death without an attempt to undergo metamorphosis. Minimal viable weight should not be mistaken for critical weight, which is the size where starvation no longer delays metamorphosis. In *Drosophila* minimal viable weight and critical weight are attained almost simultaneously, making the distinction difficult.

Several classical studies using lepidopteran insects suggest that attainment of critical weight is followed by a drop in juvenile hormone (JH) levels, which is permissive for prothoracicotropic hormone (PTTH) release at a certain time defined by the photoperiod. According to this scheme, the release of PTTH from the brain is the principal event committing the PG to ramp up ecdysone production and release to trigger metamorphosis (Nijhout & Williams, 1974a,1974b). Although the decline in circulating JH levels following attainment of critical weight is believed to be the key determinant for PTTH release in lepidopterans, a role for JH titer drop in triggering PTTH release from the PG neurons in *Drosophila* is less clear since ablation of the corpus allatum, the organ responsible for JH synthesis, does not appear to alter critical weight or accelerate the timing of pupariation in this species (Riddiford, 2011; Riddiford, Truman, Mirth, & Shen, 2010). Instead, a body of recent evidence shows that, although PTTH release is an important event, other factors converge on the PG to coordinate ecdysone synthesis and release. For example, recent studies have demonstrated the importance of insulin/TOR signaling in the PG for critical weight assessment and final body size determination (Caldwell et al., 2005; Colombani et al., 2005; Gibbens et al., 2011; Layalle et al., 2008; Mirth et al., 2005), while other factors such as myosuppression appear to act negatively to reduce PG activity (Yamanaka et al., 2005; Yamanaka et al., 2010; Yamanaka et al., 2006; see Marchal et al., 2010 for a review of factors regulating PG synthesis of ecdysone). Together these studies place the PG in a more central position compared to the classical scheme that focuses on the PTTH-producing neurons.

Starvation does not change critical size, showing that it is genetically determined and unaffected by environmental factors (Beadle et al., 1938). For example, genetic reduction of insulin signaling before critical weight slows larval growth rate and delays the attainment of critical weight and pupariation (Shingleton, Das, Vinicius, & Stern, 2005). However, it does not change the critical weight, which means that it does not alter final adult body size. On the other end, reducing insulin signaling after attainment of the critical weight reduces final adult size but does not delay pupariation.

This defines another concept important for attainment of final body size which is the "terminal growth period" (also known as the interval to cessation of growth). The terminal growth period is the time interval between attainment of critical weight and the cessation of feeding that occurs when larvae transition to the wandering stage (Edgar, 2006; Warren et al., 2006). It is important to note that after critical weight is achieved, development proceeds on a fixed temporal schedule that is independent of nutritional status. However, nutrients can still have a strong influence on final body size during this time. If nutrient levels are high for instance, a normal body size is produced, while if they are low a small adult is formed.

At least two size-related checkpoints, one mediated by growth and energy status of the larval-specific tissues and the other mediated by the maturation and patterning of adult precursor tissues, known as the imaginal disks, regulate timing of metamorphosis (Edgar, 2006; Poodry & Woods, 1990). Starvation delays development in animals that lack imaginal disks, suggesting that the mechanism which coordinates growth and ecdysone release with the nutritional status resides in the nonmitotic larval cells.

2.2. Linking nutrition to developmental timing

Insects undergo a tremendous amount of growth from the embryo to the final larval stage that is ready to initiate metamorphosis (Britton & Edgar, 1998). Because the environment affects growth rate, but not critical weight, the larva must coordinate cellular growth and nutrient storage with the timing of metamorphosis. Nutrients absorbed by the gut are not the only signals that trigger cell growth in most organs. Instead, systemic growth in *Drosophila* is regulated by seven insulin-like peptides known as Dilp1–7 that signal through a single conserved insulin receptor (InR) (Rulifson, Kim, & Nusse, 2002; Wu & Brown, 2006). The insulin-producing cells (IPCs) in the brain are major sites for Dilp production and release. However, insulin signaling must be coordinated with nutrient availability which is sensed by the fat body (analogous to adipose and liver tissues of vertebrates). In a growth permissive, nutrient-rich environment, the fat body non-cell-autonomously regulates growth through the release of fat body-derived signals (FDSs) (Colombani et al., 2003; Geminard, Rulifson, & Leopold, 2009; Rideout, Marshall, & Grewal, 2012). In the fat body, TOR acts as a sensor of nutrients from the gut and responds cell-autonomously to amino acids and ATP levels. Amino acids are crucial dietary components for growth and developmental progression. Although growth of some tissues is largely decoupled from nutrients, the fat body responds rapidly to changes in the internal milieu (Britton & Edgar, 1998). Inhibiting amino acid transport, and thus, TOR signaling in the fat body is sufficient to delay metamorphosis and nonautonomously reduce insulin signaling and systemic growth (Colombani et al., 2003). Recently, it was found that nutrient mediated inhibition of Maf1, a repressor of RNA polymerase III-dependent tRNA transcription, in the fat body nonautonomously increases growth and accelerates pupariation (Rideout et al., 2012). Inhibition of Maf1 is TOR dependent and promotes transcription of $tRNA^{Met}$ which is a limiting factor for protein synthesis in the fat body that controls organismal growth and timing of pupariation. Together, these results suggest that normal timing of metamorphosis is dependent on a positive signal from that fat body which is a central tissue for coordinating nutrient availability with organismal growth and timing of maturation.

The FDS appears to be a humoral factor(s) that conveys the fat body amino acid status to the IPCs in the brain to regulate Dilp release (Geminard et al., 2009; Rideout et al., 2012). Combined with observations that insulin regulates ecdysone release from the PG (Caldwell et al., 2005; Colombani et al., 2005; Mirth et al., 2005), it provides a mechanism for coupling nutrient status information with ecdysone release (Fig. 1.2). Consistent with this, increasing insulin signaling from the IPCs is sufficient to induce premature ecdysone release and pupariation (Walkiewicz & Stern, 2009). These studies agree with the fact that bombyxin, an insulin-like peptide from the silkworm *Bombyx mori*, was originally characterized for its ability to stimulate ecdysone synthesis (Kiriishi, Nagasawa, Kataoka, Suzuki, & Sakurai, 1992; Nagasawa et al., 1986; Rybczynski, 2005). Although insulin stimulates PG cell growth, the effect on ecdysone synthesis appears to be specific as growth inducers such as dMyc and cyclin D/Cdk4 increase PG cell size, but not ecdysone synthesis (Colombani et al., 2005). Ecdysone produced in the PG also feeds back on the fat body to regulate organismal growth (Colombani et al., 2005; Delanoue, Slaidina, & Leopold, 2010).

The TOR-dependent amino acid sensor that resides in the larval fat body cells may ensure that after each molt a certain amount of nutrient-dependent growth is required to reset the "developmental timer." This checkpoint ensures a minimal period of feeding before the PG becomes competent to produce an ecdysone pulse. Considering that insulin probably plays a central role in relaying the critical weight signal, the fat body sensor may be a key parameter in critical weight assessment. Accumulating evidence suggests that the fat body with its metabolic and endocrine activities (Britton & Edgar,

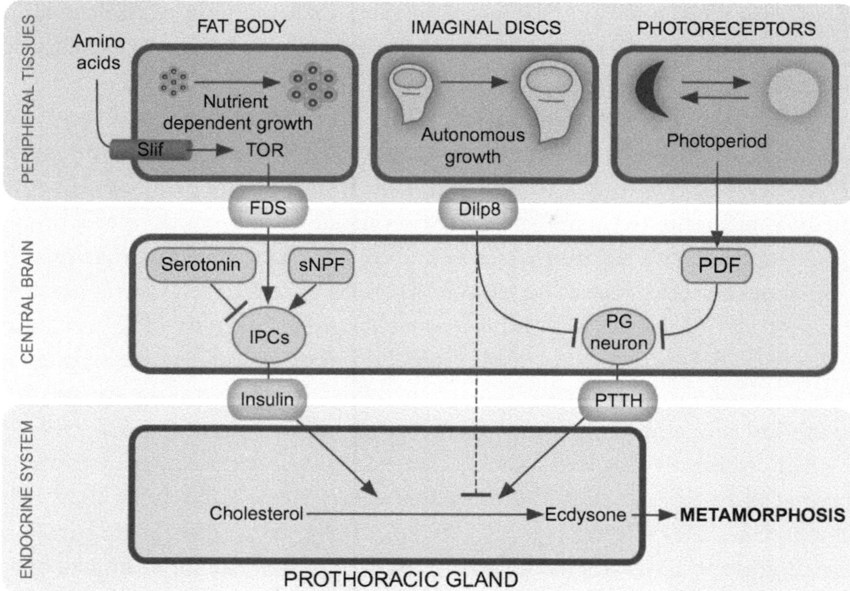

Figure 1.2 The developmental timing system monitors environmental (nutrient status and photoperiod) and developmental (disks maturation) cues. The fat body acts as a nutrient sensor that coordinates nutrient uptake with systemic growth and developmental timing. In growth permissive environments, the fat body secretes an unknown fat body-derived signal (FDS), in response to dietary amino acids, that stimulates release of insulin from the insulin producing cells (IPCs) of the brain which acts on the PG and stimulates ecdysone release. Serotonin and short neuropeptide F (sNPF) also impinge on the IPCs and regulate insulin release. In addition to insulin, the developmental timing program checks the status of the imaginal disks. Disk growth and maturation is controlled by a tissue-autonomous program that via Dilp8 crosstalks with the neuroendocrine system. Disks secrete Dilp8 which suppresses ecdysone release presumably by inhibition of PTTH release from the PG neurons until they have completed a certain amount of growth or regenerated from tissue damage. Superimposed on this, the clock neurons producing the pigment dispersing factor (PDF) impinge on the PG neurons and regulate PTTH release according to the photoperiod. (For color version of this figure, the reader is referred to the online version of this chapter.)

1998; Colombani et al., 2003; Geminard et al., 2009; Kamakura, 2011) will be the key to understanding how physiological parameters associated with mass and energy balance are integrated with the endocrine system to time maturation. In addition to relaying a signal via insulin, the FDSs may also act directly on the PG itself or indirectly through stimulation of PTTH release that then acts on the PG. It is also intriguing to note that recent evidence implicates oxygen limitation during growth of the tobacco hornworm *Manduca sexta* as a key determinator for timing of molting and metamorphosis (Callier & Nijhout, 2011), and hypoxia affects TOR activity (Wullschleger, Loewith, & Hall, 2006).

Another interesting possibility for how fat cells might sense size is that they may monitor endoreplication number. Endoreplicative tissues such as the fat body rely on polytene replication to increase the size of cells that are fully differentiated at the end of embryonic development. In the fat body, dietary amino acids are intimately linked to the endoreplicative machinery, which means that starvation results in rapid arrest of DNA replication (Britton & Edgar, 1998). However, changing fat body cell size and endoreplication number via insulin signaling does not affect developmental timing or systemic growth (Colombani et al., 2003), arguing against the possibility that endoreplication is the key for measuring cell size.

Although previous work shows that insulin regulates ecdysone biosynthesis (Caldwell et al., 2005; Colombani et al., 2005; Gibbens et al., 2011; Mirth et al., 2005; Walkiewicz & Stern, 2009), it is unlikely that insulin signaling alone generates the series of ecdysone pulses that are produced during the third instar (Warren et al., 2006). More likely, insulin is part of the size-sensing system involved in producing the first of three low-level ecdysone peaks in the third instar that coincides with the critical weight checkpoint (Rewitz & O'Connor, 2011). Passing this first checkpoint, corresponding to the accumulation of a certain amount of mass and nutrients, could allow the integration of other signals leading to production of subsequent peaks. Among other checkpoints, one that likely needs to be satisfied is that the imaginal disks must have developed sufficiently so that differentiation can take place during metamorphosis.

2.3. Signaling between the neuroendocrine timing system and a tissue-autonomous size assessment program

It is well known that lesions to imaginal disks, induced by either physical damage, radiation or genetic manipulation, also delay metamorphosis (Bryant & Levinson, 1985; Hussey, Thompson, & Calhoun, 1927; Poodry & Woods, 1990; Sehnal & Bryant, 1993; Simpson, Berreur, & Berreur-Bonnenfant, 1980; Simpson & Scheinderman, 1975; Stieper, Kupershtok, Driscoll, & Shingleton, 2008). The fact that transplantation of damaged disks delays metamorphosis suggests that they produce a secreted signal that acts on the endocrine system to suppress ecdysone release (Dewes, 1973). This developmental delay allows extra time for damaged or growth perturbed disks to regenerate and reach their target size before metamorphosis. Since these processes require cell proliferation, dividing disk cells presumably produce a signal that suppresses ecdysone release. Consistent with this view, pupariation is delayed in disk overgrowth mutants (Sehnal & Bryant, 1993). Importantly, disk growth is determined by an autonomous genetic program that allows them to stop their growth at a specific size even in culture conditions (Bryant & Levinson, 1985). The preprogrammed target size thus functions as a checkpoint that verifies that sufficient growth of disk tissue has occurred before development can progress (Fig. 1.2).

Recently, two elegant genetic studies identified the elusive disk signal as Dilp8, a molecule evolutionarily related to the insulin-like peptides (Colombani et al., 2012; Garelli et al., 2012). Genetic manipulations that (a) prolong the growth period before disks reach their correct target size, (b) give rise to neoplastic disk growth, or (c) damage tissue all result in upregulation of *Dilp8* expression. Reducing the expression of *Dilp8* does not lead to premature pupariation which agrees with previous studies showing that the disk signal is a permissive factor for developmental progression. However, it is not clear how the Dilp8 signal converges on the endocrine system to regulate ecdysone production or release. Potentially, it could work either directly on the PG, through PTTH, or both to suppress ecdysone production. Genetic disruption of disk growth mainly prolongs the duration of the third instar (L3) up to 5 days (Colombani et al., 2012), similar to the phenotype observed in larvae lacking either PTTH or its receptor Torso (McBrayer et al., 2007; Rewitz, Yamanaka, et al. 2009). Like larvae with reduced PTTH signaling, the second (L2) to L3 transition is only a few hours delayed under conditions that affect disk growth (Colombani et al., 2012). These observations suggest that Dilp8 prevents ecdysone production by inhibition of PTTH signaling. This also agrees with the fact that disk damage, like loss of PTTH, delays but does not prevent pupariation (Colombani et al., 2012; Garelli et al., 2012; Halme, Cheng, & Hariharan, 2010; McBrayer et al., 2007; Rewitz, Yamanaka, et al. 2009). Dilp8 may act at one or more levels in the PTTH signaling pathway, including production of PTTH, release of PTTH from varicosities on the PG, or by limiting processing of PTTH to its active form. An alternative but not mutually exclusive model is that Dilp8 may inhibit the capacity of the PG to respond to PTTH, perhaps by affecting Torso production or activity. Based on an earlier study which showed that disk damage reduces PTTH expression in the PG neurons (Halme et al., 2010), it seems likely that Dilp8 acts at least in part to suppress ecdysone production by downregulation of PTTH synthesis in the PG neurons.

Determining the precise temporal timing of Dilp8 activity and whether it correlates with one of the other small ecdysone peaks is an important area for future investigation. Interestingly, normal *Dilp8* expression peaks early during the third instar and drops from 80 to 96 h AEL (after egg laying) (Colombani et al., 2012). The drop in *Dilp8* expression coincides with the time when larvae reach critical weight at about 80 h AEL. This may suggest that the endocrine system is reading the Dilp8 drop as the critical weight checkpoint is reached. However, disk damage mainly delays pupariation by increasing the terminal growth period (Stieper et al., 2008), which indicates that the disks exert their suppressive influence postcritical weight, whereas nutritional restriction has no effect beyond this point. Moreover, tissue damage in the imaginal disks 96 h AEL, well after passing the critical weight checkpoint, delays pupariation (Wells & Johnston, 2012). Thus, maturation of larval-specific and adult precursor tissues may not coincide, and critical weight and imaginal disk maturation are likely different checkpoints, each of which may independently regulate the small fluctuations in the ecdysone titer prior to the larger molting peak. Although passing the critical weight checkpoint leads to nutrient-independent patterning of the disks (Mirth, Truman, & Riddiford, 2009), overexpression of *Dilp8*, which delays pupariation, does not influence disk patterning (Colombani et al., 2012), suggesting that *Dilp8* decrease is the result, and not the cause, of disk maturation. The fact that *Dilp8* mutants pupariate almost on time further supports the view that two independent timing mechanisms exist, one in the adult progenitor tissues and one in the larval-specific tissues. A detailed analysis of the exact time for the drop in *Dilp8* expression and the ecdysone profile may help unraveling if Dilp8 is responsible for the critical weight peak or one of the subsequent low-level ecdysone peaks. Taken together, these data together with the fact that manipulating insulin signaling in the disks does not affect timing (Stieper et al., 2008) show that ecdysone production is controlled by a developmental checkpoint run by an intrinsic genetic program that monitors disk maturation and patterning largely independent of nutrients.

2.4. The brain relays internal and external cues to the endocrine system

Some checkpoint signals from peripheral tissues such as fat body and disks are likely communicated through the central brain to the endocrine system. For example, nutritional assessment must ensure that shutdown of feeding only occurs when enough mass has been accumulated to survive the nonfeeding metamorphic stage. The brain regulates insulin release from the IPCs in response to the fat body-derived nutritional signal. If this signal directly regulates the IPCs or other classes of neurons that control the activity of the IPCs is not known. Cell-autonomous nutrient sensing that controls the release of adipokinetic hormone, the functional equivalent of the mammalian glucagon, does not seem to operate in the insulin release control (Kim & Rulifson, 2004). Instead, insulin secretion is controlled by PKA/CREB and ERK activity stimulated by short neuropeptide F (sNPF) or other neuropeptides and neurotransmitters such as serotonin and GABA (Birse, Soderberg, Luo, Winther, & Nassel, 2011; Enell, Kapan, Soderberg, Kahsai, & Nassel, 2010; Kaplan, Zimmermann, Suyama, Meyer, & Scott, 2008; Lee et al., 2008; Luo, Becnel, Nichols, & Nassel, 2012; Walkiewicz & Stern, 2009). Inhibition of PKA/CREB in the IPCs increases insulin signaling and results in premature ecdysone production and metamorphosis. Conversely, animals with ablated IPCs and mutants with increased serotonin levels that inhibit insulin release from the IPCs have reduced growth rates and delayed pupariation (Kaplan et al., 2008; Rulifson et al., 2002). This shows that neural circuits as well as endocrine mechanisms, such as the FDS, that regulate insulin release from the IPCs are important for the control of ecdysone production. Insulin likely acts directly on the PG (see below); in fact, the axons of the IPCs exit the brain and release insulin from terminals on the anterior aorta and corpus cardiacum close to the PG portion of the ring gland. It is also possible, however, that systemic insulin acts on the nerve endings of the PG neurons to send a retrograde signal that affects PTTH production. In general, the identity and mechanisms by which both internal and environmental inputs control PTTH activity is poorly understood, with perhaps the exception of photoperiodic inputs.

As found for lepidopterans, PTTH release in *Drosophila* appears to be controlled by the photoperiod (Fig. 1.2), since the circadian clock neurons, producing the pigment dispersing factor (PDF), impinge on the PG neurons (McBrayer et al., 2007; Siegmund & Korge, 2001). The photoperiodic gating of PTTH recurs daily but is presumably not "read" by the system until passage through other checkpoints have verified that the larva has completed enough growth (Nijhout, 1981). In *pdf* mutant larvae the periodicity of PTTH transcription is disrupted and clock mutations alter developmental timing (Kyriacou, Oldroyd, Wood, Sharp, & Hill, 1990; McBrayer et al., 2007), suggesting a circadian checkpoint regulating PTTH release. In lepidopterans, the PTTH release event is downstream of critical weight and the period between critical weight and PTTH secretion (determined by the time required for JH clearance) is called the PTTH delay period (Nijhout & Williams, 1974a,1974b). Interestingly, rearing *Drosophila* larvae with increased insulin signaling in the PG under constant light decreases time to pupariation (Mirth et al., 2005). One possible interpretation is that the PG of these larvae is prematurely estimating that critical weight has been reached because of the enhanced insulin signaling and the constant light condition then reduces the PTTH delay period.

Future studies should examine the mechanism(s) regulating PG neuron activity including the importance of transcriptional control, as well as processing and release of PTTH. Presently, the mechanism controlling PTTH release is only known for some insects in the group Hemiptera, where activation of abdominal stretch receptors in response to the attainment of a certain size results in PTTH release (Nijhout, 1981, 2003). Although distension of stretch receptors is not sufficient for PTTH release and initiation of molting in insects outside this order, peripheral sensory neurons tiling the larval body wall have been linked with changes in feeding behavior associated with the progression of development from the feeding to wandering stage in *Drosophila* (Ainsley, Kim, Wegman, Pettus, & Johnson, 2008; Wegman, Ainsley, & Johnson, 2010).

As discussed above, timing of pupariation involves checkpoints that ensure coordination of growth between larval and adult precursor tissues. However, it is less clear how each of these signals contributes to each of the discrete pulses of ecdysone seen during the third instar (Warren et al., 2006). One possibility is that clearance of all checkpoints is required to produce the first low-level, critical weight associated ecdysone peak and the successive subsequent peaks are generated autonomously by feedback circuits. However, the ecdysone peaks do not occur with a regular ultradian rhythm; the time between peaks being 8, 12, or 16 h (Warren et al., 2006). Another, and perhaps more likely, explanation is that the endocrine system monitors growth, energy status, and photoperiod checkpoints successively, each of which results in a low-level ecdysone peak that drives a subsequent part of the developmental program, ultimately resulting in pupariation and the initiation of metamorphosis.

3. SIGNALS CONVERGING ON THE PG

3.1. PTTH and insulin/TOR coordinate PG cell growth and ecdysone production

Like in peripheral tissues, the insulin/TOR pathway promotes growth of PG cells and therefore can have an indirect effect on total levels of ecdysone simply by controlling the PG cell size. However, several studies have shown

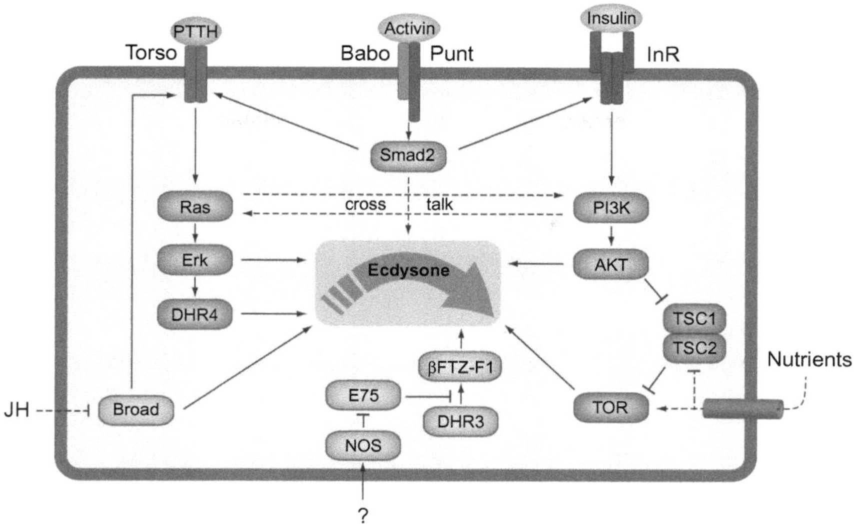

Figure 1.3 Mechanisms converging on the PG time ecdysone production. TGFβ/Activin is required for normal expression of the *insulin receptor* (*InR*) and *torso*, which provides glandular competence to PTTH (developmental) and insulin (nutrient) cues. Furthermore, the PG harbors a TOR-dependent nutrients sensor that presumably allows compensation for poor nutrient environments. Under such conditions, ecdysone release is delayed which prolongs the growth period. Broad is required for normal expression of the Halloween genes encoding the enzymes mediating ecdysone synthesis, although it is not clear if JH regulates *broad* in the PG. Nitric oxide (NO) generated by the nitric oxide synthase (NOS) regulates nuclear receptor signaling in the PG. In turn, NO inhibits E75, a repressor of DHR3 which then activates expression of *βFTZ-F1*, a nuclear receptor required for expression of at least two key ecdysone biosynthetic enzymes, Phantom and Disembodied. (See Color Insert.)

that insulin/TOR also has a direct role in regulating ecdysone synthesis in the PG (Caldwell et al., 2005; Colombani et al., 2005; Gibbens et al., 2011; Layalle et al., 2008; Mirth et al., 2005). Under this view, the developmental timer in the PG is informed of the nutritional status via insulin/TOR to adjust ecdysone synthesis (Fig. 1.3). Although the role insulin plays in the production of the ecdysone peaks is not clear, as described insulin is likely involved in setting the critical weight parameter that commits the PG to ecdysone biosynthesis. Assessment of the temporal requirement for insulin during development has demonstrated that reduced insulin signaling delays timing of pupariation before, but not after, the critical weight checkpoint (Shingleton et al., 2005). This implies that insulin signaling is not important for stimulation of ecdysone biosynthesis in the PG during the postcritical weight period, and so may not influence the two low-level ecdysone peaks at 20 and 28 h after ecdysis to the third instar nor the subsequent high-level peak at 44 h that is thought to trigger pupariation (Warren et al., 2006). In fact, starvation after reaching critical weight, which is assumed to reduce insulin signaling in the PG, actually slightly accelerates pupariation in *Drosophila* (Mirth & Riddiford, 2007; Stieper et al., 2008). Thus, insulin signaling seems to be required for the first critical weight peak approximately 8 h after the L2/L3 transition. Increase in insulin signaling possibly informs the endocrine system of the nutritional status of the larva. But how does the system sense critical size? One possibility is that some threshold level of insulin stimulates the biosynthetic pathway directly to produce the critical weight ecdysone peak. Another possibility is that insulin makes the PG commit to ecdysone synthesis by providing competence to other signals such as PTTH which then dictates the timing of some, or perhaps all, of the subsequent ecdysone peaks during the third instar. Under this view, insulin signaling would not directly produce the critical weight peak. This is consistent with the observation that insulin does not stimulate ecdysone biosynthesis in *Manduca* (Walsh & Smith, 2011).

From one perspective, it seems more intuitive that pulses of PTTH generate the precisely timed ecdysone peaks, instead of insulin. For example, PTTH is a more potent inducer of ecdysone synthesis than insulin and is released from neurons that directly innervate the PG (McBrayer et al., 2007; Rybczynski, 2005). Transcription of *PTTH* exhibits ultradian periodicity during the third instar (McBrayer et al., 2007), also making PTTH a likely candidate for a periodic ecdysone pulse generator. Recently, nucleocytoplasmic trafficking of the nuclear receptor DHR4 in the PG was shown to be PTTH dependent (Ou et al., 2011). Lack of PTTH or *torso* disrupts the first nucleocytoplasmic transition 4 h after the L2/L3 transition. This implies that PTTH signaling indeed occurs in advance to the critical weight peak. A second low-level ecdysone peak 90 h AEL is associated with a change in glue gene expression that allows the puparium to adhere to its substrate. The third low-level ecdysone peak 98 h AEL is associated with wandering behavior and the cessation of feeding. Consistent with this, reduced ecdysone levels results in a prolonged feeding period, delayed wandering and asynchrony of the ecdysone gene response (McBrayer et al., 2007; Rewitz, Yamanaka, et al. 2009).

If the role of insulin is to endow the PG with competence to produce ecdysone in response to PTTH, it would have to interact with a central component in the PTTH signaling machinery or ecdysone biosynthetic pathway. Interestingly, nutrient-dependent *torso* expression has been observed in *Bombyx* (Young, Yeh, & Gu, 2012), perhaps involving insulin. Moreover, the transcription factor Broad is required for expression of *torso* (Xiang, Liu, & Huang, 2010) and insulin induces expression of *broad* in the wing disks of *Manduca*, overcoming the inhibition by JH (Koyama, Syropyatova, & Riddiford, 2008). Assuming a similar scenario in the PG, insulin might increase *broad* levels to upregulate *torso* expression, providing competence to PTTH that in turn leads to commitment for metamorphosis. In agreement with a role in regulating ecdysone synthesis, *broad* null mutants fail to pupariate and *broad* is expressed in the PG after the L2/L3 transition with a rapid increase during wandering as the ecdysone titer rises (Kiss, Beaton, Tardiff, Fristrom, & Fristrom, 1988; Zhou, Zhou, Truman, & Riddiford, 2004).

In the previous section, we considered how TOR in the fat body might help generate FDSs that act either on the PG or PTTH activity. However TOR signaling in the PG itself is also important for determining timing since recent evidence suggests that TOR plays a role in the PG in the regulation of ecdysone synthesis (Layalle et al., 2008). In contrast to moderate changes in insulin signaling, reducing TOR activity in the PG prolongs larval development after the critical weight checkpoint (Colombani et al., 2005; Layalle et al., 2008). Although reducing TOR signaling delays pupariation, activation of TOR in the PG under normal food conditions does not accelerate development (Layalle et al., 2008). This only occurs under conditions with restricted nutrition, suggesting an upper limit for nutrient activation of ecdysone signaling. Like TOR, increasing insulin signaling in the PG only accelerates metamorphosis in larvae reared without yeast supplement (Mirth et al., 2005). This again supports the idea that nutrients provide a positive signal that is limited by the maximum growth rate. In contrast, increasing Ras signaling in the PG dramatically accelerates metamorphosis even under standard food conditions (Rewitz, Yamanaka, et al. 2009). The MAPK pathway is the major pathway relaying the signal downstream of the PTTH receptor Torso (Fig. 1.3). Like insulin signaling, activating the MAPK pathway in the PG results in increased PG cell growth, and the larvae pupariate at a smaller size than normal. On the other hand, decreasing MAPK signaling in the PG results in developmental delay and overgrowth similar to larvae lacking PTTH or Torso. Together with insulin, the importance of TOR signaling in the PG may help explain why overgrown larvae lacking PTTH eventually pupariate (McBrayer et al., 2007).

In addition to activating the MAPK pathway, studies in lepidopterans have shown that PTTH also activates a complex network of pathways that include PKA/cAMP, calcium and PKC (Rybczynski, 2005; Rybczynski & Gilbert, 2006). Whether Torso is required for activation of these alternative pathways and whether their activation is conserved in *Drosophila* and how they contribute to either the production and/or release of ecdysone from the PG requires additional study.

The finding that under low nutrient conditions TOR informs the PG to extend the growth period (Layalle et al., 2008) is difficult to reconcile with the fact that starvation after critical weight accelerates metamorphosis in *Drosophila* (Mirth & Riddiford, 2007; Stieper et al., 2008). One explanation might be a cross talk between the PTTH and the TOR pathway in the PG. Previous work has shown that rapamycin inhibits PTTH-stimulated ecdysone synthesis (Rybczynski, 2005). Indeed, PTTH has been shown to regulate 4E-BP and S6 kinase (S6K), a major target of TOR (Gu, Yeh, Young, Lin, & Li, 2012; Rybczynski, 2005). In addition to its stimulatory effect on ecdysone synthesis, PTTH has a non-steroidogenic trophic effect on PG growth and protein synthesis (Rybczynski, 2005; Rybczynski & Gilbert, 1994) and lack of PTTH reduces PG cell size (Ghosh, McBrayer, & O'Connor, 2010). PTTH-induced TOR signaling seems to rely on PI3K (Gu et al., 2012), suggesting that PTTH and insulin converge on PI3K to regulate PG cell growth and ecdysone biosynthesis. On the other hand, insulin stimulation of ecdysone synthesis may, at least in part, involve cross talk with the MAPK (Rewitz, Yamanaka, et al. 2009).

How do PTTH and insulin increase activity of the ecdysone biosynthetic pathway? PTTH-stimulated ecdysone synthesis occurs within minutes and requires translation and likely posttranslational protein modifications (Rewitz, Larsen, et al., 2009; Rybczynski, 2005). The rapid increase in the flux through the biosynthetic pathway that converts cholesterol to ecdysone possibly involves regulation of a rate-limiting molecule(s) that acts in, the so-called Black Box reaction(s), an early step in the biosynthetic pathway which is not completely understood (Gilbert et al., 2002). Consistent with this view, in lepidopterans PTTH stimulates the rapid increase in translation and phosphorylation of Spook, a cytochrome P450 enzyme involved in the Black Box (Ono, Morita, Asakura, & Nishida, 2012; Ono et al., 2006; Rewitz, Larsen, et al., 2009). However, it is not clear if the Black Box is the rate-limiting reaction in all insects.

In addition to the acute regulation, PTTH elicits a long-term transcriptional effect that involves upregulation of the Halloween genes encoding the ecdysone biosynthetic enzymes (Keightley, Lou, & Smith, 1990; Niwa et al., 2005; Yamanaka et al., 2007). High expression of the ecdysone biosynthetic enzymes coincides with the ecdysone peaks, suggesting that upregulation of the components in the pathway is required to increase the steroidogenic capacity of the gland (Rewitz, Rybczynski, Warren, & Gilbert, 2006; Yamanaka et al., 2007). The importance of the transcriptional regulation is highlighted by evidence suggesting that low expression of these genes, including *spook*, presumably prevents PTTH-stimulated ecdysone synthesis despite PTTH being able to induce a glandular phosphorylation response in the early fifth instar *Bombyx* (Gu & Chow, 2005; Lin & Gu, 2007; Yamanaka et al., 2007). Reduced expression of the ecdysone biosynthetic enzymes in PTTH-ablated larvae indicates that PTTH plays a role in the transcriptional up-regulation of these genes in *Drosophila* (McBrayer et al., 2007). However, the moderate effect of PTTH does not fully explain the dramatic increase in expression of these genes during development (Yamanaka et al., 2007). Thus, other mechanisms must be involved in the transcriptional regulation of the Halloween genes. Consistent with this notion, the transcriptional repressor DHR4, a key target of PTTH signaling, has only minor effect on Halloween gene expression (Ou et al., 2011). Other mechanisms might involve feedforward regulation by the action of ecdysone itself, believed to autonomously activate the PG in culture, or perhaps involve input from TGFβ/Activin and/or nitric oxide signaling (see discussion below: Caceres et al., 2011; Gibbens et al., 2011; Yamanaka et al., 2007).

In addition to the enzymes involved in ecdysone biosynthesis, *npc1* is also expressed specifically in the PG where it is thought to be involved in supplying cholesterol for steroidogenesis (Huang, Suyama, Buchanan, Zhu, & Scott, 2005). Normal expression of *npc1* and the Halloween genes requires Broad, which is interesting considering that JH, which suppresses *broad* expression, also inhibits *spook* expression in *Bombyx* (Xiang et al., 2010; Yamanaka et al., 2007).

3.2. TGFβ/Activin and nitric oxide signaling are essential for steroidogenesis

Although several factors have been shown to regulate PG activity (see Marchal et al., 2010 for a recent review), recent reports have highlighted the essential role of nitric oxide (NO) and TGFβ/Activin (Caceres et al., 2011; Gibbens et al., 2011). In peripheral tissues, ecdysone elicits a genetic response through a set of primary ecdysone-inducible nuclear receptors (King-Jones & Thummel, 2005). However, some of these nuclear receptors, such as E75, DHR3, and βFTZ-F1, are also involved in the production of ecdysone in the PG (Caceres et al., 2011; Parvy et al., 2005). Although little is known about the factors involved in the transcriptional regulation of the Halloween genes, βFTZ-F1 is required for expression of the genes *phantom* and *disembodied* (Parvy et al., 2005). Expression of *βFTZ-F1* requires activation by DHR3, which is repressed by E75. Activity of the NO synthase (NOS), catalyzing the production of NO, is required for NO-mediated inhibition of E75 activity in the PG (Caceres et al., 2011). Reducing expression of NOS in the PG results in a failure to produce ecdysone and an inability to undergo metamorphosis. These larvae have enlarged PG cells with increased endoreplication number, consistent with the view that increased PG cell size alone is not sufficient for high level ecdysone production (Colombani et al., 2005). Together, these results show that NO signaling has an essential role in the regulation of ecdysone production that controls developmental transitions, although the mechanism regulating NO signaling in the PG remains unknown.

In addition to NO, recent work in *Drosophila* has shown that TGFβ/Activin signaling is essential for ecdysone synthesis and developmental transitions (Gibbens et al., 2011). Interestingly, TGFβ and insulin signaling also converge in *Caenorhabditis elegans* on the synthesis of the steroid hormone dafachronic acid (DA), which interacts with the nuclear receptor DAF-12 to determine whether larvae undergo reproductive development or arrest in a dauer state (Tennessen & Thummel, 2011). In *Drosophila*, reduced TGFβ/Activin signaling in the PG, mediated by knockdown of *dSmad2*, the major downstream mediator of TGFβ signals, causes developmental arrest in the third instar (Gibbens et al., 2011). Disrupting TGFβ/Activin reduces expression of *torso* and the *InR* in the PG, presumably blocking the ability to synthesize ecdysone in response to PTTH and insulin. Moreover, restoring Torso/MAPK or InR/Akt in the PG rescues the phenotype of *dSmad2*-RNAi knockdown. Interestingly, resupplying the insulin receptor in the PG of larvae with reduced PG expression of *dSmad2* restores protein, but not transcript levels, of key ecdysone biosynthetic enzymes. Conversely, activating Torso/Ras signaling in the PG of these larvae rescues the reduced expression of these genes. Thus, potentially insulin and PTTH regulate the ecdysone biosynthetic machinery by distinct mechanisms one via transcription and the other via translational control. Together, these observations indicate that TGFβ/Activin signaling is important for the PG to develop competence to respond to additional developmental (PTTH) and nutritional (insulin) cues. Presently, it is not known whether TGFβ/Activin signaling is developmentally regulated to provide stage-specific competence or is required for constitutive expression of *torso* and *InR*. Another interesting possibility is that TGFβ/Activin signaling is linked to the effect of the SUMOylation pathway, since it has also been demonstrated that knockdown of the SUMO gene, *smt3*, in the PG arrests development in the L3 (Talamillo et al., 2008). In this regard, it is interesting to note that Medea, the co-Smad that forms a complex with dSmad2, is a primary target of SUMOylation (Miles et al., 2008).

4. SHAPING DISCRETE ECDYSONE PULSES

4.1. Feedback control of ecdysone production

In addition to the mechanisms that increase ecdysone synthesis, there must be ways to shut down PG activity as well as inactivate and remove ecdysone from circulation in order to produce the temporally defined pulses of ecdysone. Thus, the endocrine system must be under feedback regulation and receive inputs from its own target tissues (Fig. 1.4). This idea is consistent with studies showing that ecdysone inhibits ecdysone synthesis in the PG of lepidopterans (Beydon & Lafont, 1983; Gilbert et al., 2002; Sakurai & Williams, 1989). Furthermore, *EcR* is expressed in the PG of *Drosophila* during the time when the larvae initiate pupariation (Talbot, Swyryd, & Hogness, 1993). Long-term incubation of the PG in the presence of ecdysone reduces the sensitivity to PTTH (Gilbert, Song, & Rybczynski, 1997; Song & Gilbert, 1998), which indicates that ecdysone likely downregulates *torso* or other components mediating PTTH signaling or ecdysteroidogenesis.

Conversely, when the activity of the PG is low, ecdysone appears to stimulate its own synthesis, a mechanism that can increase and amplify the signal. Further amplification of the signal may occur through the EcR autoregulatory loop, through which *EcR* induces its own expression (Koelle et al., 1991). These observations suggest that ecdysone synthesis in the PG is controlled by feedforward and feedback loops that rapidly modulate the ecdysone titer and determine the temporal boundaries for the pulses. Interestingly, *torso* expression is induced by ecdysone, indicating a potential feedforward regulatory mechanism (Young et al., 2012). On the other hand, PTTH stimulation decreases *torso* mRNA levels, suggesting a desensitizing mechanism similar to that observed for InR (Puig, Marr, Ruhf, & Tjian, 2003). Although some controversies exist about the role of ecdysone in the PG (Kozlova & Thummel, 2002; Yamanaka &

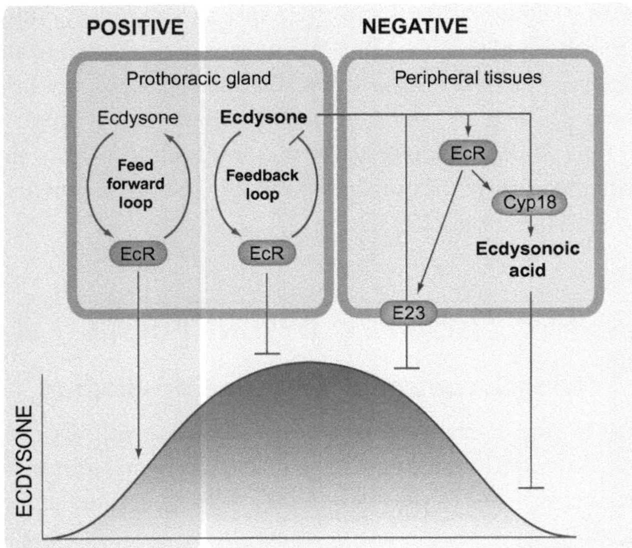

Figure 1.4 Feedback control shapes the ecdysone pulses. A short positive and a long negative feedback loop are believed to operate in the PG to control synthesis of ecdysone. The short positive feedforward loop presumably amplifies the ecdysone signal causing a fast increase of the titer. On the other hand, a long negative feedback loop shuts off the PG which allows peripheral mechanisms to clear ecdysone from the system. Ecdysone induces two feedback mechanisms, in tissues peripheral to the PG, which are eventually responsible for lowering cellular levels (E23) and the decline of the titer (Cyp18). Ecdysone induces Cyp18 that eliminates ecdysone by converting it into the inactive ecdysonoic acid. The ecdysone-inducible *E23* encodes an ABC transporter believed to pump ecdysone out of the cells. Together these feedback mechanisms determine the duration of the ecdysone pulses. (For color version of this figure, the reader is referred to the online version of this chapter.)

O'Connor, 2011), such feedback regulation makes intuitive sense and the physiological significance awaits further investigations.

Feedback control of PG activity through EcR may involve other ecdysone-inducible genes required for the ecdysone response in peripheral tissues. It is interesting to note that in a recent study DHR4, previously characterized as an ecdysone-inducible gene, was shown to be necessary to control the duration of the ecdysone pulses (King-Jones, Charles, Lam, & Thummel, 2005; Ou et al., 2011; Rewitz & O'Connor, 2011). When located in the nucleus, DHR4 represses ecdysone production, which helps establish the duration of the ecdysone pulses. Although PTTH-regulated nucleocytoplasmic shuttling of the DHR4 protein could account for this, the rapid effect of DHR4 RNAi argues that there is more to this than simple shuttling of a stable

DHR4 protein. Cycling of *DHR4* transcript in concert with PTTH-stimulated ecdysone peaks would presumably be a way to help terminate ecdysone production and regulate the duration of the pulse.

4.2. Termination of ecdysone pulses by feedback regulated degradation

Although ecdysone production in the PG must be turned off to bring levels back to basal, circulating ecdysone must also be removed to terminate the pulse. Two elegant feedback mechanisms have evolved to rapidly decrease circulating levels of ecdysone following a peak (Fig. 1.4). The first mechanism involves Cyp18a1, a cytochrome P450 enzyme required for the metabolic inactivation of ecdysone (Guittard et al., 2011; Rewitz et al., 2010). Cyp18a1 is required for the rapid decline of the ecdysone titer after the peak that triggers pupariation. Loss of *Cyp18a1* results in elevated ecdysone levels that disrupt the mid-prepupal expression of *βFTZ-F1*, which is necessary for the genetic response to ecdysone that drives unidirectional progression of development (Rewitz et al., 2010). *Cyp18a1* expression is induced by ecdysone, providing a mechanism for generating a pulse, where elevated ecdysone levels are responsible for its eventual decline. In addition to metabolic inactivation, another mechanism ensures that cellular levels of ecdysone are reduced following a peak. One of the early ecdysone-inducible genes, *E23*, encodes an ABC transporter protein (Hock, Cottrill, Keegan, & Garza, 2000). *E23* is one of the last of the early genes to be induced which makes physiological sense because the function of E23 is to pump ecdysone out of the cells and reduce cellular concentration. Together with feedback regulation of the PG activity, these peripheral mechanisms provide an autonomous regulatory system that determines the duration of the ecdysone pulses.

5. SUMMARY AND PERSPECTIVES

Steroids synthesized in response to signaling pathways including insulin-like peptides, TGFβ and other neuropeptides control the developmental transition leading to maturation in worms, insects and mammals. Recent progress from *Drosophila* research shows that nutrient sensitive insulin signaling and tissue-autonomous size determination are part of the underlying size-monitoring mechanisms that activate the endocrine system and initiate maturation once a characteristic size is achieved. An autonomous genetic size determining program that is dependent on the number of cell divisions and relatively insensitive to nutrient uptake controls disk growth. In contrast, growth of endoreplicative larval tissues is tightly coupled with nutrient availability and relies on the increase in cell size rather than number. Because larvae pupariate even in the absence of disks, Dilp8 is a permissive signal, rather than a positive "trigger" of pupariation, which provides a checkpoint to ensure that the adult precursor tissues are ready for metamorphosis. Conversely, nutrient sensitive growth influenced by the fat body likely provides positive signals to the endocrine system.

Based on recent insights discussed here, we propose that developmental timing is regulated by a series of checkpoints that verify a certain amount of nutrient stores and disk growth before a circadian input allows developmental progression (Fig. 1.5). The first checkpoint, which probably corresponds to critical weight, is the key decision whether to enter or delay pupariation if nutrient becomes limited. Another checkpoint verifies that the imaginal disks are ready, and finally the photoperiod is monitored to impose a circadian gate on the developmental transition. Although the mechanisms responsible for generating each peak of ecdysone are poorly understood,

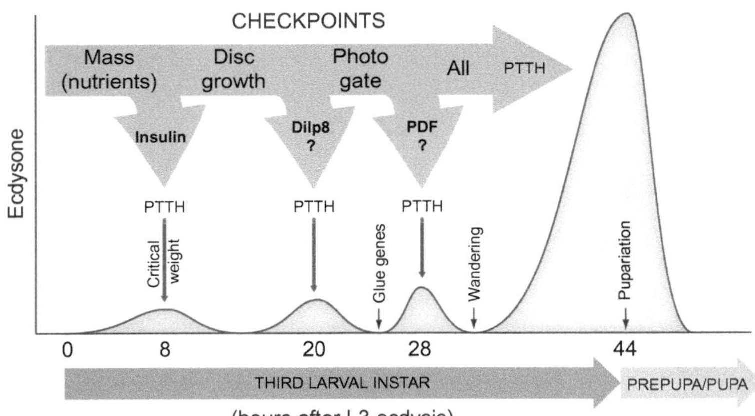

Figure 1.5 Checkpoint controls of developmental timing of metamorphosis. The figure shows a proposed model for checkpoint control of ecdysone release and timing of pupariation. Accumulation of nutrients corresponding to the critical weight presumably results in the insulin dependent production of the critical weight ecdysone peak 8 h after L3 ecdysis. Maturation and patterning of the imaginal disks and the photoperiod are also verified by the endocrine system before development can proceed. These checkpoints may potentially translate into to the other low-level ecdysone peaks 20 and 28 h after L3 ecdysis. When all checkpoints are cleared, developmental transition to the metamorphic stage can take place. (For color version of this figure, the reader is referred to the online version of this chapter.)

these peaks likely coordinate developmental and behavioral transitions during the third instar. Potentially each checkpoint control mechanism may be translated into one of the low-level ecdysone peaks which ensure that the molecular events are correctly ordered for unidirectional developmental progression. Alternatively, passing all checkpoints may activate a neuroendocrine switch to an autonomous program that produces the successive pulses of ecdysone. While it is obvious that the onset of metamorphosis is regulated by a number of cues from both the external and internal environment, further studies are needed to decipher how these cues are integrated to produce the sequential pulses of ecdysone.

Despite recent advances, the secreted FDSs have not been identified. The identification of the FDSs is the key to a more comprehensive understanding of how peripheral nutritional signals coordinate timing of ecdysone production. In this regard, the increasing prevalence of childhood obesity in humans has been associated with premature onset of puberty and reproductive dysfunction (Ahmed, Ong, & Dunger, 2009; Kaplowitz, 2008). Like developmental timing in insects, nutritional regulation of human pubertal timing is influenced by peripheral nutritional signals from adipose tissues similar to the fat body. We note a study by Norbert Perrimon's group identifying Unpaired 2 as a secreted FDS, functionally equivalent to human leptin, that regulates systemic growth presumably through its effect on insulin release from the IPCs (note added in proof).

The insulin-like molecules seem to be part of a conserved genetic mechanism that controls timing of maturation. In *C. elegans*, mutations disrupting insulin signaling result in dauer formation and arrest before the sexual maturation independent of the environmental conditions (Kimura, Tissenbaum, Liu, & Ruvkun, 1997). Moreover, puberty is delayed by mutations affecting insulin-like growth factor (IGF) signaling and in children diagnosed with certain types of diabetes (Divall et al., 2010; Domene et al., 2009; Kjaer, Hagen, Sando, & Eshoj, 1992; Messina et al., 2011). The discovery that Dilp8 coordinates tissue growth with developmental timing further highlights the central role for these conserved molecules in the control of body size, maturation, and metabolism.

How insulin/TOR and PTTH signaling interact in the PG to coordinate the timing of ecdysone pulses remain poorly understood. Insulin may work upstream of PTTH either by producing the critical weight ecdysone peak that allows subsequent PTTH release or by mediating glandular growth that provides competence to respond to PTTH. Because the PG neurons are anatomically developed earlier, regulating the PG competence might be important to prevent a premature response to the PTTH. Alternatively, these two pathways may work together to control ecdysone production in the PG. Perhaps signaling through both pathways is necessary to fully activate ecdysone biosynthesis. Investigating the role of transcriptional and posttranscriptional regulation of components in the ecdysone biosynthetic pathway as well as the underlying mechanisms will be the key to understanding the changes in the PG activity required to produce pulses of ecdysone.

In most systems, mechanisms responsible for timing the juvenile-adult transition are poorly understood, in part, because it involves complex interorgan communication to monitor numerous internal and external cues such as organ size, nutritional status, and photoperiod. As illustrated here, the number of factors that converge on the neuroendocrine system to control such transition in insects has expanded substantially during the recent years. However, these studies have also indicated, that as found for many other aspects of development, the overall architecture of the system that coordinates juvenile-adult transitions is conserved. Therefore, future insights from *Drosophila* and other insects will continue to serve as a general paradigm for understanding how metazoans coordinate growth and developmental timing.

ACKNOWLEDGMENTS

K.F.R. is supported by grant 11-105446 from the Danish Council for Independent Research, Natural Sciences. N.Y. is supported by NIH grant K99 HD073239 and M.B.O by NIH R01 GM093301.

REFERENCES

Ahmed, M. L., Ong, K. K., & Dunger, D. B. (2009). Childhood obesity and the timing of puberty. *Trends in Endocrinology and Metabolism, 20*, 237–242.

Ainsley, J. A., Kim, M. J., Wegman, L. J., Pettus, J. M., & Johnson, W. A. (2008). Sensory mechanisms controlling the timing of larval developmental and behavioral transitions require the *Drosophila* DEG/ENaC subunit, Pickpocket1. *Developmental Biology, 322*, 46–55.

Baehrecke, E. H. (2000). Steroid regulation of programmed cell death during *Drosophila* development. *Cell Death and Differentiation*, 7, 1057–1062.

Beadle, G., Tatum, E., & Glancy, C. (1938). Food level in relation to rate of development and eye pigmentation in *Drosophila melanogaster*. *The Biological Bulletin, 75*, 447–462.

Beydon, P., & Lafont, R. (1983). Feedback inhibition of ecdysone production by 20-hydroxyecdysone in *Pieris brassicae* pupae. *Journal of Insect Physiology, 29*, 529–533.

Birse, R. T., Soderberg, J. A., Luo, J., Winther, A. M., & Nassel, D. R. (2011). Regulation of insulin-producing cells in the adult *Drosophila* brain via the tachykinin peptide receptor DTKR. *The Journal of Experimental Biology, 214*, 4201–4208.

Britton, J. S., & Edgar, B. A. (1998). Environmental control of the cell cycle in *Drosophila*: Nutrition activates mitotic and endoreplicative cells by distinct mechanisms. *Development, 125*, 2149–2158.

Bryant, P. J., & Levinson, P. (1985). Intrinsic growth control in the imaginal primordia of *Drosophila*, and the autonomous action of a lethal mutation causing overgrowth. *Developmental Biology*, *107*, 355–363.

Caceres, L., Necakov, A. S., Schwartz, C., Kimber, S., Roberts, I. J., & Krause, H. M. (2011). Nitric oxide coordinates metabolism, growth, and development via the nuclear receptor E75. *Genes & Development*, *25*, 1476–1485.

Caldwell, P. E., Walkiewicz, M., & Stern, M. (2005). Ras activity in the *Drosophila* prothoracic gland regulates body size and developmental rate via ecdysone release. *Current Biology*, *15*, 1785–1795.

Callier, V., & Nijhout, H. F. (2011). Control of body size by oxygen supply reveals size-dependent and size-independent mechanisms of molting and metamorphosis. *Proceedings of the National Academy of Sciences of the United States of America*, *108*, 14664–14669.

Colombani, J., Andersen, D. S., & Leopold, P. (2012). Secreted peptide Dilp8 coordinates *Drosophila* tissue growth with developmental timing. *Science*, *336*, 582–585.

Colombani, J., Bianchini, L., Layalle, S., Pondeville, E., Dauphin-Villemant, C., Antoniewski, C., et al. (2005). Antagonistic actions of ecdysone and insulins determine final size in *Drosophila*. *Science*, *310*, 667–670.

Colombani, J., Raisin, S., Pantalacci, S., Radimerski, T., Montagne, J., & Leopold, P. (2003). A nutrient sensor mechanism controls *Drosophila* growth. *Cell*, *114*, 739–749.

D'Avino, P. P., & Thummel, C. S. (2000). The ecdysone regulatory pathway controls wing morphogenesis and integrin expression during *Drosophila* metamorphosis. *Developmental Biology*, *220*, 211–224.

Delanoue, R., Slaidina, M., & Leopold, P. (2010). The steroid hormone ecdysone controls systemic growth by repressing dMyc function in *Drosophila* fat cells. *Developmental Cell*, *18*, 1012–1021.

Dewes, E. (1973). Regeneration in transplanted halves of male genital disks and its influence upon duration of development in *Ephestia kühniella* Z. *Rouxs Archives of Developmental Biology*, *172*, 349–354.

Divall, S. A., Williams, T. R., Carver, S. E., Koch, L., Bruning, J. C., Kahn, C. R., et al. (2010). Divergent roles of growth factors in the GnRH regulation of puberty in mice. *The Journal of Clinical Investigation*, *120*, 2900–2909.

Domene, H. M., Hwa, V., Argente, J., Wit, J. M., Camacho-Hubner, C., Jasper, H. G., et al. (2009). Human acid-labile subunit deficiency: Clinical, endocrine and metabolic consequences. *Hormone Research*, *72*, 129–141.

Edgar, B. A. (2006). How flies get their size: Genetics meets physiology. *Nature Reviews. Genetics*, *7*, 907–916.

Edgar, B. A., & Orr-Weaver, T. L. (2001). Endoreplication cell cycles: More for less. *Cell*, *105*, 297–306.

Enell, L. E., Kapan, N., Soderberg, J. A., Kahsai, L., & Nassel, D. R. (2010). Insulin signaling, lifespan and stress resistance are modulated by metabotropic GABA receptors on insulin producing cells in the brain of *Drosophila*. *PLoS One*, *5*, e15780.

Garelli, A., Gontijo, A. M., Miguela, V., Caparros, E., & Dominguez, M. (2012). Imaginal discs secrete insulin-like peptide 8 to mediate plasticity of growth and maturation. *Science*, *336*, 579–582.

Geminard, C., Rulifson, E. J., & Leopold, P. (2009). Remote control of insulin secretion by fat cells in *Drosophila*. *Cell Metabolism*, *10*, 199–207.

Ghosh, A., McBrayer, Z., & O'Connor, M. B. (2010). The *Drosophila* gap gene giant regulates ecdysone production through specification of the PTTH-producing neurons. *Developmental Biology*, *347*, 271–278.

Gibbens, Y. Y., Warren, J. T., Gilbert, L. I., & O'Connor, M. B. (2011). Neuroendocrine regulation of *Drosophila* metamorphosis requires TGFbeta/Activin signaling. *Development*, *138*, 2693–2703.

Gilbert, L. I., Rybczynski, R., & Warren, J. T. (2002). Control and biochemical nature of the ecdysteroidogenic pathway. *Annual Review of Entomology*, *47*, 883–916.

Gilbert, L. I., Song, Q., & Rybczynski, R. (1997). Control of ecdysteroidogenesis: Activation and inhibition of prothoracic gland activity. *Invertebrate Neuroscience*, *3*, 205–216.

Gu, S. H., & Chow, Y. S. (2005). Analysis of ecdysteroidogenic activity of the prothoracic glands during the last larval instar of the silkworm, *Bombyx mori*. *Archives of Insect Biochemistry and Physiology*, *58*, 17–26.

Gu, S. H., Yeh, W. L., Young, S. C., Lin, P. L., & Li, S. (2012). TOR signaling is involved in PTTH-stimulated ecdysteroidogenesis by prothoracic glands in the silkworm, *Bombyx mori*. *Insect Biochemistry and Molecular Biology*, *42*, 296–303.

Guittard, E., Blais, C., Maria, A., Parvy, J. P., Pasricha, S., Lumb, C., et al. (2011). CYP18A1, a key enzyme of *Drosophila* steroid hormone inactivation, is essential for metamorphosis. *Developmental Biology*, *349*, 35–45.

Halme, A., Cheng, M., & Hariharan, I. K. (2010). Retinoids regulate a developmental checkpoint for tissue regeneration in *Drosophila*. *Current Biology*, *20*, 458–463.

Hock, T., Cottrill, T., Keegan, J., & Garza, D. (2000). The E23 early gene of *Drosophila* encodes an ecdysone-inducible ATP-binding cassette transporter capable of repressing ecdysone-mediated gene activation. *Proceedings of the National Academy of Sciences of the United States of America*, *97*, 9519–9524.

Huang, X., Suyama, K., Buchanan, J., Zhu, A. J., & Scott, M. P. (2005). A *Drosophila* model of the Niemann-Pick type C lysosome storage disease: dnpc1a is required for molting and sterol homeostasis. *Development*, *132*, 5115–5124.

Hussey, R. G., Thompson, W. R., & Calhoun, E. T. (1927). The influence of X-rays on the development of *Drosophila* larvae. *Science*, *66*, 65–66.

Kamakura, M. (2011). Royalactin induces queen differentiation in honeybees. *Nature*, *473*, 478–483.

Kaplan, D. D., Zimmermann, G., Suyama, K., Meyer, T., & Scott, M. P. (2008). A nucleostemin family GTPase, NS3, acts in serotonergic neurons to regulate insulin signaling and control body size. *Genes & Development*, *22*, 1877–1893.

Kaplowitz, P. B. (2008). Link between body fat and the timing of puberty. *Pediatrics*, *121* (Suppl. 3), S208–S217.

Keightley, D. A., Lou, K. J., & Smith, W. A. (1990). Involvement of translation and transcription in insect steroidogenesis. *Molecular and Cellular Endocrinology*, *74*, 229–237.

Kim, S. K., & Rulifson, E. J. (2004). Conserved mechanisms of glucose sensing and regulation by *Drosophila* corpora cardiaca cells. *Nature*, *431*, 316–320.

Kimura, K. D., Tissenbaum, H. A., Liu, Y., & Ruvkun, G. (1997). daf-2, an insulin receptor-like gene that regulates longevity and diapause in *Caenorhabditis elegans*. *Science*, *277*, 942–946.

King-Jones, K., Charles, J. P., Lam, G., & Thummel, C. S. (2005). The ecdysone-induced DHR4 orphan nuclear receptor coordinates growth and maturation in *Drosophila*. *Cell*, *121*, 773–784.

King-Jones, K., & Thummel, C. S. (2005). Nuclear receptors—A perspective from *Drosophila*. *Nature Reviews. Genetics*, *6*, 311–323.

Kiriishi, S., Nagasawa, H., Kataoka, H., Suzuki, A., & Sakurai, S. (1992). Comparison of in vivo and in vitro effects of Bombyxin and prothoracicotropic hormone on the prothoracic glands of the silkworm, *Bombyx mori*. *Zoological Science*, *9*, 149–155.

Kiss, I., Beaton, A. H., Tardiff, J., Fristrom, D., & Fristrom, J. W. (1988). Interactions and developmental effects of mutations in the Broad-Complex of *Drosophila melanogaster*. *Genetics*, *118*, 247–259.

Kjaer, K., Hagen, C., Sando, S. H., & Eshoj, O. (1992). Epidemiology of menarche and menstrual disturbances in an unselected group of women with insulin-dependent diabetes mellitus compared to controls. *The Journal of Clinical Endocrinology and Metabolism*, *75*, 524–529.

Koelle, M. R., Talbot, W. S., Segraves, W. A., Bender, M. T., Cherbas, P., & Hogness, D. S. (1991). The *Drosophila* EcR gene encodes an ecdysone receptor, a new member of the steroid receptor superfamily. *Cell, 67*, 59–77.

Koyama, T., Syropyatova, M. O., & Riddiford, L. M. (2008). Insulin/IGF signaling regulates the change in commitment in imaginal discs and primordia by overriding the effect of juvenile hormone. *Developmental Biology, 324*, 258–265.

Kozlova, T., & Thummel, C. S. (2002). Spatial patterns of ecdysteroid receptor activation during the onset of *Drosophila* metamorphosis. *Development, 129*, 1739–1750.

Kyriacou, C. P., Oldroyd, M., Wood, J., Sharp, M., & Hill, M. (1990). Clock mutations alter developmental timing in *Drosophila. Heredity (Edinb), 64*(Pt 3), 395–401.

Layalle, S., Arquier, N., & Leopold, P. (2008). The TOR pathway couples nutrition and developmental timing in *Drosophila. Developmental Cell, 15*, 568–577.

Lee, H. O., Davidson, J. M., & Duronio, R. J. (2009). Endoreplication: Polyploidy with purpose. *Genes & Development, 23*, 2461–2477.

Lee, K. S., Kwon, O. Y., Lee, J. H., Kwon, K., Min, K. J., Jung, S. A., et al. (2008). *Drosophila* short neuropeptide F signalling regulates growth by ERK-mediated insulin signalling. *Nature Cell Biology, 10*, 468–475.

Lin, J. L., & Gu, S. H. (2007). In vitro and in vivo stimulation of extracellular signal-regulated kinase (ERK) by the prothoracicotropic hormone in prothoracic gland cells and its developmental regulation in the silkworm, *Bombyx mori. Journal of Insect Physiology, 53*, 622–631.

Luo, J., Becnel, J., Nichols, C. D., & Nassel, D. R. (2012). Insulin-producing cells in the brain of adult *Drosophila* are regulated by the serotonin 5-HT1A receptor. *Cellular and Molecular Life Sciences, 69*, 471–484.

Marchal, E., Vandersmissen, H. P., Badisco, L., Van de Velde, S., Verlinden, H., Iga, M., et al. (2010). Control of ecdysteroidogenesis in prothoracic glands of insects: A review. *Peptides, 31*, 506–519.

McBrayer, Z., Ono, H., Shimell, M., Parvy, J. P., Beckstead, R. B., Warren, J. T., et al. (2007). Prothoracicotropic hormone regulates developmental timing and body size in *Drosophila. Developmental Cell, 13*, 857–871.

Messina, M. F., Arrigo, T., Valenzise, M., Ghizzoni, L., Caruso-Nicoletti, M., Zucchini, S., et al. (2011). Long-term auxological and pubertal outcome of patients with hereditary insulin-like growth factor-I deficiency (Laron and growth hormone-gene deletion syndrome) treated with recombinant human insulin-like growth factor-I. *Journal of Endocrinological Investigation, 34*, 292–295.

Miles, W. O., Jaffray, E., Campbell, S. G., Takeda, S., Bayston, L. J., Basu, S. P., et al. (2008). Medea SUMOylation restricts the signaling range of the Dpp morphogen in the *Drosophila* embryo. *Genes & Development, 22*, 2578–2590.

Mirth, C., & Shingleton, A. W. (2012). Integrating body and organ size in *Drosophila*: Recent advances and outstanding problems. *Frontiers in Endocrinology, 3*, 49.

Mirth, C., Truman, J. W., & Riddiford, L. M. (2005). The role of the prothoracic gland in determining critical weight for metamorphosis in *Drosophila melanogaster. Current Biology, 15*, 1796–1807.

Mirth, C. K., & Riddiford, L. M. (2007). Size assessment and growth control: How adult size is determined in insects. *Bioessays, 29*, 344–355.

Mirth, C. K., Truman, J. W., & Riddiford, L. M. (2009). The ecdysone receptor controls the post-critical weight switch to nutrition-independent differentiation in *Drosophila* wing imaginal discs. *Development, 136*, 2345–2353.

Nagasawa, H., Kataoka, H., Isogai, A., Tamura, S., Suzuki, A., Mizoguchi, A., et al. (1986). Amino acid sequence of a prothoracicotropic hormone of the silkworm *Bombyx mori. Proceedings of the National Academy of Sciences of the United States of America, 83*, 5840–5843.

Nijhout, H. F. (1981). Physiological control of molting in insects. *American Zoologist, 21*, 631–640.
Nijhout, H. F. (2003). The control of body size in insects. *Developmental Biology, 261*, 1–9.
Nijhout, H. F., & Williams, C. M. (1974a). Control of moulting and metamorphosis in the tobacco hornworm, Manduca sexta (L.): Cessation of juvenile hormone secretion as a trigger for pupation. *The Journal of Experimental Biology, 61*, 493–501.
Nijhout, H. F., & Williams, C. M. (1974b). Control of moulting and metamorphosis in the tobacco hornworm, *Manduca sexta* (L.): Growth of the last-instar larva and the decision to pupate. *The Journal of Experimental Biology, 61*, 481–491.
Niwa, R., Namiki, T., Ito, K., Shimada-Niwa, Y., Kiuchi, M., Kawaoka, S., et al. (2010). Non-molting glossy/shroud encodes a short-chain dehydrogenase/reductase that functions in the 'Black Box' of the ecdysteroid biosynthesis pathway. *Development, 137*, 1991–1999.
Niwa, R., Sakudoh, T., Namiki, T., Saida, K., Fujimoto, Y., & Kataoka, H. (2005). The ecdysteroidogenic P450 Cyp302a1/disembodied from the silkworm, *Bombyx mori*, is transcriptionally regulated by prothoracicotropic hormone. *Insect Molecular Biology, 14*, 563–571.
Okamoto, N., Yamanaka, N., Yagi, Y., Nishida, Y., Kataoka, H., O'Connor, M. B., et al. (2009). A fat body-derived IGF-like peptide regulates postfeeding growth in *Drosophila*. *Developmental Cell, 17*, 885–891.
Ono, H., Morita, S., Asakura, I., & Nishida, R. (2012). Conversion of 3-oxo steroids into ecdysteroids triggers molting and expression of 20E-inducible genes in *Drosophila* melanogaster. *Biochemical and Biophysical Research Communications, 421*, 561–566.
Ono, H., Rewitz, K. F., Shinoda, T., Itoyama, K., Petryk, A., Rybczynski, R., et al. (2006). Spook and Spookier code for stage-specific components of the ecdysone biosynthetic pathway in Diptera. *Developmental Biology, 298*, 555–570.
Ou, Q., Magico, A., & King-Jones, K. (2011). Nuclear receptor DHR4 controls the timing of steroid hormone pulses during *Drosophila* development. *PLoS Biology, 9*, e1001160.
Parvy, J. P., Blais, C., Bernard, F., Warren, J. T., Petryk, A., Gilbert, L. I., et al. (2005). A role for betaFTZ-F1 in regulating ecdysteroid titers during post-embryonic development in *Drosophila melanogaster. Developmental Biology, 282*, 84–94.
Petryk, A., Warren, J. T., Marques, G., Jarcho, M. P., Gilbert, L. I., Kahler, J., et al. (2003). Shade is the *Drosophila* P450 enzyme that mediates the hydroxylation of ecdysone to the steroid insect molting hormone 20-hydroxyecdysone. *Proceedings of the National Academy of Sciences of the United States of America, 100*, 13773–13778.
Poodry, C. A., & Woods, D. F. (1990). Control of the developmental timer for *Drosophila* pupariation. *Roux's Archives of Developmental Biology, 199*, 219–227.
Puig, O., Marr, M. T., Ruhf, M. L., & Tjian, R. (2003). Control of cell number by Drosophila FOXO: Downstream and feedback regulation of the insulin receptor pathway. *Genes & Development, 17*, 2006–2020.
Rewitz, K., & O'Connor, M. B. (2011). Timing is everything: PTTH mediated DHR4 nucleocytoplasmic trafficking sets the tempo of *Drosophila* steroid production. *Frontiers in Endocrinology, 2*, 108.
Rewitz, K. F., Larsen, M. R., Lobner-Olesen, A., Rybczynski, R., O'Connor, M. B., & Gilbert, L. I. (2009). A phosphoproteomics approach to elucidate neuropeptide signal transduction controlling insect metamorphosis. *Insect Biochemistry and Molecular Biology, 39*, 475–483.
Rewitz, K. F., Rybczynski, R., Warren, J. T., & Gilbert, L. I. (2006). The Halloween genes code for cytochrome P450 enzymes mediating synthesis of the insect moulting hormone. *Biochemical Society Transactions, 34*, 1256–1260.

Rewitz, K. F., Yamanaka, N., Gilbert, L. I., & O'Connor, M. B. (2009). The insect neuropeptide PTTH activates receptor tyrosine kinase torso to initiate metamorphosis. *Science*, *326*, 1403–1405.

Rewitz, K. F., Yamanaka, N., & O'Connor, M. B. (2010). Steroid hormone inactivation is required during the juvenile-adult transition in *Drosophila*. *Developmental Cell*, *19*, 895–902.

Riddiford, L. M. (2011). When is weight critical? *The Journal of Experimental Biology*, *214*, 1613–1615.

Riddiford, L. M., Truman, J. W., Mirth, C. K., & Shen, Y. C. (2010). A role for juvenile hormone in the prepupal development of *Drosophila melanogaster*. *Development*, *137*, 1117–1126.

Rideout, E. J., Marshall, L., & Grewal, S. S. (2012). *Drosophila* RNA polymerase III repressor Maf1 controls body size and developmental timing by modulating tRNAiMet synthesis and systemic insulin signaling. *Proceedings of the National Academy of Sciences of the United States of America*, *109*, 1139–1144.

Rulifson, E. J., Kim, S. K., & Nusse, R. (2002). Ablation of insulin-producing neurons in flies: Growth and diabetic phenotypes. *Science*, *296*, 1118–1120.

Rybczynski, R. (2005). The prothoracicotropic hormone. In L. I. Gilbert, K. Iatrou & S. Gill (Eds.), *Comprehensive molecular insect science* (pp. 61–123). Oxford: Elsevier.

Rybczynski, R., & Gilbert, L. I. (1994). Changes in general and specific protein synthesis that accompany ecdysteroid synthesis in stimulated prothoracic glands of *Manduca sexta*. *Insect Biochemistry and Molecular Biology*, *24*, 175–189.

Rybczynski, R., & Gilbert, L. I. (2006). Protein kinase C modulates ecdysteroidogenesis in the prothoracic gland of the tobacco hornworm, *Manduca sexta*. *Molecular and Cellular Endocrinology*, *251*, 78–87.

Sakurai, S., & Williams, C. M. (1989). Short-loop negative and positive feedback on ecdysone secretion by prothoracic gland in the tobacco hornworm, *Manduca sexta*. *General and Comparative Endocrinology*, *75*, 204–216.

Sehnal, F., & Bryant, P. J. (1993). Delayed pupariation in *Drosophila* imaginal disc overgrowth mutants is associated with reduced ecdysteroid titer. *Journal of Insect Physiology*, *39*, 1051–1059.

Shingleton, A. W., Das, J., Vinicius, L., & Stern, D. L. (2005). The temporal requirements for insulin signaling during development in *Drosophila*. *PLoS Biology*, *3*, e289.

Siaussat, D., Porcheron, P., & Debernard, S. (2009). The Ecdysteroids' effects in the control of cell proliferation and differentiation. In G. Smagghe (Ed.), *Ecdysone: Structures and functions* (pp. 185–204). Netherlands: Springer.

Siegmund, T., & Korge, G. (2001). Innervation of the ring gland of *Drosophila melanogaster*. *The Journal of Comparative Neurology*, *431*, 481–491.

Simpson, P., Berreur, P., & Berreur-Bonnenfant, J. (1980). The initiation of pupariation in *Drosophila*: Dependence on growth of the imaginal discs. *Journal of Embryology and Experimental Morphology*, *57*, 155–165.

Simpson, P., & Scheinderman, H. A. (1975). Isolation of temperature sensitive mutations blocking clone development in *Drosophila melanogaster*, and effects of a temperature sensitive cell lethal mutation on pattern formation in imaginal disks. *Wilhelm Rouxs Archives of Developmental Biology*, *178*, 247–275.

Slaidina, M., Delanoue, R., Gronke, S., Partridge, L., & Leopold, P. (2009). A *Drosophila* insulin-like peptide promotes growth during nonfeeding states. *Developmental Cell*, *17*, 874–884.

Smith, A. V., & Orr-Weaver, T. L. (1991). The regulation of the cell cycle during *Drosophila* embryogenesis: The transition to polyteny. *Development*, *112*, 997–1008.

Song, Q., & Gilbert, L. I. (1998). Alterations in ultraspiracle (USP) content and phosphorylation state accompany feedback regulation of ecdysone synthesis in the insect prothoracic gland. *Insect Biochemistry and Molecular Biology, 28*, 849–860.

Stieper, B. C., Kupershtok, M., Driscoll, M. V., & Shingleton, A. W. (2008). Imaginal discs regulate developmental timing in *Drosophila melanogaster*. *Developmental Biology, 321*, 18–26.

Talamillo, A., Sanchez, J., Cantera, R., Perez, C., Martin, D., Caminero, E., et al. (2008). Smt3 is required for *Drosophila melanogaster* metamorphosis. *Development, 135*, 1659–1668.

Talbot, W. S., Swyryd, E. A., & Hogness, D. S. (1993). *Drosophila* tissues with different metamorphic responses to ecdysone express different ecdysone receptor isoforms. *Cell, 73*, 1323–1337.

Tennessen, J. M., & Thummel, C. S. (2011). Coordinating growth and maturation - insights from *Drosophila*. *Current Biology, 21*, R750–R757.

Walkiewicz, M. A., & Stern, M. (2009). Increased insulin/insulin growth factor signaling advances the onset of metamorphosis in *Drosophila*. *PLoS One, 4*, e5072.

Walsh, A. L., & Smith, W. A. (2011). Nutritional sensitivity of fifth instar prothoracic glands in the tobacco hornworm, *Manduca sexta*. *Journal of Insect Physiology, 57*, 809–818.

Warren, J. T., Petryk, A., Marques, G., Jarcho, M., Parvy, J. P., Dauphin-Villemant, C., et al. (2002). Molecular and biochemical characterization of two P450 enzymes in the ecdysteroidogenic pathway of *Drosophila melanogaster*. *Proceedings of the National Academy of Sciences of the United States of America, 99*, 11043–11048.

Warren, J. T., Petryk, A., Marques, G., Parvy, J. P., Shinoda, T., Itoyama, K., et al. (2004). Phantom encodes the 25-hydroxylase of *Drosophila melanogaster* and *Bombyx mori*: A P450 enzyme critical in ecdysone biosynthesis. *Insect Biochemistry and Molecular Biology, 34*, 991–1010.

Warren, J. T., Yerushalmi, Y., Shimell, M. J., O'Connor, M. B., Restifo, L. L., & Gilbert, L. I. (2006). Discrete pulses of molting hormone, 20-hydroxyecdysone, during late larval development of *Drosophila melanogaster*: Correlations with changes in gene activity. *Developmental Dynamics, 235*, 315–326.

Wegman, L. J., Ainsley, J. A., & Johnson, W. A. (2010). Developmental timing of a sensory-mediated larval surfacing behavior correlates with cessation of feeding and determination of final adult size. *Developmental Biology, 345*, 170–179.

Wells, B. S., & Johnston, L. A. (2012). Maintenance of imaginal disc plasticity and regenerative potential in *Drosophila* by p53. *Developmental Biology, 361*, 263–276.

Wu, Q., & Brown, M. R. (2006). Signaling and function of insulin-like peptides in insects. *Annual Review of Entomology, 51*, 1–24.

Wullschleger, S., Loewith, R., & Hall, M. N. (2006). TOR signaling in growth and metabolism. *Cell, 124*, 471–484.

Xiang, Y., Liu, Z., & Huang, X. (2010). br regulates the expression of the ecdysone biosynthesis gene npc1. *Developmental Biology, 344*, 800–808.

Yamanaka, N., Honda, N., Osato, N., Niwa, R., Mizoguchi, A., & Kataoka, H. (2007). Differential regulation of ecdysteroidogenic P450 gene expression in the silkworm, *Bombyx mori*. *Bioscience, Biotechnology, and Biochemistry, 71*, 2808–2814.

Yamanaka, N., Hua, Y. J., Mizoguchi, A., Watanabe, K., Niwa, R., Tanaka, Y., et al. (2005). Identification of a novel prothoracicostatic hormone and its receptor in the silkworm *Bombyx mori*. *The Journal of Biological Chemistry, 280*, 14684–14690.

Yamanaka, N., Hua, Y. J., Roller, L., Spalovska-Valachova, I., Mizoguchi, A., Kataoka, H., et al. (2010). *Bombyx* prothoracicostatic peptides activate the sex peptide receptor to regulate ecdysteroid biosynthesis. *Proceedings of the National Academy of Sciences of the United States of America, 107*, 2060–2065.

Yamanaka, N., & O'Connor, M. B. (2011). Nitric oxide directly regulates gene expression during *Drosophila* development: Need some gas to drive into metamorphosis? *Genes & Development, 25*, 1459–1463.

Yamanaka, N., Rewitz, K., & O'Connor, M. B. (2013). Ecdysone control of developmental transitions: Lesson from *Drosophila* research. *Annual Review of Entomology, 58*, 497–516.

Yamanaka, N., Zitnan, D., Kim, Y. J., Adams, M. E., Hua, Y. J., Suzuki, Y., et al. (2006). Regulation of insect steroid hormone biosynthesis by innervating peptidergic neurons. *Proceedings of the National Academy of Sciences of the United States of America, 103*, 8622–8627.

Yoshiyama-Yanagawa, T., Enya, S., Shimada-Niwa, Y., Yaguchi, S., Haramoto, Y., Matsuya, T., et al. (2011). The conserved Rieske oxygenase DAF-36/Neverland is a novel cholesterol-metabolizing enzyme. *The Journal of Biological Chemistry, 286*, 25756–25762.

Yoshiyama, T., Namiki, T., Mita, K., Kataoka, H., & Niwa, R. (2006). Neverland is an evolutionally conserved Rieske-domain protein that is essential for ecdysone synthesis and insect growth. *Development, 133*, 2565–2574.

Young, S. C., Yeh, W. L., & Gu, S. H. (2012). Transcriptional regulation of the PTTH receptor in prothoracic glands of the silkworm, *Bombyx mori*. *Journal of Insect Physiology, 58*, 102–109.

Zhou, X., Zhou, B., Truman, J. W., & Riddiford, L. M. (2004). Overexpression of broad: A new insight into its role in the *Drosophila* prothoracic gland cells. *The Journal of Experimental Biology, 207*, 1151–1161.

CHAPTER TWO

What Goes Up Must Come Down: Transcription Factors Have Their Say in Making Ecdysone Pulses

Qiuxiang Ou, Kirst King-Jones[1]
Department of Biological Sciences, University of Alberta, Edmonton, Alberta, Canada
[1]Corresponding author: e-mail address: kirst.king-jones@ualberta.ca

Contents

Abstract

Insect metamorphosis is one of the most fascinating biological processes in the animal kingdom. The dramatic transition from an immature juvenile to a reproductive adult is under the control of the steroid hormone ecdysone, also known as the insect molting hormone. During *Drosophila* development, periodic pulses of ecdysone are released from the prothoracic glands, upon which the hormone is rapidly converted in peripheral tissues to its biologically active form, 20-hydroxyecdysone. Each hormone pulse has a unique profile and causes different developmental events, but we only have a

Current Topics in Developmental Biology, Volume 103
ISSN 0070-2153
http://dx.doi.org/10.1016/B978-0-12-385979-2.00002-2

rudimentary understanding of how the timing, amplitude, and duration of a given pulse are controlled. A key component involved in the timing of ecdysone pulses is PTTH, a brain-derived neuropeptide. PTTH stimulates ecdysone production through a Ras/Raf/ERK signaling cascade; however, comparatively little is known about the downstream targets of this pathway. In recent years, it has become apparent that transcriptional regulation plays a critical role in regulating the synthesis of ecdysone, but only one transcription factor has a well-defined link to PTTH. Interestingly, many of the ecdysteroidogenic transcription factors were originally characterized as primary response genes in the ecdysone signaling cascade that elicits the biological responses to the hormone in target tissues. To review these developments, we will first provide an overview of the transcription factors that act in the *Drosophila* ecdysone regulatory hierarchy. We will then discuss the roles of these transcriptional regulators in controlling ecdysone synthesis. In the last section, we will briefly outline transcription factors that likely have roles in regulating ecdysone synthesis but have not been formally identified as downstream effectors of ecdysone.

1. INTRODUCTION

Metamorphosis translates into "change in form." While most organisms display some changes in form during development, one typically uses the term "metamorphosis" in the context of the dramatic remodeling that occurs during the transition from a larval to an adult stage, as seen in holometabolous insects or amphibians. Insect metamorphosis is arguably one of the most striking developmental processes in the animal kingdom, allowing us to witness the transformation of an entire body plan into another. The change in body plans is not just a developmental necessity. In insects, for example, the use of different body plans serves as a remarkable adaptation to different habitats and food sources while undergoing development (Truman & Riddiford, 1999).

Typically, an insect's life cycle comprises embryonic, larval, pupal, and adult stages, where growth occurs exclusively during larval development. The succession of different stages and body plans begs the question as to how these transitions are regulated. Key components responsible for the dramatic reprogramming of body plans are small lipophilic hormones, such as ecdysteroids in insects and thyroid hormones in amphibians (Galton, 1992; Riddiford, 1993). In either case, these hormones act as ligands for members of the nuclear receptor superfamily that, upon ligand binding, initiate a cascade of gene programs that drive forward the remodeling process. In *Drosophila*, the ligand-bound ecdysteroid receptor activates a small group of early response genes, which mostly encode transcription factors. These early (or primary) regulatory proteins will then coordinate the expression of late (or secondary) response genes that ultimately effectuate the required developmental changes (Fig. 2.1B). These downstream effects of steroid and thyroid hormones are complex events; however, they represent just one side of the coin. Clearly, we may also ask what happens upstream of the hormone to understand how the synthesis and release of these hormones are controlled in the first place. As it turns out, the release of these developmentally important hormones is strictly controlled, resulting in precisely timed pulses that control the onset of developmental transitions such as hatching, molting, and metamorphosis itself (Thummel, 1995).

In this review, we will focus on transcriptional regulators acting downstream and upstream of the steroid hormone ecdysone. After a brief historical perspective, we will review the classic ecdysone hierarchy and its role in metamorphosis, and then turn our attention to the more recent finding that many of the transcription factors known to act directly downstream of ecdysone also act upstream of the hormone, thus regulating the production of the hormone itself. In the last part, we will discuss transcription factors that have not yet been linked to the ecdysone hierarchy but have been shown to function in the prothoracic gland, the principal source of ecdysone synthesis in the developing insect.

2. HISTORICAL PERSPECTIVE

The identification of ecdysone as the key molting hormone in the 1950s is a milestone in the history of insect endocrinology. In 1954, Peter Karlson and colleagues purified 25 mg of ecdysone crystals from 500 kg of silk moth pupae using the *Calliphora* bioassay to track the activity of the hormone (Butenandt & Karlson, 1954). In a series of chemical experiments and the analysis of the crystals, ecdysone was later shown to be a steroid hormone (Huber & Hoppe, 1965; Karlson, Hoffmeister, Hummel, Hocks, & Spiteller, 1965). The first evidence that ecdysone has a direct role in regulating gene expression was based on the puffing of the salivary gland polytene chromosomes. Puffs are enlargements of specific loci on these giant chromosomes and were interpreted as local transcriptional activity. In particular, it was found that some of these puffs were induced rapidly after the addition of ecdysone to cultured salivary glands of the midge *Chironomus* (Clever & Karlson, 1960). Curiously, some of the puffs were responding rapidly to the hormone (early puffs), while others were delayed (late puffs). To test

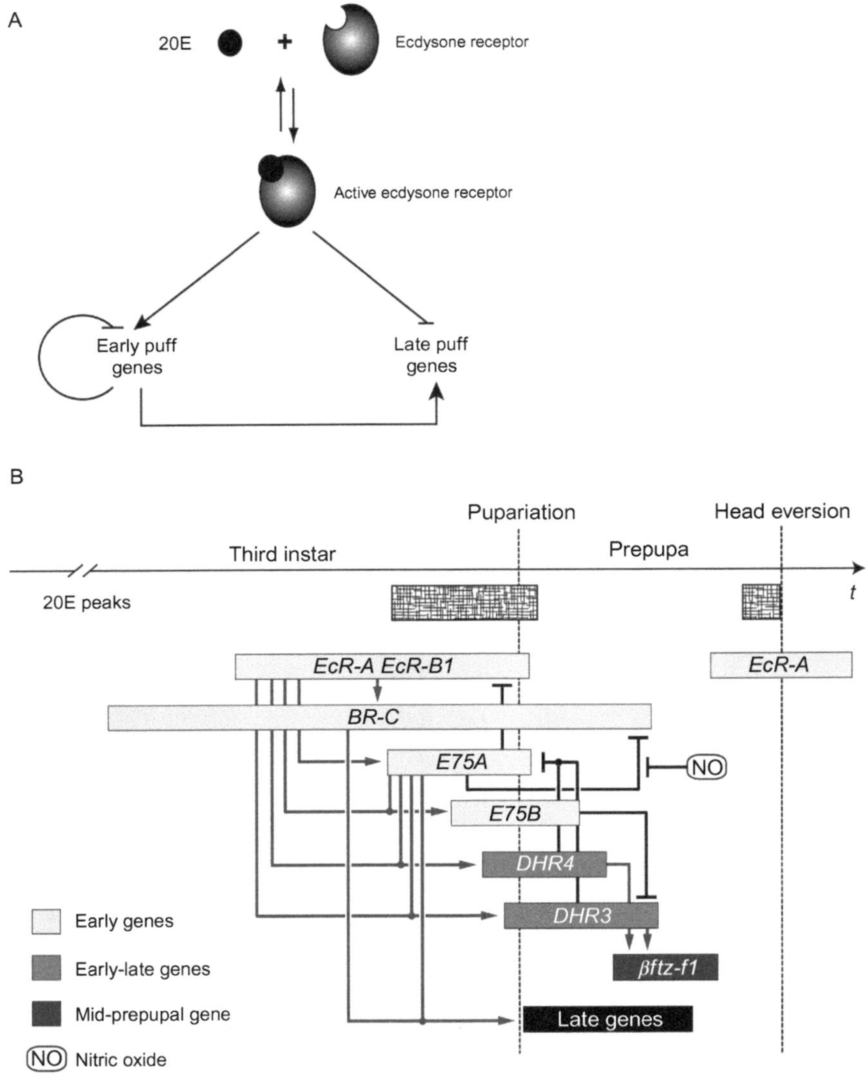

Figure 2.1 Ecdysone hierarchy. (A) Ashburner model. The hormone-bound and thus active ecdysone receptor directly induces the expression of early puff genes and represses late puff genes. A small set of early puff genes repress their own expression and are required for the induction of a large set of late puff genes. (B) Overview of *Drosophila* ecdysone hierarchy genes at the onset of metamorphosis. The expression of genes is shown in bars with different shades of gray representing different gene categories (see inset), and the length of the bars indicates the approximate duration of their expression. Positive and inhibitory interactions are shown. Ecdysone peaks are shown in dotted boxes at the top, and approximate timing of puparium formation and head eversion are indicated by dotted lines.

whether puffing was a direct consequence of ecdysone activity, a series of elegant studies by Clever (in *Chironomus*) and later by Ashburner (in *Drosophila*) tested whether protein synthesis was a requirement for the induction of puffs by ecdysone. These studies showed that the early puffs were still induced in the presence of protein synthesis inhibitors, while the late puffs were not. Ashburner also found that these early puffs are autoregulated because they failed to regress when protein synthesis was inhibited. Ashburner correctly predicted that the early puffs are direct targets of the ecdysone-bound receptor and that the corresponding early genes encode regulatory proteins that are required for inducing the late puffs (Fig. 2.1A). To this day, this conceptual framework is referred to as the Ashburner model (Ashburner, 1974; Ashburner, Chihara, Meltzer, & Richards, 1974).

Ultimately, the studies performed by Karlson, Clever, Ashburner, and colleagues not only offered some of the first insights into how genes may be regulated but also produced the key ingredients for establishing an elegant and versatile model for how steroid hormones coordinate complex developmental processes.

3. THE ECDYSONE HIERARCHY I: HORMONE ACTION AT THE ONSET OF METAMORPHOSIS

In *Drosophila*, all major developmental transitions, including the molts and the onset of metamorphosis, are triggered by major pulses of ecdysone (Riddiford, 1993). Each of these pulses has its own characteristics, such as amplitude and duration, which are largely determined by the rate and duration of hormone synthesis, how efficiently the hormone is converted to its biologically active form, and how fast it is degraded. Ecdysone is produced and released from the prothoracic gland cells, which are part of a composite endocrine organ called the ring gland. Once taken up by its target tissues, ecdysone is converted to the biologically active form 20-hydroxyecdysone (hereafter referred to as 20E) (Gilbert, Rybczynski, & Warren, 2002). Like vertebrate steroid hormones, 20E acts by binding to members of the nuclear receptor superfamily. These are ligand-dependent transcription factors that harbor a highly conserved DNA-binding domain (DBD) as well as less conserved ligand-binding domain (LBD) (King-Jones & Thummel, 2005). The identification of the *Drosophila* ecdysone receptor gene (*EcR*) and the discovery of several early ecdysone response genes established the molecular era of ecdysone biology in the early 1990s (Burtis, Thummel, Jones, Karim, & Hogness, 1990; DiBello, Withers, Bayer, Fristrom, & Guild, 1991; Koelle et al., 1991; Segraves & Hogness, 1990).

EcR requires heterodimerization with another nuclear receptor, Ultraspiracle (USP) to form a functional ecdysteroid receptor capable of binding to 20E with high affinity (Thomas, Stunnenberg, & Stewart, 1993; Yao et al., 1993). *EcR* encodes three protein isoforms, EcR-A, EcR-B1, and EcR-B2, as a result of two promoters and alternative splicing (Talbot, Swyryd, & Hogness, 1993). All three EcR isoforms are able to interact with USP, and all can bind to 20E with similar affinity. The crystal structure of the EcR LBD suggested that USP is required for forming a ligand-binding conformation, corroborating the observation that EcR alone cannot transcriptionally activate genes (Billas et al., 2003; Hu, Cherbas, & Cherbas, 2003). Likewise, structural studies demonstrated that the LBD of dipteran and lepidopteran USP is locked in an inactive conformation, consistent with the idea that ecdysteroids achieve transcriptional activation through binding to EcR (Billas, Moulinier, Rochel, & Moras, 2001; Clayton, Peak-Chew, Evans, & Schwabe, 2001).

The EcR/USP heterodimer functions at the top of ecdysone regulatory cascade and triggers the transcription of primary and secondary response genes in ecdysone target tissues that play more direct functions during development (Fig. 2.1B). Mutations affecting the region common to all isoforms of *EcR* are embryonic lethal, consistent with the finding that ecdysone signaling plays a critical role during germ-band retraction in the developing *Drosophila* embryo (Bender, Imam, Talbot, Ganetzky, & Hogness, 1997; Kozlova & Thummel, 2003). *EcR-B1* is predominantly expressed in larval tissues that do not contribute to adult structures, and loss of *EcR-B1* function blocks the ecdysone responses in these tissues, resulting in a failure to complete metamorphosis (Bender et al., 1997; Schubiger, Wade, Carney, Truman, & Bender, 1998). In contrast, the *EcR-A* isoform is expressed in imaginal disks and the ring gland, and animals mutant for *EcR-A* arrest development during late stages of pupal development (Davis, Carney, Robertson, & Bender, 2005; Talbot et al., 1993), indicating that the different EcR isoforms have distinct functions during development.

The EcR dimerization partner USP is the fly homolog of vertebrate RXR (Henrich, Sliter, Lubahn, MacIntyre, & Gilbert, 1990; Oro, McKeown, & Evans, 1990). Like EcR, USP is required during embryogenesis and metamorphosis, consistent with the idea that USP acts as a key partner for EcR throughout development (Hall & Thummel, 1998; Oro, McKeown, & Evans, 1992; Perrimon, Engstrom, & Mahowald, 1985). USP also dimerizes with the nuclear receptors DHR38 and Seven-up (Baker et al., 2003; Zelhof, Yao, Chen, Evans, & McKeown, 1995), and a recent report found that EcR forms functional dimers with DHR38 as well (Zoglowek et al., 2012). In addition, genetic evidence shows that *usp* is not required for the ecdysone-dependent induction of the larval glue genes, raising the possibility that EcR requires a different partner for this response (Costantino et al., 2008). The ability of nuclear receptors to form multiple heterodimers adds another layer of regulatory complexity that will be fascinating to unravel in the future.

4. THE ECDYSONE HIERARCHY II: EARLY RESPONSE GENES

The molecular characterization of three early ecdysone-inducible genes *BR-C*, *E74*, and *E75* revealed that all of them encode transcription factors, albeit belonging to different DNA-binding protein families (Thummel, 1990). These primary ecdysone response genes are key regulators of the ecdysone genetic hierarchy, which induce the transcription of secondary response genes that in turn execute the appropriate biological effects in response to an ecdysone pulse at the onset of metamorphosis (Fig. 2.1B).

Mutations that disrupt all *BR-C* functions (*npr1* alleles) result in larval lethality, indicating that *BR-C* is an essential gene for entry into metamorphosis (Kiss, Beaton, Tardiff, Fristrom, & Fristrom, 1988). The *broad* gene (hereafter referred to as *Broad-Complex* or *BR-C*) maps to the 2B5 early puff and is undoubtedly the most complex of the early genes. FlyBase currently acknowledges 14 transcript isoforms (McQuilton, St Pierre, & Thurmond, 2012), and genetically the locus contains up to four complementation groups (DiBello et al., 1991). *BR-C* produces four protein classes, dependent on which zinc finger module, designated Z1 to Z4, is incorporated into a given isoform. The common N-terminal region comprises a BTB/POZ domain, which is a protein–protein interaction domain commonly found in chromatin and transcription factors. The zinc fingers are believed to confer target specificity (DiBello et al., 1991; Zollman, Godt, Prive, Couderc, & Laski, 1994). However, high-affinity DNA binding was never established for BR-C proteins, and existing EMSA (Xiang, Liu, & Huang, 2010) and footprinting (von Kalm, Crossgrove, Von Seggern, Guild, & Beckendorf, 1994) studies all used uncommonly high BR-C concentrations to achieve DNA binding. Future studies will have to address whether BR-C recognizes its target genes via binding to DNA elements or through interactions with other chromatin-bound proteins, in which case the zinc finger domains may have a less direct role in target gene recognition.

Like BR-C, *E74* is directly induced by ecdysone and responsible for the 74EF early puff. Mutations in *E74* confer pupal lethality, indicating that this gene plays essential roles during metamorphosis. *E74* produces two protein isoforms, E74A and E74B, which share a C-terminal ETS DBD (Burtis et al., 1990). Both isoforms are precisely controlled by changes in ecdysone titers and display complementary profiles. *E74A* is induced when hormone concentrations are high, while *E74B* is abundant when ecdysone concentrations have fallen to intermediate or lower levels. Correspondingly, *E74A* transcript levels fall when ecdysone concentrations start to decline, and *E74B* mRNA is repressed by rising hormone titers. This behavioral link between the two isoforms is critical for the proper timing of secondary gene responses (Fletcher, D'Avino, & Thummel, 1997; Karim & Thummel, 1991; Urness & Thummel, 1995).

The *E75* early gene maps to the 75B early puff and encodes a member of the nuclear receptor superfamily. E75 forms at least three protein isoforms (E75A–C) (Segraves & Hogness, 1990). Like all *Drosophila* nuclear receptor genes, alternative splicing tends to produce protein isoforms that differ in their N-terminal sequences but share a common LBD in the C-terminus. This is not any different for E75; however, the E75B isoform represents an unusual nuclear receptor protein: While E75A and E75C both have a complete DBD and LBD domain, splicing of *E75B* removes a part of the DBD domain, which abolishes its ability to bind to DNA. This splice form appears to be a fairly ancient invention, as its closest fly homolog, *E78*, also generates a protein isoform (E78B) with a truncated DBD domain (Stone & Thummel, 1993). Mutations specific for *E75B* are viable, however, molecular data demonstrated that E75B binds to another nuclear receptor, DHR3, in an inhibitory fashion to delay the induction of a third nuclear receptor, βFTZ-F1 (White, Hurban, Watanabe, & Hogness, 1997). It should be noted that *E75B* null mutants do not display defects in the timing of *βftz-f1* expression, raising the possibility that E75B and E78B are functionally redundant (Russell, Heimbeck, Goddard, Carpenter, & Ashburner, 1996; Stone & Thummel, 1993).

In contrast to *E75B*, animals mutant for *E75A* display larval lethality, molting defects, and developmental delays, while *E75C* is required for late pupal development and adult viability (Bialecki, Shilton, Fichtenberg, Segraves, & Thummel, 2002). In 2005, the Krause lab published the surprising finding that E75 binds with high affinity to heme (Reinking et al., 2005). This led to the suggestion that the protein either acts as a heme or gas sensor. A recent study from the same lab showed that E75 is a sensor for the signaling molecule nitric oxide (NO), and we will discuss these findings in a later section (Caceres et al., 2011).

As briefly alluded to above, many components of the ecdysone regulatory hierarchy are—like EcR and USP—members of the nuclear receptor superfamily. These include *Drosophila* hormone receptor 3 (DHR3), DHR4, DHR39, E75, E78, and FTZ-F1 (King-Jones & Thummel, 2005; Woodard, Baehrecke, & Thummel, 1994). Below, we will briefly review the roles of DHR3, DHR4, and FTZ-F1 in the context of early metamorphic stages.

5. THE ECDYSONE HIERARCHY III: EARLY-LATE RESPONSE GENES AND *βftz-f1*

Early-late genes can be operationally defined as genes that require both the 20E-bound EcR/USP heterodimer and an early gene product for maximal transcriptional induction (Fig. 2.1B). This is typically shown in organ culture assays using protein synthesis inhibitors to block the translation of early gene mRNAs. Two early-late genes with very similar temporal expression profiles are the nuclear receptor genes *DHR3* and *DHR4*. DHR3 is orthologous to the vertebrate retinoid-related orphan receptor (ROR), while DHR4 is represented by the germ-cell nuclear factor (GCNF) in vertebrates. DHR3 and DHR4 expression profiles show a peak at the beginning of prepupal stage, when the expression of early genes such as *BR-C*, *E74A*, and *E75A* is receding, and *βftz-f1* expression is about to be induced. Both DHR3 and DHR4 are sufficient to repress the early genes and are required for maximal *βftz-f1* expression in mid-prepupae (King-Jones, Charles, Lam, & Thummel, 2005; Lam, Jiang, & Thummel, 1997), strongly suggesting that these two factors act in concert to regulate the early genes and *βftz-f1*. Interestingly, *DHR4* mutants display precocious wandering behavior followed by premature onset of metamorphosis, resulting in a small body size due to a shortened feeding period, a phenotype not observed in any other mutants associated with the ecdysone hierarchy (King-Jones et al., 2005). This peculiar phenotype was eventually tracked to a role for DHR4 in the prothoracic gland, which we will discuss in the next section.

The *Drosophila ftz-f1* gene encodes yet another nuclear receptor acting in the ecdysone cascade and is orthologous to vertebrate steroidogenic factor 1 (SF-1). Two protein isoforms have been described, αFTZ-F1 and βFTZ-F1 (Lavorgna, Ueda, Clos, & Wu, 1991; Ueda, Sonoda, Brown, Scott, & Wu, 1990). While αFTZ-F1 is maternally supplied and has critical roles in embryogenesis, βFTZ-F1 is also expressed in the early stages of puparium formation (Yamada et al., 2000; Yu et al., 1997). Mutations in *βftz-f1* severely perturb the ecdysone signaling pathway at the onset of metamorphosis and consequently result in prepupal lethality. Later, it was shown that *βftz-f1* functions as a competence factor during prepupal development, ensuring that the responses to the late larval ecdysone pulse are different from the prepupal pulse 12 h later (Broadus, McCabe, Endrizzi, Thummel, & Woodard, 1999) (Fig. 2.1B). Taken together, the interplay between nuclear receptors E75, DHR3, and DHR4 controls the expression of *βftz-f1* during the prepupal stage, thereby safeguarding the appropriate sequence of programs necessary for the progression of pupal development.

6. UPSTREAM OF THE HORMONE: ECDYSONE SYNTHESIS IN THE PROTHORACIC GLANDS

It is well established that the prothoracic glands are the principal source of ecdysone during larval development (Fig. 2.2A). In total, six major ecdysone pulses control embryogenesis, the molts, the onset of metamorphosis, and the differentiation of adult tissues. In addition, three minor pulses of ecdysone that occur in the third larval instar are critical for physiological and behavioral changes prior to metamorphosis, including the commitment of a larva to a pupal fate (critical weight checkpoint), the induction of the glue genes that allow the pupa to attach to a solid substrate, and the transition from feeding to wandering behavior (Davidowitz, D'Amico, & Nijhout, 2003; Lehmann, 1996; Mirth, Truman, & Riddiford, 2005; Riddiford, 1993; Sokolowski, 2001; Warren et al., 2006) (Fig. 2.2B). While these ecdysone pulses are well defined, it remains rather unclear as to how the onset, amplitude, and duration of these hormone peaks are regulated. Our current understanding is based on two cornerstones: The PTTH (prothoracicotropic hormone) signaling pathway and the enzymes involved in ecdysone biosynthesis. PTTH is a brain-derived peptide that is rhythmically released from two pairs of neurons and acts as a temporal stimulus for ecdysone production (Fig. 2.4A). Ultimately, PTTH must impinge on ecdysone biosynthesis for which currently eight enzymes and their corresponding genes have been identified; however, how this is accomplished is largely unclear. In this section, we will summarize our current understanding of the PTTH signaling pathway, and how it might regulate ecdysone biosynthesis.

The prothoracic gland has to import and transport dietary cholesterol, which serves as a precursor for the synthesis of ecdysone (Gilbert, 2004;

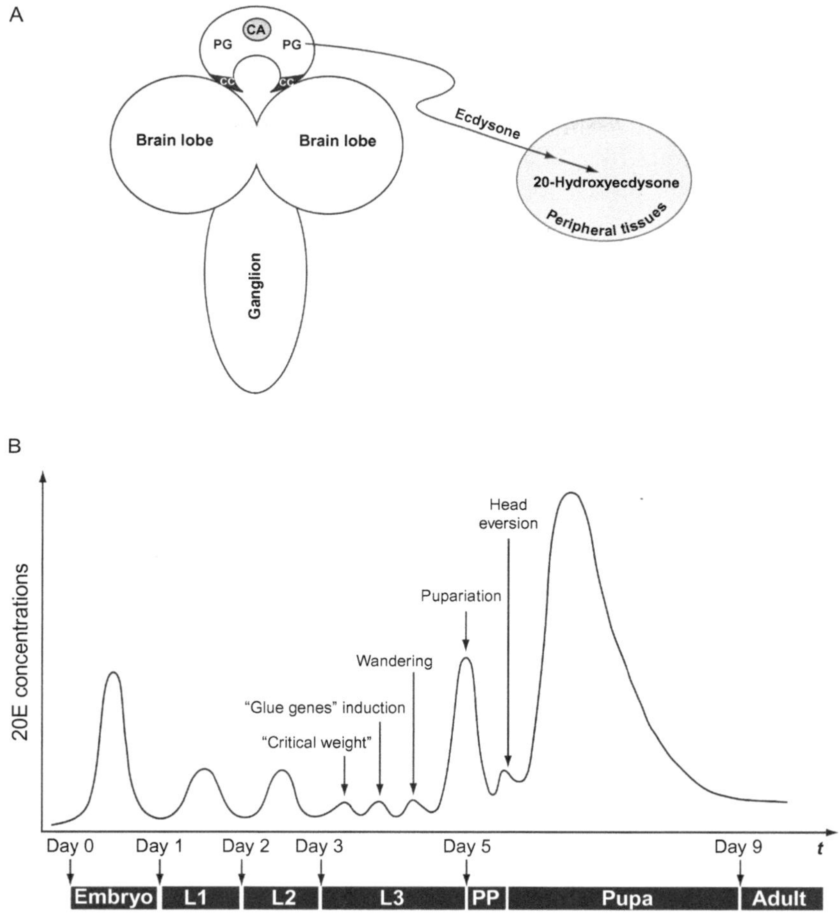

Figure 2.2 Ecdysone directs developmental transitions in *Drosophila*. (A) Ecdysteroids are produced in prothoracic gland cells of the ring gland, which is attached to the anterior side of the brain. Hormones are then released into the hemolymph and subsequently converted to the biologically active form 20-hydroxyecdysone (20E) in peripheral tissues. (B) Schematic representation of whole-body concentrations of 20E during *Drosophila* development. Three minor ecdysone pulses occur during the third instar and have been suggested to trigger the physiological and behavioral changes indicated by the arrows. PG, prothoracic gland; CA, corpus allatum; CC, corpora cardiaca; L1/L2/L3, first/second/third instar; PP, prepupa.

Gilbert et al., 2002). The conversion from cholesterol to ecdysone starts with Neverland, a Rieske electron oxygenase, which generates 7-dehydrocholesterol (Yoshiyama, Namiki, Mita, Kataoka, & Niwa, 2006; Yoshiyama-Yanagawa et al., 2011). The last four steps toward 20E are also characterized and are carried out by the cytochrome P450 enzymes Phantom, Disembodied, Shadow, and

Shade, collectively referred to as the Halloween enzymes (Chavez et al., 2000; Niwa et al., 2005; Nusslein-Volhard & Wieschaus, 1980; Petryk et al., 2003; Warren et al., 2002, 2004). By comparison, relatively little is known about the reaction steps converting 7-dehydrocholesterol to 5β-ketodiol, commonly known as the "Black Box." The black box is believed to harbor the rate-limiting step(s) of ecdysone production, and enzymes believed to act in this part of the pathway are Shroud (a short-chain dehydrogenase/reductase), Spookier and Cyp6t3 (both cytochrome P450 enzymes) (Gilbert, 2004; Niwa et al., 2010; Ono et al., 2006; Ou, Magico, & King-Jones, 2011) (Fig. 2.3).

It has long been known that a small brain-derived peptide, now known as PTTH, stimulates the production and release of ecdysone

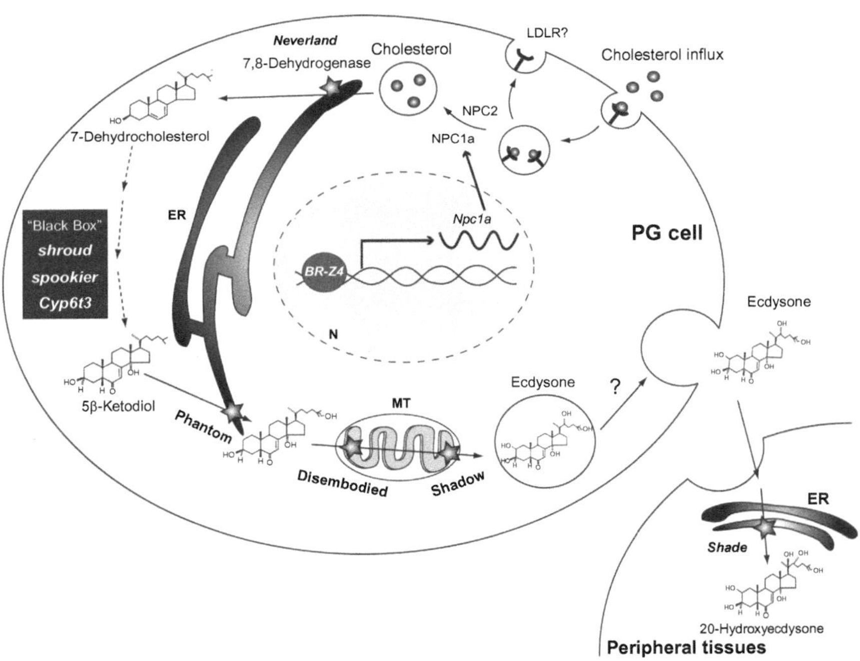

Figure 2.3 A schematic diagram of ecdysone biosynthesis in the *Drosophila* prothoracic gland. Cholesterol is converted into the prohormone ecdysone via a series of reactions that occur in the ER, cytosol (suggested for other arthropod species) (Blais et al., 1996), and mitochondria, represented here by stars, and exported possibly by secretory vesicles into hemolymph (O'Connor, 2011). The early gene *broad* (isoform BR-Z4) positively regulates the expression of *Npc1a*, which encodes a cholesterol transporter mediating the trafficking of cholesterol out of late endosomes into the ER, a critical step toward the synthesis of ecdysone. LDLR, low-density lipoprotein receptor; MT, mitochondria; ER, endoplasmic reticulum; N, nucleus; PG, prothoracic gland.

(Rybczynski, 2005). Studies in *Manduca* and *Bombyx* have shown that the PTTH prohormone is synthesized in a pair of neurosecretory cells in the brain, and released into the hemolymph, where it is processed into its mature form before it reaches the prothoracic glands (Kawakami et al., 1990; Westbrook & Bollenbacher, 1990). In contrast, *Drosophila* PTTH-producing neurons innervate the prothoracic gland directly, resulting in the release of PTTH in the immediate vicinity of prothoracic gland cells (Fig. 2.4A). The gene encoding *Drosophila* PTTH was only recently identified, which opened the door to genetically dissecting the PTTH signaling pathway. Unfortunately, no mutants for the *Drosophila ptth* gene currently exist; however, genetic ablation of PTTH-producing neurons provided valuable insights into this signaling pathway (McBrayer et al., 2007). Surprisingly, removing PTTH function through ablation is not lethal to the animal; rather, it results in prolonged larval stages and significantly larger animals compared to controls. Remarkably, not all larval stages are equally delayed, and it appears that larvae without functional PTTH reach the third instar with comparatively small delays but then remain in this stage considerably longer than the normal 2 days. PTTH-ablation does lower the expression of the Halloween genes, especially *disembodied* and *phantom*, suggesting that the developmental delay in these animals is caused by a reduced rate of ecdysone biosynthesis (McBrayer et al., 2007). The fact that loss-of-PTTH function hardly results in delays of the first and second instars indicates that ecdysone production occurs efficiently enough at these stages, even without PTTH stimulation. This is interesting for several reasons: First, it seems to suggest that PTTH acts as a booster of the Halloween genes transcription toward the end of the third instar when ecdysone titers rise dramatically, but that the previous molts do not require such a stimulus. This then raises two questions: How are the previous molts temporally regulated in the absence of PTTH? Are Halloween transcripts expressed at high levels throughout larval development, even without PTTH stimulation? A recent study published from our lab agrees with this notion, based on the finding that ring gland-specific microarrays show very high transcript levels of the classic Halloween genes in the early third instar, long before PTTH stimulates the late third instar peak (Ou et al., 2011).

How is the PTTH signal transduced in prothoracic gland cells? A recent study identified a receptor for *Drosophila* PTTH, a receptor tyrosine kinase called Torso, and showed that it activates a MAP kinase pathway (Rewitz, Yamanaka, Gilbert, & O'Connor, 2009). In particular, upon binding to PTTH, Torso activates a small GTPase, Ras, which in turn triggers

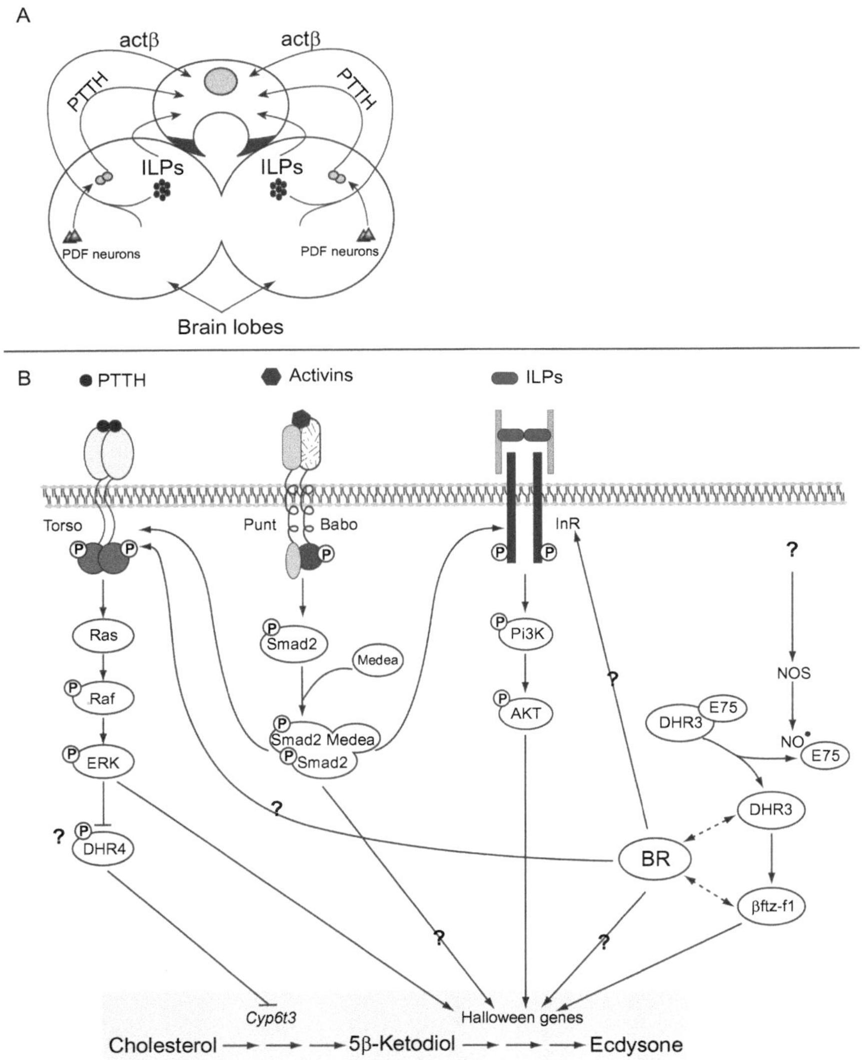

Figure 2.4 An overview of proposed mechanisms underlying regulation of ecdysone biosynthesis in the *Drosophila* prothoracic gland. (A) A range of developmental factors exert effects on the prothoracic gland to control the production of ecdysone pulses, including PTTH, TGFβ/Activins, ILPs, and circadian signals. (B) Multiple signaling cascades that regulate ecdysone biosynthesis in the prothoracic gland. Inactivation of these pathways could compromise ecdysone production and results in developmental defects. Arrows represent positive regulations, and cross lines indicate inhibitory effects. P, phosphorylation; NO, nitric oxide radical.

Raf/ERK phosphorylation (Fig. 2.4B). The main question, however, still remains: What are downstream effectors of PTTH signaling? Biochemical studies in *Manduca* have demonstrated that the ribosomal protein S6 (RpS6) is a target of PTTH, suggesting that upregulation of translation is a requirement for maximal ecdysone production (Song & Gilbert, 1995, 1997). It will be interesting to determine whether RpS6 phosphorylation is genetically dependent on the PTTH/Ras/ERK pathway. However, there is no doubt that many other PTTH targets exist, and in a separate section, we will later discuss the recent discovery that the nuclear receptor DHR4 is regulated by PTTH signaling.

Ultimately, many questions remain: How are the periodic PTTH signals regulated in the first place? Is PTTH required for all pulses, major and minor, or can other pathways compensate for a loss of PTTH signaling? Does PTTH stimulate increased enzyme activity or transcriptional upregulation of the Halloween genes? Is the withdrawal of PTTH enough to terminate a pulse? Answering these and other questions will greatly advance our understanding of how neuropeptide signaling pathways regulate the formation of steroid hormone pulses.

7. ROLE REVERSAL: THE ECDYSONE HIERARCHY REGULATES ITS OWN HORMONE

The ecdysone regulatory cascade is best understood at the onset of metamorphosis, and the finding that some components of the hierarchy might also play important roles in the production of ecdysone is not entirely new. One of the first indicators was that *EcR-A* is expressed in the prothoracic gland but not the other two isoforms encoded by *EcR* (Talbot et al., 1993). Another study reported that USP modulates PTTH-dependent ecdysone synthesis in *Manduca* prothoracic glands (Song & Gilbert, 1998). Together, these results suggested that EcR and USP may have roles in ecdysteroidogenesis, possibly through negative feedback regulators in response to rising levels of 20E. Finally, a null mutation in *E75A* causes a dramatic decrease in ecdysone levels, indicating that E75A plays a dual role, acting both downstream of ecdysone as an 20E target during the onset of metamorphosis and also upstream of ecdysone as a regulator of ecdysone production in the prothoracic gland (Bialecki et al., 2002) (Table 2.1).

Below, we will summarize recent findings on the novel functions of several classic ecdysone hierarchy genes that have roles important for three

Table 2.1 Transcription factors with roles in the prothoracic gland

Transcription factor	Domain information	Mutant phenotype[a]	PG expression[b]	Signaling pathway	Putative target	Reference
EcR (*EcR*)	Znf-C4, LBD (NR family)	Embryonic lethal	Immunostaining (EcR-A)	Negative feedback?	?	Talbot et al. (1993)
USP (*ultraspiracle*)	Znf-C4, LBD (NR family)	L1 lethal	WB on *Manduca* PG	Negative feedback?	?	Song & Gilbert (1998)
E75A (*Eip75B*)	Znf-C4, LBD (NR family)	Larval/pupal lethal	Immunostaining	NO signaling	*ftz-f1*	Caceres et al. (2011)
E75B (*Eip75B*)	Znf-C4, LBD (NR family)	Viable	Immunostaining	NO signaling	*ftz-f1*	Caceres et al. (2011)
DHR3 (*DHR3*)	Znf-C4, LBD (NR family)	Embryonic lethal	Immunostaining	NO signaling	*ftz-f1*	Caceres et al. (2011)
DHR4 (*DHR4*)	Znf-C4, LBD (NR family)	Prepupal lethal	Immunostaining	PTTH signaling	*Cyp6t3*	King-Jones et al. (2005)
βFTZ-F1 (*ftz-f1*)	Znf-C4, LBD (NR family)	Embryonic lethal	ISH, immunostaining	NO signaling	*dib* and *phm*	Parvy et al. (2005), Caceres et al. (2011)
WOC (*without children*)	Znf-C4	Larval lethal	ISH	?	7,8-Dehydrogenase activity	Wismar et al. (2000)
MLD (*molting defective*)	Znf-C2H2	L1 lethal	Immunostaining	?	*spok*	Neubueser, Warren, Gilbert, and Cohen (2005)
DRE4 (*dre4*)	FACT/DUF1747 domain	Larval and pupal lethal	Genetic evidence	?	?	Sliter & Gilbert (1992)

Smad2 (*smox*)	SMAD domain	Pupal lethal	Genetic evidence	TGFβ/ Activin signaling	*torso, inR*	Gibbens, Warren, Gilbert, and O'Connor (2011)
Med (*medea*)	SMAD domain	Embryonic lethal	Genetic evidence	TGFβ/ Activin signaling	*torso, InR*	Gibbens et al. (2011)
Per (*period*)	PAS domain	Viable	Immunostaining	?	?	Emery, Noveral, Jamison, and Siwicki (1997) Myers, Yu, and Sehgal (2003)
Tim (*timeless*)	DNA-binding	Viable	Immunostaining	?	?	Myers et al. (2003)
HLH54F (*HLH54F/ CG5005*)	bHLH domain (MyoD family)	Lethal	FISH	?	?	Namiki et al. (2009)
Ttk (*tramtrack*)	Znf-C2H2, BTB/ POZ domain	Lethal	Enhancer traps	?	?	Harvie, Filippova, and Bryant (1998)
BR (*broad complex*)	Znf-C2H2, BTB/ POZ domain	Early larval lethal	Immunostaining (BR Z2-Z4)	?	*Npc1a*	Zhou et al. (2004)
CRC (*cryptocephal*)	bZIP domain (CREB/ATF family)	Lethal	ISH	?	?	Hewes, Schaefer, and Taghert (2000)

[a]Phenotype for the strongest mutant allele reported.
[b]PG expression was originally reported in the reference listed.
Abbreviations: Znf, Zinc finger; bHLH, basic helix–loop–helix; bZIP, basic leucine zipper; NO, nitric oxide radical.

aspects of ecdysone synthesis: (a) coordinating cholesterol availability in prothoracic glands, (b) integration of PTTH signaling and ecdysone biosynthesis, and (c) modulation of the expression of the ecdysone biosynthetic enzymes (Figs. 2.3 and 2.4; Table 2.1).

8. *broad* CONTROLS A KEY COMPONENT OF CELLULAR CHOLESTEROL TRANSPORT

A recent study has shown that *broad* serves as a key regulator for coordinating available cholesterol levels required for ecdysone synthesis in *Drosophila* prothoracic glands (Xiang et al., 2010). *Drosophila*, like all other insects, is a cholesterol auxotroph and must obtain cholesterol or other suitable sterols directly from a dietary source (Carvalho et al., 2010; Clayton, 1964). Therefore, cholesterol uptake and transport by the prothoracic gland represent critical steps for the synthesis of steroid hormones in insects.

The study published by Xiang and colleagues shows that *broad* positively regulates the expression of *Npc1a* in the prothoracic gland, a cholesterol transporter gene (Xiang et al., 2010). In humans, mutations in the *NPC1* gene cause Niemann-Pick Type C disease, a fatal neurodegenerative disorder where patients accumulate free cholesterol in lysosomes and display defects in glycolipid sorting (Sturley, Patterson, Balch, & Liscum, 2004). NPC1 proteins are highly conserved and have a sterol-sensing domain that presumably recognizes free sterols and assists cholesterol to move out of lysosomes (Davies & Ioannou, 2000). The *Drosophila* genome encodes two NPC1 homologs, NPC1a and NPC1b, which exhibit 42% and 35% identity to the human NPC1 protein, respectively (Carstea et al., 1997; Fluegel, Parker, & Pallanck, 2006; Loftus et al., 1997). In contrast to NPC1b, which is restricted to the gut, the *Drosophila Npc1a* gene is ubiquitously expressed, and a null allele of *Npc1a* results in early larval lethality (Fluegel et al., 2006; Huang, Suyama, Buchanan, Zhu, & Scott, 2005). Interestingly, the lethal phenotype of *Npc1a* mutants can be partially rescued by rearing animals on a high cholesterol diet or by adding the insect steroid hormone 20E to the medium, suggesting that *Npc1a* mutants are ecdysone deficient (Huang et al., 2005).

Npc1a transcripts are significantly downregulated in the ring gland of both *broad* mutants (*npr* alleles) and ring gland-specific *broad* RNAi animals, but not affected by mutations in other transcription factor genes including *ftz-f1*, *woc*, *mld*, and *ecd1* (see Table 2.1), suggesting that cholesterol uptake is transcriptionally controlled by *broad*. This raised the possibility that *npr* mutants are unable to initiate metamorphosis because low levels of *Npc1a* result in reduced uptake of cholesterol and consequently insufficient ecdysone production. However, *broad* mutants could not be rescued with *Npc1a* overexpression, indicating that the reason for the lethality is not just due to the lack of NPC1a. The authors also report the identification of a 220-bp DNA region within the *Npc1a* regulatory region that is necessary and sufficient to drive the ring gland-specific expression of a reporter gene, which constitutes the first *cis*-regulatory module (CRM) reported for the ring gland. Studies in vertebrates have shown that the regulatory regions of the human, porcine, and mouse *NPC1* genes harbor a conserved consensus element recognized by sterol-regulatory element binding proteins (SREBPs) (Gevry, Schoonjans, Guay, & Murphy, 2008). SREBP proteins induce *NPC1* expression in steroidogenic cells under low intracellular cholesterol concentrations. While this is an intriguing parallel, *Drosophila* SREBP (HLH106) has not been linked to ecdysone or cholesterol pathways in prothoracic gland cells, but rather plays a role in regulating fatty acid biosynthesis (Kunte, Matthews, & Rawson, 2006).

The authors further suggest that the Z4 isoform of Broad directly binds to a conserved 20-bp region within the ring gland CRM, based on published binding data for the Z4 isoform we mentioned earlier. Experimentally, they show that $broad^{npr\text{-}3}$, an allele of *broad* with defects in Z4, has very low *Npc1a* transcript levels, and that overexpression of Z4, but not Z1–Z3 leads to a concomitant increase of *Npc1a* transcript levels. While these data convincingly tie BR-Z4 to the regulation of *Npc1a*, more experiments are needed to show that the Z4 isoform is indeed binding the proposed ring gland CRM. As the authors point out, *broad* expression is not restricted to the ring gland, while the identified enhancer activity appears to be restricted to this tissue. It is possible that other ring gland-specific transcription factors are required to facilitate BR-Z4 binding to the ring gland CRM, but more data, such as chromatin IP, are needed to demonstrate direct binding of BR-Z4 to this DNA region.

Finally, a range of genes with known ring gland expression was severely downregulated in a $broad^{npr\text{-}3}$ mutant background, including several Halloween genes and the PTTH receptor Torso (Fig. 2.4B). This raises the interesting question as to whether *broad* acts as a general regulator of these critical ecdysone biosynthetic genes, or whether this observation reflects a secondary response to the loss of *Npc1a*. It seems plausible that the reduced availability of cholesterol caused by the loss of functional NPC1a in turn triggers a widespread feedback response that downregulates ecdysone production by throttling the expression of ecdysone biosynthetic genes. It is reasonable to assume that the expression of ecdysone biosynthetic enzymes is somehow tied into cellular cholesterol levels, as it would be rather wasteful to express these genes at high levels when cholesterol is in short supply. To test this possibility, one could examine whether ring gland-specific knockdown of *Npc1a* causes downregulation of ecdysone biosynthetic genes. Previous studies have shown that NPC2 family members NPC2a and NPC2b play essential roles in the prothoracic gland. *Drosophila Npc2a* and *Npc2b* double mutants display similar phenotype as the *Npc1a* mutant, and were fully rescued by supplying 20E, suggesting that these animals lack ecdysone (Huang, Warren, Buchanan, Gilbert, & Scott, 2007). Therefore, it will be intriguing to see whether *Npc2a*/*Npc2b* double mutants affect the expression of genes involved in ecdysteroidogenesis. Experiments along these lines would test whether cellular cholesterol levels are sensed in the ring gland, and whether changes in available cholesterol affect the expression of the Halloween genes.

9. *DHR4* IS A KEY MEDIATOR OF PTTH SIGNALING

As mentioned earlier, PTTH-mediated stimulation of ecdysone synthesis is best established for the last major larval ecdysone pulse that triggers the onset of metamorphosis. However, whether the three minor ecdysone pulses during the last larval instar are also controlled in this manner remained unexplored. Recently, our group has demonstrated that *DHR4*, which we discussed earlier as an ecdysone hierarchy gene, is a target of PTTH signaling in the *Drosophila* prothoracic gland cells and plays a key role in the proper timing of ecdysone pulses.

DHR4 is expressed in three major larval tissues: the prothoracic gland, the salivary glands, and the fat body. A hypomorphic mutation in *DHR4* causes two distinct phenotypes. First, larvae stop feeding prematurely, which results in small animals, and second, animals arrest development in the early prepupal stage, indicating that DHR4 has critical roles in larvae as well as during the prepupal stage (King-Jones et al., 2005). This duality was later confirmed by tissue-specific knockdown of *DHR4* (Ou et al., 2011). When *DHR4* function is reduced specifically in the ring gland, larvae stop feeding prematurely and enter early into metamorphosis, while knockdown of *DHR4* in the fat body results in prepupal lethality, but timing is normal. Closer inspection of *DHR4* expression in the ring gland revealed that the protein oscillates with 8–16 h ultradian cycle times between cytoplasm and nucleus of prothoracic gland cells. This behavior closely matched the oscillations that were reported for *Drosophila* PTTH mRNA levels (McBrayer et al., 2007), which suggested a functional link between PTTH and DHR4. Indeed, when PTTH neurons were ablated, DHR4 remained primarily in the nucleus, but when the PTTH pathway was hyperactivated, the protein was predominantly cytoplasmic. Consistent with this observation, hyperactivation of the PTTH pathway via expressing constitutively active *Ras* (Ras^{V12}) in the prothoracic gland (Caldwell, Walkiewicz, & Stern, 2005; Karim & Rubin, 1998) also results in accelerated entry into metamorphosis, suggesting that this is achieved precisely because active Ras prevents DHR4 from entering the nucleus (Fig. 2.4B).

If loss-of-*DHR4* function causes shortened feeding times and accelerated entry into metamorphosis, one would predict that this might arise from a premature occurrence of ecdysone pulses. This appears to be the case, as depleting DHR4 in the ring gland via RNAi causes a faster rise of ecdysone levels that fail to regress when hormone levels normally drop. We suspect that the animal interprets these higher ecdysone levels as an early ecdysone pulse, resulting in developmental acceleration. Our model therefore proposes that DHR4 acts as a repressor of ecdysone pulses by counteracting the PTTH-stimulated rise of ecdysone levels. This idea is further corroborated by the finding that prothoracic gland-specific overexpression of *DHR4* blocks molting, a phenotype that can be partially rescued by feeding 20E. If DHR4 is a target of the PTTH pathway, then one would expect that ERK phosphorylates DHR4 to trigger its removal from the nucleus when the PTTH signaling is active. While this has not been demonstrated directly, it appears that there is an inverse correlation between the subcellular localization of ERK and DHR4, suggesting that there is a functional link (Fig. 2.4B). Indeed, DHR4 is predicted to have several clusters of ERK target sites, and future studies such as mutational analysis of putative ERK phosphorylation sites of DHR4 will have to be carried out in order to determine whether these sites affect the subcellular localization of the protein.

What are the downstream targets of DHR4? Does DHR4 regulate the expression of the Halloween genes? In *Drosophila*, the expression levels of *phantom*, *disembodied*, and *shadow* are transcriptionally upregulated roughly 8 h prior to the onset of metamorphosis, consistent with the idea that PTTH induces the Halloween genes at the transcriptional level (Parvy et al., 2005). However, the expression levels of these three Halloween genes are relatively constant (and fairly high as we pointed out earlier) in the last larval instar before the major PTTH pulses, suggesting that the three minor ecdysone pulses are not simply a consequence of modulating mRNA levels of the Halloween genes. In line with this, none of the Halloween genes appeared to be affected in ring glands of *DHR4* mutants or *DHR4* RNAi animals. Through ring gland-specific microarrays, a hitherto uncharacterized cytochrome P450 gene, *Cyp6t3*, was identified as a candidate downstream target of DHR4 (Fig. 2.4B). Several lines of evidence link *Cyp6t3* to *DHR4* and ecdysone synthesis: (1) decreasing *Cyp6t3* expression in the prothoracic gland results in molting defects and developmental delays, which can be rescued by 20E, suggesting that these animals are ecdysone deficient; (2) Cyp6t3 activity can be provisionally placed into the "Black Box" of the pathway: The ecdysone intermediate 5β-ketodiol can partially rescue the *Cyp6t3* RNAi phenotype, however, precursors upstream of the "Black Box" cannot; (3) *Cyp6t3* expression is significantly increased in the ring gland of both *DHR4* mutant and *DHR4* RNAi animals, indicating the gene is normally repressed by DHR4; (4) knockdown of *Cyp6t3* expression in a *DHR4* mutant background can reverse the precocious pupariation phenotype; (5) *Cyp6t3* mRNA levels oscillate, where lower levels correlate with times when DHR4 is nuclear. Taken together, these data demonstrate that *Cyp6t3* is negatively regulated by DHR4 to downregulate ecdysone production, but whether the gene is a direct target of *DHR4* needs to be determined by further experiments.

10. DHR3/E75-MEDIATED NO SIGNALING IS A CRITICAL PLAYER OF ECDYSTEROIDOGENESIS

The recent report by Caceres et al. (2011) that NO plays a key role in the production of ecdysone comes undoubtedly as a surprise. In brief, it was shown that NO signaling regulates the interaction of DHR3 and E75 to control the transcriptional activation of *βftz-f1*, all well-characterized players of the ecdysone hierarchy (Fig. 2.4B).

NO is produced by nitric oxide synthase (NOS) and is a short-lived diatomic molecule that acts as an intra- and transcellular messenger that regulates many physiological functions in both vertebrates and invertebrates (Bredt & Snyder, 1994). The *Drosophila* genome harbors only one nitric oxide synthase gene called *NOS*, which is expressed throughout *Drosophila* development (Stasiv, Regulski, Kuzin, Tully, & Enikolopov, 2001). Previous studies have shown that NO plays important roles in imaginal disk development, synaptogenesis, formation of retinal projection patterns, response to hypoxia, and behavioral responses (Suman, Seth, & Chandna, 2008).

Two papers reporting the effects of a null mutation in *NOS* come to different conclusions. One paper reports that a *NOS* null mutation results in early larval lethality (Regulski, Stasiv, Tully, & Enikolopov, 2004), while a second publication finds that *NOS* is not required for *Drosophila* development and results in viable flies (Yakubovich, Silva, & O'Farrell, 2010). The paper discussed here (Caceres et al., 2011) finds that two nonoverlapping *NOS* RNAi lines have similar phenotypes and cause lethality when expressed in the prothoracic gland. These seemingly contradictory results may arise when a tissue-specific loss-of-gene-function is more detrimental than when the same gene is mutated in all cells of the organism, possibly due to compensatory mechanisms. A similar case was reported for Smad2, where two nonoverlapping RNAi transgenes cause lethality when expressed in the prothoracic gland, while the published null *Smad2* mutant is viable (Gibbens et al., 2011). Future research has to address whether this phenomenon is real, or if a more mundane explanations exist.

A knockdown of *NOS* in the prothoracic gland via RNAi causes animals to remain third instars, and the inability to undergo metamorphosis is likely caused by insufficient amounts of ecdysone. How does NO signaling affect ecdysone synthesis? NO often exerts its function by binding to heme groups, and Henry Krause's lab showed in 2005 (Reinking et al., 2005) that the nuclear receptor E75 contains a heme moiety that, upon binding to NO, controls its ability to interact with DHR3. This exciting discovery was soon followed by a series of studies on the REV-ERB proteins, the vertebrate homologs of E75. Like E75, the REV-ERBs are transcriptional repressors that regulate circadian rhythms, lipid, and glucose metabolism (Ramakrishnan & Muscat, 2006). The REV-ERBs also bind heme, but unlike E75, heme-binding is not essential for protein stability, rather, it appears that heme binds to REV-ERBs in a reversible manner. However, the transcriptional repression mediated by heme-bound REV-ERBs and RORs, the vertebrate homolog of DHR3, is reversed by the addition of NO (Marvin et al., 2009; Raghuram et al., 2007). Taken together, these studies show that heme is structurally required for E75, and NO is a ligand for this nuclear receptor.

During prepupal development, the expression of *βftz-f1* is controlled by the interaction of the nuclear receptors DHR3 and E75B. In particular, DHR3 induces the *βftz-f1* gene; however, E75B blocks this inductive function by binding to DHR3 on the *βftz-f1* promoter, until the disappearance of E75B relieves this repression (White et al., 1997). Is it possible that NO regulates the DHR3/E75B interaction in the prothoracic gland via binding to heme? This concept is based on the idea that βFTZ-F1 regulates ecdysone biosynthesis, as it was shown that similar to its vertebrate homolog SF-1, βFTZ-F1 also has critical roles in the regulation steroidogenic genes (Broadus et al., 1999; Parker & Schimmer, 1997; Parvy et al., 2005; Yamada et al., 2000). SF-1 expression is restricted to steroidogenic tissues in mice, and the protein controls the expression of several steroid hydroxylase genes. Likewise, βFTZ-F1 was shown to regulate the expression of at least two ecdysone biosynthetic enzymes, Phantom and Disembodied, in prothoracic gland cells (Parvy et al., 2005) (Fig. 2.4B).

Several lines of evidence support the idea that NO signaling modulates the DHR3/E75-mediated regulation of βFTZ-F1, which would in turn affect ecdysone production (Caceres et al., 2011). First, the *NOS* RNAi knockdown in prothoracic glands results in a failure to induce the expression of *βftz-f1*, consistent with the idea that the late third instar ecdysone peak relies on proper *βftz-f1* upregulation. Second, *βftz-f1* expression can be restored by either providing a constitutively active form of the NOS enzyme *in vivo* or by coculturing ring glands with an NO donor *in vitro*. These data strongly suggest that NO is sufficient to activate *βftz-f1* expression at this time in development. Third, the inhibition of DHR3 activity by E75 binding can be reversed by NO. Taken together, the report by Caceres and colleagues proposes a novel mechanism by which ecdysteroidogenesis is controlled in the *Drosophila* prothoracic gland. To further consolidate this model, one could determine whether a ring gland-specific knockdown of *βftz-f1* results in low ecdysone titers, and whether the expression or upregulation of the two reported βFTZ-F1 targets, *phantom* and *disembodied*, are abolished in the absence of NO signaling.

It is of interest to note that the prothoracic gland-specific knockdown of *NOS* triggers a massive overgrowth of the ring gland. NO is a well-known regulator of cell proliferation (Villalobo, 2006), which raises the interesting question as to whether this overgrowth phenotype is due to an increase in cell number or cell size. The *Drosophila* ring gland cells undergo endoreplication where the cell number is determined during embryogenesis, which appears to be consistent with the idea that NO acts as a negative effector of cell growth. However, it was recently suggested that DHR3 plays a role in maintaining cell-autonomous growth (Montagne et al., 2010; Villalobo, 2006), raising the possibility that hyperactivation of DHR3 due to low NO levels could contribute to the overgrowth of the gland in *NOS* RNAi animals.

11. OTHER TRANSCRIPTION FACTORS WITH KNOWN EXPRESSION IN THE PROTHORACIC GLAND

In this last section, we will briefly review other transcription factors that have been directly or indirectly linked to the ecdysone synthesis pathway (Table 2.1). Below, we will highlight seven of them and discuss their putative functions in the *Drosophila* prothoracic gland.

11.1. WOC (*without children*)

The *Drosophila woc* gene encodes a transcription factor that harbors eight C_2C_2 zinc finger repeats. The *woc* gene is expressed in many tissues, including the ring gland, imaginal disks, the brain–ganglion complex, and others. Homozygous *woc* mutants have overgrown ring glands and fail to pupariate; however, feeding 20E allows mutants to undergo metamorphosis (Klose, Gateff, Emmerich, & Beikirch, 1980; Wismar et al., 2000), suggesting WOC plays a role in regulating ecdysone synthesis. This was supported by the finding that ring glands of *woc* mutants are not able to convert radiolabelled cholesterol into 7-dehydrocholesterol. Further, feeding mutants with 7-dehydrocholesterol, but not cholesterol, leads to a significant increase in the ecdysone titers (Warren, Wismar, Subrahmanyam, & Gilbert, 2001). Together, these studies indicate that WOC regulates the first step of ecdysone biosynthesis in *Drosophila*, likely through regulating the activity of 7,8-cholesterol dehydrogenase (now known to be Neverland); however, the mechanism by which WOC achieves this remains unaddressed.

11.2. MLD (*molting defective*)

The *Drosophila mld* gene encodes a nuclear protein that is predicted to have seven C_2H_2 zinc fingers in two clusters in the C-terminal region. Immunodetection showed that MLD protein is present in larval tissues including the ring gland, salivary gland, fat body, and imaginal disks. Mutations of *mld* result in an arrest of development in the first larval instar. This defect can be rescued by administration of excess E or 20E, suggesting that *mld* participates in modulating ecdysone synthesis (Neubueser et al., 2005). Semiquantitative PCR showed that homozygous *mld* mutant first instar larvae have decreased transcript levels of the ecdysone biosynthetic gene *spookier*, while other known enzymatic genes remain unaffected or upregulated, suggesting that *spookier* is a target of *mld* in the steroidogenic pathway (Ono et al., 2006). However, restoration of *spookier* expression in the *mld* mutant ring gland is not sufficient to rescue animals to develop beyond the first instar, suggesting that *spookier* is not the only target of MLD. Similar to the *woc* mutant, mutations in the *mld* gene also result in overgrowth of the ring gland, reflecting a possibility that increasing the gland size serves as a common strategy to compensate for the inability of the tissue to synthesize ecdysone (Neubueser et al., 2005). However, it may be still worthwhile to examine whether either WOC or MLD plays a role in the regulation of cell size/growth.

11.3. DRE4 (*dre4*)

Loss-of-function mutations of the *Drosophila dre4* gene cause stage-specific developmental arrests due to a reduction of ecdysone levels. Studies on a temperature-sensitive *dre4* allele showed that mutant ring glands failed to produce ecdysone shortly after a shift to the restrictive temperature, indicating a cell-autonomous function of the *dre4* gene in the ring gland (Sliter & Gilbert, 1992). Some evidence suggests that the DRE4 functions by remodeling chromatin. A heterodimer of DRE4 and SSPR1 (known as FACT) associates with the GAGA factor, a well-established chromatin remodeling factor (Shimojima et al., 2003). While these results have not been obtained for the prothoracic gland, it is still quite likely that DRE4 functions by remodeling chromatin structure in the prothoracic gland, possibly facilitating local access for transcription factors to regulate ecdysteroidogenic genes.

11.4. Smad2/Med (*smox/medea*)

Smad2, along with its heteromeric partner Medea (Smad4), serves as the core downstream mediator of the TGFβ/Activin signaling in *Drosophila* (Brummel et al., 1999; Schmierer & Hill, 2007), and a recent report shows that this pathway plays an important role in ecdysone synthesis pathways (Gibbens et al., 2011) (Fig. 2.4). Knocking down *Smad2* via RNAi specifically in the prothoracic gland causes larvae to arrest in the third instar, and this defect can be rescued by providing 20E in the medium. Knocking down other key components of the TGFβ/Activin pathway gave rise to the same nonpupariating phenotype, suggesting that TGFβ/Activin signaling is required for the proper onset of metamorphosis. This study has also shown that transcript levels of *torso* (Strecker, Yip, & Lipshitz, 1991) and *Insulin Receptor* (*InR*) (Chen, Jack, & Garofalo, 1996) are dramatically reduced when *Smad2* is silenced via RNAi. In addition, two ecdysone biosynthetic genes, *dib* and *spok*, show severely reduced transcript levels in *Smad2* RNAi ring glands. These results suggest that the TGFβ/Activin pathway plays a critical role in ecdysteroidogenesis through controlling the competence of the prothoracic gland to respond to metamorphic factors, in particular, PTTH and insulin-like peptides (ILPs) (Wu & Brown, 2006).

There are several questions that would be worth addressing in future studies to further dissect roles of the TGFβ/Activin signaling in the *Drosophila* prothoracic gland. First, it is unclear in which manner the TGFβ/Activin pathway is required for the expression of *torso* and *InR*. For instance, is the requirement dynamic and therefore only needed at specific times in development or is the requirement more general to keep *InR* and *torso* levels up at all times? Second, are *torso* and *InR* direct targets of the Smad2/Med complex? In vertebrates, SMAD-binding sites have been elucidated (SMAD-binding element or SBDs) (Zawel et al., 1998), however, corresponding data for *Drosophila* Smad2/Med complexes are unavailable.

11.5. Per/Tim (*period/timeless*)

The release of ecdysone from the prothoracic glands is under circadian control in many insect species (Giebultowicz, 1999; Schwartz & Truman, 1983). Studies in *Rhodnius* have shown that the last larval instar displays daily fluctuations of ecdysone in the hemolymph, which was attributed to the rhythmic release of PTTH being controlled by the central clock in the brain. In particular, four neurons in the dorsal brain form the central pacemaker that releases a peptide hormone, PDF (pigmentdispersing factor), in a circadian manner. Interestingly, not only do these PDF neurons innervate the prothoracic gland but also does a mutation in the PDF gene affect the periodicity of *ptth* transcript levels, strongly suggesting that ecdysone synthesis is under circadian control. However, further experiments are needed to examine whether PTTH synthesis and/or release of the peptide are affected in PDF mutants (McBrayer et al., 2007; Renn, Park, Rosbash, Hall, & Taghert, 1999). Interestingly, cultured prothoracic glands are able to properly time the release of ecdysone, suggesting that the gland itself contains an independent clock (Ampleford & Steel, 1985; Vafopoulou & Steel, 1996a, 1996b). The circadian genes *period* and *timeless* encode transcription factors that regulate the expression of clock-controlled genes (Bae, Lee, Sidote, Chuang, & Edery, 1998). A previous report showed that *period* is rhythmically expressed in the prothoracic gland of *Drosophila* pupae, supporting the notion of circadian clock components in the tissue, at least at the stage that was examined (Emery et al., 1997). A more recent study demonstrated that a peripheral clock is required for normal eclosion in addition to the central pacemaker in the brain (Myers, 2003; Myers et al., 2003). Taken together, these data support the idea of a direct role for timeless and period in the regulation of ecdysone synthesis.

11.6. HLH54F (*HLH54F*)

Drosophila HLH54F has roles in muscle development and encodes a basic helix–loop–helix transcription factor that is highly conserved between flies and vertebrates (Ismat et al., 2010; Robb, Hartley, Wang, Harvey, & Begley, 1998). *HLH54F* is also expressed in prothoracic glands of wandering *Bombyx* larvae and *Drosophila* embryos, but its expression in the post-embryonic stages of *Drosophila* development has not been reported (Namiki et al., 2009). The expression in prothoracic glands could indicate a role in ecdysteroid biosynthesis or the development of the gland, but we have to await genetic data before more can be said about the roles of this transcription factor.

11.7. Ttk (*tramtrack*)

The *Drosophila tramtrack* gene encodes two protein isoforms, Ttk69 and Ttk88, which are transcription factors that share an N-terminal BTB/POZ domain for protein–protein interaction but have different sets of zinc fingers for distinct DNA-binding specificities (Giesen et al., 1997; Read & Manley, 1992; Zollman et al., 1994). *ttk* is strongly expressed during mid-embryogenesis and is present in various larval tissues during postembryonic development. Based on enhancer traps, *ttk* expression was detected in the prothoracic gland at all three larval stages (Harvie et al., 1998). However, it remains unknown whether *ttk* is important for any aspect of prothoracic gland development and/or function. It was shown that Ttk69 represses gene expression mediated by the GAGA factor through attenuating the interaction between GAGA and the general transcription machinery (Pagans, Pineyro, Kosoy, Bernues, & Azorin, 2004). More recent evidence suggested that the GAGA/Hsf/Ttk69 complex acts downstream of Trunk/Torso signaling in regulating gene expression via remodeling local chromatin structure (Chen, Lin, Chen, Chiang, & Liaw, 2009), raising the interesting possibility that *ttk* may function as a target of the PTTH/Torso in the prothoracic gland as well. It is interesting to note that Ttk and Broad share the same type of protein–protein interaction domain (BTB/POZ domains). Another protein harboring this domain, Batman (a.k.a. Lola like) was reported to interact physically with Broad (Faucheux et al., 2003). Interestingly, genome-wide protein–protein interaction data suggest that Batman can interact with Ttk (Murali et al., 2011; Pacifico et al., 2006; Yu & Finley, 2009; Yu, Pacifico, Liu, & Finley, 2008). Taken together, these interesting anecdotal data raise the possibility that Ttk could form multimers not just with itself but also with other proteins such as Broad and Batman (Bonchuk, Denisov, Georgiev, & Maksimenko, 2011).

11.8. CRC/ATF4 (*cryptocephal*)

The *cryptocephal* gene encodes the *Drosophila* homolog of vertebrate ATF4, which belongs to the CREB/ATF family of basic leucine zipper (bZIP) proteins (Hewes et al., 2000; Persengiev & Green, 2003). Mutations in *crc* result in a range of developmental abnormalities, including molting defects, the failure of the gas bubble to translocate during pupal development, the absence of head eversion, and defects in imaginal disks elongation. Together, these data show that *crc* is crucial for controlling molting and metamorphosis in *Drosophila* (Hewes et al., 2000). Further, *crc* is required for the expression of the ecdysis triggering hormone (Gauthier & Hewes, 2006). *crc* is strongly expressed in the larval ring gland, the CNS, and the imaginal disks in third instar larvae at the wandering stage, raising the possibility that *crc* might exert its control over metamorphic processes through controlling ecdysone biosynthesis as well as acting downstream of the hormone in target tissues.

REFERENCES

Ampleford, E. J., & Steel, C. G. (1985). Circadian control of a daily rhythm in hemolymph ecdysteroid titer in the insect Rhodnius prolixus (Hemiptera). *General and Comparative Endocrinology*, *59*, 453–459.

Ashburner, M. (1974). Sequential gene activation by ecdysone in polytene chromosomes of Drosophila melanogaster. II. The effects of inhibitors of protein synthesis. *Developmental Biology*, *39*, 141–157.

Ashburner, M., Chihara, C., Meltzer, P., & Richards, G. (1974). Temporal control of puffing activity in polytene chromosomes. *Cold Spring Harbor Symposia on Quantitative Biology*, *38*, 655–662.

Bae, K., Lee, C., Sidote, D., Chuang, K. Y., & Edery, I. (1998). Circadian regulation of a Drosophila homolog of the mammalian Clock gene: PER and TIM function as positive regulators. *Molecular and Cellular Biology*, *18*, 6142–6151.

Baker, K. D., Shewchuk, L. M., Kozlova, T., Makishima, M., Hassell, A., Wisely, B., et al. (2003). The Drosophila orphan nuclear receptor DHR38 mediates an atypical ecdysteroid signaling pathway. *Cell*, *113*, 731–742.

Bender, M., Imam, F. B., Talbot, W. S., Ganetzky, B., & Hogness, D. S. (1997). Drosophila ecdysone receptor mutations reveal functional differences among receptor isoforms. *Cell*, *91*, 777–788.

Bialecki, M., Shilton, A., Fichtenberg, C., Segraves, W. A., & Thummel, C. S. (2002). Loss of the ecdysteroid-inducible E75A orphan nuclear receptor uncouples molting from metamorphosis in Drosophila. *Developmental Cell, 3*, 209–220.

Billas, I. M., Iwema, T., Garnier, J. M., Mitschler, A., Rochel, N., & Moras, D. (2003). Structural adaptability in the ligand-binding pocket of the ecdysone hormone receptor. *Nature, 426*, 91–96.

Billas, I. M., Moulinier, L., Rochel, N., & Moras, D. (2001). Crystal structure of the ligand-binding domain of the ultraspiracle protein USP, the ortholog of retinoid X receptors in insects. *The Journal of Biological Chemistry, 276*, 7465–7474.

Blais, C., Dauphin-Villemant, C., Kovganko, N., Girault, J. P., Descoins, C., Jr., & Lafont, R. (1996). Evidence for the involvement of 3-oxo-delta 4 intermediates in ecdysteroid biosynthesis. *The Biochemical Journal, 320*(Pt 2), 413–419.

Bonchuk, A., Denisov, S., Georgiev, P., & Maksimenko, O. (2011). Drosophila BTB/POZ domains of "ttk group" can form multimers and selectively interact with each other. *Journal of Molecular Biology, 412*, 423–436.

Bredt, D. S., & Snyder, S. H. (1994). Nitric oxide: A physiologic messenger molecule. *Annual Review of Biochemistry, 63*, 175–195.

Broadus, J., McCabe, J. R., Endrizzi, B., Thummel, C. S., & Woodard, C. T. (1999). The Drosophila beta FTZ-F1 orphan nuclear receptor provides competence for stage-specific responses to the steroid hormone ecdysone. *Molecular Cell, 3*, 143–149.

Brummel, T., Abdollah, S., Haerry, T. E., Shimell, M. J., Merriam, J., Raftery, L., et al. (1999). The Drosophila activin receptor baboon signals through dSmad2 and controls cell proliferation but not patterning during larval development. *Genes & Development, 13*, 98–111.

Burtis, K. C., Thummel, C. S., Jones, C. W., Karim, F. D., & Hogness, D. S. (1990). The Drosophila 74EF early puff contains E74, a complex ecdysone-inducible gene that encodes two ets-related proteins. *Cell, 61*, 85–99.

Butenandt, A., & Karlson, P. (1954). Über die Isolierung eines Metamorphose-Hormons der Insekten in kristallisierter Form. *Zeitschrift für Naturforschung, 9b*, 389–391.

Caceres, L., Necakov, A. S., Schwartz, C., Kimber, S., Roberts, I. J., & Krause, H. M. (2011). Nitric oxide coordinates metabolism, growth, and development via the nuclear receptor E75. *Genes & Development, 25*, 1476–1485.

Caldwell, P. E., Walkiewicz, M., & Stern, M. (2005). Ras activity in the Drosophila prothoracic gland regulates body size and developmental rate via ecdysone release. *Current Biology, 15*, 1785–1795.

Carstea, E. D., Morris, J. A., Coleman, K. G., Loftus, S. K., Zhang, D., Cummings, C., et al. (1997). Niemann-Pick C1 disease gene: Homology to mediators of cholesterol homeostasis. *Science (New York, NY), 277*, 228–231.

Carvalho, M., Schwudke, D., Sampaio, J. L., Palm, W., Riezman, I., Dey, G., et al. (2010). Survival strategies of a sterol auxotroph. *Development (Cambridge, England), 137*, 3675–3685.

Chavez, V. M., Marques, G., Delbecque, J. P., Kobayashi, K., Hollingsworth, M., Burr, J., et al. (2000). The Drosophila disembodied gene controls late embryonic morphogenesis and codes for a cytochrome P450 enzyme that regulates embryonic ecdysone levels. *Development (Cambridge, England), 127*, 4115–4126.

Chen, C., Jack, J., & Garofalo, R. S. (1996). The Drosophila insulin receptor is required for normal growth. *Endocrinology, 137*, 846–856.

Chen, Y. C., Lin, S. I., Chen, Y. K., Chiang, C. S., & Liaw, G. J. (2009). The Torso signaling pathway modulates a dual transcriptional switch to regulate tailless expression. *Nucleic Acids Research, 37*, 1061–1072.

Clayton, R. B. (1964). The utilization of sterols by insects. *Journal of Lipid Research, 15*, 3–19.

Clayton, G. M., Peak-Chew, S. Y., Evans, R. M., & Schwabe, J. W. (2001). The structure of the ultraspiracle ligand-binding domain reveals a nuclear receptor locked in an inactive

conformation. *Proceedings of the National Academy of Sciences of the United States of America, 98*, 1549–1554.
Clever, U., & Karlson, P. (1960). Induktion von Puff-Veränderungen in den Speicheldrüsenchromosomen von *Chironomus tentans* durch Ecdyson Exp. *Cell Research, 20*, 623–626.
Costantino, B. F., Bricker, D. K., Alexandre, K., Shen, K., Merriam, J. R., Antoniewski, C., et al. (2008). A novel ecdysone receptor mediates steroid-regulated developmental events during the mid-third instar of Drosophila. *PLoS Genetics, 4*, e1000102.
Davidowitz, G., D'Amico, L. J., & Nijhout, H. F. (2003). Critical weight in the development of insect body size. *Evolution & Development, 5*, 188–197.
Davies, J. P., & Ioannou, Y. A. (2000). Topological analysis of Niemann-Pick C1 protein reveals that the membrane orientation of the putative sterol-sensing domain is identical to those of 3-hydroxy-3-methylglutaryl-CoA reductase and sterol regulatory element binding protein cleavage-activating protein. *The Journal of Biological Chemistry, 275*, 24367–24374.
Davis, M. B., Carney, G. E., Robertson, A. E., & Bender, M. (2005). Phenotypic analysis of EcR-A mutants suggests that EcR isoforms have unique functions during Drosophila development. *Developmental Biology, 282*, 385–396.
DiBello, P. R., Withers, D. A., Bayer, C. A., Fristrom, J. W., & Guild, G. M. (1991). The Drosophila broad-complex encodes a family of related proteins containing zinc fingers. *Genetics, 129*, 385–397.
Emery, I. F., Noveral, J. M., Jamison, C. F., & Siwicki, K. K. (1997). Rhythms of Drosophila period gene expression in culture. *Proceedings of the National Academy of Sciences of the United States of America, 94*, 4092–4096.
Faucheux, M., Roignant, J. Y., Netter, S., Charollais, J., Antoniewski, C., & Theodore, L. (2003). batman Interacts with polycomb and trithorax group genes and encodes a BTB/POZ protein that is included in a complex containing GAGA factor. *Molecular and Cellular Biology, 23*, 1181–1195.
Fletcher, J. C., D'Avino, P. P., & Thummel, C. S. (1997). A steroid-triggered switch in E74 transcription factor isoforms regulates the timing of secondary-response gene expression. *Proceedings of the National Academy of Sciences of the United States of America, 94*, 4582–4586.
Fluegel, M. L., Parker, T. J., & Pallanck, L. J. (2006). Mutations of a Drosophila NPC1 gene confer sterol and ecdysone metabolic defects. *Genetics, 172*, 185–196.
Galton, V. A. (1992). The role of thyroid hormone in amphibian metamorphosis. *Trends in Endocrinology and Metabolism, 3*, 96–100.
Gauthier, S. A., & Hewes, R. S. (2006). Transcriptional regulation of neuropeptide and peptide hormone expression by the Drosophila dimmed and cryptocephal genes. *The Journal of Experimental Biology, 209*, 1803–1815.
Gevry, N., Schoonjans, K., Guay, F., & Murphy, B. D. (2008). Cholesterol supply and SREBPs modulate transcription of the Niemann-Pick C-1 gene in steroidogenic tissues. *Journal of Lipid Research, 49*, 1024–1033.
Gibbens, Y. Y., Warren, J. T., Gilbert, L. I., & O'Connor, M. B. (2011). Neuroendocrine regulation of Drosophila metamorphosis requires TGFbeta/Activin signaling. *Development (Cambridge, England), 138*, 2693–2703.
Giebultowicz, J. M. (1999). Insect circadian clocks: Is it all in their heads? *Journal of Insect Physiology, 45*, 791–800.
Giesen, K., Hummel, T., Stollewerk, A., Harrison, S., Travers, A., & Klambt, C. (1997). Glial development in the Drosophila CNS requires concomitant activation of glial and repression of neuronal differentiation genes. *Development (Cambridge, England), 124*, 2307–2316.
Gilbert, L. I. (2004). Halloween genes encode P450 enzymes that mediate steroid hormone biosynthesis in Drosophila melanogaster. *Molecular and Cellular Endocrinology, 215*, 1–10.

Gilbert, L. I., Rybczynski, R., & Warren, J. T. (2002). Control and biochemical nature of the ecdysteroidogenic pathway. *Annual Review of Entomology*, *47*, 883–916.

Hall, B. L., & Thummel, C. S. (1998). The RXR homolog ultraspiracle is an essential component of the Drosophila ecdysone receptor. *Development (Cambridge, England)*, *125*, 4709–4717.

Harvie, P. D., Filippova, M., & Bryant, P. J. (1998). Genes expressed in the ring gland, the major endocrine organ of Drosophila melanogaster. *Genetics*, *149*, 217–231.

Henrich, V. C., Sliter, T. J., Lubahn, D. B., MacIntyre, A., & Gilbert, L. I. (1990). A steroid/thyroid hormone receptor superfamily member in Drosophila melanogaster that shares extensive sequence similarity with a mammalian homologue. *Nucleic Acids Research*, *18*, 4143–4148.

Hewes, R. S., Schaefer, A. M., & Taghert, P. H. (2000). The cryptocephal gene (ATF4) encodes multiple basic-leucine zipper proteins controlling molting and metamorphosis in Drosophila. *Genetics*, *155*, 1711–1723.

Hu, X., Cherbas, L., & Cherbas, P. (2003). Transcription activation by the ecdysone receptor (EcR/USP): Identification of activation functions. *Molecular Endocrinology (Baltimore, MD)*, *17*, 716–731.

Huang, X., Suyama, K., Buchanan, J., Zhu, A. J., & Scott, M. P. (2005). A Drosophila model of the Niemann-Pick type C lysosome storage disease: dnpc1a is required for molting and sterol homeostasis. *Development (Cambridge, England)*, *132*, 5115–5124.

Huang, X., Warren, J. T., Buchanan, J., Gilbert, L. I., & Scott, M. P. (2007). Drosophila Niemann-Pick type C-2 genes control sterol homeostasis and steroid biosynthesis: A model of human neurodegenerative disease. *Development (Cambridge, England)*, *134*, 3733–3742.

Huber, R., & Hoppe, W. (1965). Die Kristall- und Molekülstrukturanalyse des Insekten Verpupungshormons Ecdyson mit der automatisierten Faltmolekülmethode. *Chemische Berichte*, *98*, 2403–2424.

Ismat, A., Schaub, C., Reim, I., Kirchner, K., Schultheis, D., & Frasch, M. (2010). HLH54F is required for the specification and migration of longitudinal gut muscle founders from the caudal mesoderm of Drosophila. *Development (Cambridge, England)*, *137*, 3107–3117.

Karim, F. D., & Rubin, G. M. (1998). Ectopic expression of activated Ras1 induces hyperplastic growth and increased cell death in Drosophila imaginal tissues. *Development (Cambridge, England)*, *125*, 1–9.

Karim, F. D., & Thummel, C. S. (1991). Ecdysone coordinates the timing and amounts of E74A and E74B transcription in Drosophila. *Genes & Development*, *5*, 1067–1079.

Karlson, P., Hoffmeister, H., Hummel, H., Hocks, P., & Spiteller, G. (1965). On the chemistry of ecdysone. VI. Reactions of ecdysone molecules. *Chemische Berichte*, *98*, 2394–2402.

Kawakami, A., Kataoka, H., Oka, T., Mizoguchi, A., Kimura-Kawakami, M., Adachi, T., et al. (1990). Molecular cloning of the Bombyx mori prothoracicotropic hormone. *Science (New York, NY)*, *247*, 1333–1335.

King-Jones, K., Charles, J. P., Lam, G., & Thummel, C. S. (2005). The ecdysone-induced DHR4 orphan nuclear receptor coordinates growth and maturation in Drosophila. *Cell*, *121*, 773–784.

King-Jones, K., & Thummel, C. S. (2005). Nuclear receptors—A perspective from Drosophila. *Nature Reviews. Genetics*, *6*, 311–323.

Kiss, I., Beaton, A. H., Tardiff, J., Fristrom, D., & Fristrom, J. W. (1988). Interactions and developmental effects of mutations in the broad-complex of Drosophila melanogaster. *Genetics*, *118*, 247–259.

Klose, W., Gateff, E., Emmerich, H., & Beikirch, H. (1980). Developmental studies on two ecdysone deficient mutants of Drosophila melanogaster. *Roux's Archives of Developmental Biology*, *189*, 57–67.

Koelle, M. R., Talbot, W. S., Segraves, W. A., Bender, M. T., Cherbas, P., & Hogness, D. S. (1991). The Drosophila EcR gene encodes an ecdysone receptor, a new member of the steroid receptor superfamily. *Cell, 67*, 59–77.

Kozlova, T., & Thummel, C. S. (2003). Essential roles for ecdysone signaling during Drosophila mid-embryonic development. *Science (New York, NY), 301*, 1911–1914.

Kunte, A. S., Matthews, K. A., & Rawson, R. B. (2006). Fatty acid auxotrophy in Drosophila larvae lacking SREBP. *Cell Metabolism, 3*, 439–448.

Lam, G. T., Jiang, C., & Thummel, C. S. (1997). Coordination of larval and prepupal gene expression by the DHR3 orphan receptor during Drosophila metamorphosis. *Development (Cambridge, England), 124*, 1757–1769.

Lavorgna, G., Ueda, H., Clos, J., & Wu, C. (1991). FTZ-F1, a steroid hormone receptor-like protein implicated in the activation of fushi tarazu. *Science (New York, NY), 252*, 848–851.

Lehmann, M. (1996). Drosophila Sgs genes: Stage and tissue specificity of hormone responsiveness. *Bioessays, 18*, 47–54.

Loftus, S. K., Morris, J. A., Carstea, E. D., Gu, J. Z., Cummings, C., Brown, A., et al. (1997). Murine model of Niemann-Pick C disease: Mutation in a cholesterol homeostasis gene. *Science (New York, NY), 277*, 232–235.

Marvin, K. A., Reinking, J. L., Lee, A. J., Pardee, K., Krause, H. M., & Burstyn, J. N. (2009). Nuclear receptors homo sapiens Rev-erbbeta and Drosophila melanogaster E75 are thiolate-ligated heme proteins which undergo redox-mediated ligand switching and bind CO and NO. *Biochemistry, 48*, 7056–7071.

McBrayer, Z., Ono, H., Shimell, M., Parvy, J. P., Beckstead, R. B., Warren, J. T., et al. (2007). Prothoracicotropic hormone regulates developmental timing and body size in Drosophila. *Developmental Cell, 13*, 857–871.

McQuilton, P., St Pierre, S. E., & Thurmond, J. (2012). FlyBase 101—The basics of navigating FlyBase. *Nucleic Acids Research, 40*, D706–D714.

Mirth, C., Truman, J. W., & Riddiford, L. M. (2005). The role of the prothoracic gland in determining critical weight for metamorphosis in Drosophila melanogaster. *Current Biology, 15*, 1796–1807.

Montagne, J., Lecerf, C., Parvy, J. P., Bennion, J. M., Radimerski, T., Ruhf, M. L., et al. (2010). The nuclear receptor DHR3 modulates dS6 kinase-dependent growth in Drosophila. *PLoS Genetics, 6*, e1000937.

Murali, T., Pacifico, S., Yu, J., Guest, S., Roberts, G. G., 3rd, & Finley, R. L., Jr. (2011). DroID 2011: A comprehensive, integrated resource for protein, transcription factor, RNA and gene interactions for Drosophila. *Nucleic Acids Research, 39*, D736–D743.

Myers, E. M. (2003). The circadian control of eclosion. *Chronobiology International, 20*, 775–794.

Myers, E. M., Yu, J., & Sehgal, A. (2003). Circadian control of eclosion: Interaction between a central and peripheral clock in Drosophila melanogaster. *Current Biology, 13*, 526–533.

Namiki, T., Niwa, R., Higuchi, A., Yoshiyama, T., Mita, K., & Kataoka, H. (2009). A basic-HLH transcription factor, HLH54F, is highly expressed in the prothoracic gland in the silkworm Bombyx mori and the fruit fly Drosophila melanogaster. *Bioscience, Biotechnology, and Biochemistry, 73*, 762–765.

Neubueser, D., Warren, J. T., Gilbert, L. I., & Cohen, S. M. (2005). Molting defective is required for ecdysone biosynthesis. *Developmental Biology, 280*, 362–372.

Niwa, R., Namiki, T., Ito, K., Shimada-Niwa, Y., Kiuchi, M., Kawaoka, S., et al. (2010). Non-molting glossy/shroud encodes a short-chain dehydrogenase/reductase that functions in the 'Black Box' of the ecdysteroid biosynthesis pathway. *Development (Cambridge, England), 137*, 1991–1999.

Niwa, R., Sakudoh, T., Namiki, T., Saida, K., Fujimoto, Y., & Kataoka, H. (2005). The ecdysteroidogenic P450 Cyp302a1/disembodied from the silkworm, Bombyx mori, is transcriptionally regulated by prothoracicotropic hormone. *Insect Molecular Biology, 14*, 563–571.

Nusslein-Volhard, C., & Wieschaus, E. (1980). Mutations affecting segment number and polarity in Drosophila. *Nature, 287*, 795–801.

O'Connor, M. B. (2011). Abstract: Vesicle mediated secretion of ecdysone from the Drosophila prothoracic gland. In *NASCE 2011: The inaugural meeting of the North American Society for Comparative Endocrinology*.

Ono, H., Rewitz, K. F., Shinoda, T., Itoyama, K., Petryk, A., Rybczynski, R., et al. (2006). Spook and Spookier code for stage-specific components of the ecdysone biosynthetic pathway in Diptera. *Developmental Biology, 298*, 555–570.

Oro, A. E., McKeown, M., & Evans, R. M. (1990). Relationship between the product of the Drosophila ultraspiracle locus and the vertebrate retinoid X receptor. *Nature, 347*, 298–301.

Oro, A. E., McKeown, M., & Evans, R. M. (1992). The Drosophila retinoid X receptor homolog ultraspiracle functions in both female reproduction and eye morphogenesis. *Development (Cambridge, England), 115*, 449–462.

Ou, Q., Magico, A., & King-Jones, K. (2011). Nuclear receptor DHR4 controls the timing of steroid hormone pulses during Drosophila development. *PLoS Biology, 9*, e1001160.

Pacifico, S., Liu, G., Guest, S., Parrish, J. R., Fotouhi, F., & Finley, R. L., Jr. (2006). A database and tool, IM Browser, for exploring and integrating emerging gene and protein interaction data for Drosophila. *BMC Bioinformatics*, 7, 195.

Pagans, S., Pineyro, D., Kosoy, A., Bernues, J., & Azorin, F. (2004). Repression by TTK69 of GAGA-mediated activation occurs in the absence of TTK69 binding to DNA and solely requires the contribution of the POZ/BTB domain of TTK69. *The Journal of Biological Chemistry, 279*, 9725–9732.

Parker, K. L., & Schimmer, B. P. (1997). Steroidogenic factor 1: A key determinant of endocrine development and function. *Endocrine Reviews, 18*, 361–377.

Parvy, J. P., Blais, C., Bernard, F., Warren, J. T., Petryk, A., Gilbert, L. I., et al. (2005). A role for betaFTZ-F1 in regulating ecdysteroid titers during post-embryonic development in Drosophila melanogaster. *Developmental Biology, 282*, 84–94.

Perrimon, N., Engstrom, L., & Mahowald, A. P. (1985). Developmental genetics of the 2C-D region of the Drosophila X chromosome. *Genetics, 111*, 23–41.

Persengiev, S. P., & Green, M. R. (2003). The role of ATF/CREB family members in cell growth, survival and apoptosis. *Apoptosis, 8*, 225–228.

Petryk, A., Warren, J. T., Marques, G., Jarcho, M. P., Gilbert, L. I., Kahler, J., et al. (2003). Shade is the Drosophila P450 enzyme that mediates the hydroxylation of ecdysone to the steroid insect molting hormone 20-hydroxyecdysone. *Proceedings of the National Academy of Sciences of the United States of America, 100*, 13773–13778.

Raghuram, S., Stayrook, K. R., Huang, P., Rogers, P. M., Nosie, A. K., McClure, D. B., et al. (2007). Identification of heme as the ligand for the orphan nuclear receptors REV-ERBalpha and REV-ERBbeta. *Nature Structural & Molecular Biology, 14*, 1207–1213.

Ramakrishnan, S. N., & Muscat, G. E. (2006). The orphan Rev-erb nuclear receptors: A link between metabolism, circadian rhythm and inflammation? *Nuclear Receptor Signaling, 4*, e009.

Read, D., & Manley, J. L. (1992). Alternatively spliced transcripts of the Drosophila tramtrack gene encode zinc finger proteins with distinct DNA binding specificities. *The EMBO Journal, 11*, 1035–1044.

Regulski, M., Stasiv, Y., Tully, T., & Enikolopov, G. (2004). Essential function of nitric oxide synthase in Drosophila. *Current Biology, 14*, R881–R882.

Reinking, J., Lam, M. M., Pardee, K., Sampson, H. M., Liu, S., Yang, P., et al. (2005). The Drosophila nuclear receptor e75 contains heme and is gas responsive. *Cell, 122*, 195–207.

Renn, S. C., Park, J. H., Rosbash, M., Hall, J. C., & Taghert, P. H. (1999). A pdf neuropeptide gene mutation and ablation of PDF neurons each cause severe abnormalities of behavioral circadian rhythms in Drosophila. *Cell, 99*, 791–802.

Rewitz, K. F., Yamanaka, N., Gilbert, L. I., & O'Connor, M. B. (2009). The insect neuropeptide PTTH activates receptor tyrosine kinase torso to initiate metamorphosis. *Science (New York, NY), 326*, 1403–1405.

Riddiford, L. M. (1993). Hormone receptors and the regulation of insect metamorphosis. *Receptor, 3*, 203–209.

Robb, L., Hartley, L., Wang, C. C., Harvey, R. P., & Begley, C. G. (1998). Musculin: A murine basic helix-loop-helix transcription factor gene expressed in embryonic skeletal muscle. *Mechanisms of Development, 76*, 197–201.

Russell, S. R., Heimbeck, G., Goddard, C. M., Carpenter, A. T., & Ashburner, M. (1996). The Drosophila Eip78C gene is not vital but has a role in regulating chromosome puffs. *Genetics, 144*, 159–170.

Rybczynski, R. (2005). Prothoracicotropic hormone. In L. I. Gilbert, K. Latrou & S. Gill (Eds.), *Comprehensive molecular insect science*, Vol. 3, (pp. 61–123). Pergamon: Elsevier.

Schmierer, B., & Hill, C. S. (2007). TGFbeta-SMAD signal transduction: Molecular specificity and functional flexibility. *Nature Reviews. Molecular Cell Biology, 8*, 970–982.

Schubiger, M., Wade, A. A., Carney, G. E., Truman, J. W., & Bender, M. (1998). Drosophila EcR-B ecdysone receptor isoforms are required for larval molting and for neuron remodeling during metamorphosis. *Development (Cambridge, England), 125*, 2053–2062.

Schwartz, L. M., & Truman, J. W. (1983). Hormonal control of rates of metamorphic development in the tobacco hornworm Manduca sexta. *Developmental Biology, 99*, 103–114.

Segraves, W. A., & Hogness, D. S. (1990). The E75 ecdysone-inducible gene responsible for the 75B early puff in Drosophila encodes two new members of the steroid receptor superfamily. *Genes & Development, 4*, 204–219.

Shimojima, T., Okada, M., Nakayama, T., Ueda, H., Okawa, K., Iwamatsu, A., et al. (2003). Drosophila FACT contributes to Hox gene expression through physical and functional interactions with GAGA factor. *Genes & Development, 17*, 1605–1616.

Sliter, T. J., & Gilbert, L. I. (1992). Developmental arrest and ecdysteroid deficiency resulting from mutations at the dre4 locus of Drosophila. *Genetics, 130*, 555–568.

Sokolowski, M. B. (2001). Drosophila: Genetics meets behaviour. *Nature Reviews, 2*, 879–890.

Song, Q., & Gilbert, L. I. (1995). Multiple phosphorylation of ribosomal protein S6 and specific protein synthesis are required for prothoracicotropic hormone-stimulated ecdysteroid biosynthesis in the prothoracic glands of Manduca sexta. *Insect Biochemistry and Molecular Biology, 25*, 591–602.

Song, Q., & Gilbert, L. I. (1997). Molecular cloning, developmental expression, and phosphorylation of ribosomal protein S6 in the endocrine gland responsible for insect molting. *The Journal of Biological Chemistry, 272*, 4429–4435.

Song, Q., & Gilbert, L. I. (1998). Alterations in ultraspiracle (USP) content and phosphorylation state accompany feedback regulation of ecdysone synthesis in the insect prothoracic gland. *Insect Biochemistry and Molecular Biology, 28*, 849–860.

Stasiv, Y., Regulski, M., Kuzin, B., Tully, T., & Enikolopov, G. (2001). The Drosophila nitric-oxide synthase gene (dNOS) encodes a family of proteins that can modulate NOS activity by acting as dominant negative regulators. *The Journal of Biological Chemistry, 276*, 42241–42251.

Stone, B. L., & Thummel, C. S. (1993). The Drosophila 78C early late puff contains E78, an ecdysone-inducible gene that encodes a novel member of the nuclear hormone receptor superfamily. *Cell, 75*, 307–320.

Strecker, T. R., Yip, M. L., & Lipshitz, H. D. (1991). Zygotic genes that mediate torso receptor tyrosine kinase functions in the Drosophila melanogaster embryo. *Proceedings of the National Academy of Sciences of the United States of America, 88*, 5824–5828.

Sturley, S. L., Patterson, M. C., Balch, W., & Liscum, L. (2004). The pathophysiology and mechanisms of NP-C disease. *Biochimica et Biophysica Acta, 1685*, 83–87.

Suman, S., Seth, R. K., & Chandna, S. (2008). Role of nitric oxide synthase in insect cell radioresistance: An in-silico analysis. *Bioinformation, 3*, 8–13.

Talbot, W. S., Swyryd, E. A., & Hogness, D. S. (1993). Drosophila tissues with different metamorphic responses to ecdysone express different ecdysone receptor isoforms. *Cell, 73*, 1323–1337.

Thomas, H. E., Stunnenberg, H. G., & Stewart, A. F. (1993). Heterodimerization of the Drosophila ecdysone receptor with retinoid X receptor and ultraspiracle. *Nature, 362*, 471–475.

Thummel, C. S. (1990). Puffs and gene regulation—Molecular insights into the Drosophila ecdysone regulatory hierarchy. *Bioessays, 12*, 561–568.

Thummel, C. S. (1995). From embryogenesis to metamorphosis: The regulation and function of Drosophila nuclear receptor superfamily members. *Cell, 83*, 871–877.

Truman, J. W., & Riddiford, L. M. (1999). The origins of insect metamorphosis. *Nature, 401*, 447–452.

Ueda, H., Sonoda, S., Brown, J. L., Scott, M. P., & Wu, C. (1990). A sequence-specific DNA-binding protein that activates fushi tarazu segmentation gene expression. *Genes & Development, 4*, 624–635.

Urness, L. D., & Thummel, C. S. (1995). Molecular analysis of a steroid-induced regulatory hierarchy: The Drosophila E74A protein directly regulates L71-6 transcription. *The EMBO Journal, 14*, 6239–6246.

Vafopoulou, X., & Steel, C. G. (1996a). Circadian regulation of a daily rhythm of release of prothoracicotropic hormone from the brain retrocerebral complex of Rhodnius prolixus (hemiptera) during larval-adult development. *General and Comparative Endocrinology, 102*, 123–129.

Vafopoulou, X., & Steel, C. G. (1996b). The insect neuropeptide prothoracicotropic hormone is released with a daily rhythm: Re-evaluation of its role in development. *Proceedings of the National Academy of Sciences of the United States of America, 93*, 3368–3372.

Villalobo, A. (2006). Nitric oxide and cell proliferation. *The FEBS Journal, 273*, 2329–2344.

von Kalm, L., Crossgrove, K., Von Seggern, D., Guild, G. M., & Beckendorf, S. K. (1994). The Broad-Complex directly controls a tissue-specific response to the steroid hormone ecdysone at the onset of Drosophila metamorphosis. *The EMBO Journal, 13*, 3505–3516.

Warren, J. T., Petryk, A., Marques, G., Jarcho, M., Parvy, J. P., Dauphin-Villemant, C., et al. (2002). Molecular and biochemical characterization of two P450 enzymes in the ecdysteroidogenic pathway of Drosophila melanogaster. *Proceedings of the National Academy of Sciences of the United States of America, 99*, 11043–11048.

Warren, J. T., Petryk, A., Marques, G., Parvy, J. P., Shinoda, T., Itoyama, K., et al. (2004). Phantom encodes the 25-hydroxylase of Drosophila melanogaster and Bombyx mori: A P450 enzyme critical in ecdysone biosynthesis. *Insect Biochemistry and Molecular Biology, 34*, 991–1010.

Warren, J. T., Wismar, J., Subrahmanyam, B., & Gilbert, L. I. (2001). Woc (without children) gene control of ecdysone biosynthesis in Drosophila melanogaster. *Molecular and Cellular Endocrinology, 181*, 1–14.

Warren, J. T., Yerushalmi, Y., Shimell, M. J., O'Connor, M. B., Restifo, L. L., & Gilbert, L. I. (2006). Discrete pulses of molting hormone, 20-hydroxyecdysone, during late larval development of Drosophila melanogaster: Correlations with changes in gene activity. *Developmental Dynamics, 235*, 315–326.

Westbrook, A. L., & Bollenbacher, W. E. (1990). The development of identified neurosecretory cells in the tobacco hornworm, Manduca sexta. *Developmental Biology, 140*, 291–299.

White, K. P., Hurban, P., Watanabe, T., & Hogness, D. S. (1997). Coordination of Drosophila metamorphosis by two ecdysone-induced nuclear receptors. *Science (New York, NY)*, *276*, 114–117.

Wismar, J., Habtemichael, N., Warren, J. T., Dai, J. D., Gilbert, L. I., & Gateff, E. (2000). The mutation without children(rgl) causes ecdysteroid deficiency in third-instar larvae of Drosophila melanogaster. *Developmental Biology*, *226*, 1–17.

Woodard, C. T., Baehrecke, E. H., & Thummel, C. S. (1994). A molecular mechanism for the stage specificity of the Drosophila prepupal genetic response to ecdysone. *Cell*, *79*, 607–615.

Wu, Q., & Brown, M. R. (2006). Signaling and function of insulin-like peptides in insects. *Annual Review of Entomology*, *51*, 1–24.

Xiang, Y., Liu, Z., & Huang, X. (2010). br regulates the expression of the ecdysone biosynthesis gene npc1. *Developmental Biology*, *344*, 800–808.

Yakubovich, N., Silva, E. A., & O'Farrell, P. H. (2010). Nitric oxide synthase is not essential for Drosophila development. *Current Biology*, *20*, R141–R142.

Yamada, M., Murata, T., Hirose, S., Lavorgna, G., Suzuki, E., & Ueda, H. (2000). Temporally restricted expression of transcription factor betaFTZ-F1: Significance for embryogenesis, molting and metamorphosis in Drosophila melanogaster. *Development (Cambridge, England)*, *127*, 5083–5092.

Yao, T. P., Forman, B. M., Jiang, Z., Cherbas, L., Chen, J. D., McKeown, M., et al. (1993). Functional ecdysone receptor is the product of EcR and Ultraspiracle genes. *Nature*, *366*, 476–479.

Yoshiyama, T., Namiki, T., Mita, K., Kataoka, H., & Niwa, R. (2006). Neverland is an evolutionally conserved Rieske-domain protein that is essential for ecdysone synthesis and insect growth. *Development (Cambridge, England)*, *133*, 2565–2574.

Yoshiyama-Yanagawa, T., Enya, S., Shimada-Niwa, Y., Yaguchi, S., Haramoto, Y., Matsuya, T., et al. (2011). The conserved Rieske oxygenase DAF-36/Neverland is a novel cholesterol-metabolizing enzyme. *The Journal of Biological Chemistry*, *286*, 25756–25762.

Yu, J., & Finley, R. L., Jr. (2009). Combining multiple positive training sets to generate confidence scores for protein-protein interactions. *Bioinformatics*, *25*, 105–111.

Yu, Y., Li, W., Su, K., Yussa, M., Han, W., Perrimon, N., et al. (1997). The nuclear hormone receptor Ftz-F1 is a cofactor for the Drosophila homeodomain protein Ftz. *Nature*, *385*, 552–555.

Yu, J., Pacifico, S., Liu, G., & Finley, R. L., Jr. (2008). DroID: The Drosophila Interactions Database, a comprehensive resource for annotated gene and protein interactions. *BMC Genomics*, *9*, 461.

Zawel, L., Dai, J. L., Buckhaults, P., Zhou, S., Kinzler, K. W., Vogelstein, B., et al. (1998). Human Smad3 and Smad4 are sequence-specific transcription activators. *Molecular Cell*, *1*, 611–617.

Zelhof, A. C., Yao, T. P., Chen, J. D., Evans, R. M., & McKeown, M. (1995). Seven-up inhibits ultraspiracle-based signaling pathways in vitro and in vivo. *Molecular and Cellular Biology*, *15*, 6736–6745.

Zhou, X., Zhou, B., Truman, J. W., & Riddiford, L. M. (2004). Overexpression of broad: a new insight into its role in the Drosophila prothoracic gland cells. *The Journal of Experimental Biology*, *207*, 1151–1161.

Zoglowek, A., Orlowski, M., Pakula, S., Dutko-Gwozdz, J., Pajdzik, D., Gwozdz, T., et al. (2012). The composite nature of the interaction between nuclear receptors EcR and DHR38. *Biological Chemistry*, *393*, 457–471.

Zollman, S., Godt, D., Prive, G. G., Couderc, J. L., & Laski, F. A. (1994). The BTB domain, found primarily in zinc finger proteins, defines an evolutionarily conserved family that includes several developmentally regulated genes in Drosophila. *Proceedings of the National Academy of Sciences of the United States of America*, *91*, 10717–10721.

CHAPTER THREE

Hormonal Regulation of Insect Metamorphosis with Special Reference to Juvenile Hormone Biosynthesis

Kiyoshi Hiruma[*,†,1], **Yu Kaneko**[*,†]
[*]Faculty of Agriculture and Life Sciences, Hirosaki University, Hirosaki, Japan
[†]Graduate School of Agricultural Sciences, Iwate University, Morioka, Japan
[1]Corresponding author: e-mail address: hiruma@cc.hirosaki-u.ac.jp

Contents

Abstract

Ecdysteroids and juvenile hormones (JHs) are key hormones that are responsible for insect molting and metamorphosis. JH maintains the larval state and the decline of its level in the hemolymph is crucial to elicit transformation to the pupal stage; therefore, the precise control of JH biosynthesis is necessary for normal development and the initiation of metamorphosis. This chapter summarizes mechanisms of the regulation of JH biosynthesis by the corpora allata and shows that several factors such as ecdysteroids, neurotransmitters, and peptides act together in the stage-specific manner to guarantee the accurate production of JH in each stage, in particular, in the last larval stage when metamorphosis is initiated with the transformation of the larva to the pupa. In addition, recent progress in understanding the JH signaling pathway is briefly discussed, including the identification of a long elusive JH receptor.

Current Topics in Developmental Biology, Volume 103
ISSN 0070-2153
http://dx.doi.org/10.1016/B978-0-12-385979-2.00003-4

1. INTRODUCTION

Insect molting and metamorphosis are primarily under the control of two hormone classes; ecdysteroids, which cause molting, and juvenile hormones (JHs), which modulate ecdysteroid action. In most insects, ecdysone (E) is synthesized by the prothoracic glands (PGs) and converted systemically to the active ecdysteroid, 20-hydroxyecdysone (20E). JHs are synthesized by the corpora allata (CA). The chemical identity of JH varies among different insect groups; in most insects, the main JH is Juvenile Hormone III (JH-III), but additional forms are present in Lepidoptera (JH-0, JH-I, 4-methyl JH-I, and JH-II) and Diptera (JH-III bisepoxide; see review by Goodman & Granger, 2005).

In hemimetabolous insects, metamorphosis occurs during a single molt; ecdysteroid secretion repeatedly elicits larval molts as long as JH is present in the hemolymph. In the last larval stage, however, JH declines after ecdysis and adult development is initiated by ecdysteroid in the absence of JH (Hiruma, 2003; Riddiford, 1985). In holometabolous insects, by contrast, metamorphosis is a more complicated process that takes two successive molts to complete. In these insects, as in Hemimetabola, repeated larval molting is induced by ecdysteroid in the presence of JH, and metamorphosis begins after several larval ecdyses. When a holometabolous larva enters the last instar larval stadium, the level of JH in the hemolymph gradually declines and becomes undetectable (as in *Manduca sexta*—Baker, Tsai, Reuter, & Schooley, 1987) or reaches trace levels (as in *Bombyx mori*—Niimi & Sakurai, 1997), then a small amount of ecdysteroid in the absence of JH induces larval–pupal commitment of the epidermis. Pupal commitment is an irreversible event, representing the first crucial step of metamorphosis; it precedes any externally observable change in the insect's form, being followed only later by the actual formation of pupal cuticle, an event that is induced by the secretion of a larger amount of ecdysteroid (Muramatsu, Kinjoh, Shinoda, & Hiruma, 2008; Riddiford, 1976, 1985; Fig. 3.1B). Metamorphosis continues after the formation of the pupa, which only then begins to transform into the imago. Adult development is caused by ecdysteroid in the absence of JH during the pupal stage. The decline of the JH level in the last larval stadium is one of the most important events required for initiation and progress of metamorphosis; the continued presence of JH prevents metamorphosis so that JH preserves the *status quo*. Thus, it is the regulation of JH level in the hemolymph that determines whether and when metamorphosis begins and the manner of its continuation.

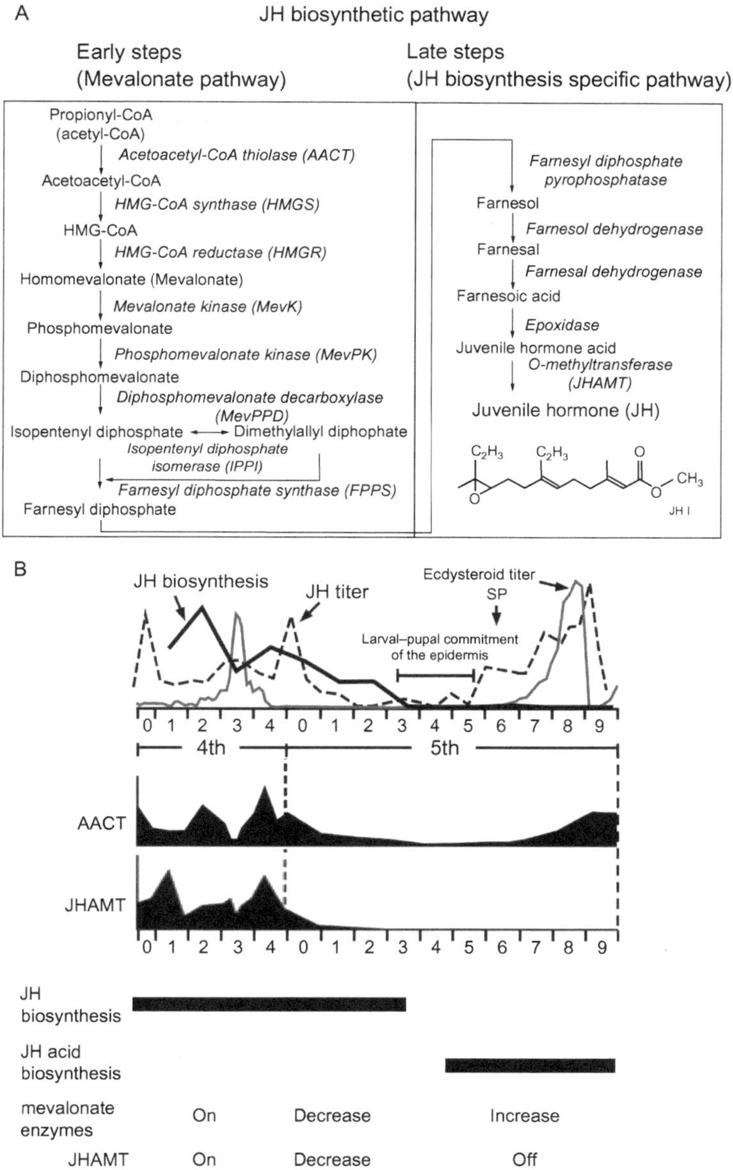

Figure 3.1 The JH biosynthetic pathway and the relationship between insect hormone titers and JH biosynthesis in *Bombyx mori*. (A) Diagram of the mevalonate pathway up to FPP (early step) and the JH-specific pathway (late step). (B) Fluctuations of JH synthetic activity by CC–CA and the mRNA expression of two representative JH synthetic enzymes in the CA, and the summary for the expression levels of the JH biosynthetic enzymes during the fourth and fifth instar stadium (Kinjoh et al., 2007). SP, spinneret pigmentation. *Panel (A): Modified from Bellés, Martín, & Piulachs (2005) and Kinjoh et al. (2007).*

The concentration of JH in the hemolymph is regulated not only by its biosynthesis but also by its degradation, secretion, sequestration, etc. It is thought that synthesis is the most important among these regulated processes (Gilbert, Granger, & Roe, 2000). Much progress has been made in the past 25 years in understanding the regulation of JH biosynthesis by various factors; peptide hormones such as allatotropins (ATs) and allatostatins (ASTs) (Stay & Tobe, 2007; Weaver & Audsley, 2009), neurotransmitters (see review, Goodman & Granger, 2005), cAMP (Granger et al., 1994), and ecdysteroids (Gu & Chow, 1996; Kaneko, Kinjoh, Kiuchi, & Hiruma, 2011; Whisenton, Bowen, Granger, Gilbert, & Bollenbacher, 1985; Whisenton, Watsona, Granger, & Bollenbacher, 1987) are all regulators of JH biosynthesis (see reviews, Goodman & Granger, 2005; Stay & Tobe, 2007). Most studies of JH-regulatory peptides, however, have concentrated on the gene structures of the peptides, and their actions have been studied on relatively few different developmental stages and different insect species. The stage-specific regulatory mechanisms and roles of these factors on JH synthesis in relation to growth, molting, and metamorphosis are as yet poorly understood.

Studies on JH lag far behind in those on ecdysteroids, which is mainly due to the minute size of the CA and the "sticky" chemical nature of JHs, which make experimental manipulation and measurement of these hormones challenging. The sesquiterpenoid nature of these hormones, which are not found in mammals, also makes it hard to incorporate knowledge from other better-studied organisms. Moreover, the effects of JH on one of the most prominent model insects, *Drosophila melanogaster*, are less easily monitored than those on other insect groups such as lepidopterans.

For these reasons, it is helpful to study JH and its actions in a well-studied lepidopteran model in which JH has strong, clear effects. The silkworm *B. mori* is suitable for this, genome information having been available for some time (The International Silkworm Genome Consortium, 2008). The various biological actions of JH have also been well studied, so that *Bombyx* allows us to study JH action at the molecular level much more easily than in other non-*Drosophila* insect species. This chapter summarizes the regulatory mechanisms of JH biosynthesis by three major factors, ecdysteroids, catecholamines, and peptides, during molting and metamorphosis mainly in *Bombyx* larvae as a model case. The precise stage-specific regulation of JH synthesis is shown to be achieved by the coordinated action of several factors, which activate or inactivate JH synthesis in different ways at different life stages, so that JH synthesis is properly adjusted to allow progress through the normal life cycle. In addition, we briefly describe how JH acts on cells, mainly emphasizing the progress that has occurred since the excellent review by Riddiford (2008), including the recent finding of a JH receptor (Charles et al., 2011).

2. REGULATION OF JH BIOSYNTHESIS

2.1. Genes responsible for JH biosynthesis

There are two main components for the JH biosynthetic pathway: the early steps belong to the mevalonate pathway up to farnesyl diphosphate (FPP), while the late steps are responsible for the conversion of FPP to JH (Bellés et al., 2005; Fig. 3.1A). Genes encoding the enzymes involved in the early steps (Kinjoh et al., 2007) and JH acid *O*-methyltransferase (JHAMT) in the late steps (Shinoda & Itoyama, 2003) of *Bombyx* have been isolated. mRNAs for these enzymes occur only in the CA and not in the corpora cardiaca (CC) in the cephalic organs (Kaneko, Shinoda, & Hiruma, 2011; Ueda, Shinoda, & Hiruma, 2009), and the expression patterns in the CA are well correlated to the patterns of JH synthetic activity by the CA (Kinjoh et al., 2007; Fig. 3.1B). Therefore, JH synthesis by the CA is considered to be controlled at the level of expression of these biosynthetic enzymes; among them, JHAMT apparently functions as a rate-limiting enzyme (Kinjoh et al., 2007).

2.2. Regulation by ecdysteroid

Regulation of JH synthesis by ecdysteroids had not been well studied until our *Bombyx* studies were performed, in particular, at the molecular level (Kaneko, Kinjoh, et al., 2011). It was known that ecdysteroid regulated JH synthesis (Goodman & Granger, 2005; Gu & Chow, 1993, 1996; Whisenton et al., 1985, 1987). The action appears to be indirect with ecdysteroid stimulating JH (or JH acid in the case of day 3 fifth instar) synthesis on day 2 fourth and day 3 fifth instar *M. sexta* larvae via the brain–CC complex *in vitro* (Whisenton et al., 1985, 1987). In *Bombyx*, feeding mulberry leaves sprayed with 20E to newly ecdysed fifth (last) instar larvae caused activation of JH synthesis by the CA; *in vitro* studies showed that this activation was direct on the CA (Gu & Chow, 1996). Yet these studies were not designed to explain the mechanisms of changing JH synthesis during the development, which is essential to understand the roles of JH on the normal development including molting and metamorphosis; rather they only showed that ecdysteroid was a stimulator of JH biosynthesis at a particular stage.

2.2.1 Fourth (penultimate) instar stadium

JH synthesis by a CC–CA complex during the fourth instar stadium fluctuates in *Bombyx* (Kinjoh et al., 2007; Fig. 3.1B). When the ecdysteroid titer in the hemolymph begins to increase on the second day of the instar, JH synthesis increases dramatically, but then declines at the time of the peak of ecdysteroid titer on day 3. In addition, mRNA levels of JH synthetic enzymes belonging to the early steps and *JHAMT* show very similar patterns to JH synthesis itself (Kinjoh et al., 2007; Fig. 3.1B). The close correspondence between these patterns suggests that JH synthesis is controlled by ecdysteroid through the regulation of the JH biosynthetic enzymes.

The concentration of ecdysteroid in the hemolymph increases in larvae fed on a diet containing 20E (Kamimura, Shimura, & Kiuchi, 2003; Kaneko, Kinjoh, et al., 2011); therefore, it is an ideal system to study ecdysteroid action *in vivo* by changing an ecdysteroid titer artificially. When fourth instar larvae are fed on diet containing 400 ppm 20E, the ecdysteroid titer in the hemolymph begins to increase shortly after the start of the feeding. JH synthetic activity by CC–CA also increases, but its activity then declines when the ecdysteroid titer exceeds 250 ng/ml (Kaneko, Kinjoh, et al., 2011). These relationships between ecdysteroid concentration and JH synthesis are very similar to that seen in larvae fed on normal diet without 20E (Fig. 3.1B), indicating that in both cases a low level of ecdysteroid stimulates JH synthesis but at high concentration ecdysteroid suppresses JH synthesis (Fig. 3.3A). The stimulation and suppression of JH synthesis are due to a change in the expression of a subset of JH biosynthetic enzymes; activation of JH synthesis *in vivo* is caused by increased expression of *FPPS1* and *JHAMT*, while suppression *in vivo* is due to inhibition of *IPPI*, *FPPS1*, and *JHAMT* expression (Kaneko, Kinjoh, et al., 2011). Similar regulation of the JH biosynthetic enzymes by 20E has also been observed *in vitro*. The inhibitory action of high concentrations of ecdysteroid on *JHAMT* expression is mediated by the action of a transcription factor E75, expression of which was stimulated by ecdysteroid, as RNAi suppression of E75 expression by the addition of dsE75, abolished ecdysteroid-induced inhibition of *JHAMT* expression *in vitro* (T. Kinjoh & K. Hiruma, unpublished data).

After the decline of JH synthesis induced by high ecdysteroid titers, JH synthesis increases again around the time of the last larval ecdysis (Fig. 3.1B). This is partly due to the falling ecdysteroid titer at this time, but additional factors also seem to be involved in this increase (Kaneko, Kinjoh, et al., 2011) (see Section 2.4.5).

The fluctuation of JH biosynthesis during the fourth larval stage is thus mainly under the control of ecdysteroid, which both stimulates and prevents the expression of individual JH biosynthetic enzymes, and this regulation is apparently both stage specific and dose dependent.

2.2.2 Fifth (last) instar stadium

After the last larval ecdysis, JH titer in the hemolymph declines gradually and then becomes a trace level by 3 days in *Bombyx* (Niimi & Sakurai, 1997; Fig. 3.1B). It is thought that the decline of JH titer is primarily based on the decline in JH biosynthesis, but the proper timing of JH esterase synthesis is also essential for normal larval molt in both *Manduca* and *Bombyx* (Hammock, Bonning, Possee, Hanzlik, & Maeda, 1990; Tan, Tanaka, Tamura, & Shiotsuki, 2005) as well as for pupal commitment by removing the remaining traces of JH in the hemolymph in *Manduca* (Browder, D'Amico, & Nijhout, 2001). If the JH level does not decline, pupal metamorphosis is prevented; the larvae attempt another larval ecdysis to form supernumerary larvae or remain as dauer larvae in *Manduca* (Reynolds, Brown, Seth, Riddiford, & Hiruma, 2009) and *Bombyx* (Muramatsu et al., 2008; Sakurai, 1984), respectively.

JH synthetic activity by *Bombyx* CA at the time of the last larval ecdysis is quite low (about 0.3 fmol/pair CC–CA/h), and this low activity shuts down completely 60 h after this ecdysis (Kaneko, Kinjoh, et al., 2011; Kinjoh et al., 2007). Ecdysteroid titer declines to low levels shortly before the last larval ecdysis, becomes 31 ng/ml at the time of ecdysis, and then gradually declines to undetectable levels by 60 h (Kaneko, Kinjoh, et al., 2011). Therefore, the decline of ecdysteroid appears to be the key factor for the shutdown of JH biosynthesis during the last larval stage. To verify this hypothesis, the same feeding approach used in the fourth instar stage studies mentioned in the previous section was used to change ecdysteroid titers in the fifth stage. Kamimura et al. (2003) and Kaneko, Kinjoh, et al. (2011) showed that feeding high concentration of 20E (400 ppm) to fifth instar larvae induced an increase in ecdysteroid titer shortly after feeding began, which reached 400 ng/ml 36 h later. Even feeding a low concentration (20 ppm) of 20E in diet maintained the low level of ecdysteroid (30 ng/ml) in the hemolymph. Under this condition, the JH synthetic activity of the CA increased, reaching maximum activity (3.5 fmol/pair/CC–CA/h) 48 and 36 h after feeding the diet with 20 and 400 ppm 20E, respectively, rather than declining as occurs in larvae fed on normal diet.

The responsiveness of the CA to ecdysteroid declines as a function of development after the last larval ecdysis and disappears completely by 60 h when synthetic activity is shut off (Kaneko, Kinjoh, et al., 2011). It must be noted that concentrations of ecdysteroid as low as 30 ng/ml strongly stimulate JH biosynthesis, indicating that the presence of even a low concentration of ecdysteroid is enough to maintain JH biosynthesis; therefore, the decline of ecdysteroid titer after the last larval ecdysis is one of the key events leading to the cessation of JH biosynthesis at this time (Fig. 3.3A).

At the time of the declining ecdysteroid titer, the mRNA levels of most of the JH synthetic enzymes in the CA gradually declined (Kinjoh et al., 2007; Fig. 3.1B). The presence of ecdysteroid prevented these declines, while mRNAs for both *JHAMT* and HMG-CoA reductase (*HMGR*) were stimulated as JH synthesis became high, so that the stimulation of JH synthesis was primarily due to the activation of JHAMT, one of the rate-limiting enzymes in *Bombyx* (T. Kinjoh & K. Hiruma, unpublished data; Fig. 3.4A), and also HMGR.

The activation of the CA by 20E requires nervous connections with the brain, as 20E only activates JH synthesis by the CA when a brain–CC–CA complex was intact (Kaneko, Kinjoh, et al., 2011). How the brain acts on the CA is unknown, but whatever the mechanisms involved, the expression of JH biosynthetic enzymes appears to be indirectly regulated by 20E. Unlike the inhibitory action of a high concentration of ecdysteroid on JH synthesis in the fourth instar larvae, RNAi suppression of E75 had little effect on the stimulative action of ecdysteroid on *JHAMT* expression in the early fifth stage larvae (Kinjoh & Hiruma, unpublished data), supporting the hypothesis of indirect ecdysteroid action on the CA (Fig. 3.4A).

Thus, in very early fifth stage *Bombyx* larvae, the brain responds to the postecdysial decline of ecdysteroid titer and signals to the CA via nervous connections to change JH biosynthetic enzymes so as to shut off JH biosynthesis (Figs. 3.3A and 3.4A).

2.3. Regulation by neurotransmitters

Neurotransmitters, in particular biogenic amines, have been thought to regulate JH, as maintaining intact nervous connections between the brain and the CA affects JH synthesis (Goodman & Granger, 2005; Granger, Sturgis, Ebersohl, Geng, & Sparks, 1996). Some biogenic amines such as octopamine, L-glutamate, and dopamine have been shown to stimulate JH synthesis by the isolated CA *in vitro*. Here, we summarize how two well-studied amines, dopamine and glutamate, act to regulate JH synthesis in the CA.

2.3.1 Dopamine

In *Manduca*, dopamine is the only biogenic amine present in the CA (Granger et al., 1996) and is detected also in the brain (Goodman & Granger, 2005). *In vitro* studies show that dopamine stimulates JH production by the CA in the early fifth instar, but inhibits its production on day 3 fifth instar (Granger et al., 1996).

In *Bombyx*, dopamine levels in the hemolymph increase after the last larval ecdysis (Noguchi & Hayakawa, 2001); therefore, it is possible that dopamine is responsible for the decline of JH synthesis in the early fifth instar stadium (Kaneko & Hiruma, 2007). *In vitro* studies showed that dopamine had little effect on JH synthetic activity by the CA in fourth stage larvae, but it suppressed JH synthesis in the fifth stage at concentrations higher than 5 μM (Fig. 3.2B and C). This strong suppressive action was observed shortly after the last larval ecdysis and the sensitivity of the CA to dopamine declined with development. Dopamine is produced by neurosecretory cells in the brain and transferred to the CA through nerve axons in fifth stage *Bombyx* larvae (Fig. 3.2A). Thus, it is thought that both dopamine from the brain and hemolymph dopamine may act directly on the CA to suppress JH synthesis (Fig. 3.4B).

Dopamine receptors have been isolated in insects and are classified as D1 and D2 types (Mustard, Beggs, & Mercer, 2005); the *Bombyx* D1 receptor has two isoforms: *BmDopR1* and *BmDopR2* (Mitsumasu et al., 2008). As mentioned above, dopamine suppresses JH synthesis by the CA of fifth instar larvae *in vitro*. Yet addition of a D1 receptor-specific antagonist, R(+)-SCH23390, inhibits this suppressive action (Fig. 3.2D), suggesting that dopamine action is through the D1 receptor. The expression patterns of *BmDopR1* and *BmDopR2* in the CA show similar profiles (Fig. 3.2E), with both exhibiting a sharp peak shortly before the last larval ecdysis, but the amount of *BmDopR2* mRNA was much higher than that of *BmDopR1*, especially just after ecdysis when only trace amounts of DopR1 mRNA are seen, although DopR2 transcripts are present at noticeably higher levels (Fig. 3.2E). Thus, most likely the suppressive action of JH biosynthesis by dopamine is through the arthropod-type BmDopR2 receptor (Fig. 3.4B). The reason dopamine has little effect on the CA in the fourth instar larvae is unknown. Presumably, other factors such as cofactor(s) are required for dopamine action and fourth instar larvae lack these factors. Unlike 20E action in the fifth instar stadium, dopamine does not affect the expressions of any of the JH synthetic enzymes (data not shown). How dopamine regulates JH synthesis is

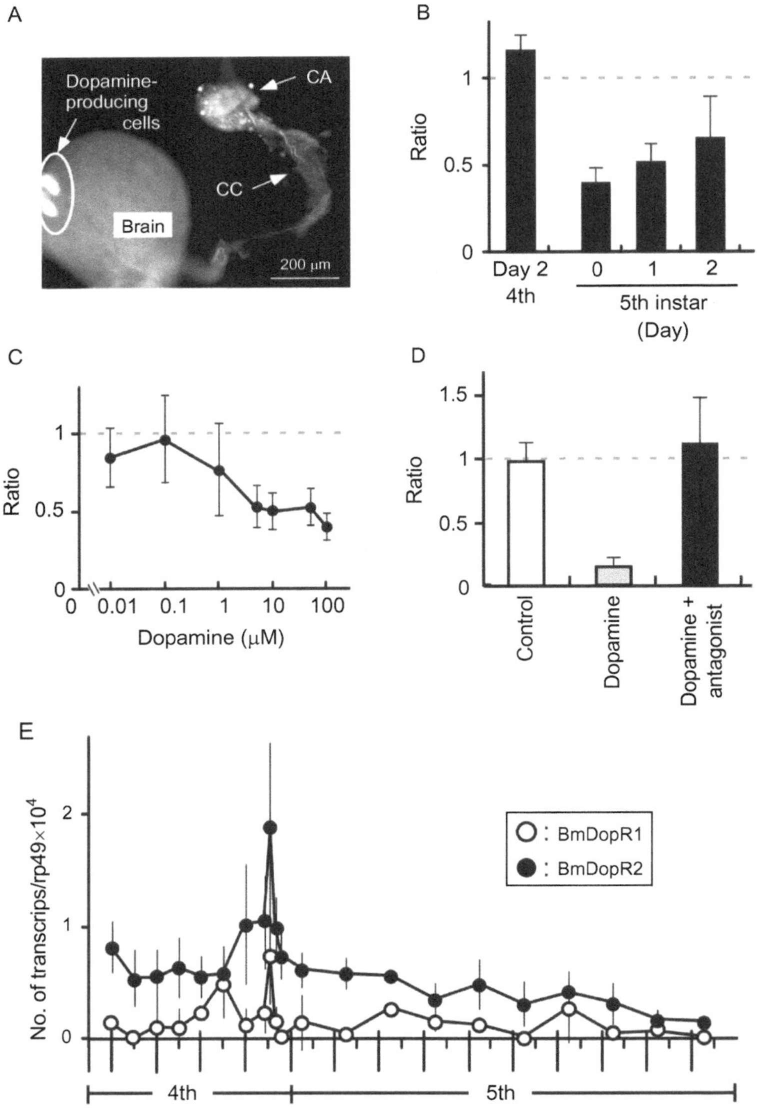

Figure 3.2 Inhibition of JH biosynthesis by dopamine in the last (fifth) instar stadium in *Bombyx*. (A) Dopamine-producing cells located in the medial portion of the day 0 fifth instar *Bombyx* brain detected by a dopamine antiserum. (B) Stage-dependent inhibition of JH synthesis by dopamine. CC–CA complexes from larvae of various stages were cultured with or without 100 μM dopamine for 6 h followed by the determination of JH synthetic activity ($N=3$–5, ±SE). Ratio represents the quantity of the amount of JH synthesized by the CA cultured with dopamine divided by the amount without dopamine. Note that the CA from only fifth instar larvae were suppressed to synthesize JH.

unknown, and further studies will be necessary to elucidate its mode of action.

Thus, declining levels of ecdysteroid and an increase in dopamine released from nerve endings in the CA act together to shut down JH biosynthesis by the CA after the last larval ecdysis (Fig. 3.3A). These factors apparently target different steps on the JH biosynthetic pathway to ensure the cessation of JH synthesis, allowing the initiation of metamorphosis.

2.3.2 Glutamate

The neurotransmitter glutamate has been shown to affect JH biosynthesis in the cockroach, *Diploptera punctata* (Pszczolkowski, Lee, Liu, & Chiang, 1999). Glutamate causes an increase in cytosolic calcium concentration in CA cells, which stimulates JH synthesis. Glutamate has also been shown to stimulate JH synthesis in *Drosophila* larvae (Huang et al., 2011), and the molecular mechanisms of this effect on JH synthesis have been studied well. Glutamate is the natural stimulatory ligand of the *N*-methyl-D-aspartate (NMDA) subtype of glutamate receptor, and in *Drosophila* a mutation of this receptor influences JH biosynthesis; the expression of the regulatory gene *Decapentaplegic* (*Dpp*) and the JH biosynthetic enzyme gene *JHAMT* declines. In addition, a mutant of the gene *Mothers against decapentaplegic* (*mad*), which is located downstream of Dpp in the pathway that regulates the transforming growth factor beta (TGF-β) receptor signaling pathway, decreased *JHAMT* expression so that JH biosynthesis declined. Based on these results, it is suggested that glutamate produced by the brain binds to NMDA receptors in CA cells to induce *Dpp* expression, in turn activating the TGF-β pathway, which enhances *JHAMT* expression. As a result, JH biosynthesis is activated. It is not known whether this signaling pathway is involved in the control of JH synthesis in other insect species, including *Bombyx*.

(C) Dose-dependent inhibition of JH synthesis. CC–CA complexes of day 0 fifth instar larvae were cultured with various concentrations of dopamine for 6 h followed by the determination of JH synthetic activity ($N=3$–4, ±SE). Ratio was calculated as in (B). (D) Prevention of the inhibitory action of dopamine by a dopamine D1 receptor antagonist. CC–CA complexes from day 0 fifth instar larvae were cultured with or without 0.1 μM antagonist, R(+)-SCH23390, and 100 μM dopamine for 6 h and then determined the JH synthetic activity ($N=3$, ±SE). Ratio was calculated as in (B). (E) Fluctuation of the mRNAs of two isoforms of dopamine D1 receptors, *BmDopR1* and *BmDopR2*, in the CA during development ($N=3$, ±SD).

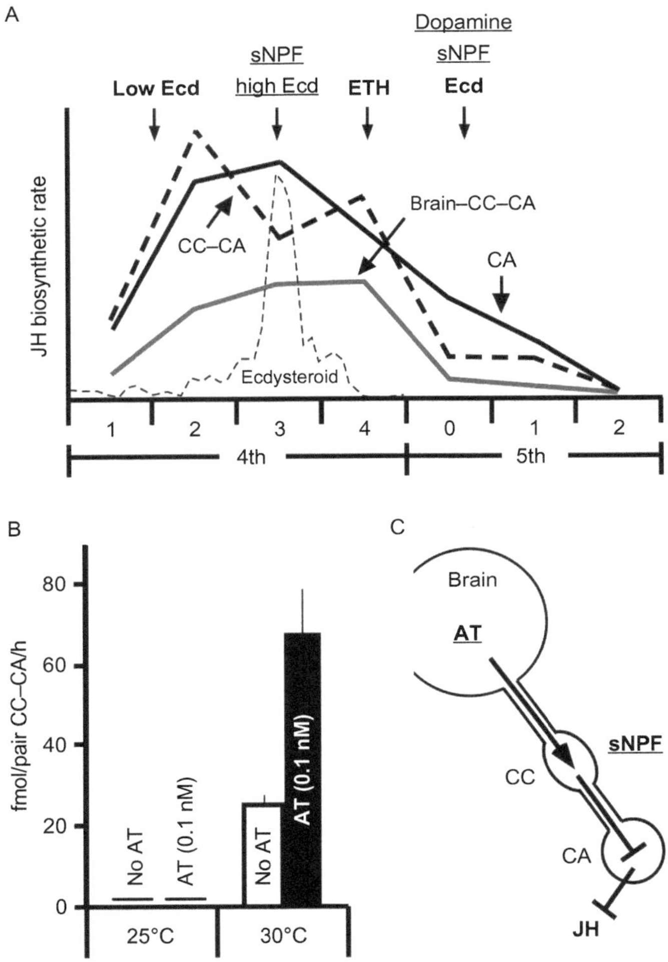

Figure 3.3 (A) Synthetic activity of JH by a brain–CC–CA complex, CC–CA complex, and CA alone in *Bombyx*. Summary of the action of ecdysteroid, sNPF, ETH, and dopamine on JH biosynthesis is shown on top. Note that the action of these factors are well corresponded to the synthetic activity by CC–CA complexes in the fourth instar stadium (see text for discussion). Bolds: stimulative action, underlines: inhibitory action. (B) AT activates the CA of *Agrius convolvuli*, a close relative of *Manduca*, only at high temperature such as 30 °C, but not at 25 °C. CC–CA complexes from female day 0 adult were cultured with or without 0.1 nM AT for 6 h followed by the measurement of JH synthetic activity ($N=5$, $\pm$SE). The male CC–CA were not activated by AT even at 30 °C. (C) The mode of suppressive action of AT in *Bombyx*. The receptor for AT is expressed in the same cells as sNPF cells in the CC, AT stimulates sNPF expression, and then the translated sNPF in the CC is transferred to the CA, where the sNPF receptor presents, through the nerve axon to suppress JH biosynthesis.

2.4. Regulation by peptide hormones

JH biosynthetic activity by the CA can only be measured *in vitro*; the extent of JH synthesis depends on how this is assayed *in vitro*. When a brain–CC–CA complex is dissected intact and its JH synthetic activity is measured *in vitro*, the amount of JH produced by the CA in each stage is lower than would be the case without the brain (Fig. 3.3A). The JH synthetic activity of the CA alone is approximately doubled compared to the intact brain–CC–CA, but the pattern of JH synthesis during development is similar to that of the brain–CC–CA complex. Although the CC–CA complex without the brain produces a similar amount of JH to the CA alone, the fluctuating pattern of JH synthetic activity is distinct; there are two periods of depressed synthetic activity: on day 3 of the fourth larval stage and again on day 0 of the fifth stage just after the final larval ecdysis (Fig. 3.3A). JH synthetic activity by the CC–CA complex in the fourth instar stadium thus parallels the JH titer in the hemolymph, although synthesis is (as expected) shifted forward by several hours (Niimi & Sakurai, 1997; Fig. 3.1B); therefore, several factors supplied by the cephalic organs regulate JH synthesis under *in vivo* conditions.

Recent studies show that many kinds of peptides can stimulate or inhibit JH production by the CA (Goodman & Granger, 2005), but most of these studies concern the isolation, characterization, and distribution of these peptides (Stay & Tobe, 2007; Weaver & Audsley, 2009). In this section, we concentrate on four peptides, short neuropeptide F (sNPF), AT, AST, and ecdysis triggering hormone (ETH), and discuss how these peptides cooperate to stimulate and/or suppress JH synthesis in stage-specific manners.

2.4.1 sNPF suppresses JH biosynthesis in a stage-specific fashion

sNPF is a short peptide [(A/L)R(P/L)RFamide] present in *Bombyx* for which the mRNA is expressed primarily in the CC and a small amount in the brain; the translated peptide in the CC is transferred to the CA through nerve axons to suppress JH synthesis in early last stage larvae (Yamanaka et al., 2008). Detailed analysis of the action of sNPF showed that its inhibitory action on the CA was effective only from the day 3 fourth instar and in early fifth stage larvae just after the last larval ecdysis; these are the stages when CC–CA complexes produce a smaller amount of JH *in vitro* than that is produced by the CA alone (Fig. 3.3A). The two depressions of the JH synthesis by the CC–CA complexes at these times are most likely due to the release of sNPF from the CC where this peptide is synthesized. Although the fluctuation of sNPF in the hemolymph is unknown, mRNA levels of sNPF in the CC are relatively high with some fluctuations during the fourth instar and in the feeding stage of the fifth instar stadium (Hiruma, Kaneko, & Kinjoh, 2010; unpublished data).

There are two isoforms of sNPF receptors: isoform 1 (*BNGR-A10*) and isoform 2 (*BNGR-A11*) (Yamanaka et al., 2008) both are expressed in the CA. Receptor mRNA levels in the CA, in particular of isoform 1, become high on days 2–3 of the fourth larval stage, just when sNPF is effective in inhibiting the JH synthesis, but decline sharply at the time (day 3 of the fourth larval stage) when suppressive action of sNPF disappears. This indicates that the pattern of sNPF receptor expression is one of the factors that causes JH biosynthesis suppression by sNPF to be stage specific, at least in the fourth larval stage. *In vitro* CA culture experiments with 20E suppress receptor mRNA levels and indicate that the premolting surge of ecdysteroid is among the factors regulating receptor levels. In addition, insulin also suppressed *sNPF* receptor expression at this time (unpublished data). Similar downregulation of the *sNPF* receptor by insulin signaling has also been found in *Drosophila* odorant receptor neurons (Root, Ko, Jafari, & Wang, 2011). It will be necessary to measure insulin concentration in the hemolymph to establish whether or not insulin functions to regulate sNPF receptor levels *in vivo*.

Thus, the suppressive action of sNPF on JH synthesis after the final larval ecdysis also contributes, in addition to the actions of ecdysteroid and dopamine (see above), to the cessation of JH synthesis that is required for the initiation of metamorphosis (Figs. 3.3A and 3.4).

2.4.2 AT is a regulator for sNPF expression

AT was the first peptide to be isolated that affects JH synthesis by the CA, but the allatotropic action of AT evidently does not extend to insects from other families. AT only activates the CA of female adult *Manduca*, the CA of other insects being unaffected (Kataoka et al., 1989). We have attempted to elucidate the strange function of this tridecapeptide using another Sphingid moth, *Agrius convolvuli*, a relative of *Manudca* [*Agrius* AT is identical to that of *Manduca* (H. Kataoka, personal communication)]. We found that AT did activate JH synthesis by the CA from adult female *A. convolvuli in vitro*, but this was only seen when we increased the temperature to 30 °C, which was the temperature that the original paper reported, from 25 °C (Fig. 3.3B). AT did not activate the CA from other stages or from males of *Agrius* even at

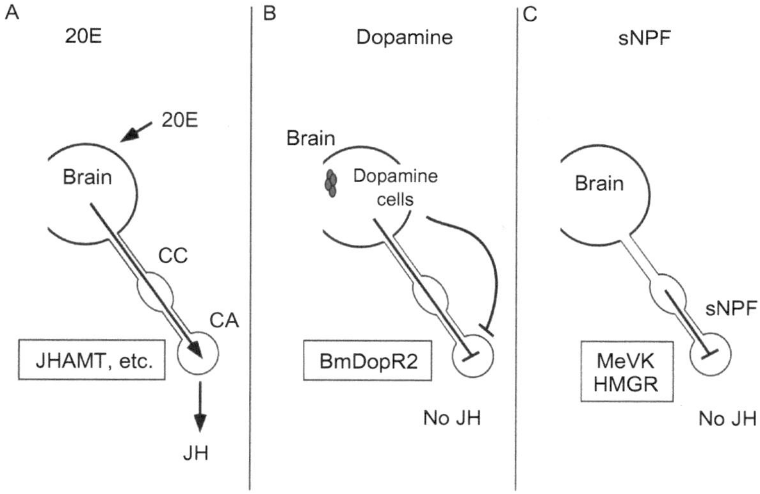

Figure 3.4 Model for the mode of action of 20E (A), dopamine (B), and sNPF (C) on the CA of the early fifth (last) instar stadium in *Bombyx*. See text for detailed explanations.

30 °C; therefore, the trophic action of AT is sex-, stage-, and temperature dependent, which might be the reason that many reports have failed to find the activation of CA by AT. The reason for the temperature dependence of the allatotropic action is unknown, and we have not determined this AT action in *Bombyx*. AT has been found in other insect species such as *Aedes aegypti* (Li, Unnithan, Veenstra, Feyereisen, & Noriega, 2003) and others (Stay, 2000), but the roles of these peptides on insect molting and metamorphosis need to be studied.

Bombyx AT is the same peptide as that of *Manduca*, and the receptor for AT in *Bombyx* is not expressed in the CA but unexpectedly in the CC (Yamanaka et al., 2008). This might be the reason that AT does not have allatotropic action on *Bombyx*, *Manduca*, and other lepidopterans. As the AT receptor is produced in the same cells as the sNPF-producing cells in the CC (based on the *in situ* hybridization and immunohistochemistry experiments—Yamanaka et al., 2008) and the developmental transcript fluctuations of *sNPF* in the CC mirror those of the *AT* receptor (unpublished data), it is possible that interactions exist between AT and sNPF. Surprisingly, *in vitro* experiments showed that 1 μM AT stimulated sNPF expression in the CC of day 1 fifth instar larvae (unpublished data), indicating that AT is one of the stimulatory regulators for sNPF (Fig. 3.3C).

2.4.3 AST suppresses JH biosynthesis without stage specificity

Among the peptides, ASTs are the most well-studied JH-regulatory peptides which reversibly inhibit JH biosynthesis by the CA (see reviews by Goodman & Granger, 2005; Stay & Tobe, 2007; Weaver & Audsley, 2009). ASTs have three different types by their structures. AST-A peptides were the first to be isolated from the cockroach *D. punctata* and have been found in many different insect species, but some AST-A peptides stimulate JH biosynthesis in a stage-specific fashion (Clark, Lange, Zhang, & Tobe, 2008; Stay, Zhang, & Tobe, 2002). AST-B family peptides have been identified in crickets and locusts, and AST-C peptides are members of the PISCF family found in Lepidoptera (Goodman & Granger, 2005). ASTs are produced in the brain and are considered to be transferred to the CA through nerve axons. AST-C suppressed JH biosynthesis by 50–80% in *Bombyx* in all the stages we studied, from the fourth to fifth instar stadium, and did not show any stage-specific variation in effect (Hiruma et al., 2010; unpublished data). The lower amount of JH synthesis by an intact brain–CC–CA complex (Fig. 3.3A) is probably due to the inhibitory action of AST-C synthesized in the brain and supplied to the CA via the CC. The role of AST-C might be to adjust the total amount of JH synthesis by preventing the overproduction of JH, although the timing of the release of this peptide is unknown.

2.4.4 Mode of action of sNPF and AST

Both sNPF and AST-C suppress JH biosynthesis by the CA, but the stage-dependent suppressive action of sNPF is distinct from that of AST-C. When the CA of day 0 fifth instar larvae were incubated with these peptides, *MevK* and *HMGR* expressions were prevented by sNPF, although not only *MevK* and *HMGR* but also *IPPI* mRNA levels were depressed by AST-C (unpublished data); therefore, the modes of action of these peptides must be different from each other. AST-C was initially believed to inhibit JH synthesis by the transfer of citrate to the cytoplasm across the mitochondrial membrane or by the cleavage of citrate to yield cytoplasmic acetyl-CoA rather than by the inhibition of the enzymes in the mevalonate pathway (Sutherland & Feyereisen, 1996), but it is apparent that inhibition of some JH synthetic enzymes is also involved.

2.4.5 ETH activates JH biosynthesis at the time of the final larval ecdysis

ETH is a peptide first found in *Manduca* which is responsible for initiating and regulating insect ecdysis, along with eclosion hormone (EH), corazonin, and crustacean cardioactive peptide (CCAP). ETH is produced by epitracheal glands (Inka cells) that are located in the epitracheal endocrine system in each spiracle (Zitnan & Adams, 2005; Zitnan, Kingan, Hermesman, & Adams, 1996). When the ecdysteroid titer declines during the last larval molt, corazonin is released from the brain, which stimulates ETH release from the epitracheal glands. The ETH then acts on ventromedial cells in the brain to release EH which acts to stimulate CCAP cells in the central nervous system (Zitnan & Adams, 2005).

There are two isoforms of the ETH receptor in *Bombyx*, both of which are expressed in the CA shortly before the last larval ecdysis (Yamanaka et al., 2008), which suggests that ETH might modulate JH synthesis at this time. Detailed analyses of ETH receptor expression in the CA showed that the maximum expression is observed 5 h before the last larval ecdysis, at the time when JH biosynthesis ceases to decline and increases again. ETH receptor is not expressed at the time just before pupation when the CA are not synthesizing JH (Hiruma et al., 2010). At the time of adult ecdysis, the female CA (but not males) that synthesize JH (Kinjoh et al., 2007) also express the ETH receptor (Zitnan, personal communication). When CA from larvae 5–6 h before the final larval ecdysis are cultured with ETH, JH synthesis by the CA is stimulated. ETH does not have this effect at other stages, indicating that ETH is responsible for the increase in JH synthesis that occurs at the time of the final larval ecdysis along with the decline of ecdysteroid titer (see above). Since the decline of ecdysteroid titer induced JH biosynthesis only under *in vivo* conditions, and not *in vitro* (Kaneko, Kinjoh, et al., 2011), the induction of JH synthesis by declining ecdysteroid might be through the induction of ETH release. The detailed action of ETH, such as the mode of action and the mechanisms of the stage specificity, needs to be studied. The increase in JH titer that occurs at the time of the final larval ecdysis is important to determine the length of feeding period (Hiruma, 1980; Hiruma, Shinoda, Malone, & Riddiford, 1999; Sakurai, 1984) before spinning and gut purge, a factor that in turn affects pupal weight.

2.5. Regulation by nutrition

JH synthesis is also influenced by nutrition (Bhaskaran, Jones, & Jones, 1980; Cymborowski, Bogus, Beckage, Williams, & Riddiford, 1982; Lee & Horodyski, 2006; Maestro, Cobo, & Bellés, 2009; Noriega, 2004). In *Manduca*, starvation after the final larval ecdysis stimulates JH biosynthesis and also suppresses the degradation of JH in the hemolymph, so that JH titer in the hemolymph increases (Cymborowski et al., 1982; Lee &

Horodyski, 2006). Nutritional effects in insects are mediated by insulin signaling (Britton, Lockwood, Li, Cohen, & Edgar, 2002), and these observations suggest involvement of the insulin signaling pathway in the control of JH synthesis. In the insulin receptor mutant (*InR*) in *Drosophila*, JH synthesis is reduced compared to wild-type flies (Tatar et al., 2001; Tu, Yin, & Tatar, 2005), an effect that is due to decreased *HMGR* expression (Belgacem & Martin, 2007). Since insulin causes the reduction of sNPF expression in *Drosophila* (Root et al., 2011), and considering that sNPF also reduces HMGR expression in *Bombyx*, the reduced JH synthesis seen in *InR* flies is probably due to the upregulation of sNPF, so that HMGR expression is suppressed followed by the decrease in JH synthesis. In addition, the target of rapamycin pathway also regulates the expression of both HMGR and HMGS in the cockroach, *Blattella germanica*, and this pathway is involved in the control of JH biosynthesis in that insect (Maestro et al., 2009).

3. MOLECULAR ACTION OF JH ON INSECT MOLTING AND METAMORPHOSIS

Studies of the action of JH at the molecular level have been hampered due to, in particular, the marked tendency of the hormone to bind to surfaces, but notable advances have been made in the last few years; the most important results have been summarized by Riddiford (2008). In this review, we briefly discuss the most recent findings including the long-awaited discovery of a JH receptor.

3.1. Methoprene-tolerant (Met) as a JH receptor

The *methoprene-tolerant* (*Met*) gene, first found in *Drosophila*, is a member of the basic helix–loop–helix (bHLH)–Per–Arnt–Sim (PAS) family of transcriptional regulators (Ashok, Turner, & Wilson, 1998). *Met* null mutants are resistant to the morphogenetic effect of the JH analog methoprene and are viable, although their fecundity is reduced (Wilson & Ashok, 1998). JH is necessary for egg maturation so that yolk protein uptake is controlled (Soller, Bownes, & Kubli, 1999); therefore, it has been long thought that the Met protein is involved in the JH signaling pathway. In *Drosophila*, *germ cell-expressed* (*gce*) is known as a paralog of *Met* (Godlewski, Wang, & Wilson, 2006), but only *Met* has been found in non-*Drosophila* insects (Charles et al., 2011; Wang, Baumann, & Wilson, 2007). In non-*Drosophila* insects such as *Tribolium*, knockout of *Met* RNA expression by the injection of dsRNA caused precocious metamorphosis, and those individuals are unresponsive to methoprene and JH-III (Konopova & Jindra, 2007; Parthasarathy, Tan, & Palli, 2008). In *Drosophila*, the removal of CA causes the formation of smaller pupae and death at head eversion (Riddiford, Truman, Mirth, & Shen, 2010), and the *Met/gce* double mutant dies during the larval–pupal transition, although both *Met* and *gce* null single mutants are viable because of their redundancy (Abdou et al., 2011).

Met protein specifically binds JH-III and other biologically active JH mimics at physiologically relevant concentrations [K_d = 5.3 nM for JH-III in *Drosophila* (Miura, Oda, Makita, & Chinzei, 2005); K_d = 2.94 nM for JH-III in *Tribolium castaneum* (Charles et al., 2011)]. This binding is through the C-terminal PAS-B domain (K_d = 12.3 nM) (Charles et al., 2011). Met forms Met/Met homodimers [also Met/GCE heterodimers in *Drosophila* (Godlewski et al., 2006)] in the absence of JH, and dimer formation was prevented by JH (Charles et al., 2011; Godlewski et al., 2006). In the mosquito, *A. aegypti*, a transcriptional coactivator of the ecdysteroid receptor complex FISC (Li, Mead, & Zhu, 2011) and a steroid receptor coactivator SRC (also known as Taiman) (Zhang, Xu, Sheng, Sui, & Palli, 2011), both of which belong to the member of the bHLH–PAS family, act as functional partners of Met in mediating JH action on target genes such as *Krüppel-homolog* 1 (Kr-h1). Unlike in the case of the Met/Met formation, Met/FISC and Met/SRC form a complex in the presence of JH (Charles et al., 2011; Li et al., 2011). Mutations of *Tribolium* Met within the ligand-binding pocket which disrupt JH binding did not affect the formation of the Met/Met dimer complex, but prevented the ligand-dependent dissociation of the Met/Met homodimer and the ligand-dependent interaction of Met with its partner SRC (Taiman) (Charles et al., 2011). The evidence of both biological actions and the characteristic nature of Met such as direct and specific binding to JH strongly suggest that Met is a JH receptor with SRC (Taiman) as a partner (Fig. 3.5A).

3.2. JH responsive genes and the JH signaling pathway

Kr-h1 is an evolutionarily conserved JH-induced gene that seems to be a crucial factor for molting and metamorphosis in both holometabolous and hemimetabolous insects (Duportets et al., 2012, Lozano & Belles, 2011; Minakuchi, Namiki, & Shinoda, 2009; Minakuchi, Zhou, & Riddiford, 2008; Zhu, Busche, & Zhang, 2010). RNAi knockout of the *Kr-h1* gene induces precocious metamorphosis in *Tribolium* (Minakuchi et al., 2009) and *Blattella* (Lozano & Belles, 2011), but the JH analog methoprene was unable to rescue this effect so that Kr-h1 is considered to mediate the JH

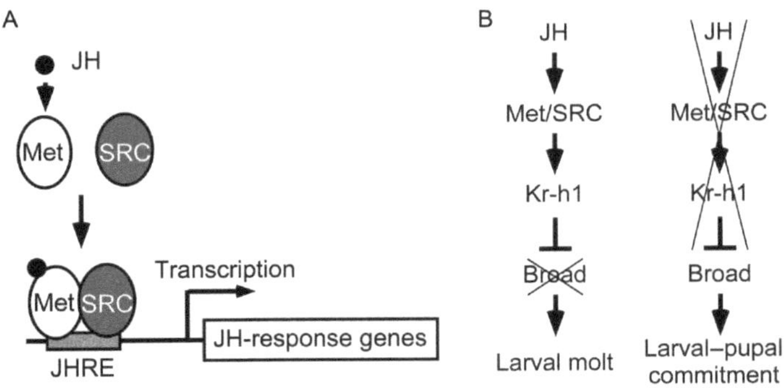

Figure 3.5 Model for Met as a JH receptor in insects (A) and JH signaling pathway during larval–pupal commitment (B). JHRE, JH response element. See text for detailed explanations.

signaling pathway downstream of the hormone's interaction with its receptor. Consistent with this, when *Met* gene expression was suppressed, *Kr-h1* gene expression was also suppressed (Li et al., 2011; Minakuchi et al., 2009; Zhu et al., 2010). Kr-h1 is not only a crucial factor for JH control of larval molting but is also induced by ecdysteroid and interacts with other genes such as *broad* (Beck, Pecasse, & Richards, 2004), SRC (Zhang et al., 2011), and FISC (Li et al., 2011).

The induction of *broad* expression by ecdysteroid does not occur until pupal commitment has been performed (Zhou, Hiruma, Shinoda, & Riddiford, 1998), and its ectopic expression in early second instar *Drosophila* larvae before the rise of ecdysteroid titer prevented molting to the third instar, but caused precocious pupae (Zhou, Zhou, Truman, & Riddiford, 2004). These results show that *broad* is one of the ecdysteroid-induced genes that both specify pupal development and mediate the prevention of the pupal–adult transformation (Zhou & Riddiford, 2002; Zhou et al., 2004). RNAi suppression of *Met* or *Kr-h1* expression by dsRNA stimulates *broad* expression so that precocious metamorphosis results (Konopova & Jindra, 2008; Minakuchi et al., 2009); it, therefore, appears that Met bound to JH stimulates Kr-h1 expression, thereby suppressing *broad* expression so that the larval state is maintained (Fig. 3.5B).

Met/SRC and Met/FISC are required for the expression of JH responsive genes including Kr-h1 (Li et al., 2011; Zhang et al., 2011), and SRC is required for the expression of ecdysteroid-responsive genes. Therefore, these factors as well as *broad* might mediate cross talk between JH and ecdysteroid to prevent the ecdysteroid-induced switch necessary for metamorphosis.

E75A is also induced by JH in *Drosophila* S2 cells (Dubrovsky, Dubrovskaya, & Berger, 2004). This induction of E75A is mediated by an intercellular pathway utilizing GCE/FTZ-F1 and Met/FTZ-F1 that forms transcriptionally active heterodimers, so that the removal of FTZ-F1 prevents JH activation of E75A (Bernardo & Dubrovsky, 2012; Dubrovsky et al., 2011). Although about 20% of E75A mutants undergo precocious metamorphosis in *Drosophila* (Bialecki, Shilton, Fichtenberg, Segraves, & Thummel, 2002), this transcription factor does not seem to be involved in the JH signaling pathway; rather the phenotype of the E75A mutant involves a feed-forward pathway that amplifies or maintains the ecdysteroid level, so that there is submaximal ecdysteroid induction of FTZ-F1 expression (Bialecki et al., 2002) in turn reducing the latter's ability to act as a competence factor facilitating JH (Dubrovsky et al., 2011) and ecdysteroid (Broadus, McCabe, Endrizzi, Thummel, & Woodard, 1999) activation of gene expression.

4. CONCLUSION

After entering the last larval stadium, lepidopteran larvae prepare for pupal metamorphosis, which is characterized by the larval–pupal commitment of various tissues such as epidermal cells (Muramatsu et al., 2008; Riddiford, 1985, 1996). The indispensable event for this process is the removal of JH from the hemolymph, and one of the most important contributors to this endocrine event is the shut down of JH synthesis by the CA. Here, we have shown that at least three factors, ecdysteroid, dopamine, and sNPF, work together to guarantee cessation of JH synthesis; all of these factors have different mode of actions (Figs. 3.3A and 3.4), so that the cessation is definitely accomplished.

Many additional factors are involved in JH synthesis (see Goodman & Granger, 2005), but the roles of these other factors on JH synthesis in association with insect development have not been well studied. One of the interesting peptides among them is allatinhibin found in *Manduca*, released from the brain shortly before the larvae empty their gut content, and inhibits JH synthesis (Bhaskaran et al., 1990). Therefore, allatinhibin may also be involved in the cessation of the JH synthesis in the last larval stage, although we have not determined the presence of allatinhibin in *Bombyx* and other insect species. This peptide is distinct from AST (Bhaskaran et al., 1990) and apparently from sNPF as well (Yamanaka et al., 2008), but no other studies have been done and we do not know much about this peptide.

Another candidate for JH regulation is prothoracicotropic hormone (PTTH) which stimulates PGs to secrete ecdysteroid. This peptide is produced by brain neurosecretory cells and is released into the hemolymph from the CA in lepidopterans, so that PTTH might also affect JH synthesis, but our study showed little effect of this nature (unpublished data).

Ecdysteroid is known to have a feedback action, both positively and negatively, on its own synthetic organ, the prothoracic gland (Sakurai, 2005). It has been suggested that JH too might have feedback loops (Goodman & Granger, 2005). However, when CA of day 1 fourth instar larvae were cultured with or without 0.01 or 1 μg/ml methoprene for 8 h, JH synthesis was little affected. In addition, when we cultured CC–CA complexes from day 0 fifth instar *Bombyx* larvae with 100 μM pravastatin, a derivative compound of a competitive inhibitor of HMGR, compactin (Endo, Kuroda, & Tanzawa, 1976) for 24 h, JH biosynthesis was completely shut down, but the expression of none of the JH synthetic enzymes in the CA, including HMGR itself, was affected (Kinjoh, Kaneko, & Hiruma, unpublished data). Therefore, there is little evidence of the presence of feedback regulation of JH synthesis by JH both at the level of JH synthesis and at the JH synthetic enzymes.

Finally, a JH receptor has been found, and cumulative information on the JH signaling pathway has increased tremendously during the past few years; therefore, we can expect the components of the JH signaling network to be fully described within the near future. This will allow fuller understanding of not only the cellular actions of JH in regulating metamorphosis but also the interendocrine signaling that plays an important role in determining the timing of metamorphic developmental events.

ACKNOWLEDGMENTS

We thank Professor Stuart E. Reynolds for critically, and patiently, reading the manuscript. This work was supported by grants from Japanese Society for the Promotion of Science (23658044 and 22380032) and the Program for Promotion of Basic Research Activities for Innovative Biosciences (PROBRAIN).

REFERENCES

Abdou, M. A., He, Q., Wen, D., Zyaan, O., Wang, J., Xu, J., et al. (2011). *Drosophila* Met and Gce are partially redundant in transducing juvenile hormone action. *Insect Biochemistry and Molecular Biology, 41*, 938–945.

Ashok, M., Turner, C., & Wilson, T. G. (1998). Insect juvenile hormone resistance gene homology with the bHLH-PAS family of transcriptional regulators. *Proceedings of the National Academy of Sciences of the United States of America, 95*, 2761–2766.

Baker, F. C., Tsai, L. W., Reuter, C. C., & Schooley, D. A. (1987). *In vivo* fluctuation of JH, JH acid, and ecdysteroid titer, and JH esterase activity, during development of fifth stadium *Manduca sexta*. *Insect Biochemistry*, *17*, 989–996.

Beck, Y., Pecasse, F., & Richards, G. (2004). *Krüppel-homolog* is essential for the coordination of regulatory gene hierarchies in early *Drosophila* development. *Developmental Biology*, *268*, 64–75.

Belgacem, Y. H., & Martin, J. R. (2007). Hmgcr in the *corpus allatum* controls sexual dimorphism of locomotor activity and body size via the insulin pathway in *Drosophila*. *PLoS One*, *2*, e187. http://dx.doi.org/10.1371/journal.pone.0000187.

Bellés, X., Martín, D., & Piulachs, M. D. (2005). The mevalonate pathway and the synthesis of juvenile hormone in insects. *Annual Review of Entomology*, *50*, 181–199.

Bernardo, T. J., & Dubrovsky, E. B. (2012). The *Drosophila* juvenile hormone receptor candidates methoprene-tolerant (MET) and germ cell-expressed (GCE) utilize a conserved LIXXL motif to bind the FTZ-F1 nuclear receptor. *The Journal of Biological Chemistry*, *287*, 7821–7833.

Bhaskaran, G., Dahm, K. H., Barrera, P., Pacheco, J. L., Peck, K. E., & Muszynska-Pytel, M. (1990). Allatinhibin, a neurohormonal inhibitor of juvenile hormone biosynthesis in *Manduca sexta*. *General and Comparative Endocrinology*, *78*, 123–136.

Bhaskaran, G., Jones, G., & Jones, D. (1980). Neuroendcrine regulation of corpus allatum activity in *Manduca sexta*: Sequential neurohormonal and nervous inhibition in the last-instar larva. *Proceedings of the National Academy of Sciences of the United States of America*, *77*, 4407–4411.

Bialecki, M., Shilton, A., Fichtenberg, C., Segraves, W. A., & Thummel, C. S. (2002). Loss of the ecdysteroid-inducible E75A orphan nuclear receptor uncouples molting from metamorphosis in *Drosophila*. *Developmental Cell*, *3*, 209–220.

Britton, J. S., Lockwood, W. K., Li, L., Cohen, S. M., & Edgar, B. A. (2002). *Drosophila*'s insulin/P13-kinase pathway coordinates cellular metabolism with nutritional conditions. *Developmental Cell*, *2*, 239–249.

Broadus, J., McCabe, J. R., Endrizzi, B., Thummel, C. S., & Woodard, C. T. (1999). The *Drosophila* beta FTZ-F1 orphan nuclear receptor provides competence for stage-specific responses to the steroid hormone ecdysone. *Molecular Cell*, *3*, 143–149.

Browder, M. H., D'Amico, L. J., & Nijhout, H. F. (2001). The role of low levels of juvenile hormone esterase in the metamorphosis of *Manduca sexta*. *Journal of Insect Science*, *1*, 1–4.

Charles, J. P., Iwema, T., Epa, V. C., Takaki, K., Rynes, J., & Jindra, M. (2011). Ligand-binding properties of a juvenile hormone receptor, Methoprene-tolerant. *Proceedings of the National Academy of Sciences of the United States of America*, *108*, 21128–21133.

Clark, L., Lange, A. B., Zhang, J. R., & Tobe, S. S. (2008). The roles of Dippu-allatostatin in the modulation of hormone release in *Locusta migratoria*. *Journal of Insect Physiology*, *54*, 949–958.

Cymborowski, B., Bogus, M., Beckage, N. E., Williams, C. M., & Riddiford, L. M. (1982). Juvenile hormone titres and metabolism during starvation-induced supernumerary larval moulting of the tobacco hornworm, *Manduca sexta* (L.). *Journal of Insect Physiology*, *28*, 129–135.

Dubrovsky, E. B., Dubrovskaya, V. A., & Berger, E. M. (2004). Hormonal regulation and functional role of *Drosophila* E75A orphan nuclear receptor in the juvenile hormone signaling pathway. *Developmental Biology*, *268*, 258–270.

Dubrovsky, E. B., Dubrovskaya, V. A., Bernardo, T., Otte, V., DiFilippo, R., & Bryan, H. (2011). The *Drosophila* FTZ-F1 nuclear receptor mediates juvenile hormone activation of *E75A* gene expression through an intracellular pathway. *The Journal of Biological Chemistry*, *286*, 33689–33700.

Duportets, L., Bozzolan, F., Abrieux, A., Maria, A., Gadenne, C., & Debernard, S. (2012). The transcription factor *Krüppel homolog 1* is linked to the juvenile hormone-dependent maturation of sexual behavior in the male moth, *Agrotis ipsilon. General and Comparative Endocrinology, 176*, 158–166.

Endo, A., Kuroda, M., & Tanzawa, K. (1976). Competitive inhibition of 3-hydroxy-3-methylglutaryl coenzyme A reductase by ML-236A and ML-236B fungal metabolites, having hypocholesterolemic activity. *FEBS Letters, 72*, 323–326.

Gilbert, L. I., Granger, N. A., & Roe, R. M. (2000). The juvenile hormones: Historical facts and speculations on future research directions. *Insect Biochemistry and Molecular Biology, 30*, 617–644.

Godlewski, J., Wang, S., & Wilson, T. G. (2006). Interaction of bHLH-PAS proteins involved in juvenile hormone reception in *Drosophila. Biochemical and Biophysical Research Communications, 342*, 1305–1311.

Goodman, W. G., & Granger, N. A. (2005). The juvenile hormones. In: L. I. Gilbert, K. Iatrou & S. S. Gill (Eds.), *Comprehensive molecular insect science*, Vol. 3, (pp. 319–408). Oxford: Elsevier.

Granger, N. A., MacDonald, J. D., Menold, M., Ebersohl, R., Hiruma, K., Cann, C. L., et al. (1994). Evidence of a stimulatory effect of cyclic AMP on corpus allatum activity in *Manduca sexta. Molecular and Cellular Endocrinology, 103*, 73–80.

Granger, N. A., Sturgis, S. L., Ebersohl, R., Geng, C., & Sparks, T. C. (1996). Dopaminergic control of corpora allata activity in the larval tobacco hornworm, Manduca sexta. *Archives of Insect Biochemistry and Physiology, 32*, 449–466.

Gu, S. H., & Chow, Y. S. (1993). Role of ecdysteroid levels in the early last larval instar of *Bombyx mori. Experientia, 49*, 806–809.

Gu, S. H., & Chow, Y. S. (1996). Regulation of juvenile hormone biosynthesis by ecdysteroid levels during the early stages of the last two larval instars of *Bombyx mori. Journal of Insect Physiology, 42*, 625–632.

Hammock, B. D., Bonning, B. C., Possee, R. D., Hanzlik, T. N., & Maeda, S. (1990). Expression and effects of the juvenile hormone esterase in a baculovirus vector. *Nature, 344*, 458–461.

Hiruma, K. (1980). Possible roles of juvenile hormone in the prepupal stage of *Mamestra brassicae. General and Comparative Endocrinology, 41*, 392–399.

Hiruma, K. (2003). Juvenile hormone action in insect development. In H. L. Henry & A. W. Norman (Eds.), *Encyclopedia of hormones*, Vol. 2, (pp. 528–535). San Diego: Academic Press.

Hiruma, K., Kaneko, Y., & Kinjoh, T. (2010). Regulation of juvenile hormone synthesis by corpora allata. In *The 18th international ecdysone workshop 2010, Ceske Budejovice, Czech Republic* (p. 10).

Hiruma, K., Shinoda, T., Malone, F., & Riddiford, L. M. (1999). Juvenile hormone modulates 20-hydroxyecdysone-inducible ecdysone receptor and Ultraspiracle gene expression in the tobacco hornworm, *Manduca sexta. Development Genes and Evolution, 209*, 18–30.

Huang, J., Tian, L., Peng, C., Abdou, M., Wen, D., Wang, Y., et al. (2011). DPP-mediated TGF beta signaling regulates juvenile hormone biosynthesis by activating the expression of juvenile hormone acid methyltransferase. *Development, 138*, 2283–2291.

Kamimura, M., Shimura, S., & Kiuchi, M. (2003). Simple manipulation of silkworm molting by an artificial diet cotaining plant-derived 20-hydroxyecdysone. *Journal of Insect Biotechnology and Sericology, 72*, 197–201.

Kaneko, Y., & Hiruma, K. (2007). Dopamine is a regulator of juvenile hormone biosynthesis by the corpora allata in *Bombyx mori*. In *9th International conference on juvenile hormones, York, UK* (p. 67).

Kaneko, Y., Kinjoh, T., Kiuchi, M., & Hiruma, K. (2011). Stage-specific regulation of juvenile hormone biosynthesis by ecdysteroid in *Bombyx mori. Molecular and Cellular Endocrinology, 335*, 204–210.

Kaneko, Y., Shinoda, T., & Hiruma, K. (2011). Remodeling of the corpora cardiaca and the corpora allata during adult metamorphosis in *Bombyx mori*: Identification of invisible corpora cardiaca by the expression of adipokinetic hormone. *Applied Entomology and Zoology*, *46*, 87–93.

Kataoka, H., Toschi, A., Li, J. P., Carney, R. L., Schooley, D. A., & Kramer, S. J. (1989). Identification of an allatotropin from adult *Manduca sexta*. *Science*, *243*, 1481–1483.

Kinjoh, T., Kaneko, Y., Itoyama, K., Mita, K., Hiruma, K., & Shinoda, T. (2007). Control of juvenile hormone biosynthesis in *Bombyx mori*: Cloning of the enzymes in the mevalonate pathway and assessment of their developmental expression in the corpora allata. *Insect Biochemistry and Molecular Biology*, *37*, 808–818.

Konopova, B., & Jindra, M. (2007). Juvenile hormone resistance gene *Methoprene-tolerant* controls entry into metamorphosis in the beetle *Tribolium castaneum*. *Proceedings of the National Academy of Sciences of the United States of America*, *104*, 10488–10493.

Konopova, B., & Jindra, M. (2008). Broad-Complex acts downstream of Met in juvenile hormone signaling to coordinate primitive holometabolan metamorphosis. *Development*, *135*, 559–568.

Lee, K. Y., & Horodyski, F. M. (2006). Effects of starvation and mating on corpora allata activity and allatotropin (Manse-AT) gene expression in *Manduca sexta*. *Peptides*, *27*, 567–574.

Li, M., Mead, E. A., & Zhu, J. (2011). Heterodimer of two bHLH-PAS proteins mediates juvenile hormone-induced gene expression. *Proceedings of the National Academy of Sciences of the United States of America*, *108*, 638–643.

Li, Y., Unnithan, G. C., Veenstra, J. A., Feyereisen, R., & Noriega, F. G. (2003). Stimulation of JH biosynthesis by the corpora allata of adult female *Aedes aegypti in vitro*: Effect of farnesoic acid and *Aedes* allatotropin. *The Journal of Experimental Biology*, *206*, 1825–1832.

Lozano, J., & Belles, X. (2011). Conserved repressive function of Krüppel homolog 1 on insect metamorphosis in hemimetabolous and holometabolous species. *Science Reports*, *1*, 1–7. http://dx.doi.org/10.1038/srep00163.

Maestro, J. L., Cobo, J., & Bellés, X. (2009). Target of rapamycin (TOR) mediates the transduction of nutritional signals into juvenile hormone production. *The Journal of Biological Chemistry*, *284*, 5506–5513.

Minakuchi, C., Namiki, T., & Shinoda, T. (2009). *Krüppel homolog* 1, an early juvenile hormone-response gene downstream of *Methoprene-tolerant*, mediates its anti-metamorphic action in the red flour beetle *Tribolium castaneum*. *Developmental Biology*, *325*, 341–350.

Minakuchi, C., Zhou, X., & Riddiford, L. M. (2008). Krüppel homolog 1 (Kr-h1) mediates juvenile hormone action during metamorphosis of *Drosophila melanogaster*. *Mechanisms of Development*, *125*, 91–105.

Mitsumasu, K., Ohta, H., Tsuchihara, K., Asaoka, K., Ozoe, Y., Niimi, T., et al. (2008). Molecular cloning and characterization of cDNAs encoding dopamine receptor-1 and -2 from brain-suboesophageal ganglion of the silkworm, *Bombyx mori*. *Insect Molecular Biology*, *17*, 185–195.

Miura, K., Oda, M., Makita, S., & Chinzei, Y. (2005). Characterization of the *Drosophila Methoprene-tolerant* gene product. Juvenile hormone binding and ligand-dependent gene regulation. *FEBS Journal*, *272*, 1169–1178.

Muramatsu, D., Kinjoh, T., Shinoda, T., & Hiruma, K. (2008). The role of 20-hydroxyecdysone and juvenile hormone in pupal commitment of the epidermis of the silkworm, *Bombyx mori*. *Mechanisms of Development*, *125*, 411–420.

Mustard, J. A., Beggs, K. T., & Mercer, A. R. (2005). Molecular biology of the invertebrate dopamine receptors. *Archives of Insect Biochemistry and Physiology*, *59*, 103–117.

Niimi, S., & Sakurai, S. (1997). Developmental changes in juvenile hormone and juvenile hormone acid titers in the hemolymph and *in vitro* juvenile hormone synthesis by corpora allata of the silkworm, *Bombyx mori*. *Journal of Insect Physiology*, *43*, 875–884.

Noguchi, H., & Hayakawa, Y. (2001). Dopamine is a key factor for the induction of egg diapause of the silkworm, *Bombyx mori*. *European Journal of Biochemistry*, *268*, 774–780.
Noriega, F. G. (2004). Nutritional regulation of JH synthesis: A mechanism to control reproductive maturation in mosquitoes? *Insect Biochemistry and Molecular Biology*, *34*, 687–693.
Parthasarathy, R., Tan, A., & Palli, S. R. (2008). bHLH-PAS family transcription factor methoprene-tolerant plays a key role in JH action in preventing the premature development of adult structures during larval-pupal metamorphosis. *Mechanisms of Development*, *125*, 601–616.
Pszczolkowski, M. A., Lee, W. S., Liu, H. P., & Chiang, A. S. (1999). Glutamate-induced rise in cytosolic calcium concentration stimulates *in vitro* rates of juvenile hormone biosynthesis in corpus allatum of *Diploptera punctata*. *Molecular and Cellular Endocrinology*, *158*, 163–171.
Reynolds, S. E., Brown, A. M., Seth, R. K., Riddiford, L. M., & Hiruma, K. (2009). Induction of supernumerary larval moulting in the tobacco hornworm *Manduca sexta*: Interaction of bisacylhydrazine ecdysteroid agonists with endogenous juvenile hormone. *Physiological Entomology*, *34*, 30–38.
Riddiford, L. M. (1976). Hormonal control of insect epidermal cell commitment *in vitro*. *Nature*, *259*, 115–117.
Riddiford, L. M. (1985). Hormone action at the cellular level. In G. A. Kerkut & L. I. Gilbert (Eds.), *Comprehensive insect physiology, biochemistry and pharmacology*, Vol. 8, (pp. 37–84). Oxford: Pergamon Press.
Riddiford, L. M. (1996). Juvenile hormone: The status of its "status quo" action. *Archives of Insect Biochemistry and Physiology*, *32*, 271–286.
Riddiford, L. M. (2008). Juvenile hormone action: A 2007 perspective. *Journal of Insect Physiology*, *54*, 895–901.
Riddiford, L. M., Truman, J. W., Mirth, C. K., & Shen, Y. C. (2010). A role for juvenile hormone in the prepupal development of *Drosophila melanogaster*. *Development*, *137*, 1117–1126.
Root, C. M., Ko, K. I., Jafari, A., & Wang, J. W. (2011). Presynaptic facilitation by neuropeptide signaling mediates odor-driven food search. *Cell*, *145*, 133–144.
Sakurai, S. (1984). Temporal organization of endocrine events underlying larval-pupal metamorphosis in the silkworm, *Bombyx mori*. *Journal of Insect Physiology*, *30*, 657–664.
Sakurai, S. (2005). Feedback regulation of prothoracic gland activity. In L. I. Gilbert, K. Iatrou & S. Gill (Eds.), *Comprehensive molecular insect science*, Vol. 3, (pp. 409–431). Oxford: Elsevier.
Shinoda, T., & Itoyama, K. (2003). Juvenile hormone acid methyltransferase: A key regulatory enzyme for insect metamorphosis. *Proceedings of the National Academy of Sciences of the United States of America*, *100*, 11986–11991.
Soller, M., Bownes, M., & Kubli, E. (1999). Control of oocyte maturation in sexually mature *Drosophila* females. *Developmental Biology*, *208*, 337–351.
Stay, B. (2000). A review of the role of neurosecretion in the control of juvenile hormone synthesis: A tribute to Berta Scharrer. *Insect Biochemistry and Molecular Biology*, *30*, 653–662.
Stay, B., & Tobe, S. S. (2007). The role of allatostatins in juvenile hormone synthesis in insects and crustaceans. *Annual Review of Entomology*, *52*, 277–299.
Stay, B., Zhang, J. R., & Tobe, S. S. (2002). Methyl farnesoate and juvenile hormone production in embryos of *Diploptera punctata* in relation to innervation of corpora allata and their sensitivity to allatostatin. *Peptides*, *23*, 1981–1990.
Sutherland, T. D., & Feyereisen, R. (1996). Target of cockroach allatostatin in the pathway of juvenile hormone biosynthesis. *Molecular and Cellular Endocrinology*, *120*, 115–123.
Tan, A., Tanaka, H., Tamura, T., & Shiotsuki, T. (2005). Precocious metamorphosis in transgenic silkworms overexpressing juvenile hormone esterase. *Proceedings of the National Academy of Sciences of the United States of America*, *102*, 11751–11756.

Tatar, M., Kopelman, A., Epstein, D., Tu, M. P., Yin, C. M., & Garofalo, R. S. (2001). A mutant *Drosophila* insulin receptor homolog that extends life-span and impairs neuroendocrine function. *Science, 292*, 107–110.

The International Silkworm Genome Consortium, (2008). The genome of a lepidopteran model insect, the silkworm *Bombyx mori. Insect Biochemistry and Molecular Biology, 38*, 1036–1045.

Tu, M. P., Yin, C. M., & Tatar, M. (2005). Mutations in insulin signaling pathway alter juvenile hormone synthesis in *Drosophila melanogaster. General and Comparative Endocrinology, 142*, 347–356.

Ueda, H., Shinoda, T., & Hiruma, K. (2009). Spatial expression of the mevalonate enzymes involved in juvenile hormone biosynthesis in the corpora allata in *Bombyx mori. Journal of Insect Physiology, 55*, 798–804.

Wang, S., Baumann, A., & Wilson, T. G. (2007). *Drosophila melanogaster* Methoprene-tolerant (Met) gene homologs from three mosquito species: Members of PAS transcriptional factor family. *Journal of Insect Physiology, 53*, 246–253.

Weaver, R. J., & Audsley, N. (2009). Neuropeptide regulators of juvenile hormone synthesis: Structures, functions, distribution, and unanswered questions. *Annals of the New York Academy of Sciences, 1163*, 316–329.

Whisenton, L. R., Bowen, M. F., Granger, N. A., Gilbert, L. I., & Bollenbacher, W. E. (1985). Brain-mediated 20-hydroxyecdysone regulation of juvenile hormone synthesis by the corpora allata of the tobacco hornworm, *Manduca sexta. General and Comparative Endocrinology, 58*, 311–318.

Whisenton, L. R., Watson, D. R., Granger, N. A., & Bollenbacher, W. E. (1987). Regulation of juvenile hormone biosynthesis by 20-hydroxyecdysone during the fourth larval instar of the tobacco hornworm, *Manduca sexta. General and Comparative Endocrinology, 66*, 62–70.

Wilson, T. G., & Ashok, M. (1998). Insecticide resistance resulting from an absence of target-site gene product. *Proceedings of the National Academy of Sciences of the United States of America, 95*, 14040–14044.

Yamanaka, N., Yamamoto, S., Zitnan, D., Watanabe, K., Kawada, T., Satake, H., et al. (2008). Neuropeptide receptor transcriptome reveals unidentified neuroendocrine pathways. *PLoS One, 3*, e3048. http://dx.doi.org/10.1371/journal.pone.0003048.

Zhang, Z., Xu, J., Sheng, Z., Sui, Y., & Palli, S. R. (2011). Steroid receptor co-activator is required for juvenile hormone signal transduction through a bHLH-PAS transcription factor, methoprene tolerant. *The Journal of Biological Chemistry, 286*, 8437–8447.

Zhou, B., Hiruma, K., Shinoda, T., & Riddiford, L. M. (1998). Juvenile hormone prevents ecdysteroid-induced expression of broad complex RNAs in the epidermis of the tobacco hornworm, *Manduca sexta. Developmental Biology, 203*, 233–244.

Zhou, X., & Riddiford, L. M. (2002). Broad specifies pupal development and mediates the 'status quo' action of juvenile hormone on the pupal-adult transformation in *Drosophila* and *Manduca. Development, 129*, 2259–2269.

Zhou, X., Zhou, B., Truman, J. W., & Riddiford, L. M. (2004). Overexpression of broad: A new insight into its role in the *Drosophila* prothoracic gland cells. *The Journal of Experimental Biology, 207*, 1151–1161.

Zhu, J., Busche, J. M., & Zhang, X. (2010). Identification of juvenile hormone target genes in the adult female mosquitoes. *Insect Biochemistry and Molecular Biology, 40*, 23–29.

Zitnan, D., & Adams, M. E. (2005). Neuroendocrine regulation of insect ecdysis. In L. I. Gilbert, K. Iatrou & S. S. Gill (Eds.), *Comprehensive molecular insect science*, Vol. 3, (pp. 1–60). Oxford: Elsevier.

Zitnan, D., Kingan, T. G., Hermesman, J. L., & Adams, M. E. (1996). Identification of ecdysis-triggering hormone from an epitracheal endocrine system. *Science, 271*, 88–91.

NOTE ADDED IN PROOF

After submission of the manuscript, the following excellent reviews regarding the JH signaling pathway were published

Jindra, M., Palli, S. R., & Riddiford, L. M. (2012). The juvenile hormone signaling pathway in insect metamorphosis. *Annual Review of Entomology*, in press. http://dx.doi.org/10.1146/annurev-ento-120811-153700.

Riddiford, L. M. (2012). How does juvenile hormone control insect metamorphosis and reproduction? *General and Comparative Endocrinology*, in press. http://dx.doi.org/10.1016/j.ygcen.2012.06.001.

CHAPTER FOUR

The Role of Autophagy in *Drosophila* Metamorphosis

Kirsten Tracy, Eric H. Baehrecke[1]
Department of Cancer Biology, University of Massachusetts Medical School, Worcester, Massachusetts, USA
[1]Corresponding author: e-mail address: eric.baehrecke@umassmed.edu

Contents

Abstract

Macroautophagy (autophagy) is a conserved catabolic process that targets cytoplasmic components to lysosomes for degradation. Autophagy is required for cellular homeostasis and cell survival in response to starvation and stress, and paradoxically, it also plays a role in programmed cell death during development. The mechanisms that regulate the relationship between autophagy, cell survival, and cell death are poorly understood. Here we review research in *Drosophila* that has provided insights into the regulation of autophagy by steroid hormones and nutrient restriction and discuss how autophagy influences cell growth, nutrient utilization, cell survival, and cell death.

Current Topics in Developmental Biology, Volume 103
ISSN 0070-2153
http://dx.doi.org/10.1016/B978-0-12-385979-2.00004-6

1. INTRODUCTION

All animals transition through several different stages during their development. The first stage is embryonic development, followed by a juvenile growth phase, then sexual maturation, and finally reproductive adulthood. Some animals, such as mammals, exhibit few changes in their body plan during development except for growth; human babies look like miniature adults. Other animals go through drastic changes, such as the frog, starting its juvenile phase as a tadpole and developing into a frog. Progression through the developmental stages requires the coordination of cell and tissue growth, cell survival, and cell death. In the example of the frog, cell growth allows the limbs to develop, and cell death causes the regression of the tadpole's tail. The balance of cell growth, survival, and death is critical to maintaining homeostasis of the organism.

Autophagy is a catabolic process that functions at the crossroads of the different cell fates. Autophagy is predominately associated with cell survival in response to cellular stress; however, mounting evidence suggests that it also plays a role in programmed cell death. Additionally, autophagy is regulated by the same pathways that control cell growth. Autophagy and the pathways that regulate it have been studied extensively in the fruit fly *Drosophila melanogaster*. These studies have provided insights into the relationship between cell growth, cell death, and autophagy, but the questions of how and why these signals are integrated remain poorly understood.

2. AUTOPHAGY

Autophagy is an important catabolic process in all eukaryotic cells. There are three known types of autophagy: macroautophagy, microautophagy, and chaperone-mediated autophagy (Klionsky, 2005). Macroautophagy (hereafter referred to as autophagy) is the best characterized of the three types, and it involves the sequestration of cytoplasmic components and long-lived proteins into lysosomes for degradation. During autophagy, an isolation membrane sequesters cytoplasmic material, and it elongates to form a double-membrane vesicle, the autophagosome (Fig. 4.1). The autophagosome traffics to the lysosomal compartment where its outer membrane fuses with lysosomes and releases the inner cargo for degradation. Lysosomal permeases then recycle the degradation products back to the cytoplasm (Mizushima & Komatsu, 2011). Autophagy is an

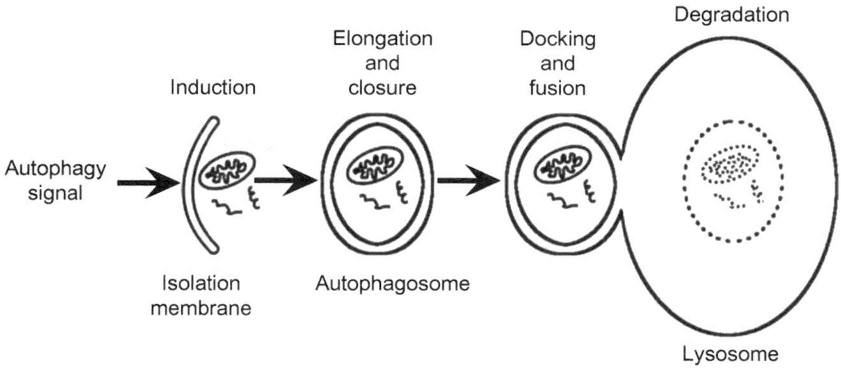

Figure 4.1 Diagram of the steps in autophagy. Autophagy is a catabolic process by which organelles and cytoplasmic proteins are degraded. Induction of autophagy results in the formation of an isolation membrane, which expands and closes around cytoplasmic material, forming the double-membraned autophagosome. The autophagosome traffics to the lysosome where it docks and fuses, releasing its inner membrane and its contents. The autophagosome contents are degraded by lysosomal enzymes and are recycled back to the cytoplasm through permeases.

important process for maintaining cell homeostasis, responding to stress, and surviving nutrient starvation.

2.1. Regulatory pathways

Several metabolic regulatory factors affect autophagy induction, including nutrient availability, insulin signaling, and ATP levels (Meijer & Codogno, 2004). The mechanistic target of rapamycin (TOR) plays a central role in autophagy by integrating the class I phosphatidylinositol-3-kinase (PI3K) and amino acid signaling pathways (Wullschleger, Loewith, & Hall, 2006). When nutrients are available, class I PI3K activates TOR, which represses autophagy by phosphorylating Atg13. This hyperphosphorylation reduces the affinity of Atg13 for Atg1, decreasing the kinase activity of Atg1 and inhibiting autophagy (Kamada et al., 2000; Noda & Ohsumi, 1998). During nutrient starvation, TOR activity is reduced, relieving its repression of Atg1, and autophagy is induced. Increased autophagy contributes to cell survival by producing amino acids and fatty acids that are used by the tricarboxylic acid cycle to generate ATP (Lum et al., 2005).

The origin of the autophagic membrane is not completely understood and remains a subject of debate (Juhasz & Neufeld, 2006). In yeast, autophagy proteins gather at the pre-autophagosomal structure (PAS) near the vacuole (Mizushima, 2007). In animal cells, a PAS-like structure has never been observed. Some studies suggest that in mammalian cells, the autophagosomal membrane originates from the endoplasmic reticulum (Axe et al., 2008; Dunn, 1990). In addition, more recent research suggests that autophagosome formation involves membrane derived from the mitochondria or the plasma membrane (Hailey et al., 2010; Ravikumar, Moreau, Jahreiss, Puri, & Rubinsztein, 2010).

Formation of the autophagosomal membrane requires phosphorylation of phosphatidylinositol. In yeast, this is accomplished by a class III PI3K complex consisting of Vps30/Atg6 /Beclin1, Vps34/ class III PI3K, Atg14, and Vps15 (Kametaka, Okano, Ohsumi, & Ohsumi, 1998; Kihara, Noda, Ishihara, & Ohsumi, 2001; Suzuki et al., 2001). Atg6 also forms a complex required for the vacuolar protein sorting (VPS) pathway in yeast, which consists of Atg6, Vps35, Vps15, and Vps38 (Kihara et al., 2001). The Beclin1–Vps34 complex is similar in mammalian cells; however, it contains additional regulators, including UVRAG, Bif1, Ambra1, and Barkor (Fimia et al., 2007; Liang et al., 2006; Sun et al., 2008; Takahashi et al., 2007). As in yeast, it has been suggested that Beclin1 forms at least two distinct complexes in animal cells that play different roles in membrane trafficking (Itakura, Kishi, Inoue, & Mizushima, 2008).

2.2. Autophagosome formation

Genetic studies in yeast have identified several *Atg* genes that are required for autophagy (Harding, Hefner-Gravink, Thumm, & Klionsky, 1996; Harding, Morano, Scott, & Klionsky, 1995; Klionsky et al., 2003; Thumm et al., 1994; Tsukada & Ohsumi, 1993). Many of these genes are involved in two conserved ubiquitin-like conjugation systems that are required for autophagosome formation, Atg12 and Atg8 (LC3 in mammals) (Klionsky & Emr, 2000; Ohsumi, 2001). Atg12 and Atg8 are both activated by the E1-like enzyme Atg7. Atg12 is then transferred to the E2-like enzyme Atg10. Finally, Atg12 is conjugated to Atg5 and forms a complex with Atg16 on the isolation membrane (Kuma, Mizushima, Ishihara, & Ohsumi, 2002; Mizushima, Noda, & Ohsumi, 1999; Mizushima et al., 1998; Shintani et al., 1999; Tanida et al., 1999). Atg8 is transferred to the E2-like enzyme Atg3 and is then conjugated to the phospholipid anchor phosphatidylethanolamine (PE) (Ichimura et al., 2000). This final conjugation results in the anchoring of Atg8-PE to the isolation membrane and is thought to regulate the elongation of the isolation membrane (Nakatogawa, Ichimura, & Ohsumi, 2007). In addition to Atg7 and Atg3, Atg8 modification requires Atg4, a cysteine protease that processes Atg8 before conjugation and cleaves Atg8 from PE once the autophagosome has fused with the lysosome (Ichimura et al., 2000). Since Atg8 remains on the membrane throughout autophagosome maturation, it is a useful marker of autophagosomes (Klionsky et al., 2008).

3. *DROSOPHILA* AS A MODEL FOR STUDYING THE INTERFACE BETWEEN STEROID SIGNALING, NUTRITION, AND GROWTH DURING DEVELOPMENT

Drosophila development provides a useful system for studying the coordination of cell growth, division, and death that is necessary for the animal to reach its proper size. Fly development is regulated by the steroid 20-hydroxyecdysone (ecdysone), and insulin and insulin-like growth factor signaling. These pathways are also known to regulate autophagy in different contexts; however, the coordination of steroid, insulin signaling, and autophagy is poorly understood. Recent studies have investigated the relationship between ecdysone and growth factor signaling in flies (Colombani et al., 2005; Layalle, Arquier, & Léopold, 2008), and understanding how these two pathways coordinate with each other may provide insight into how autophagy fits into this dynamic to facilitate animal homeostasis.

3.1. Steroid signaling

During development, *Drosophila* transitions through many different stages, and these transitions are signaled by pulses of the steroid hormone, ecdysone (Riddiford, Cherbas, & Truman, 2000; Thummel, 2001). *Drosophila* begins life as an embryo, and approximately 1 day after egg lay, they hatch as first instar larvae. The larvae feed and grow for approximately 3.5 days, and they molt twice during this period to become second instar larvae 24 h after hatching and third instar larvae 48 h after hatching. After the larval period, the animal stops feeding and a high titer pulse of ecdysone triggers puparium formation. This ecdysone pulse also induces the programmed cell death of the larval midgut (Lee, Cooksey, & Baehrecke, 2002). Prepupal development lasts for 12 h, and another peak in ecdysone titer triggers the prepupal–pupal transition and initiates programmed cell death of the larval salivary glands (Lee et al., 2003). Pupal development lasts for 3.5 days, after which the adult animal ecloses. A remarkable transformation occurs during this final developmental period; the tissues necessary to the feeding larva

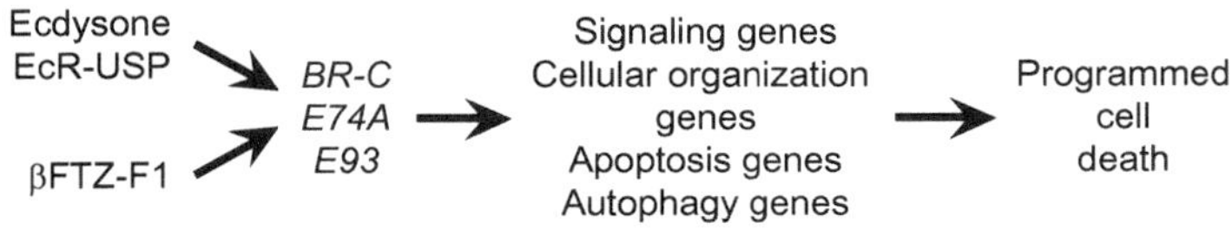

Figure 4.2 Genetic regulation of ecdysone-induced autophagy in *Drosophila* salivary glands. At 10 h after puparium formation, there is a rise in ecdysone titer, and ecdysone binds to its heterodimeric receptor which consists of EcR and USP. The ecdysone receptor complex functions together with βFTZ-F1 to induce transcription of the early genes; *BR-C, E74A,* and *E93*. The early genes activate transcription of many late genes involved in signaling, cellular organization, apoptosis, and autophagy.

degrade through histolysis and are replaced by growing tissues that will be necessary to the walking, flying, and reproducing adult.

Ecdysone signaling has been studied extensively in the larval salivary glands of *Drosophila*. The pulses of ecdysone regulate stage- and tissue-specific developmental pathways through a transcriptional hierarchy (Thummel, 1995) (Fig. 4.2). Ecdysone signals by binding its receptor which is a heterodimer of two nuclear receptors, ecdysone receptor (EcR) and ultraspiracle (USP) (Koelle et al., 1991; Thomas, Stunnenberg, & Stewart, 1993; Yao, Segraves, Oro, McKeown, & Evans, 1992). The EcR complex activates transcription of the early genes; these include *Broad Complex* (*BR-C*), *E74A*, *E75*, and *E93* (Baehrecke & Thummel, 1995; Burtis, Thummel, Jones, Karim, & Hogness, 1990; DiBello, Withers, Bayer, Fristrom, & Guild, 1991; Segraves & Hogness, 1990). The early genes then activate transcription of the late genes, which are thought to function more directly in the regulation of developmental processes. In the salivary glands, the βFTZ-F1 orphan nuclear receptor is expressed during the mid-prepupal dip in ecdysone titer (Lavorgna, Karim, Thummel, & Wu, 1993). During the ecdysone peak that triggers salivary gland degradation, the EcR complex and βFTZ-F1 function together to reinduce transcription of *BR-C*, *E74A*, and *E75* and to activate transcription of the stage-specific early gene, *E93* (Baehrecke & Thummel, 1995; Broadus, McCabe, Endrizzi, Thummel, & Woodard, 1999; Woodard, Baehrecke, & Thummel, 1994). *βFTZ-F1*, *BR-C*, *E74A*, and *E93* are all necessary for the proper degradation of larval salivary glands (Broadus et al., 1999; Jiang, Lamblin, Steller, & Thummel, 2000; Lee et al., 2000; Restifo and White, 1991). *E93* may have a more prominent role in autophagic cell death than the other early genes as it also appears to be required for autophagosome formation in the dying larval midgut (Lee, Cooksey, & Baehrecke, 2002).

3.2. Growth and nutrient utilization

Growth regulation at the cellular, tissue, and organismal level is critical for proper size development in all multicellular organisms, and it is affected by several environmental factors including nutrient availability (Mirth & Riddiford, 2007). In *Drosophila*, the feeding larva grows an astounding amount, increasing its size by ~200-fold during the 3.5-day period (Church & Robertson, 1966). Without this accumulation of body mass, the fly may have reduced reproductive success as an adult or it may not even be able to survive metamorphosis from the larva to adult.

For the adult fly to reach its proper size, the larva must pass three weight checkpoints. The first checkpoint occurs near the second instar to third instar molt and is called the threshold size for metamorphosis (Zhou, Zhou, Truman, & Riddiford, 2004). This size assessment determines whether the next molt will be a larval or metamorphic molt (Nijhout, 1975). The second checkpoint is the minimal viable weight which is the minimum body mass that is necessary to complete larval and pupal development in the absence of nutrients (Bakker, 1959). The final checkpoint, critical weight, occurs during the last larval stage (Nijhout, 2003; Nijhout & Williams, 1974). Reaching critical weight ensures that the animal will pupate within a certain amount of time regardless of nutrient availability (Bakker, 1959; Mirth & Riddiford, 2007; Nijhout, 2003; Robertson, 1963). Of these three size assessment checkpoints, critical weight is the most studied and best understood in *Drosophila*.

Once larvae reach their critical weight, environmental factors have a large impact on adult size. Larvae that starve before they achieve critical weight will delay their development until the nutrient supply improves. If nutrients are still abundant after larvae reach critical weight, they will continue to accumulate body mass (Mirth & Riddiford, 2007; Tennessen & Thummel, 2011). On the other hand, if postcritical weight larvae starve, they will stop growing in size. Since these starved larvae have reached their critical weight, they will enter metamorphosis within a similar time frame as fed larvae, but they will be smaller and will mature into smaller adults than the fed animals. This suggests that the mechanisms that regulate development and puparium formation must coordinate with nutrient utilization.

The endocrine cascade that follows critical weight achievement was originally described in the tobacco hookworm, *Manduca sexta* (Nijhout & Williams, 1974; Truman & Riddiford, 1974). Briefly, once larvae reach critical weight, juvenile hormone (JH) titers drop, causing a release of prothoracicotropic hormone, which signals to the prothoracic gland (PG) to produce ecdysone. However, this function of JH does not seem to be conserved in *Drosophila*, suggesting that critical weight is determined through another mechanism (Nijhout, 2003; Stern & Emlen, 1999).

Recent studies have elucidated some of the mechanisms required for critical weight assessment in *Drosophila*. One study showed that the *Drosophila* insulin receptor (InR), which has a conserved role in nutrition-dependent growth in animals, affects growth differently in precritical weight and postcritical weight larvae (Shingleton, Das, Vinicius, & Stern, 2005). Before larvae reach critical weight, InR signaling influences developmental timing but not larval growth. In contrast, InR activity affects final body size but not developmental timing in postcritical weight larvae. This is consistent with the observations in starved larvae discussed above. Several other studies showed that in *Drosophila* the size of the PG affects developmental rate and body size (Caldwell, Walkiewicz, & Stern, 2005; Colombani et al., 2005; Mirth, Truman, & Riddiford, 2005). They did this by manipulating insulin-dependent growth in the PG. When PG growth was suppressed by the expression of PTEN, a phosphatase that antagonizes class I PI3K activity, dominant negative class I PI3K, or dominant negative Ras, the larvae were larger than controls and had a longer developmental period. Conversely, larvae with an enlarged PG due to either class I PI3K or Ras activation initiated metamorphosis earlier than controls and thus the adults were smaller. Interestingly, the effects of growth in the PG appear to be specific to the insulin signaling pathway and not to cell size increase in general. In the study done by Colombani et al., they increased PG size by manipulating two other growth pathways in addition to PI3K: Myc and cyclin D/Cdk4. Although activation of these two genes increased the size of the PG, they had no effect on pupal or adult size (Colombani et al., 2005).

It is clear from these studies that tissue growth coordinates with developmental timing through InR signaling; however, the signals that regulate this have not been well studied. Recently, two independent groups performed screens to identify molecules that couple tissue growth with developmental timing and identified a novel *Drosophila* insulin-like peptide (dilp), *dilp8* (Colombani, Andersen, & Léopold, 2012; Garelli, Gontijo, Miguela, Caparros, & Dominguez, 2012). Perturbing growth of larval imaginal disks through either damage or tumor promotion causes a delay in the time to pupariation, allowing the imaginal disks to reach their correct size (Menut et al., 2007; Poodry & Woods, 1990; Simpson, Berreur, & Berreur-Bonnenfant, 1980; Smith-Bolton, Worley, Kanda, & Hariharan, 2009). *dilp8* is highly induced in imaginal disks with growth perturbations (Colombani et al., 2012; Garelli et al., 2012). Importantly, knockdown of *dilp8* in tissues with abnormal growth prevents the delay in pupariation, suggesting that it is required for the coupling of tissue growth and developmental timing. Expression of *dilp8* in imaginal disks is also sufficient to delay the onset of metamorphosis, which can be overcome by feeding larvae ecdysone (Garelli et al., 2012). Additionally, coculture experiments reveal that ecdysone production in the ring gland is suppressed in response to Dilp8 produced by imaginal disks (Colombani et al., 2012). Taken together, these results suggest that Dilp8 is secreted by the imaginal disks and remotely acts on the ring gland to suppress ecdysone production and delay development. How Dilp8 suppresses ecdysone is not known, but it may signal through the InR pathway.

It has been shown that insulin signaling and ecdysone regulate each other antagonistically (Caldwell et al., 2005; Colombani et al., 2005; Mirth et al., 2005). A recent study has demonstrated a role for the nuclear cofactor, dDOR, in the relationship between insulin signaling and ecdysone. They show that dDOR is a coactivator of EcR, and that its expression is downregulated by insulin signaling via the inhibition of FOXO activity (Francis, Zorzano, & Teleman, 2010). In addition, ecdysone induces translocation of dFOXO into the nucleus, promoting *dDOR* expression, which further activates EcR and initiates a feed-forward loop. Intriguingly, *dDOR* knockout flies have a salivary gland degradation defect, and DOR has been shown to regulate autophagy in both mammalian and *Drosophila* cells (Francis et al., 2010; Mauvezin et al., 2010). These results provide one of the few clues to how the relationship between insulin signaling, ecdysone, and autophagy functions (Fig. 4.3).

4. AUTOPHAGY AND *DROSOPHILA* DEVELOPMENT

Most autophagy studies have been done using either yeast or mammalian cell culture. While these studies have been essential to our understanding of the genetic mechanisms that regulate autophagy, there is little known about the impact of autophagy on the homeostasis of multicellular organisms. It would be interesting to understand how autophagy in different cell contexts, such as cell growth, cell survival, and cell death, affects the organism as a whole.

Drosophila is an ideal system for studying autophagy in a multicellular organism. The steroid and growth factor signaling pathways that regulate autophagy are similar in flies and humans. Importantly, *Atg* genes and their regulators are highly conserved between flies and humans (Baehrecke, 2003).

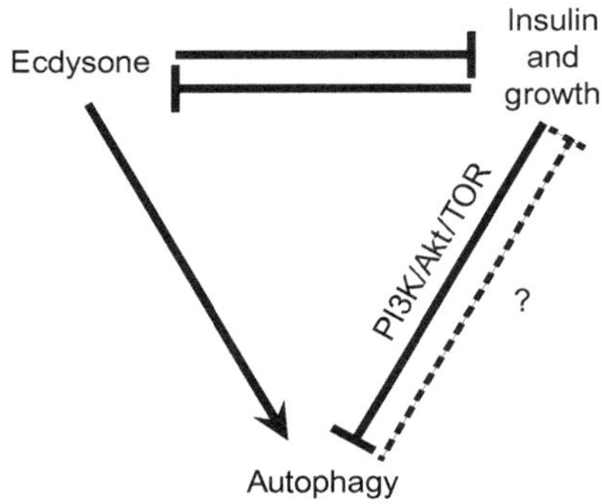

Figure 4.3 Relationship between ecdysone, autophagy, and insulin signaling and growth. Ecdysone and insulin have opposing effects on autophagy; ecdysone induces autophagy, and insulin inhibits autophagy through class I PI3K and TOR signaling. Ecdysone and insulin also antagonize each other, suggesting that a balance between these two hormones may be required to regulate autophagy. For example, increased ecdysone would inhibit insulin signaling, releasing insulin's inhibition on autophagy and further inducing autophagy. Whether autophagy itself regulates ecdysone and insulin is unclear; however, in some contexts autophagy may be a negative regulator of growth.

In contrast to mammalian systems, *Drosophila* has little genetic redundancy and has single copies for most genes in the autophagic pathway and its regulatory pathways. In addition, autophagy is induced in *Drosophila* tissues in response to either nutrient starvation or the steroid hormone ecdysone (Lee & Baehrecke, 2001; Lee, Cooksey, & Baehrecke, 2002; Rusten et al., 2004).

4.1. Autophagy in growth and nutrient utilization

Autophagy is critical for proper nutrient utilization during *Drosophila* larval development. In the fly, the major storage site for glycogen, lipids, and proteins is the fat body, an organ that shares attributes with both mammalian adipose tissue and liver. The fat body provides an excellent model for studying the mechanisms that regulate autophagy. When larvae are deprived of amino acids, autophagy is induced in the fat body, and this starvation-induced autophagy is regulated by TOR signaling (Scott, Schuldiner, & Neufeld, 2004). It has been shown that inactivation of TOR signaling either by a *TOR* null mutant or by manipulating upstream regulators of TOR induces autophagy in the fat body of feeding larvae. On the other hand, activation of either TOR or class I PI3K suppresses starvation-induced autophagy in the fat body (Scott et al., 2004). These results, taken together with the result that constitutive expression of PI3K in the fat body causes reduced viability during starvation (Britton, Lockwood, Li, Cohen, & Edgar, 2002), suggest that proper regulation of the class I PI3K signaling pathway is necessary for autophagy to promote survival during starvation.

In addition to being necessary for survival during starvation, autophagy may have a critical role in lipid metabolism of the *Drosophila* fat body. In mammalian cells, it has been shown that there is a connection between autophagy and lipolysis as well as lipid storage. Singh et al. demonstrated that triglycerides (TGs) and lipid droplet (LD) proteins were associated with both autophagosomes and lysosomes. Moreover, inhibition of autophagy in mouse liver cells led to increased TGs and LDs *in vitro* and *in vivo*, while increased autophagy led to decreased TGs and LDs *in vitro* (Singh et al., 2009). Their data suggests that lipid accumulation during autophagy inhibition is a result of blocked lipolysis. By contrast, it has been shown that loss of either *Atg5* or *Atg7* in mouse adipocytes leads to reduced lipid accumulation and impaired adipocyte differentiation (Baerga, Zhang, Chen, Goldman, & Jin, 2009; Zhang et al., 2009). Similar results were obtained in a recent study of *Drosophila* larval fat body. *Atg7* loss-of-function mutants had smaller LDs in the fat body, indicating a lipid accumulation defect (Wang et al., 2012). One possible explanation for the discrepancies between these studies is that autophagy may affect lipid metabolism in a tissue-specific manner. It would be interesting to further investigate the relationship between autophagy and lipid metabolism and how it is regulated in different tissues.

Wang et al. (2012) provided insight into the relationship between lipid metabolism and autophagy. Members of the Rab small GTPase family have been associated with LDs, and are known to participate in many cellular processes, including endocytosis, exocytosis, autophagosome formation, lysosome formation, and signaling transduction (Liu et al., 2007; Stenmark, 2009; Zehmer et al., 2009). In a screen for Rab proteins that affect LD size, Wang et al. found 18 Rab proteins that either increased or decreased LD size (Wang et al., 2012). They focused on Rab32 and showed that as well as having smaller LDs, *Rab32* mutants have impaired autophagy in the fat body. Importantly, Rab32 localized on autophagosomes, but not LDs, suggesting that its effect on LD size is due to regulation of autophagy rather than a direct effect on LDs. Since different Rab proteins have different effects on LD size, investigating the remaining Rab proteins might shed some light on the regulation of the relationship between autophagy and lipid metabolism.

Autophagy is also induced in the fat body and other tissues, including the salivary glands and mid gut during development in response to rises in ecdysone titer. This developmental autophagy is induced during the wandering larval stage and metamorphosis at times when the animal is not feeding, suggesting that autophagy may play an important role in survival and even tissue growth during nonfeeding periods (Lee & Baehrecke, 2001; Lee, Cooksey, & Baehrecke, 2002; Rusten et al., 2004). In the fat body, programmed autophagy is induced in response to ecdysone late during the third larval stage. This induction requires the downregulation of class I PI3K signaling (Rusten et al., 2004), suggesting that regulation of the class I PI3K pathway is involved in both starvation-induced autophagy and developmental autophagy.

Studies in the *Drosophila* fat body have identified other genes that are necessary for autophagy induced in response to ecdysone. SNF4Aγ, the *Drosophila* homologue of the AMP-activated protein kinase (AMPK) γ subunit, was identified in a fat body screen for mutants that fail to induce autophagy in response to ecdysone (Lippai et al., 2008). AMPK is an evolutionarily conserved enzyme that maintains cellular energy balance and is an inhibitor of TOR signaling (Shaw, 2009). In mammalian cells, AMPK has been implicated in the induction of autophagy in response to stimuli other than starvation, including growth factor withdrawal and increased calcium signaling (Hoyer-Hansen et al., 2007; Liang et al., 2007). Importantly, several recent studies in mammalian cells have shown that AMPK may directly control ULK1, the mammalian homologue of Atg1, via phosphorylation; however, the exact sites of phosphorylation are still debated (Egan et al., 2011; Kim, Kundu, Viollet, & Guan, 2011; Lee, Park, Takahashi, & Wang, 2010; Shang et al., 2011). AMPK has also been shown to suppress cell proliferation in *Drosophila* (Mandal, Guptan, Owusu-Ansah, & Banerjee, 2005). Taken together, these studies suggest that AMPK is an important regulator of the relationship between autophagy and growth.

Studies in *Drosophila* have further investigated the relationship between autophagy and growth. TOR is a key regulator of cell growth that was first implicated in the regulation of autophagy when rapamycin, a TOR inhibitor, was shown to induce autophagy (Blommaart, Luiken, Blommaart, van Woerkom, & Meijer, 1995). TOR represses autophagy through phosphorylation of Atg1 (Kamada et al., 2000; Scott, Juhász, & Neufeld, 2007). In *Drosophila* larval fat body, overexpression of Atg1 inhibits cell growth through a negative feedback mechanism on TOR. Conversely, *Atg1* mutant cells with reduced TOR signaling have increased growth (Scott et al., 2007). These results suggest that autophagy is a negative regulator of cell growth. Interestingly, it has been shown that inhibiting autophagy in a *TOR* null background enhances the *TOR* mutant phenotypes, including reduced growth rate, smaller cell size, and decreased survival (Scott et al., 2004). This suggests that under these conditions, in contradiction to its role as a negative regulator of growth, autophagy is necessary to promote cell survival and maintain growth.

The relationship between autophagy and growth signaling has also been studied in the context of degrading tissues during *Drosophila* metamorphosis. Growth arrest is required for the induction of autophagy in degrading salivary glands (Berry & Baehrecke, 2007). This growth arrest is regulated by the class I PI3K pathway. Maintaining growth in the salivary glands through expression of activated Ras, Akt, or the class I PI3K catalytic subunit Dp110, inhibits autophagy and gland degradation. In addition, coexpression of a dominant negative TOR with either Ras or Dp110 partially suppresses the overgrowth phenotypes and the salivary gland degradation defects (Berry & Baehrecke, 2007). These data suggest that cell growth regulators signal through TOR to inhibit autophagy and prevent salivary gland degradation. Further, overexpression of Atg1, which induces autophagy, suppresses the Dp110 persistent salivary gland phenotype, while *Atg* loss-of-function mutations cause persistent salivary glands (Berry & Baehrecke, 2007), indicating that both growth arrest and autophagy are required for proper salivary gland degradation.

A recent study has observed a similar relationship between growth arrest and autophagy during midgut programmed cell death in *Drosophila*. In the midgut, as in the salivary glands, growth arrest occurs before programmed cell death induction (Denton, Chang, et al., 2012). When cell growth in the midgut is maintained by expression of either activated Ras or Dp110, autophagy is suppressed and midgut degradation is delayed (Denton, Chang, et al., 2012). These results indicate a role for growth arrest in midgut programmed cell death. In contrast, inhibition of growth by the expression of PTEN or TSC1/TSC2, negative regulators of class I PI3K signaling, results in smaller midguts and premature autophagy induction. This growth inhibition can be suppressed by knockdown of either *Atg1* or *Atg18* in a PTEN or TSC1/TSC2 expressing background (Denton, Chang, et al., 2012). Interestingly, knockdown of *Atg* genes alone in the midgut causes persistent PI3K growth signaling and a significant delay in midgut degradation. These results suggest that in the midgut, growth and autophagy have a reciprocal relationship as in the salivary glands; however, there is also a feedback mechanism by which autophagy downregulates class I PI3K signaling. The nature of this feedback mechanism is unknown and deserves future investigation.

There has been some recent progress on the study of how cell growth arrest is regulated in dying salivary glands. The evolutionarily conserved Warts (Wts)/Hippo (Hpo) signaling pathway is an important negative regulator of cell growth that functions through the inactivation of Yorkie (Yki), a transcriptional coactivator and positive regulator of growth (Huang, Wu, Barrera, Matthews, & Pan, 2005). Loss-of-function mutations in the Wts pathway or overexpression of Yki lead to tissue overgrowth (Huang et al., 2005). Importantly, *wts* is required for growth arrest and autophagy induction in degrading salivary glands (Dutta & Baehrecke, 2008). Disruption of this pathway by mutations in *wts* and *hpo* or knockdown of *sav* and *mats* prevents salivary gland degradation (Dutta & Baehrecke, 2008). Surprisingly, overexpression of Yki fails to inhibit salivary gland degradation, suggesting that Wts regulates salivary gland growth in a Yki-independent manner. Significantly, *wts* mutants cause persistent class I PI3K signaling in salivary glands, and knockdown of *chico* or expression of dominant-negative TOR suppresses the *wts* cell death defects (Dutta & Baehrecke, 2008). These data suggest that Wts regulates salivary gland cell growth in a class I PI3K-dependent manner. However, Wts does not have a common role in programmed cell death. Despite the clear requirement for class I PI3K signaling in the regulation of cell growth and cell death in the midgut, knockdown of *wts* does not affect midgut morphology or degradation (Denton, Chang, et al., 2012).

4.2. Autophagy and cell death

Programmed cell death is a highly conserved and genetically regulated fundamental biological process. During development, cell death is required for tissue pattern formation and to maintain tissue homeostasis. Cell death also functions to remove abnormal or damaged cells. Schweichel and Merker (1973) described three major types of cell death during mammalian development based on morphology and involvement of the lysosomal compartment. Type I cell death, or apoptosis, is characterized by caspase activation, cell shrinkage, cytoplasmic blebbing, nuclear and DNA fragmentation, and engulfment by a phagocyte where the lysosome of the engulfing cell degrades the dying cell (Kerr, Wyllie, & Currie, 1972). In contrast to apoptosis, type II cell death, or autophagic cell death, requires little or no help from phagocytes, and the dying cell is degraded by its own lysosome. Type III cell death, or necrosis, is the least common form of cell death, and it has no known lysosomal involvement.

Type II cell death is observed in a variety of organisms. The plant, *Arabidopsis*, requires type II cell death for the formation of tracheary elements (Kwon et al., 2010). Type II cell death has also been observed in several tissues during mammalian development, including regression of the corpus luteum and involution of mammary and prostate glands (Clarke, 1990). Type II cell death is best characterized in insects and has been observed in several tissues during development, including dying flight muscles of the Hawkmoth *Manduca sexta* (Lockshin & Williams, 1965), and degrading salivary glands and midgut in *Drosophila* (Lee & Baehrecke, 2001; Lee, Cooksey, & Baehrecke, 2002). Although autophagosomes are present in dying cells with type II morphology, the role of autophagy in cell death remains controversial (Denton, Nicolson, et al., 2012; Levine & Yuan, 2005).

Studies of dying larval tissues during *Drosophila* metamorphosis have provided evidence for a role of autophagy in programmed cell death. As described above, a peak in ecdysone titer triggers salivary gland degradation during metamorphosis. Several *Atg* genes exhibit increased transcription in salivary glands in response to the rise in ecdysone, including *Atg2*, *Atg3*, *Atg4*, *Atg5*, *Atg7*, and *Atg18* (Gorski et al., 2003; Lee et al., 2003). Additionally, mutations in transcription factors downstream of the EcR inhibit transcription of *Atg*-related genes and prevent proper salivary gland cell death (Lee et al., 2003), suggesting that ecdysone-induced autophagy promotes cell death. It was not until recently though that the function of autophagy in cell death was rigorously tested *in vivo*. Mutations in *Atg8*, *Atg18*, *Atg2*, or *Atg3* or decreased function of *Atg1* all result in incomplete degradation of the larval salivary glands (Berry & Baehrecke, 2007). In addition, knockdown of *Atg3*, *Atg6*, *Atg7*, or *Atg12* specifically in the salivary glands leads to incomplete gland destruction, suggesting that autophagy functions in a tissue-autonomous manner in these dying cells (Berry & Baehrecke, 2007). Moreover, misexpression of Atg1 in the salivary glands induces autophagy and leads to premature gland degradation in a caspase-independent manner (Berry & Baehrecke, 2007). This is in contrast to previous work which showed that overexpression of Atg1 in the fat body induces cell death that depends on caspase function (Scott et al., 2007).

There is also mounting evidence for a role of autophagy during programmed cell death of the larval midgut. Similar to salivary glands, larval midgut destruction is triggered by a peak in ecdysone titer at the end of larval development. The dying midguts have increased autophagosome formation, and inhibition of autophagy by loss-of-function mutations in *Atg2* or *Atg18* or knockdown of either *Atg1* or *Atg18* severely delays midgut removal (Denton et al., 2009). Additionally, overexpression of Atg1 in the larval midgut is sufficient to induce autophagy and premature degradation (Denton, Chang, et al., 2012). Surprisingly, caspases are active, but they are not required for removal of the midgut (Denton et al., 2009; Denton, Shravage, Simin, Baehrecke, & Kumar, 2010), indicating that there is a complex relationship between autophagy and caspases in this tissue.

Autophagy and caspases have a complex relationship that may be context dependent. During salivary gland degradation, the rise in ecdysone titer triggers increased transcription of not only *Atg* genes but also the proapoptotic genes, *rpr* and *hid*, caspases, the BCL-2 family member *buffy*, and *ark*, the fly *Apaf-1* homologue (Dorstyn, Colussi, Quinn, Richardson, & Kumar, 1999; Jiang, Baehrecke, & Thummel, 1997; Lee, Simon, Woodard, & Baehrecke, 2002). Caspase activation occurs in the glands, but expression of the caspase inhibitor p35 only partially inhibits salivary gland degradation (Lee & Baehrecke, 2001). Additionally, *ark* mutants have a partial salivary gland degradation defect, but autophagy occurs normally, suggesting that *ark* may function downstream or parallel to autophagy in programmed cell death (Akdemir et al., 2006; Mills et al., 2006). Significantly, inhibiting both caspases and autophagy by expressing p35 in salivary glands of *Atg18* loss-of-function mutants or with dominant negative Atg1 results in increased persistence of the salivary glands (Berry & Baehrecke, 2007). These results suggest that autophagy and caspases function in parallel during salivary gland cell death. Many of the components of the apoptotic machinery are also upregulated in dying midguts. Despite the presence of high levels of caspase activity, p35 expression or genetic ablation of the canonical caspase activation pathway has no effect on midgut degradation (Denton et al., 2009). This is in contrast to what has been observed in salivary glands, and it would be interesting to study what causes these distinct differences between how programmed cell death is executed in these two tissues.

Although these *in vivo* studies indicate a role for autophagy in programmed cell death, the mechanistic differences that determine whether autophagy will support cell survival or cell death are not clear. Recently, Draper (Drpr), the *Drosophila* homologue of *Caenorhabditis elegans* engulfment receptor CED-1, and other components of the engulfment pathway were shown to be required for induction of autophagy during cell death (McPhee, Logan, Freeman, & Baehrecke, 2010). Null mutations in *drpr* and salivary gland-specific knockdown of *drpr* prevent induction of autophagy and cause persistent salivary glands. Expression of Atg1 in *drpr* mutants is sufficient to rescue the salivary gland degradation defect, indicating that Drpr functions upstream of autophagy. Surprisingly, clonal analysis of degrading glands reveals that Draper functions in a cell-autonomous manner, as there is only a reduction of autophagy in the *drpr* mutant cells. Interestingly, knockdown of *drpr* in the fat body does not affect starvation-induced autophagy, implicating *drpr* as the first known factor to regulate autophagy's role in cell death but not cell survival (McPhee et al., 2010). It would be interesting to further investigate how Drpr is regulated in salivary glands and why an engulfment receptor is functioning cell-autonomously.

5. CONCLUSIONS

Organisms require a balance between cell survival and cell death to maintain homeostasis, and although *in vivo* evidence supports a role for autophagy in both cell survival and cell death, many fundamental questions remain. Since autophagy is involved in both protecting and killing the cell, it is important to determine the mechanisms that decide between these cell fates. One possibility is that autophagy selectively depletes a cell survival factor or an essential organelle, which leads to cell death (Abeliovich, 2007; Nezis et al., 2010; Yu et al., 2006). Another possibility is that there is a threshold of autophagic flux that is crossed to promote cell death. Extended growth factor withdrawal in apoptotic-resistant mouse cells leads to stress-induced autophagy and eventual death by depletion of cellular resources (Lum et al., 2005). Under more physiological conditions, degradation of the *Drosophila* salivary glands and midgut is preceded by an increase in both transcription of the *Atg* genes and autophagosome levels. Additionally, misexpression of Atg1 in several tissues promotes cell demise, supporting the idea that excessive autophagy leads to cell death. However, excessive autophagy might not always be enough to kill, and other death factors may be required in addition to autophagy. Cell death induced by Atg1 misexpression in the fat body is caspase dependent. Further, salivary glands require caspases and autophagy, functioning in parallel, to fully degrade (Berry & Baehrecke, 2007).

Autophagy has been shown to be both an alternative form of cell death in nonphysiological conditions and a necessary component of cell death in physiological contexts; however, why cells die by autophagy is not understood. Apoptosis requires a phagocyte to engulf the dying cell, while autophagic cell death has little or no phagocyte involvement. One possibility is that phagocytes have restricted access to the dying cells. In *Drosophila*, the adult midgut forms around the degrading larval midgut isolating the dying cells from the rest of the tissues. Similarly, *in vitro* models of mammary lumen formation, where the dying cells are isolated from phagocytes, implicate the necessity of both caspases and autophagy for elimination of the dying cells (Debnath et al., 2002; Mills, Reginato, Debnath, Queenan, & Brugge, 2004). Alternatively, large cells and tissues, such as the giant larval salivary glands, may be too big to degrade by phagocytosis alone, and they require autophagy for the bulk degradation of their cytoplasm. Finally, autophagy may contribute to nutrient resource reallocation and survival in multicellular organisms. In yeast and mammalian cell culture, autophagy degrades cellular content to produce ATP and resources to protect the cell during starvation. Interestingly, autophagic cell death of tissues in *Drosophila* occurs during a time when the animal receives no external nutrients and must rely on its nutrient stores for survival and development of adult structures. Further, the majority of *Atg* mutants are pupal lethal, suggesting that autophagy is necessary to survive metamorphosis. Thus, although autophagy is killing individual cells and tissues, this form of cell death could be promoting organism survival. Future studies in *Drosophila* will hopefully lead to a better understanding of autophagy's dual roles in life and death.

ACKNOWLEDGMENTS

We thank the Baehrecke lab for constructive comments. Work on this subject is supported by NIH grants GM079431 and CA159314, and the Ellison Medical Foundation to E. H. B. E. H. B. is a member of the UMass DERC (DK32520).

REFERENCES

Abeliovich, H. (2007). Mitophagy: The life-or-death dichotomy includes yeast. *Autophagy*, *3*, 275–277.

Akdemir, F., Farkas, R., Chen, P., Juhasz, G., Medved'ová, L., Sass, M., et al. (2006). Autophagy occurs upstream or parallel to the apoptosome during histolytic cell death. *Development (Cambridge, England)*, *133*, 1457–1465.

Axe, E. L., Walker, S. A., Manifava, M., Chandra, P., Roderick, H. L., Habermann, A., et al. (2008). Autophagosome formation from membrane compartments enriched in phosphatidylinositol 3-phosphate and dynamically connected to the endoplasmic reticulum. *The Journal of Cell Biology*, *182*, 685–701.

Baehrecke, E. H. (2003). Autophagic programmed cell death in Drosophila. *Cell Death and Differentiation*, *10*, 940–945.

Baehrecke, E. H., & Thummel, C. S. (1995). The Drosophila E93 gene from the 93F early puff displays stage- and tissue-specific regulation by 20-hydroxyecdysone. *Developmental Biology*, *171*, 85–97.

Baerga, R., Zhang, Y., Chen, P.-H., Goldman, S., & Jin, S. (2009). Targeted deletion of autophagy-related 5 (atg5) impairs adipogenesis in a cellular model and in mice. *Autophagy*, *5*, 1118–1130.
Bakker, K. (1959). Feeding period, growth, and pupation in larvae of Drosophila melanogaster. *Entomologia Experimentalis et Applicata*, *2*, 171–186.
Berry, D. L., & Baehrecke, E. H. (2007). Growth arrest and autophagy are required for salivary gland cell degradation in Drosophila. *Cell*, *131*, 1137–1148.
Blommaart, E. F., Luiken, J. J., Blommaart, P. J., van Woerkom, G. M., & Meijer, A. J. (1995). Phosphorylation of ribosomal protein S6 is inhibitory for autophagy in isolated rat hepatocytes. *The Journal of Biological Chemistry*, *270*, 2320–2326.
Britton, J. S., Lockwood, W. K., Li, L., Cohen, S. M., & Edgar, B. A. (2002). Drosophila's insulin/PI3-kinase pathway coordinates cellular metabolism with nutritional conditions. *Developmental Cell*, *2*, 239–249.
Broadus, J., McCabe, J. R., Endrizzi, B., Thummel, C. S., & Woodard, C. T. (1999). The Drosophila beta FTZ-F1 orphan nuclear receptor provides competence for stage-specific responses to the steroid hormone ecdysone. *Molecular Cell*, *3*, 143–149.
Burtis, K. C., Thummel, C. S., Jones, C. W., Karim, F. D., & Hogness, D. S. (1990). The Drosophila 74EF early puff contains E74, a complex ecdysone-inducible gene that encodes two ets-related proteins. *Cell*, *61*, 85–99.
Caldwell, P. E., Walkiewicz, M., & Stern, M. (2005). Ras activity in the Drosophila prothoracic gland regulates body size and developmental rate via ecdysone release. *Current Biology*, *15*, 1785–1795.
Church, R. B., & Robertson, F. W. (1966). Biochemical analysis of genetic differences in the growth of Drosophila. *Genetical Research*, *7*, 383–407.
Clarke, P. G. (1990). Developmental cell death: Morphological diversity and multiple mechanisms. *Anatomy and Embryology*, *181*, 195–213.
Colombani, J., Andersen, D. S., & Léopold, P. (2012). Secreted peptide Dilp8 coordinates Drosophila tissue growth with developmental timing. *Science*, *336*, 582–585.
Colombani, J., Bianchini, L., Layalle, S., Pondeville, E., Dauphin-Villemant, C., Antoniewski, C., et al. (2005). Antagonistic actions of ecdysone and insulins determine final size in Drosophila. *Science*, *310*, 667–670.
Debnath, J., Mills, K. R., Collins, N. L., Reginato, M. J., Muthuswamy, S. K., & Brugge, J. S. (2002). The role of apoptosis in creating and maintaining luminal space within normal and oncogene-expressing mammary acini. *Cell*, *111*, 29–40.
Denton, D., Chang, T.-K., Nicolson, S., Shravage, B., Simin, R., Baehrecke, E. H., et al. (2012). Relationship between growth arrest and autophagy in midgut programmed cell death in Drosophila. *Cell Death and Differentiation*, *19*, 1299–1307.
Denton, D., Nicolson, S., & Kumar, S. (2012). Cell death by autophagy: Facts and apparent artefacts. *Cell Death and Differentiation*, *19*, 87–95.
Denton, D., Shravage, B., Simin, R., Baehrecke, E. H., & Kumar, S. (2010). Larval midgut destruction in Drosophila: Not dependent on caspases but suppressed by the loss of autophagy. *Autophagy*, *6*, 163–165.
Denton, D., Shravage, B., Simin, R., Mills, K., Berry, D. L., Baehrecke, E. H., et al. (2009). Autophagy, not apoptosis, is essential for midgut cell death in Drosophila. *Current Biology*, *19*, 1741–1746.
DiBello, P. R., Withers, D. A., Bayer, C. A., Fristrom, J. W., & Guild, G. M. (1991). The Drosophila Broad-Complex encodes a family of related proteins containing zinc fingers. *Genetics*, *129*, 385–397.
Dorstyn, L., Colussi, P. A., Quinn, L. M., Richardson, H., & Kumar, S. (1999). DRONC, an ecdysone-inducible Drosophila caspase. *Proceedings of the National Academy of Sciences of the United States of America*, *96*, 4307–4312.

Dunn, W. A., Jr. (1990). Studies on the mechanisms of autophagy: Formation of the autophagic vacuole. *The Journal of Cell Biology, 110*, 1923–1933.
Dutta, S., & Baehrecke, E. H. (2008). Warts is required for PI3K-regulated growth arrest, autophagy, and autophagic cell death in Drosophila. *Current Biology, 18*, 1466–1475.
Egan, D. F., Shackelford, D. B., Mihaylova, M. M., Gelino, S., Kohnz, R. A., Mair, W., et al. (2011). Phosphorylation of ULK1 (hATG1) by AMP-activated protein kinase connects energy sensing to mitophagy. *Science, 331*, 456–461.
Fimia, G. M., Stoykova, A., Romagnoli, A., Giunta, L., Di Bartolomeo, S., Nardacci, R., et al. (2007). Ambra1 regulates autophagy and development of the nervous system. *Nature, 447*, 1121–1125.
Francis, V. A., Zorzano, A., & Teleman, A. A. (2010). dDOR is an EcR coactivator that forms a feed-forward loop connecting insulin and ecdysone signaling. *Current Biology, 20*, 1799–1808.
Garelli, A., Gontijo, A. M., Miguela, V., Caparros, E., & Dominguez, M. (2012). Imaginal discs secrete insulin-like peptide 8 to mediate plasticity of growth and maturation. *Science, 336*, 579–582.
Gorski, S. M., Chittaranjan, S., Pleasance, E. D., Freeman, J. D., Anderson, C. L., Varhol, R. J., et al. (2003). A SAGE approach to discovery of genes involved in autophagic cell death. *Current Biology, 13*, 358–363.
Hailey, D. W., Rambold, A. S., Satpute-Krishnan, P., Mitra, K., Sougrat, R., Kim, P. K., et al. (2010). Mitochondria supply membranes for autophagosome biogenesis during starvation. *Cell, 141*, 656–667.
Harding, T. M., Hefner-Gravink, A., Thumm, M., & Klionsky, D. J. (1996). Genetic and phenotypic overlap between autophagy and the cytoplasm to vacuole protein targeting pathway. *The Journal of Biological Chemistry, 271*, 17621–17624.
Harding, T. M., Morano, K. A., Scott, S. V., & Klionsky, D. J. (1995). Isolation and characterization of yeast mutants in the cytoplasm to vacuole protein targeting pathway. *The Journal of Cell Biology, 131*, 591–602.
Hoyer-Hansen, M., Bastholm, L., Szyniarowski, P., Campanella, M., Szabadkai, G., Farkas, T., et al. (2007). Control of macroautophagy by calcium, calmodulin-dependent kinase kinase-beta, and Bcl-2. *Molecular Cell, 25*, 193–205.
Huang, J., Wu, S., Barrera, J., Matthews, K., & Pan, D. (2005). The Hippo signaling pathway coordinately regulates cell proliferation and apoptosis by inactivating Yorkie, the Drosophila Homolog of YAP. *Cell, 122*, 421–434.
Ichimura, Y., Kirisako, T., Takao, T., Satomi, Y., Shimonishi, Y., Ishihara, N., et al. (2000). A ubiquitin-like system mediates protein lipidation. *Nature, 408*, 488–492.
Itakura, E., Kishi, C., Inoue, K., & Mizushima, N. (2008). Beclin 1 forms two distinct phosphatidylinositol 3-kinase complexes with mammalian Atg14 and UVRAG. *Molecular Biology of the Cell, 19*, 5360–5372.
Jiang, C., Baehrecke, E. H., & Thummel, C. S. (1997). Steroid regulated programmed cell death during Drosophila metamorphosis. *Development (Cambridge, England), 124*, 4673–4683.
Jiang, C., Lamblin, A. F., Steller, H., & Thummel, C. S. (2000). A steroid-triggered transcriptional hierarchy controls salivary gland cell death during Drosophila metamorphosis. *Molecular Cell, 5*, 445–455.
Juhasz, G., & Neufeld, T. P. (2006). Autophagy: A forty-year search for a missing membrane source. *PLoS Biology, 4*, e36.
Kamada, Y., Funakoshi, T., Shintani, T., Nagano, K., Ohsumi, M., & Ohsumi, Y. (2000). Tor-mediated induction of autophagy via an Apg1 protein kinase complex. *The Journal of Cell Biology, 150*, 1507–1513.
Kametaka, S., Okano, T., Ohsumi, M., & Ohsumi, Y. (1998). Apg14p and Apg6/Vps30p form a protein complex essential for autophagy in the yeast, Saccharomyces cerevisiae. *The Journal of Biological Chemistry, 273*, 22284–22291.

Kerr, J. F., Wyllie, A. H., & Currie, A. R. (1972). Apoptosis: A basic biological phenomenon with wide-ranging implications in tissue kinetics. *British Journal of Cancer*, *26*, 239–257.
Kihara, A., Noda, T., Ishihara, N., & Ohsumi, Y. (2001). Two distinct Vps34 phosphatidylinositol 3-kinase complexes function in autophagy and carboxypeptidase Y sorting in Saccharomyces cerevisiae. *The Journal of Cell Biology*, *152*, 519–530.
Kim, J., Kundu, M., Viollet, B., & Guan, K.-L. (2011). AMPK and mTOR regulate autophagy through direct phosphorylation of Ulk1. *Nature Cell Biology*, *13*, 132–141.
Klionsky, D. J. (2005). The molecular machinery of autophagy: Unanswered questions. *Journal of Cell Science*, *118*, 7–18.
Klionsky, D. J., Abeliovich, H., Agostinis, P., Agrawal, D. K., Aliev, G., et al. (2008). Guidelines for the use and interpretation of assays for monitoring autophagy in higher eukaryotes. *Autophagy*, *4*, 151–175.
Klionsky, D. J., Cregg, J. M., Dunn, W. A., Jr., Emr, S. D., Sakai, Y., Sandoval, I. V., et al. (2003). A unified nomenclature for yeast autophagy-related genes. *Developmental Cell*, *5*, 539–545.
Klionsky, D. J., & Emr, S. D. (2000). Autophagy as a regulated pathway of cellular degradation. *Science*, *290*, 1717–1721.
Koelle, M. R., Talbot, W. S., Segraves, W. A., Bender, M. T., Cherbas, P., & Hogness, D. S. (1991). The Drosophila EcR gene encodes an ecdysone receptor, a new member of the steroid receptor superfamily. *Cell*, *67*, 59–77.
Kuma, A., Mizushima, N., Ishihara, N., & Ohsumi, Y. (2002). Formation of the approximately 350-kDa Apg12-Apg5.Apg16 multimeric complex, mediated by Apg16 oligomerization, is essential for autophagy in yeast. *The Journal of Biological Chemistry*, *277*, 18619–18625.
Kwon, S. I., Cho, H. J., Jung, J. H., Yoshimoto, K., Shirasu, K., & Park, O. K. (2010). The Rab GTPase RabG3b functions in autophagy and contributes to tracheary element differentiation in Arabidopsis. *The Plant Journal*, *64*, 151–164.
Lavorgna, G., Karim, F. D., Thummel, C. S., & Wu, C. (1993). Potential role for a FTZ-F1 steroid receptor superfamily member in the control of Drosophila metamorphosis. *Proceedings of the National Academy of Sciences of the United States of America*, *90*, 3004–3008.
Layalle, S., Arquier, N., & Léopold, P. (2008). The TOR pathway couples nutrition and developmental timing in Drosophila. *Developmental Cell*, *15*, 568–577.
Lee, C. Y., & Baehrecke, E. H. (2001). Steroid regulation of autophagic programmed cell death during development. *Development (Cambridge, England)*, *128*, 1443–1455.
Lee, C.-Y., Clough, E. A., Yellon, P., Teslovich, T. M., Stephan, D. A., & Baehrecke, E. H. (2003). Genome-wide analyses of steroid- and radiation-triggered programmed cell death in Drosophila. *Current Biology*, *13*, 350–357.
Lee, C.-Y., Cooksey, B. A. K., & Baehrecke, E. H. (2002). Steroid regulation of midgut cell death during Drosophila development. *Developmental Biology*, *250*, 101–111.
Lee, J. W., Park, S., Takahashi, Y., & Wang, H.-G. (2010). The association of AMPK with ULK1 regulates autophagy. *PLoS One*, *5*, e15394.
Lee, C.-Y., Simon, C. R., Woodard, C. T., & Baehrecke, E. H. (2002). Genetic mechanism for the stage- and tissue-specific regulation of steroid triggered programmed cell death in Drosophila. *Developmental Biology*, *252*, 138–148.
Lee, C. Y., Wendel, D. P., Reid, P., Lam, G., Thummel, C. S., & Baehrecke, E. H. (2000). E93 directs steroid-triggered programmed cell death in Drosophila. *Molecular Cell*, *6*, 433–443.
Levine, B., & Yuan, J. (2005). Autophagy in cell death: An innocent convict? *The Journal of Clinical Investigation*, *115*, 2679–2688.
Liang, C., Feng, P., Ku, B., Dotan, I., Canaani, D., Oh, B.-H., et al. (2006). Autophagic and tumour suppressor activity of a novel Beclin1-binding protein UVRAG. *Nature Cell Biology*, *8*, 688–699.

Liang, J., Shao, S. H., Xu, Z.-X., Hennessy, B., Ding, Z., Larrea, M., et al. (2007). The energy sensing LKB1-AMPK pathway regulates p27(kip1) phosphorylation mediating the decision to enter autophagy or apoptosis. *Nature Cell Biology*, *9*, 218–224.

Lippai, M., Csikós, G., Maróy, P., Lukácsovich, T., Juhász, G., & Sass, M. (2008). SNF4Agamma, the Drosophila AMPK gamma subunit is required for regulation of developmental and stress-induced autophagy. *Autophagy*, *4*, 476–486.

Liu, P., Bartz, R., Zehmer, J. K., Ying, Y. S., Zhu, M., Serrero, G., et al. (2007). Rab-regulated interaction of early endosomes with lipid droplets. *Biochimica et Biophysica Acta*, *1773*, 784–793.

Lockshin, R. A., & Williams, C. M. (1965). Programmed cell death I. Cytology of degeneration in the intersegmental muscles of the pernyi silkmoth. *Journal of Insect Physiology*, *11*, 123–133.

Lum, J. J., Bauer, D. E., Kong, M., Harris, M. H., Li, C., Lindsten, T., et al. (2005). Growth factor regulation of autophagy and cell survival in the absence of apoptosis. *Cell*, *120*, 237–248.

Mandal, S., Guptan, P., Owusu-Ansah, E., & Banerjee, U. (2005). Mitochondrial regulation of cell cycle progression during development as revealed by the tenured mutation in Drosophila. *Developmental Cell*, *9*, 843–854.

Mauvezin, C., Orpinell, M., Francis, V. A., Mansilla, F., Duran, J., Ribas, V., et al. (2010). The nuclear cofactor DOR regulates autophagy in mammalian and Drosophila cells. *EMBO Reports*, *11*, 37–44.

McPhee, C. K., Logan, M. A., Freeman, M. R., & Baehrecke, E. H. (2010). Activation of autophagy during cell death requires the engulfment receptor Draper. *Nature*, *465*, 1093–1096.

Meijer, A. J., & Codogno, P. (2004). Regulation and role of autophagy in mammalian cells. *The International Journal of Biochemistry & Cell Biology*, *36*, 2445–2462.

Menut, L., Vaccari, T., Dionne, H., Hill, J., Wu, G., & Bilder, D. (2007). A mosaic genetic screen for Drosophila neoplastic tumor suppressor genes based on defective pupation. *Genetics*, *177*, 1667–1677.

Mills, K., Daish, T., Harvey, K. F., Pfleger, C. M., Hariharan, I. K., & Kumar, S. (2006). The Drosophila melanogaster Apaf-1 homologue ARK is required for most, but not all, programmed cell death. *The Journal of Cell Biology*, *172*, 809–815.

Mills, K. R., Reginato, M., Debnath, J., Queenan, B., & Brugge, J. S. (2004). Tumor necrosis factor-related apoptosis-inducing ligand (TRAIL) is required for induction of autophagy during lumen formation in vitro. *Proceedings of the National Academy of Sciences of the United States of America*, *101*, 3438–3443.

Mirth, C. K., & Riddiford, L. M. (2007). Size assessment and growth control: How adult size is determined in insects. *Bioessays*, *29*, 344–355.

Mirth, C., Truman, J. W., & Riddiford, L. M. (2005). The role of the prothoracic gland in determining critical weight for metamorphosis in Drosophila melanogaster. *Current Biology*, *15*, 1796–1807.

Mizushima, N. (2007). Autophagy: Process and function. *Genes & Development*, *21*, 2861–2873.

Mizushima, N., & Komatsu, M. (2011). Autophagy: Renovation of cells and tissues. *Cell*, *147*, 728–741.

Mizushima, N., Noda, T., & Ohsumi, Y. (1999). Apg16p is required for the function of the Apg12p-Apg5p conjugate in the yeast autophagy pathway. *The EMBO Journal*, *18*, 3888–3896.

Mizushima, N., Noda, T., Yoshimori, T., Tanaka, Y., Ishii, T., George, M. D., et al. (1998). A protein conjugation system essential for autophagy. *Nature*, *395*, 395–398.

Nakatogawa, H., Ichimura, Y., & Ohsumi, Y. (2007). Atg8, a ubiquitin-like protein required for autophagosome formation, mediates membrane tethering and hemifusion. *Cell*, *130*, 165–178.

Nezis, I. P., Shravage, B. V., Sagona, A. P., Lamark, T., Bjørkøy, G., Johansen, T., et al. (2010). Autophagic degradation of dBruce controls DNA fragmentation in nurse cells during late Drosophila melanogaster oogenesis. *The Journal of Cell Biology, 190*, 523–531.
Nijhout, H. F. (1975). A threshold size for metamorphosis in the tobacco hornworm, Manduca sexta (L.). *The Biological Bulletin, 149*, 214–225.
Nijhout, H. F. (2003). The control of body size in insects. *Developmental Biology, 261*, 1–9.
Nijhout, H. F., & Williams, C. M. (1974). Control of moulting and metamorphosis in the tobacco hornworm, Manduca sexta (L.): Growth of the last-instar larva and the decision to pupate. *The Journal of Experimental Biology, 61*, 481–491.
Noda, T., & Ohsumi, Y. (1998). Tor, a phosphatidylinositol kinase homologue, controls autophagy in yeast. *The Journal of Biological Chemistry, 273*, 3963–3966.
Ohsumi, Y. (2001). Molecular dissection of autophagy: Two ubiquitin-like systems. *Nature Reviews. Molecular Cell Biology, 2*, 211–216.
Poodry, C. A., & Woods, D. F. (1990). Control of the developmental timer for Drosophila pupariation. *Roux's Archives of Developmental Biology, 199*, 219–227.
Ravikumar, B., Moreau, K., Jahreiss, L., Puri, C., & Rubinsztein, D. C. (2010). Plasma membrane contributes to the formation of pre-autophagosomal structures. *Nature Cell Biology, 12*, 747–757.
Restifo, L. L., & White, K. (1991). Mutations in a steroid hormone-regulated gene disrupt the metamorphosis of the central nervous system in Drosophila. *Developmental Biology, 148*, 174–194.
Riddiford, L. M., Cherbas, P., & Truman, J. W. (2000). Ecdysone receptors and their biological actions. *Vitamins and Hormones, 60*, 1–73.
Robertson, F. (1963). The ecological genetics of growth in Drosophila. 6. The genetic correlation between the duration of the larval period and body size in relation to larval diet. *Genetical Research, 4*, 74–92.
Rusten, T. E., Lindmo, K., Juhász, G., Sass, M., Seglen, P. O., Brech, A., et al. (2004). Programmed autophagy in the Drosophila fat body is induced by ecdysone through regulation of the PI3K pathway. *Developmental Cell, 7*, 179–192.
Schweichel, J. U., & Merker, H. J. (1973). The morphology of various types of cell death in prenatal tissues. *Teratology, 7*, 253–266.
Scott, R. C., Juhász, G., & Neufeld, T. P. (2007). Direct induction of autophagy by Atg1 inhibits cell growth and induces apoptotic cell death. *Current Biology, 17*, 1–11.
Scott, R. C., Schuldiner, O., & Neufeld, T. P. (2004). Role and regulation of starvation-induced autophagy in the Drosophila fat body. *Developmental Cell, 7*, 167–178.
Segraves, W. A., & Hogness, D. S. (1990). The E75 ecdysone-inducible gene responsible for the 75B early puff in Drosophila encodes two new members of the steroid receptor superfamily. *Genes & Development, 4*, 204–219.
Shang, L., Chen, S., Du, F., Li, S., Zhao, L., & Wang, X. (2011). Nutrient starvation elicits an acute autophagic response mediated by Ulk1 dephosphorylation and its subsequent dissociation from AMPK. *Proceedings of the National Academy of Sciences of the United States of America, 108*, 4788–4793.
Shaw, R. J. (2009). LKB1 and AMP-activated protein kinase control of mTOR signalling and growth. *Acta Physiologica (Oxford, England), 196*, 65–80.
Shingleton, A. W., Das, J., Vinicius, L., & Stern, D. L. (2005). The temporal requirements for insulin signaling during development in Drosophila. *PLoS Biology, 3*, e289.
Shintani, T., Mizushima, N., Ogawa, Y., Matsuura, A., Noda, T., & Ohsumi, Y. (1999). Apg10p, a novel protein-conjugating enzyme essential for autophagy in yeast. *The EMBO Journal, 18*, 5234–5241.
Simpson, P., Berreur, P., & Berreur-Bonnenfant, J. (1980). The initiation of pupariation in Drosophila: Dependence on growth of the imaginal discs. *Journal of Embryology and Experimental Morphology, 57*, 155–165.

Singh, R., Kaushik, S., Wang, Y., Xiang, Y., Novak, I., Komatsu, M., et al. (2009). Autophagy regulates lipid metabolism. *Nature, 458*, 1131–1135.
Smith-Bolton, R. K., Worley, M. I., Kanda, H., & Hariharan, I. K. (2009). Regenerative growth in Drosophila imaginal discs is regulated by Wingless and Myc. *Developmental Cell, 16*, 797–809.
Stenmark, H. (2009). Rab GTPases as coordinators of vesicle traffic. *Nature Reviews. Molecular Cell Biology, 10*, 513–525.
Stern, D. L., & Emlen, D. J. (1999). The developmental basis for allometry in insects. *Development (Cambridge, England), 126*, 1091–1101.
Sun, Q., Fan, W., Chen, K., Ding, X., Chen, S., & Zhong, Q. (2008). Identification of Barkor as a mammalian autophagy-specific factor for Beclin 1 and class III phosphatidylinositol 3-kinase. *Proceedings of the National Academy of Sciences of the United States of America, 105*, 19211–19216.
Suzuki, K., Kirisako, T., Kamada, Y., Mizushima, N., Noda, T., & Ohsumi, Y. (2001). The pre-autophagosomal structure organized by concerted functions of APG genes is essential for autophagosome formation. *The EMBO Journal, 20*, 5971–5981.
Takahashi, Y., Coppola, D., Matsushita, N., Cualing, H. D., Sun, M., Sato, Y., et al. (2007). Bif-1 interacts with Beclin 1 through UVRAG and regulates autophagy and tumorigenesis. *Nature Cell Biology, 9*, 1142–1151.
Tanida, I., Mizushima, N., Kiyooka, M., Ohsumi, M., Ueno, T., Ohsumi, Y., et al. (1999). Apg7p/Cvt2p: A novel protein-activating enzyme essential for autophagy. *Molecular Biology of the Cell, 10*, 1367–1379.
Tennessen, J. M., & Thummel, C. S. (2011). Coordinating growth and maturation—Insights from Drosophila. *Current Biology, 21*, R750–R757.
Thomas, H. E., Stunnenberg, H. G., & Stewart, A. F. (1993). Heterodimerization of the Drosophila ecdysone receptor with retinoid X receptor and ultraspiracle. *Nature, 362*, 471–475.
Thumm, M., Egner, R., Koch, B., Schlumpberger, M., Straub, M., Veenhuis, M., et al. (1994). Isolation of autophagocytosis mutants of Saccharomyces cerevisiae. *FEBS Letters, 349*, 275–280.
Thummel, C. S. (1995). From embryogenesis to metamorphosis: The regulation and function of drosophila nuclear receptor superfamily members. *Cell, 83*, 871–877.
Thummel, C. S. (2001). Molecular mechanisms of developmental timing in C. elegans and Drosophila. *Developmental Cell, 1*, 453–465.
Truman, J. W., & Riddiford, L. M. (1974). Physiology of insect rhythms. 3. The temporal organization of the endocrine events underlying pupation of the tobacco hornworm. *The Journal of Experimental Biology, 60*, 371–382.
Tsukada, M., & Ohsumi, Y. (1993). Isolation and characterization of autophagy-defective mutants of Saccharomyces cerevisiae. *FEBS Letters, 333*, 169–174.
Wang, C., Liu, Z., & Huang, X. (2012). Rab32 is important for autophagy and lipid storage in Drosophila. *PLoS One*, 7, e32086.
Woodard, C. T., Baehrecke, E. H., & Thummel, C. S. (1994). A molecular mechanism for the stage specificity of the Drosophila prepupal genetic response to ecdysone. *Cell, 79*, 607–615.
Wullschleger, S., Loewith, R., & Hall, M. N. (2006). TOR signaling in growth and metabolism. *Cell, 124*, 471–484.
Yao, T. P., Segraves, W. A., Oro, A. E., McKeown, M., & Evans, R. M. (1992). Drosophila ultraspiracle modulates ecdysone receptor function via heterodimer formation. *Cell, 71*, 63–72.
Yu, L., Wan, F., Dutta, S., Welsh, S., Liu, Z., Freundt, E., et al. (2006). Autophagic programmed cell death by selective catalase degradation. *Proceedings of the National Academy of Sciences of the United States of America, 103*, 4952–4957.

Zehmer, J. K., Huang, Y., Peng, G., Pu, J., Anderson, R. G. W., & Liu, P. (2009). A role for lipid droplets in inter-membrane lipid traffic. *Proteomics*, *9*, 914–921.

Zhang, Y., Goldman, S., Baerga, R., Zhao, Y., Komatsu, M., & Jin, S. (2009). Adipose-specific deletion of autophagy-related gene 7 (atg7) in mice reveals a role in adipogenesis. *Proceedings of the National Academy of Sciences of the United States of America*, *106*, 19860–19865.

Zhou, X., Zhou, B., Truman, J. W., & Riddiford, L. M. (2004). Overexpression of broad: A new insight into its role in the Drosophila prothoracic gland cells. *The Journal of Experimental Biology*, *207*, 1151–1161.

CHAPTER FIVE

Metamorphosis in Teleosts

Sarah K. McMenamin, David M. Parichy[1]
Department of Biology, University of Washington, Seattle, Washington, USA
[1]Corresponding author: e-mail address: dparichy@u.washington.edu

Contents

Abstract

Teleosts are the largest and most diverse group of vertebrates, and many species undergo morphological, physiological, and behavioral transitions, "metamorphoses," as they progress between morphologically divergent life stages. The larval metamorphosis that generally occurs as teleosts mature from larva to juvenile involves the loss of embryo-specific features, the development of new adult features, major remodeling of different organ systems, and changes in physical proportions and overall phenotype. Yet, in contrast to anuran amphibians, for example, teleost metamorphosis can entail morphological change that is either sudden and profound, or relatively gradual and subtle. Here, we review the definition of metamorphosis in teleosts, the diversity of teleost metamorphic strategies and the transitions they involve, and what is known of their underlying endocrine and genetic bases. We suggest that teleost metamorphosis offers an outstanding opportunity for integrating our understanding of endocrine mechanisms, cellular processes of morphogenesis and differentiation, and the evolution of diverse morphologies and life histories.

Current Topics in Developmental Biology, Volume 103
ISSN 0070-2153
http://dx.doi.org/10.1016/B978-0-12-385979-2.00005-8

1. INTRODUCTION AND DEFINITIONS

Nearly half of all described vertebrates are teleosts (see Volff, 2005), and the more than 23,500 known species of teleosts exhibit a vast diversity of phenotypes, ecologies, and developmental life histories. Moreover, teleosts exhibit a tremendous range of metamorphic processes, sometimes undergoing major phenotypic and physiological transitions that allow a single species to exploit different habitats and niches during different life stages. This diversity in teleost life history has long fascinated comparative embryologists and marine ecologists, but the molecular mechanisms of these transitions remain almost entirely unexplored. With the methods of modern genetics and developmental biology, we can begin to elucidate the molecular, endocrinological, and morphogenetic processes underlying this incredibly diverse array of postembryonic transitions.

Several families of teleosts undergo spectacular morphological metamorphoses, comparable to the metamorphosis of anuran amphibians. Other teleosts undergo comparatively subtle phenotypic transformations, more analogous to the fetal-to-adult transition of mammals. Accordingly, workers in this area have adopted sometimes conflicting definitions of what should rightly constitute a "metamorphosis." For example, some count only the most dramatic of transitions to be metamorphoses, requiring profound morphological and ecological differences between life stages, and rapid transitions between these stages (Balon, 1990, 1999; Just, Kraus-Just, & Check, 1981; Manzon, 2011; Youson, 1988, 2004). Such restrictive definitions dismiss more subtle morphological, physiological, or ecological changes. Alternatively, others have suggested that metamorphosis of teleosts and other chordates can be defined as a conserved period of postembryonic remodeling, how extensive it might be, that is regulated specifically by thyroid hormone (TH; Laudet, 2011; Paris et al., 2008; Paris & Laudet, 2008; Power, Silva, & Campinho, 2008). Yet, this definition precludes transformations that may rely on other hormonal axes.

For the purposes of this chapter, we define metamorphosis to be *an irreversible developmental and physiological change that affects multiple traits during postembryonic development and is brought about by one or more systematically acting endocrine mediators, but is independent of sexual maturation, sex-specific modifications, or senescence*. This definition focuses on intrinsic features of an organism, or how the environment impacts organismal functioning, and accommodates what is clearly a phylogenetic continuum in the magnitude of postembryonic remodeling. By excluding habitat changes *per se*, this definition allows for studies of how metamorphic processes have evolved in association with particular ecological conditions or life history modes. By our definition, metamorphosis is not necessarily a specific event, but, rather, any nonsexual, nonsenescent transition entailing developmental or physiological remodeling that is systemically controlled. Thus, our criteria allow one to pose testable hypotheses for whether any particular morphological or physiological transformation constitutes a metamorphosis; that is, whether or not it is controlled by one or more endocrine mediators. The definition deliberately excludes morphological changes (e.g., acquisition of nuptial coloration) that may be temporary, sex-specific, or both, and also does not specify any requirement for overall somatic growth. Moreover, as we have defined it, metamorphoses could occur over a range of time spans, from less than a day to periods of several months, and could, in principle, occur between a variety of life history phases, though metamorphosis is generally most prevalent during the larva-to-juvenile transition (Fig. 5.1). Finally, in being agnostic to mechanism, our definition acknowledges that additional factors mediating such transformations still may await discovery or characterization. In proposing this definition, and in the review that follows, our goal is to facilitate the study of postembryonic cellular, developmental, and physiological processes, how they intersect with endocrine mechanisms, and how they have evolved.

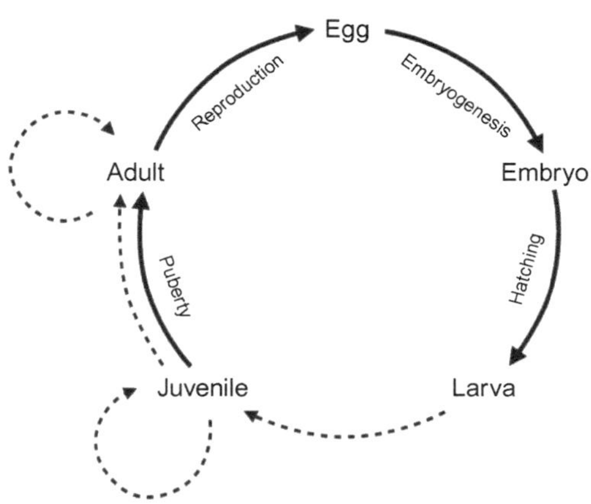

Figure 5.1 Generalized life cycle of a teleost. Dashed lines indicate potential periods of metamorphic transformations within the life cycle; the transformation from larva to juvenile represents larval metamorphosis.

2. METAMORPHIC DIVERSITY IN TELEOSTS

In this section, we briefly survey the astonishing diversity of metamorphic remodeling among teleosts, set in the context of the ecological and life history transitions that they accompany (Table 5.1). The majority of described metamorphic transitions, *sensu* our definition, are metamorphoses from larva to juvenile. These larval metamorphoses commonly include the formation of adult fins (pectoral fins generally develop during embryogenesis) and ossification of fin rays, maturation of internal organs and sensory systems, formation of scales, modifications to pigment pattern, and allometric changes in body proportions. Accompanying these alterations are a multitude of group- and species-specific morphological and physiological changes, some of which we outline below.

2.1. Marine teleosts

Marine teleosts exhibit some of the most dramatic morphological transitions seen in vertebrates. These changes in phenotype facilitate ecological transitions: most marine fish—and indeed the majority of marine organisms—exhibit a dispersive planktonic or pelagic (open water) larval stage, then undergo metamorphosis and are recruited to an adult habitat (Thorson, 1950). Pelagic larvae often posses morphological specializations that maximize survival and dispersal potential: large larval fin folds, bony plates, and long spines frequently characterize dispersive larvae. Such larval-specific features are resorbed at metamorphosis (as in the grouper, Fig. 5.2; Moser, 1981; Webb, 1999). The larval stage can last several years in some fish species, allowing larvae to disperse over long distances (Webb, 1999).

In addition to the loss of larval features, larval metamorphosis of marine fishes often involves dramatic changes in pigmentation and overall morphology, with specific alterations to head shape and body depth (de Jesus, Toledo, & Simpas, 1998; Fukuhara & Fushimi, 1988; Wittenrich et al., 2010). Flattening in the dorsoventral plane prepares many species to become "demersal" adults, which inhabit the bottom of the water column, or "benthic" adults, which live close to or in contact with the substrate or the sea floor (Webb, 1999). One remarkable group of benthic teleosts flattens laterally rather than dorsoventrally during metamorphosis (flatfish morphology citations, Table 5.1). As Pleuronectiform flatfish mature, bilaterally symmetric larvae become markedly asymmetrical as one eye translocates across the dorsal midline to the opposite side of the head, resulting in an ocular side that

Table 5.1 Characteristics of larval metamorphosis in different teleost groups

Group	Habitat	Niche shift	Duration	Additional morphological changes*	Species and references
Carp	Freshwater	None	Days–weeks	None	Zebrafish *Danio rerio*: gross morphology (Parichy, Elizondo, Mills, Gordon, & Engeszer, 2009); fin development (Cubbage & Mabee, 1996; Goldsmith, Iovine, O'Reilly-Pol, & Johnson, 2006); pigmentation (Budi, Patterson, & Parichy, 2011; Parichy & Turner, 2003a); and lateral line (Nuñez et al., 2009; Webb & Shirey, 2003)
Catfish	Freshwater	Pelagic to benthic	Days–weeks	Growth of barbels	Indian catfish *Mystus punctatus* (Ramanathan, Natarajan, & Sukumran, 1985)
Coral reef fish	Marine	Pelagic to demersal	Hours–weeks	Dramatic adult pigmentation changes in many species	Multiple species (McCormick & Makey, 1997; McCormick, Makey, et al., 2002)
Eels	Catadromous	Marine to freshwater	Months–years	Leptocephalus/glass eel to elver/silver eel transition: Salinity tolerance	American eel *Anguilla rostrata* otolith morphology (Cieri & McCleave, 2000) Japanese eel *Anguilla japonica* otolith morphology (Arai, Otake, & Tsukamoto, 1997) Australian shortfin eels *Anguilla australis* (De Silva, Gunasekera, Gooley, & Ingram, 2001) Multiple species (Raju, 1974)

Continued

Table 5.1 Characteristics of larval metamorphosis in different teleost groups—cont'd

Group	Habitat	Niche shift	Duration	Additional morphological changes	Species and references
Flatfish	Marine	Pelagic to benthic	Weeks	Cranial development and eye migration, asymmetrical pigmentation	Atlantic halibut *Hippoglossus hippoglossus* (Sæle et al., 2004) Japanese flounder *Paralichthys olivaceus* (Tanaka, Kawai, Seikai, & Burke, 1996) Senegalese sole *Solea senegalensis* (Fernández-Díaz et al., 2001) Starry flounder *Platichthys stellatus* (Policansky & Sieswerda, 1979) Summer flounder *Paralichthys dentatus* (Keefe & Able, 1993; Martinez & Bolker, 2003) Multiple species (Schreiber, 2001)
Gobies	Amphidromous	Marine to freshwater	Days–weeks	Change in mouth position and fin shape	Freshwater goby *Sicyopterus lagocephalus* (Keith et al., 2008; Taillebois et al., 2011)
Groupers	Marine	Pelagic to benthic	Months	Loss of larval spines, dorsoventral flattening	Spotted grouper *Epinephelus tauvina* (Hussain & Higuchi, 1980) Leopard grouper *Mycteroperca rosacea* (Martínez-Lagos & Gracia-López, 2009) Hong Kong grouper *Epinephelus akaara* (Fukuhara & Fushimi, 1988)
Medaka	Amphidromous	None	Days–weeks	None	Medaka *Oryzias latipes* (Iwamatsu, 1994)
Sea Breams	Marine	Pelagic to benthic	Months	None	Gilt-head sea bream *Sparus saurata* (Fukuhara, 1991)

Salmonids	Anadromous	Freshwater to marine	Days–weeks	Parr to smolt transition: Changes in lipid content and salinity tolerance; erythrocyte modifications	Atlantic salmon *Salmo salar.* Morphology (Dêbowski et al., 1999); physiology (McCormick & Saunders, 1987; Virtanen, 1987) Coho salmon *Oncorhynchus kisutch* physiology (Sullivan, Dickhoff, Mahnken, & Hersbberger, 1985) Multiple species (McCormick & Saunders, 1987; Woo, Bern, & Nishioka, 1978)
Tilapia	Freshwater	None	Days–weeks	None	Tilapia *Oreochromis mossambicus* (Ramanathan et al., 1985)
Tuna	Marine	Planktonic to pelagic	Days–weeks	None	Pacific bluefin tuna *Thunnus thynnus* (Kaji, Tanaka, Takahashi, Oka, & Ishibashi, 1996; Kawakami, Nozaki, Seoka, Kumai, & Ohta, 2008; Miyashita, Sawada, Okada, Murata, & Kumai, 2001; Tanaka, Satoh, Iwahashi, & Yamada, 2006) Yellowfin tuna *Thunnus albacares* (Kaji et al., 1999)

*At metamorphosis, all groups acquire adult fins and fin rays, internal organs and sensory systems mature, scales and adult pigment pattern form and body proportions change; additional morphological changes lists specific modifications in addition to these common changes

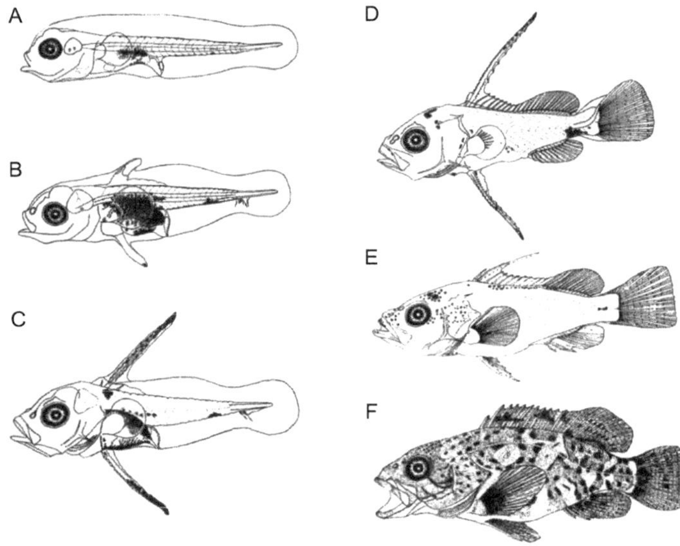

Figure 5.2 Metamorphosis of the brown-spotted grouper (*Epinephelus tauvina*). (A) Newly hatched larva, 2 mm; (B) 10-day-old larva; (C) 13-day-old larva; (D) 31-day-old metamorphosing individual; (E) 31 day old, 18 mm; and (F) 50 dpf juvenile, 31 mm. *Images reprinted with permission from (Hussain & Higuchi, 1980)*

will face the water column, and a blind side that will face the substrate (Fig. 5.3). During the transitional metamorphic period, fish swim at an increasing angle to ultimately settle onto the substrate. As in other teleosts, a new pattern of pigmentation develops during metamorphosis, but only on the ocular surface of the body (Table 5.1).

Particularly rapid and dramatic remodeling also occurs among coral reef fishes, which undergo changes in body shape and acquire conspicuous adult pigmentation (Leis & McCormick, 2006). Metamorphosis in reef fish enables the transition from a pelagic dispersal stage to a demersal, reef-associated adult form. This metamorphosis can be extremely rapid, with some species undergoing their metamorphic changes in pigmentation in as little as 6 h, or can last several weeks and include several intermediate stages; for example, goatfish and damselfish undergo 2–3 week metamorphic processes that include several intermediate shifts in habitat and morphology (McCormick & Makey, 1997; McCormick, Makey, & Dufour, 2002). Thus, marine teleosts exhibit an astonishing range of metamorphic transformations that often include substantial morphological changes allowing the exploitation of different spatial niches within an oceanic environment.

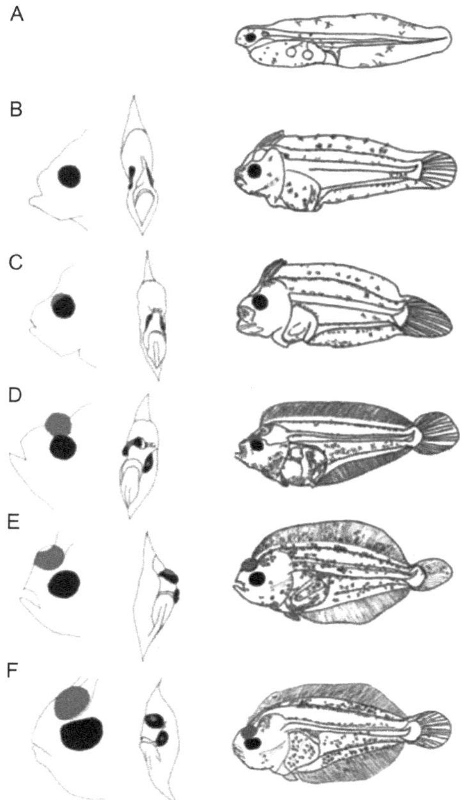

Figure 5.3 Metamorphosis of the summer flounder *Paralichthys dentatus*. (A) Hatched yolk-sac larva. (B) Pretransformation larva before eye migration commences. (C) Early metamorphosis and the beginning of eye migration. (D) Mid-metamorphosis. (E) Metamorphic climax, right eye has migrated over the dorsal midline. (F) Young juvenile. Left column in B–D shows the migration of the eye across the skull; migrating right eye is shaded in gray. Rightmost column shows whole-body morphological changes at each stage. *Images reprinted with permission from (Martinez & Bolker, 2003)*

2.2. Diadromous teleosts

Diadromous species migrate between salt water and fresh water environments; metamorphosis can prepare these fish to survive in their new habitat. The physical demands of a marine environment differ significantly from those of fresh water, and diadromous teleosts undergo substantial changes in morphology. These include changes in body shape, muscle, skin, and pigmentation; changes in the structure and function of numerous organs, including the kidneys, gut, eyes, and lateral line; and physiological changes in osmoregulation and metabolism.

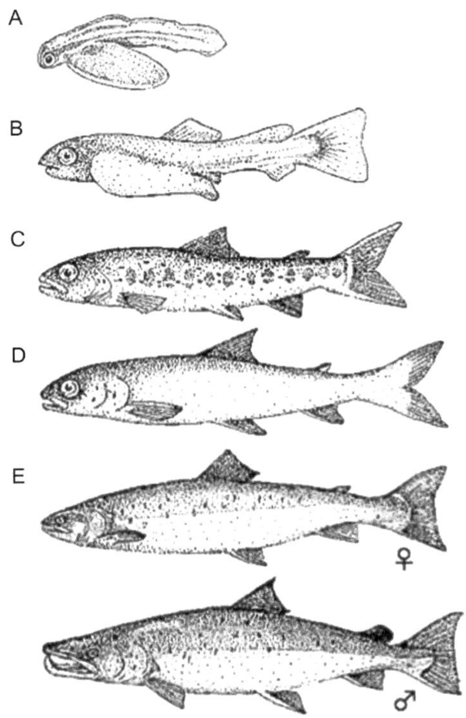

Figure 5.4 Life stages of Atlantic Salmon, *Salmo salmar*. (A) Alevin; (B) fry, 3–6 weeks old; (C) parr; (D) smolt, 1–3 years old; and (E) Breeding female and male. *Images from public domain.*

The best-studied example of a fresh-to-salt water (anadromous) migration is that of salmon. Salmon spawn in freshwater breeding grounds, where embryonic "alevin" hatch into larval "fry." After months to several years of development, larval fry develop bars of pigmentation and are called "parr" (Fig. 5.4). Swimming downstream, parr undergo smoltification, which morphologically and physiologically prepares young "smolt" juveniles for the marine habitat. By our definition, the entire parr-to-smolt period constitutes a protracted larval metamorphosis, with different morphogenetic and physiological processes stimulated by different hormones (insulin, prolactin, TH, growth hormone (GH), and cortisol) that spike at different points during the months of parr development (see Björnsson, Einarsdottir, & Power, 2012; Dickhoff, Brown, Sullivan, & Bern, 1990). As fish metamorphose into smolts, purines are deposited in the skin, producing a silvery appearance, body fat decreases, gas bladder size increases, and body shape and condition change to become leaner and sleeker (Björnsson, Einarsdottir, et al., 2012; Johnston & Saunders, 1981; McCormick & Saunders, 1987; Winans & Nishioka, 1987). In preparation for the hyperosmotic environment, smolts become able to absorb increased amounts of water through the intestine, and expression of ion channels (aquaporins) and ion pumps (Na^+/K^+-ATPase) increases, allowing active excretion of salt (D'Cotta, Valotaire, le Gac, & Prunet, 2000; Tipsmark, Sørensen, & Madsen, 2010). During smoltification, salmon also begin synthesizing additional hemoglobin isoforms with enhanced oxygen affinity (Seear et al., 2010; Sullivan et al., 1985). If smolts do not reach salt water during a critical "smolt window," they undergo desmoltification, which includes a loss of hypo-osmoregulatory abilities and metabolic adaptations, and a darkening of skin color (Stefansson, Björnsson, Ebbesson, & McCormick, 2008). Although desmoltification includes the regression of certain physiological functions, fish do not revert to an overall smolt morphology; thus, desmoltification is not a reversal of metamorphosis. Returning to freshwater environments to spawn, marine salmonids undergo yet another transformation into the life stage known as "grilse," preparing them for migration in fresh water and breeding. This transformation is controlled predominantly by the gonadotropic axes (Youngson, McLay, Wright, & Johnstone, 1988), and the majority of the morphological changes are sex-specific, so by our definition, this transformation primarily constitutes a puberty rather than a metamorphosis.

Eels represent a second well-studied instance of a diadromous life history. In contrast to salmonids, eels develop and disperse as marine larvae, then begin to metamorphose in preparation for a salt-to-fresh water (catadromous) migration. The leaf-shaped marine larvae (leptocephali) are dispersed by ocean currents and after up to 18 months of larval development, transform into transparent "glass eels" (Fig. 5.5; Wang & Tzeng, 2000). Glass eels undergo numerous morphological and physiological changes to preadapt them to freshwater environments, suggesting that the entire glass eel stage represents the metamorphic period. As glass eels are recruited to a freshwater habitat, they complete metamorphosis to become juvenile "elvers." To return to oceanic environments for spawning, eels undergo the process of silvering and regain saltwater tolerance. Silvering has been described as a secondary metamorphic event, although it is stimulated by sex hormones and involves maturation of the gonads (Aroua et al., 2005; Rousseau, Aroua, Schmitz, Elie, & Dufour, 2009). This event may represent a second metamorphosis simultaneous with puberty.

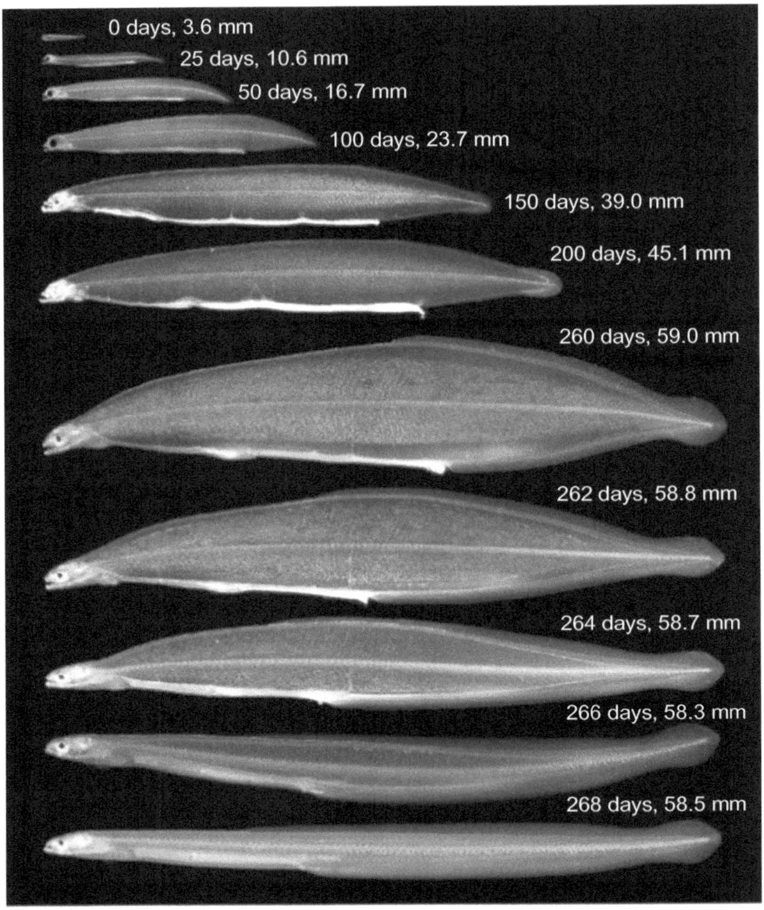

Figure 5.5 Developmental growth series of the Japanese eel *Anguilla japonica*. Larval leptocephali transform into the glass eel stage (bottom image shows a young glass eel). *Reprinted with permission from (Tsukamoto et al., 2009).* (For color version of this figure, the reader is referred to the online version of this chapter.)

2.3. Freshwater teleosts

In contrast with their marine and diadromous counterparts, freshwater teleosts generally undergo more phenotypically subtle metamorphoses. Few larva-specific adaptations are found in freshwater teleosts, reflecting the fact that freshwater fish do not generally have an explicitly dispersive larval stage. Nonetheless, freshwater teleosts undergo numerous morphological and physiological changes during larval metamorphosis, modifications that have been described most extensively in the developmental genetic model species zebrafish *Danio rerio* (Fig. 5.6; Parichy et al., 2009). Like other teleosts,

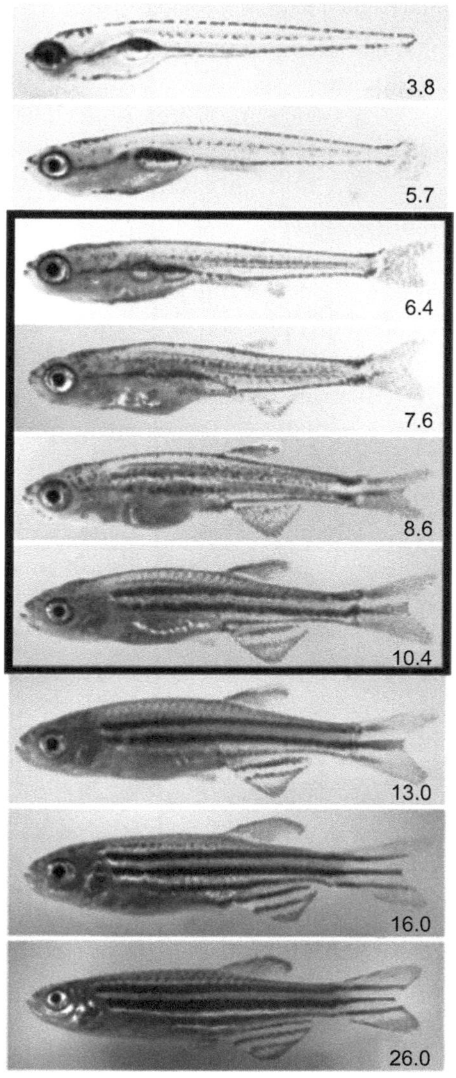

Figure 5.6 Postembryonic development in the zebrafish (*Danio rerio*). Numbers indicate standard length in mm, with the period of peak metamorphic remodeling during the larva-to-juvenile transition indicated by the black box. At standard temperature and typical rearing densities, this transition begins at about 10–14 days and takes about 2 weeks. As postembryonic fish can develop and grow at markedly different rates, days of development is an inadequate staging criterion; stages based on morphological criteria, and their size equivalents have been defined for this species and are preferable for developmental studies (Parichy et al., 2009). (See Color Insert.)

zebrafish undergo significant changes in body shape, losing larval fin folds, developing or completing development of adult fins and fin rays (Cubbage & Mabee, 1996; Goldsmith et al., 2006; Patterson, Mook, & Devoto, 2008), ossifying the axial skeleton (Bird & Mabee, 2003; Elizondo et al., 2005), forming an adult pigment pattern (Budi, Patterson, & Parichy, 2008; Budi et al., 2011; Johnson, Africa, Walker, & Weston, 1995; Parichy & Turner, 2003b), acquiring scales (Sire & Akimenko, 2004), and exhibiting maturation and remodeling of many organs, including the lateral line (Ledent, 2002; Nuñez et al., 2009; Webb & Shirey, 2003), central and peripheral nervous systems (Larson, Gordon, Lau, & Parichy, 2010), gut, kidneys, and swimbladder (Parichy et al., 2009; Robertson, McGee, Dumbarton, Croll, & Smith, 2007). Freshwater teleosts tend to have less distinct periods of larval metamorphosis than marine or diadromous species, but this postembryonic process still unequivocally occurs, transforming the larval organism into a morphologically distinct juvenile form. There are at least a few identified cases of paedomorphism in freshwater species, with miniature *Paedocypris* and *Danionella* fish failing to undergo normal metamorphosis and becoming reproductively mature while maintaining a larva-like overall morphology (Britz, Conway, & Rüber, 2009; Mayden & Chen, 2010; Rüber et al., 2007). These paedomorphic genera illustrate ways in which even relatively subtle metamorphic processes may be modified to effect major morphological change.

3. ENDOCRINE CONTROLS OF TELEOST METAMORPHOSIS

The coordinated occurrence of multiple disparate organ- and tissue-specific events during teleost life-stage transitions suggests that "local" processes are likely to be coordinated by "global" signals (and so are likely to constitute bona fide metamorphoses, as defined above). Research into the endocrine controls of postembryonic transitions in teleosts has focused almost exclusively on spectacular examples of such metamorphoses; for example, salmonid smoltification and flatfish metamorphosis. Yet, there is evidence that more subtle transitional processes are controlled by conserved endocrine mechanisms as well. Here, we outline what is known about the hormonal contributions to teleost metamorphosis. TH appears to be a key regulator of teleost metamorphosis, and most endocrine studies of metamorphosis have focused on this factor. Thus, we briefly review the metabolism of TH in teleosts, and the expression of TH-associated factors before outlining what is known about other endocrine contributions to metamorphosis.

3.1. Thyroid hormone

It became clear in the late 1980s that TH, which stimulates the metamorphosis of amphibians, also promotes metamorphosis of flatfish. Application of exogenous TH accelerates metamorphic events, including fin resorption, eye migration, and fin ray shortening (Inui & Miwa, 1985; Inui, Tagawa, Miwa, & Hirano, 1989; Klaren, Wunderink, Yüfera, Mancera, & Flik, 2008; Miwa, Yamano, & Inui, 1992). TH is also necessary for flatfish metamorphosis: treatment with TH-inhibiting goitrogens arrests metamorphosis, resulting in oversized pelagic larvae (Inui & Miwa, 1985).

Further studies showed that spikes in thyroid activity or circulating TH levels are associated with the larval metamorphosis of many teleost species, including those with spectacular transformations (salmonids, eels, flatfish) and those with more subtle metamorphoses (fatheads, greenlings, groupers, minnow, sea breams, tilapia, tuna, zebrafish; see Fig. 5.7 and hormone titer citations in Table 5.2). Treatment with exogenous TH stimulates metamorphic processes, while inhibition of TH synthesis with goitrogens inhibits metamorphosis in numerous groups: eels, gobies, greenlings, groupers, and salmon (morphological responses to TH manipulation citations, Table 5.2). Metamorphic changes in fins, pigmentation, and body size in zebrafish are inhibited by goitrogen treatment as well (Brown, 1997; D.M. Parichy, unpublished data). Evolutionary modifications of TH signaling effected by *cis*-regulatory changes at the thyroid-stimulating hormone β2 locus have also been implicated in the adaptive divergence of marine and freshwater stickleback populations (Kitano et al., 2010). Together, these studies suggest that TH involvement in metamorphosis is widespread among teleosts, and that changes in this pathway can contribute to evolutionary diversification. Interestingly, lampreys, relatives of early teleosts, undergo a metamorphosis in which TH plays an inhibitory rather than a stimulatory role (see Youson, 1997); the mechanistic bases for this divergent TH effect, and whether other factors substitute for TH in promoting metamorphosis, remain unknown.

TH synthesis and metabolism in teleosts are similar to that of tetrapods (Yen, 2001). The genes that contribute to thyroid follicle patterning and

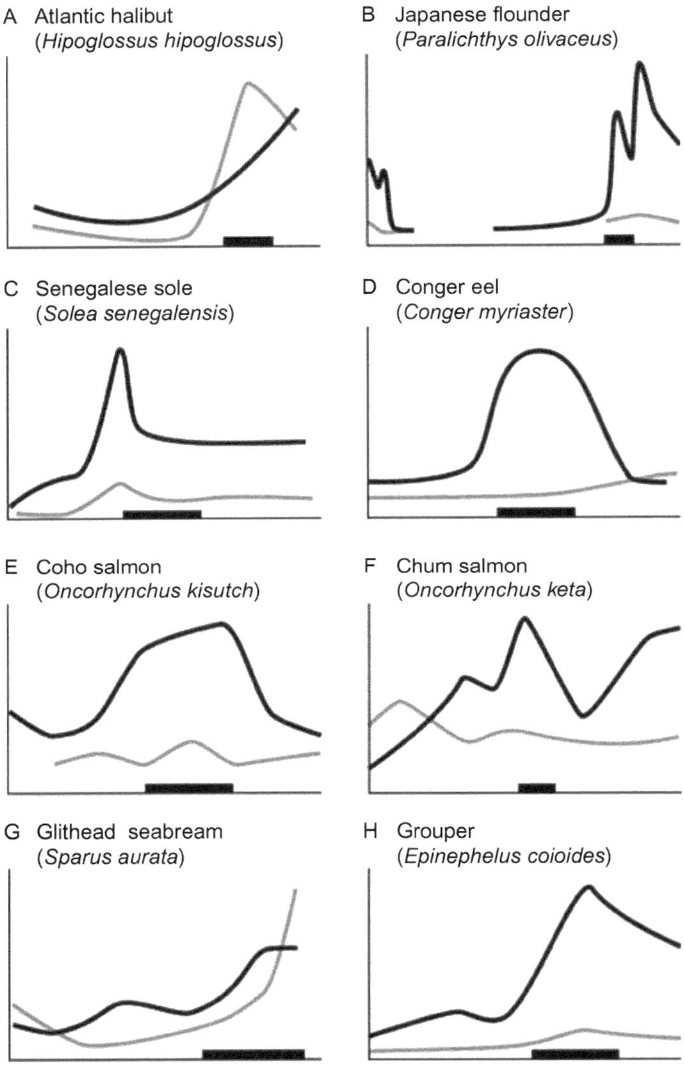

Figure 5.7 Relative whole-body concentrations of thyroid hormone in different teleosts during larval and juvenile development. Developmental stage is indicated by the *x*-axis and the horizontal black bar shows the approximate climax of the metamorphic period. Black line indicates relative level of thyroxine (T4); gray line indicates relative level of triiodothyronine (T3); lines redrawn from original figures. Relative concentrations shown for three flatfish: (A) Halibut (Galay-Burgos, Power, Llewellyn, & Sweeney, 2008); (B) Flounder (Yamano & Miwa, 1998); and (C) Sole (Klaren et al., 2008). Also shown are (D) Eel (Kawakami, Tanda, Adachi, & Yamauchi, 2003), two species of salmon: (E) Coho salmon (Harada, Yoshinaga, Ojima, & Iwata, 2008) and (F) Chum salmon (Parhar & Iwata, 1996); (G) Seabream (Szisch, Papandroulakis, Fanouraki, & Pavlidis, 2005); and (H) Grouper (de Jesus et al., 1998).

Table 5.2 Selected studies that have examined hormone involvement or gene expression during teleost larval metamorphosis

Family	Assessment	Species and references
Carp (Cyprinidae)	Hormone titer	TH (zebrafish *Danio rerio*) (Chang et al., 2012) TH (minnow *Pimephales promelas*) (Crane, Pickford, Hutchinson, & Brown, 2004)
	Morphological response to TH manipulation	Overall morphology: (Brown, 1989); Larval survival and growth: (Lam and Sharma, 1985)
Cichlids (Cichlidae)	Hormone titer	TH (Shiao, Wu, Hwang, Wu, & Hwang, 2008)
	Morphological response to TH manipulation	Overall morphology (Okimoto, Weber, & Grau, 1993) Otolith development (Shiao et al., 2008)
Eels (Anguilloidei)	Hormone titer	TH (Aroua et al., 2005; Jegstrup & Rosenkilde, 2003; Yamano, Nomura, & Tanaka, 2007; Yamano, Tagawa, et al., 1991) Cortisol (Yamano, Tagawa, et al., 1991) GH, TSH, and sex hormones (Aroua et al., 2005)
	Morphological response to TH manipulation	Overall morphology and pigmentation (Jegstrup & Rosenkilde, 2003; Kitajima, Sato, & Kawanishi, 1967)
Flatfish (Order: Pleuronectiformes)	Hormone titer	TH (de Jesus, Hirano, & Inui, 1991; Klaren et al., 2008; Miwa, Tagawa, Inui, & Hirano, 1988) GH and IGF (Hildahl, Sweeney, Galay-Burgos, Einarsdóttir, & Björnsson, 2007) Cortisol (de Jesus et al., 1991)
	Morphological response to TH manipulation	Eye migration and overall morphology (Inui & Miwa, 1985; Inui et al., 1989; Schreiber & Specker, 1998; Solbakken, Norberg, Watanabe, & Pittman, 1999; Tagawa & Aritaki, 2005) Gastric development (Miwa et al., 1992) Pigmentation (Yoo, Takeuchi, Tagawa, & Seikai, 2000) Muscle development (Yamano, Miwa, Obinata, & Inui, 1991) Fin rays (de Jesus, Hirano, & Inui, 1993; de Jesus, Inui, & Hirano, 1990) Erythrocyte development (Miwa & Inui, 1991)

Continued

Table 5.2 Selected studies that have examined hormone involvement or gene expression during teleost larval metamorphosis—cont'd

Family	Assessment	Species and references
	Gene expression	TRs (Galay-Burgos et al., 2008; Marchand, Duffraisse, Triqueneaux, Safi, & Laudet, 2004; Yamano & Miwa, 1998) Deiodinases (Itoh, Watanabe, Wu, & Suzuki, 2010) GH receptors (Hildahl, Power, Björnsson, & Einarsdóttir, 2008; Hildahl et al., 2007; Manchado, Infante, Asensio, Planas, & Cañavate, 2008; Manchado, Infante, Rebordinos, & Cañavate, 2009) CK-M1 (Chen et al., 2010) Elongation factor 1 alpha genes (Infante, Asensio, Cañavate, & Manchado, 2008) Heat shock proteins (Manchado, Salas-Leiton, et al., 2008) Trypsinogens (Manchado, Infante, Asensio, et al., 2008) Keratin (Infante, Manchado, Asensio, & Cañavate, 2007) Myogenic regulatory factors (Andersen et al., 2009)
Fatheads (Platycephalidae)	Hormone titer	TH (Taillebois et al., 2011)
	Morphological response to TH manipulation	Overall morphology (Taillebois et al., 2011)
Groupers	Hormone titer	TH (de Jesus et al., 1998)
	Morphological response to TH manipulation	Overall morphology (de Jesus et al., 1998; Trijundo, Yoseda, Hirokawa, Tagawa, & Tanaka, 2002)
Greenlings (Hexagrammidae)	Hormone titer	TH (Matsumoto & Tanaka, 1996)
	Morphological response to TH manipulation	Overall morphology and settling behavior (Matsumoto & Tanaka, 1996)

Salmonids (Salmonidae)	Hormone titer	TH (Boeuf, Le Bail, & Prunet, 1989; Dickhoff, Folmar, & Gorbman, 1978; Dickhoff, Folmar, Mighell, & Mahnken, 1982; Grau, Specker, Nishioka, & Bern, 1982; Lin et al., 1988; McCormick, Moriyama, & Björnsson, 2000; Prunet, Boeuf, Bolton, & Young, 1989; Quigley, Harvey, Hayden, Dowling, & O'Keane, 2006; Specker & Schreck, 1982; Virtanen & Soivio, 1985; Young, Bjornsson, Prunet, Lin, & Bern, 1989) GH (Ágústsson et al., 2001; Boeuf et al., 1989; McCormick et al., 2000; McCormick, Shrimpton, Moriyama, & Björnsson, 2002; Prunet et al., 1989; Sweeting & McKeown, 1989; Young et al., 1989) Cortisol (McCormick et al., 2000; McCormick, Shrimpton, Moriyama, & Björnsson, 2002; Specker & Schreck, 1982; Virtanen & Soivio, 1985; Young et al., 1989) Prolactin (Prunet et al., 1989; Young et al., 1989) IGF-I (Ágústsson et al., 2001; McCormick et al., 2000; McCormick, et al., 2002)
	Morphological response to TH manipulation	Growth and salinity tolerance (Saunders, McCormick, Henderson, Eales, & Johnston, 1985) Salinity preference (Baggerman, 1960) Adult hemoglobin expression (Sullivan et al., 1985)
	Gene expression	Microarray of multiple genes (Seear et al., 2010) Aquaporins (Tipsmark et al., 2010)
Sea Breams (Sparidae)	Hormone titer	TH (Campinho, Galay-Burgos, Sweeney, & Power, 2010; Deane & Woo, 2003) Cortisol (Deane & Woo, 2003)
	Gene expression	TRs and deiodinases (Campinho et al., 2010) HSPs (Deane & Woo, 2003)
Tuna (Scombridae)	Hormone titer	TH (Kawakami, Nozaki, et al., 2008)
	Morphological response to TH manipulation	Eye pigmentation (Kawakami, Yokoi, Kumai, & Ohta, 2008)

development are highly conserved between teleosts and mammals (Alt et al., 2006; Porazzi, Calebiro, Benato, Tiso, & Persani, 2009), as are the deiodinases that regulate TH activity (Itoh et al., 2010; Orozco & Valverde-R, 2005; Power et al., 2008). In response to pituitary thyrotropin (TSH), thyroid follicles produce TH in the form of thyroxine (T4). Circulating plasma levels of T4 increase markedly at the onset of metamorphosis in several teleost species (Fig. 5.7), suggesting that the thyroid is highly active during this period. This is further consistent with the results of radioiodide uptake assays (Brown, 1997; D.M. Parichy, unpublished) and enzyme-linked immunoassays performed in zebrafish (Chang et al., 2012). T4 is converted in peripheral tissues by deiodinase enzymes (DIO1 and DIO2) into the genomically active form, tri-iodothyronine (T3). Both T3 and T4 are inactivated by a third deiodinase, DIO3. Thus, the expression and activity of deiodinases regulate TH bioactivity and availability. In teleosts, these genes are expressed in spatially and temporally specific manners during postembryonic development, with *dio1* and *dio2* increasing in expression immediately before and during metamorphic climax of several species (Campinho et al., 2010; Itoh et al., 2010).

T3 serves as the ligand for nuclear TH receptors (TRs), which typically activate expression of target genes in the presence of the hormone, and repress expression in its absence. All teleosts examined possess at least two isoforms of THRs: TRα and TRβ, and many posses additional copies of these loci, possibly owing to an ancient, teleost-specific genome duplication (Hoegg, Brinkmann, Taylor, & Meyer, 2004; Ravi & Venkatesh, 2008); for example, zebrafish exhibit two TRα loci, *thraa* and *thrab*, though only a single TRβ locus, *thrb*. In species that have been examined so far, TRs increase in expression prior to and during metamorphic climax (Fig. 5.8). Thus, TR-mediated TH signaling appears to play a critical role in the metamorphosis of teleosts examined. Nevertheless, although TH is by far the best-studied hormonal factor in teleosts, TSH and TH titers and expression of DIOs and TRs remain undocumented for the vast majority of teleost families. Thus, the requirement or sufficiency for TH in metamorphosis remains unknown in most fish; moreover, the genetic targets and molecular mechanisms of hormonal action remain largely unclear.

3.2. Non-TH mediators of metamorphosis

In addition to TH, other endocrine factors contribute to metamorphosis as well. GH activates insulin-like growth factor (IGF) pathways, stimulating cellular proliferation and increasing basal metabolic rate. Spikes in plasma

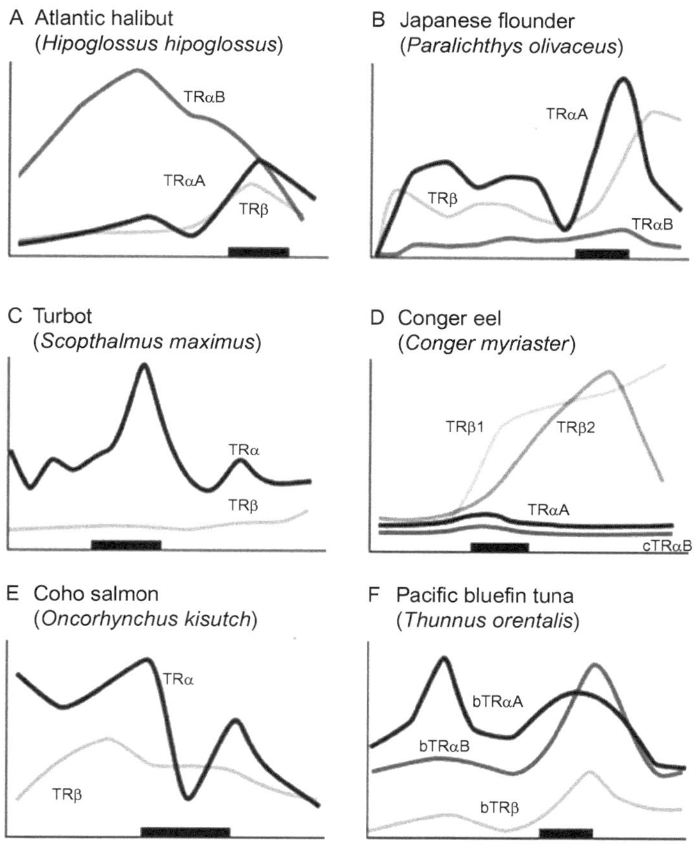

Figure 5.8 Relative different thyroid hormone receptor genes during larval and juvenile development. Expression is from whole-body assays unless otherwise noted; lines redrawn from original figures. The x-axis shows developmental progression and the thick black bar shows the approximate climax of the metamorphic period. Relative gene expression in three flatfish: (A) Halibut (Galay-Burgos et al., 2008), (B) Flounder (gene expression in head; Yamano & Miwa, 1998), and (C) Turbot (Marchand et al., 2004), and three other teleosts: (D) Eel (Kawakami et al., 2003), (E) Salmon (gene expression in brain; Harada et al., 2008), and (F) Tuna (Kawakami, Nozaki, et al., 2008). The absence of a particular receptor subtype on a graph does not necessarily mean that the species does not possess that isoform.

GH and IGF-I are observed during salmonid parr-to-smolt metamorphosis (Table 5.2). Further, treating parr with exogenous GH induces metamorphic changes, including changes in body shape, pigmentation, and seawater tolerance (Boeuf, 1993; Donaldson, Fagerlund, Higgs, & McBride, 1979; Dufour & Rousseau, 2007). Genetic loss of GH likewise delays metamorphic stage transitions in zebrafish; though this may be a secondary effect of delayed growth (McMenamin and Parichy, unpublished). In flatfish, GH receptors peak in expression immediately prior to the onset of metamorphosis, as do concentrations of IGF (Hildahl et al., 2007). The hormone prolactin antagonizes GH, and appears to have a negative influence on salinity tolerance and overall smoltification in salmonids (Björnsson, Stefansson, & McCormick, 2011; Madsen & Bern, 1992). The GH/IGF axis thus plays important roles in promoting growth and regulating metabolism during postembryonic development in teleosts (Yousefian & Shirzad, 2011), and its potential roles in directly stimulating morphogenetic events merit further investigation.

Another factor that may contribute to metamorphic progression is cortisol, known primarily for its roles in stress response. Cortisol titer increases during salmon smoltification, and may directly contribute to physiological metamorphosis (Richman, de Diaz, Nishioka, & Bern, 1985; Specker, 1982). Flatfish also exhibit peak cortisol levels at metamorphic climax, potentially indicating a stimulatory role in this process (de Jesus et al., 1991). Indeed, stress hormones are known to contribute to metamorphosis of some anuran amphibians (Denver, 1993, 1997). In eel leptocephali, however, cortisol levels decrease prior to metamorphosis and remain low throughout metamorphosis (Yamano, Tagawa, et al., 1991). Thus, stress hormones may play synergistic roles with TH during the metamorphosis of some teleosts (Dufour & Rousseau, 2007), but these roles require further investigation and particularly experimental manipulation.

In summary, it is clear that several hormonal axes are activated immediately before, during, and following metamorphic climax in the teleosts that have been studied to date. These hormonal axes, and specifically TH pathways, likely play roles in orchestrating metamorphosis, and may integrate external environmental cues into coordinated sets of disparate morphogenetic events. Nevertheless, the evolutionary conservation of these roles, the interactions between the hormonal axes, and the proximate mechanisms of endocrine response remain unclear.

4. LOCAL MECHANISMS OF MORPHOGENETIC CHANGES

The morphogenetic changes that occur at metamorphosis include major remodeling of existing features as well as the formation of entirely new tissues and organs; thus, metamorphosis requires extensive differentiation as well as the morphogenetic processes of cellular migration, proliferation, growth, and death. For the overwhelming majority of metamorphic events, however, the underlying cellular mechanics and genetic mechanisms remain unexplored by modern methods. Here, we discuss the relatively few areas in which the cellular or genetic bases underlying metamorphic changes are being thoroughly investigated, and we highlight some of the many outstanding questions that remain to be answered.

4.1. Flatfish cranial asymmetry

The most thoroughly studied teleost metamorphic process is undoubtedly the unique cranial transformation of the flatfish. Although flatfishes are exceptional in their external morphological asymmetry, the internal organs of all vertebrates are asymmetric to different degrees. This internal asymmetry originates during early embryogenesis with the asymmetric expression of genetic cascades that initiate asymmetric expression of the homeobox-containing transcription factor Pitx2 (Ryan et al., 1998). Flatfishes exploit this internal asymmetry to ultimately achieve external asymmetry at metamorphosis (Hashimoto et al., 2002; Suzuki et al., 2009). Pitx2 is reexpressed at metamorphosis in a fundamentally asymmetric portion of the brain, and this reexpression appears to initiate eye migration (Suzuki et al., 2009). Cellular proliferation in the suborbital tissue of one side of the cranium may "push" one eye across the dorsal midline (Bao et al., 2011). Whether this proliferation is stimulated by TH directly or whether intermediary signals are involved remains unknown, but GH and IGF pathways likely serve as more proximal factors (Hildahl et al., 2008). In several flatfish mutants, cranial asymmetry is decoupled from internal organ asymmetry, suggesting that the two are regulated by independent mechanisms (Hashimoto et al., 2002). Elucidating the local pattern-forming and morphogenetic cues that determine the definitive form of the adult craniofacial skeleton, and how these factors depend on global hormonal effectors, clearly represents an exciting area for future research.

4.2. Skin

Larval teleosts have a simple integumentary structure, composed primarily of epidermis; at metamorphosis, the skin becomes increasingly stratified and complex (Chang & Hwang, 2011; Hawkes, 1974; Rakers et al., 2010). Collagen fibrils are deposited in orthogonal arrays under the epidermis (Le Guellec, Morvan-Dubois, & Sire, 2004) and the acellular stroma is then invaded by fibroblasts of unknown origin (but potentially originating from the neural crest; Matsumoto et al., 1983). Later, in metamorphosis, these fibroblasts initiate scale formation, potentially regulated by expression of sonic hedgehog among other factors (Sire & Akimenko, 2004). The processes of skin stratification have been studied in greatest detail in zebrafish, but appear at least structurally similar in cichlids (Sire & Géraudie, 1983). Keratins mediate some of the skin restructuring events, and in flatfish, keratin expression is regulated directly by TH (Infante et al., 2007), decreasing after metamorphic climax (Campinho, Silva, Sweeney, & Power, 2007). The metamorphic transformation of flatfish skin resembles the metamorphic restructuring seen in amphibians (Power et al., 2008), in which keratin loci are also under TH control (Page et al., 2007). Nevertheless, further research is needed to determine the precise roles of TH in promoting integumentary metamorphosis of teleosts. In this regard, genetically tractable species such as zebrafish offer outstanding potential for studying skin metamorphosis; indeed, a variety of mutants with defects in the formation of scales and other postembryonic integumentary cell lineages have been identified (Harris et al., 2008; Lang, Patterson, Gordon, Johnson, & Parichy, 2009). As many features of skin development are conserved even with mammals, and mammalian skin undergoes a similar period of increased stratification and barrier function acquisition during fetal stages, studies of teleost skin metamorphosis may have translational relevance as well (Hoath & Maibach, 2003; Rakers et al., 2010).

4.3. Pigmentation

The formation of adult pigmentation is another common feature of teleost metamorphosis. Some of the genetic mechanisms underlying metamorphic pigmentation have been dissected in zebrafish, in which adult pigment patterns result from the spatial arrangements of neural crest-derived black melanophores, yellow xanthophores, and iridescent iridophores. During embryonic stages of neural crest migration, precursors to adult pigment cells are established in part owing to ErbB signaling (Budi et al., 2008; Hultman et al., 2009). Subsequently, these precursors are associated with peripheral nerves and, at metamorphosis, migrate to the skin along stereotypical pathways to form the adult pigment pattern, a process that can be directly visualized owing to the existence of fluorescent lineage reporters and the relative transparency of even metamorphic stage zebrafish (Budi et al., 2011). Embryonic and adult pigment cells have partially nonoverlapping genetic requirements, demonstrated by the phenotypes of several zebrafish mutants in which normal embryonic pigment cells develop yet adult precursors either fail to be established or fail to be recruited. In these mutants, pigment pattern metamorphosis is effectively decoupled from overall somatic metamorphosis (e.g., Budi et al., 2008; Larson et al., 2010; Parichy & Turner, 2003b). Once latent precursors have been recruited, a variety of genes acting both within the pigment cells and in the tissue environments through which these cells migrate are required to organize the different pigment cell classes into distinct juvenile and adult stripes (Eom et al., 2012; Iwashita et al., 2006; Lang et al., 2009; Parichy, Rawls, Pratt, Whitfield, & Johnson, 1999; Parichy & Turner, 2003a; Watanabe et al., 2006).

Phylogenetic comparisons reveal that adult pigment patterns of zebrafish relatives similarly depend on the recruitment of latent pigment cell precursors at metamorphosis. Interestingly, however, these cells have been mostly lost in the closely related *Danio nigrofasciatus*, in which the adult pigment pattern arises at metamorphosis largely through the rearrangement of embryonic/early larval melanophores that persist into the adult. Interspecific cell transplantation shows that the difference in adult melanophore development between zebrafish and *D. nigrofasciatus* lies extrinsic to the pigment cells, implicating a change in a still-unidentified tissue or cell type that influences metamorphic processes within the pigment cell lineage (Quigley et al., 2004). This example highlights the potential for comparative studies to reveal the cellular, and ultimately genetic, bases underlying evolutionary changes in metamorphic transformations.

In contrast to other teleosts, flatfish develop pigmentation in an asymmetric manner at metamorphosis, with only the ocular (upper) side normally developing adult melanophores. In flounder, pigment cell precursors migrate symmetrically to both lateral sides from the bases of fins at the dorsal and ventral margins of the flank, yet these cells differentiate as melanophores only on the ocular side (Watanabe et al., 2008; Yamada, Okauchi, & Araki, 2010). The genetic bases for this differential response of pigment cells remain unknown.

Development of metamorphic pigmentation in zebrafish is retarded by goitrogens that prevent TH synthesis (Brown, 1997). Whether TH is directly required by pigment cells or their precursors, or whether the hormone exerts an influence indirectly through other cellular intermediaries, is an active area of investigation. Treatment with high (hyper-physiological) levels of T4 actually inhibits adult melanophore development in flatfish (Yoo et al., 2000), eels (Jegstrup & Rosenkilde, 2003) and zebrafish (D.M. Parichy, unpublished data), but the biological significance of these observations remains obscure. Likewise, the genetic and cellular mechanisms of metamorphosis giving rise to the many and varied adult pigment patterns of other teleosts are largely unknown, though mutational and transgenic resources for some (Kelsh, 2004; Kelsh et al., 2004; Odenthal et al., 1996; Parichy, 2006), and evolutionary genetic strategies for others (Miller et al., 2007; Roberts, Ser, & Kocher, 2009), offer significant potential for future advances in this area.

4.4. Fin ray formation

Before and during metamorphosis, mesenchymal cells give rise to the fin endoskeleton, with rays growing from proximal to distal. Numerous pathways regulate positioning and outgrowth of these rays and the joints within the rays (Marí-Beffa & Murciano, 2010; Sims, Eble, & Iovine, 2009). Early proximo-distal patterning in the fin bud, largely controlled by Fgf signaling, is thought to establish a "prepattern" for ray positioning. Sonic hedgehog receptors are expressed in the proximal blastema, potentially inducing the differentiation of bone-forming cells (Laforest et al., 1998). Many of these morphogenetic processes are reactivated during fin regeneration (Iovine, 2007), and studying the normal metamorphic processes can lend insight into regenerative processes. Interestingly, fin regeneration in zebrafish is accompanied by enhanced expression of DIO3 and regenerative progress is retarded when DIO3 activity is blocked pharmacologically (Bouzaffour, Rampon, Ramaugé, Courtin, & Vriz, 2010), suggesting that local T3 degradation promotes regeneration. By contrast, growth cone formation of the regenerating lateral line is stimulated by TH (Bouzaffour et al., 2010). Remaining unknown is the extent to which hormonal mediators associated with normal metamorphosis are required during regeneration of other tissues and how precisely these contribute to particular cellular behaviors. The roles of these factors during normal fin development remain largely unknown, but will be interesting to uncover, particularly in light of the diversity of fin morphologies among teleosts more generally.

In summary, larval metamorphosis in teleosts involves coordinated morphological changes to many different organs; we now understand only a few of the genetic requirements for these transitions. The mechanisms by which these cascades are hormonally activated and coordinated remain largely unknown. Particularly, considering the growing number of molecular resources for nonmodel organisms (Sarropoulou, Nousdili, Magoulas, & Kotoulas, 2008; Volff, 2005), the genetics of metamorphic events and transitions in teleosts represents a promising field for future study.

5. CONCLUSIONS

Teleosts undergo a spectacular diversity of postembryonic metamorphic transitions, but most of these transitions have been described only superficially. Even in the most familiar teleosts, the proximate endocrine and genetic mechanisms for different elements of metamorphosis have yet to be described. Nonetheless, this period of profound remodeling encompasses many inherently interesting developmental phenomena involving fundamental processes of morphogenesis and differentiation. Moreover, this period of coordinated tissue modification offers the opportunity to examine differential tissue responses to global endocrine signals. Genetic modifications to these postembryonic processes have contributed to the spectacular diversity in adult morphology seen within teleost lineages, and elucidating their attendant molecular mechanisms will lend considerable additional insight into this diversification (Harris, 2012). Although the weight of evidence supports roles for TH signaling in promoting metamorphosis in teleosts, roles for other factors remain likely but uncertain, and the precise mechanisms by which TH effects particular morphogenetic or physiological outcomes remain largely unknown. Given the diversity of teleost metamorphoses and the particular tractability of some species (such as zebrafish) for analyses of developmental, genetic, and endocrine mechanisms, it seems certain that further studies of teleosts will provide important insights into the evolution of metamorphosis and, more generally, the diversification of physiology, behavior, and form.

GLOSSARY

Adult Reproductively mature form of an organism, possessing all adult organs and mature gonads. Adults are not always in breeding condition as many species are seasonal spawners.

Amphidromous Fish that migrate back and forth between fresh and salt water habitats, not necessarily for the purposes of breeding.

Anadromous Fish that spawn in fresh water, then migrate to marine habitats.

Benthic Associated with the lower layers of the water column, just above the substrate.

Catadromous Fish that spawn in marine environments, then migrate to fresh water habitats.

Demersal Associated with the very bottom of the water column, close to or in physical contact with the substrate.

Diadromous Fish that migrate between salt water and fresh water environments (catadromous and anadromous).

Embryo Developmental stage characterized by yolk sac and chorion. Non-feeding; acquires nutrients exclusively from yolk. Sometimes called *alevin*, *yolk sac fry*, or *sac fry*.

Juvenile Developmental stage at which the majority of adult organs have developed and organism is proportionally similar to an adult, but is not yet sexually mature. Salmonid juveniles are called *smolts*; eel juveniles are called *elvers* and *yellow eels*.

Larva Developmental stage characterized by free swimming and active feeding. Larvae are able to obtain and digest exogenous food and are often transparent or lightly pigmented. Generally show substantial morphological differences from adults. Larvae may also be called *fry*, salmonid larvae may be called *fingerlings* or *parr*, eel larvae are called *leptocephali*.

Metamorphosis An irreversible developmental and physiological change that affects multiple traits during postembryonic development and is brought about by one or more systematically acting endocrine mediators, but is independent of sexual maturation, sex-specific modifications, or senescence. May occur multiple times throughout during a life cycle, and may be simultaneous with puberty. The most common type of metamorphosis is larval metamorphosis, during which time a teleost transforms from larva to juvenile.

Pelagic Associated with the upper layers of the water column and open water. Many marine organisms have a pelagic open ocean larval stage of life.

Puberty The process of transforming from juvenile to adult, involving the maturation of the gonads and reproductive organs. Controlled by sex hormones in the gonadotropic axis.

REFERENCES

Ágústsson, T., Sundell, K., Sakamoto, T., Johansson, V., Ando, M., & Bjornsson, B. T. (2001). Growth hormone endocrinology of Atlantic salmon (*Salmo salar*): Pituitary gene expression, hormone storage, secretion and plasma levels during parr–smolt transformation. *The Journal of Endocrinology, 170*, 227–234.

Alt, B., Reibe, S., Feitosa, N. M., Elsalini, O. A., Wendl, T., & Rohr, K. B. (2006). Analysis of origin and growth of the thyroid gland in zebrafish. *Developmental Dynamics, 235*, 1872–1883.

Andersen, Ø., Dahle, S. W., van Nes, S., Bardal, T., Tooming-Klunderud, A., Kjørsvik, E., et al. (2009). Differential spatio-temporal expression and functional diversification of the myogenic regulatory factors MyoD1 and MyoD2 in Atlantic halibut (*Hippoglossus hippoglossus*). *Comparative Biochemistry and Physiology. Part B, Biochemistry & Molecular Biology, 154*, 93–101.

Arai, T., Otake, T., & Tsukamoto, K. (1997). Drastic changes in otolith microstructure and microchemistry accompanying the onset of metamorphosis in the Japanese eel *Anguilla japonica*. *Marine Ecology Progress Series, 161*, 17–22.

Aroua, S., Schmitz, M., Baloche, S., Vidal, B., Rousseau, K., & Dufour, S. (2005). Endocrine evidence that silvering, a secondary metamorphosis in the eel, is a pubertal rather than a metamorphic event. *Neuroendocrinology, 82*, 221–232.

Baggerman, B. (1960). Salinity preference, thyroid activity and the seaward migration of four species of Pacific salmon (*Oncorhynchus*). *Journal of the Fisheries Research Board of Canada, 17*, 295–322.

Balon, E. K. (1990). Epigenesis of an epigeneticist: The development of some alternative concepts on the early ontogeny and evolution of fishes. *Guelph Ichthyology Reviews, 1*, 1–48.

Balon, E. K. (1999). Alternative ways to become a juvenile or a definitive phenotype (and on some persisting linguistic offenses). *Environmental Biology of Fishes, 56*, 17–38.

Bao, B., Ke, Z., Xing, J., Peatman, E., Liu, Z., Xie, C., et al. (2011). Proliferating cells in suborbital tissue drive eye migration in flatfish. *Developmental Biology, 351*, 200–207.

Bird, N. C., & Mabee, P. M. (2003). Developmental morphology of the axial skeleton of the zebrafish, *Danio rerio* (Ostariophysi: Cyprinidae). *Developmental Dynamics, 228*, 337–357.

Björnsson, B. T., Einarsdottir, I. E., & Power, D. (2012). Is salmon smoltification an example of vertebrate metamorphosis? Lessons learnt from work on flatfish larval development. *Aquaculture, 28*, 264–272.

Björnsson, B. T., Stefansson, S. O., & McCormick, S. D. (2011). Environmental endocrinology of salmon smoltification. *General and Comparative Endocrinology, 170*, 290–298.

Boeuf, G. (1993). Salmonid smolting: A pre-adaptation to the oceanic environment. In J. C. Rankin & F. B. Jensen (Eds.), *Fish ecophysiology*, Vol. 9, (pp. 105–125). London: Chapman & Hall.

Boeuf, G., Le Bail, P., & Prunet, P. (1989). Growth hormone and thyroid hormones during Atlantic salmon, *Salmo salar* L., smolting, and after transfer to seawater. *Aquaculture, 82*, 257–268.

Bouzaffour, M., Rampon, C., Ramaugé, M., Courtin, F., & Vriz, S. (2010). Implication of type 3 deiodinase induction in zebrafish fin regeneration. *General and Comparative Endocrinology, 168*, 88–94.

Britz, R., Conway, K. W., & Rüber, L. (2009). Spectacular morphological novelty in a miniature cyprinid fish, *Danionella dracula* n. sp. *Proceedings of the Royal Society B: Biological Sciences, 276*, 2179.

Brown, C. L. (1989). Thyroid hormones in early development, with special reference to teleost fishes. In M. P. Schreibman & C. G. Scanes (Eds.), *Hormones in development, maturation and senescence of neuroendocrine systems* (pp. 289–306). San Diego: Academic Press.

Brown, D. D. (1997). The role of thyroid hormone in zebrafish and axolotl development. *Proceedings of the National Academy of Sciences of United States of America, 94*, 13011–13016.

Budi, E. H., Patterson, L. B., & Parichy, D. M. (2008). Embryonic requirements for ErbB signaling in neural crest development and adult pigment pattern formation. *Development (Cambridge, England), 135*, 2603–2614.

Budi, E. H., Patterson, L. B., & Parichy, D. M. (2011). Post-embryonic nerve-associated precursors to adult pigment cells: Genetic requirements and dynamics of morphogenesis and differentiation. *PLoS Genetics, 7*, e1002044.

Campinho, M. A., Galay-Burgos, M., Sweeney, G. E., & Power, D. M. (2010). Coordination of deiodinase and thyroid hormone receptor expression during the larval to juvenile transition in sea bream (*Sparus aurata*, Linnaeus). *General and Comparative Endocrinology, 165*, 181–194.

Campinho, M. A., Silva, N., Sweeney, G. E., & Power, D. M. (2007). Molecular, cellular and histological changes in skin from a larval to an adult phenotype during bony fish metamorphosis. *Cell and Tissue Research, 327*, 267–284.

Chang, W. J., & Hwang, P. P. (2011). Development of zebrafish epidermis. *Birth Defects Research. Part C, Embryo Today: Reviews, 93*, 205–214.

Chang, J., Wang, M., Gui, W., Zhao, Y., Yu, L., & Zhu, G. (2012). Changes in thyroid hormone levels during zebrafish development. *Zoological Science, 29*, 181–184.

Chen, Y., Zhang, Q., Qi, J., Wang, Z., Wang, X., Sun, Y., et al. (2010). Cloning and stage-specific expression of CK-M1 gene during metamorphosis of Japanese flounder, *Paralichthys olivaceus. Chinese Journal of Oceanology and Limnology, 28*, 558–564.

Cieri, M., & McCleave, J. (2000). Discrepancies between otoliths of larvae and juveniles of the American eel: Is something fishy happening at metamorphosis? *Journal of Fish Biology, 57*, 1189–1198.

Crane, H. M., Pickford, D. B., Hutchinson, T. H., & Brown, J. A. (2004). Developmental changes of thyroid hormones in the fathead minnow, *Pimephales promelas. General and Comparative Endocrinology, 139*, 55–60.

Cubbage, C. C., & Mabee, P. M. (1996). Development of the cranium and paired fins in the zebrafish *Danio rerio* (Ostariophysi, Cyprinidae). *Journal of Morphology, 229*, 121–160.
D'Cotta, H., Valotaire, C., le Gac, F., & Prunet, P. (2000). Synthesis of gill Na^+-K^+-ATPase in Atlantic salmon smolts: Differences in α-mRNA and α-protein levels. *American Journal of Physiology. Regulatory, Integrative and Comparative Physiology, 278*, R101–R110.
Deane, E. E., & Woo, N. (2003). Ontogeny of thyroid hormones, cortisol, hsp70 and hsp90 during silver sea bream larval development. *Life Sciences, 72*, 805–818.
Dêbowski, P., Glogowski, J., Robak, S., & Dobosz, S. (1999). Smoltification of hatchery-reared Atlantic salmon (*Salmo salar* L.)—Indices and methods of estimation. *Archives of Polish Fisheries/Archiwum Rybactwa Polskiego, 7*, 267–279.
de Jesus, E. G., Hirano, T., & Inui, Y. (1991). Changes in cortisol and thyroid hormone concentrations during early development and metamorphosis in the Japanese flounder, *Paralichthys olivaceus*. *General and Comparative Endocrinology, 82*, 369–376.
de Jesus, E. G., Hirano, T., & Inui, Y. (1993). Flounder metamorphosis: Its regulation by various hormones. *Fish Physiology and Biochemistry, 11*, 323–328.
de Jesus, E. G., Inui, Y., & Hirano, T. (1990). Cortisol enhances the stimulating action of thyroid hormones on dorsal fin-ray resorption of flounder larvae in vitro. *General and Comparative Endocrinology, 79*, 167–173.
de Jesus, E. G. T., Toledo, J. D., & Simpas, M. S. (1998). Thyroid hormones promote early metamorphosis in grouper (*Epinephelus coioides*) larvae. *General and Comparative Endocrinology, 112*, 10–16.
De Silva, S., Gunasekera, R., Gooley, G., & Ingram, B. (2001). Growth of Australian shortfin eel (*Anguilla australis*) elvers given different dietary protein and lipid levels. *Aquaculture Nutrition, 7*, 53–57.
Denver, R. J. (1993). Acceleration of anuran amphibian metamorphosis by corticotropin-releasing hormone-like peptides. *General and Comparative Endocrinology, 91*, 38–51.
Denver, R. J. (1997). Environmental stress as a developmental cue: Corticotropin-releasing hormone is a proximate mediator of adaptive phenotypic plasticity in amphibian metamorphosis. *Hormones and Behavior, 31*, 169–179.
Dickhoff, W. W., Brown, C. L., Sullivan, C. V., & Bern, H. A. (1990). Fish and amphibian models for developmental endocrinology. *The Journal of Experimental Zoology, 256*, 90–97.
Dickhoff, W. W., Folmar, L. C., & Gorbman, A. (1978). Changes in plasma thyroxine during smoltification of coho salmon, *Oncorhynchus kisutch*. *General and Comparative Endocrinology, 36*, 229–232.
Dickhoff, W. W., Folmar, L. C., Mighell, J. L., & Mahnken, C. V. W. (1982). Plasma thyroid hormones during smoltification of yearling and underyearling coho salmon and yearling chinook salmon and steelhead trout. *Aquaculture, 28*, 39–48.
Donaldson, E., Fagerlund, U., Higgs, D., & McBride, J. (1979). Hormonal enhancement of growth. In W. S. Hoar, D. J. Randall & R. J. Brett (Eds.), *Fish physiology*, Vol. 8. New York: Academic Press.
Dufour, S., & Rousseau, K. (2007). Neuroendocrinology of fish metamorphosis and puberty: Evolutionary and ecophysiological perspectives. *Journal of Marine Science and Technology, 68*, 55–68.
Elizondo, M. R., Arduini, B. L., Paulsen, J., MacDonald, E. L., Sabel, J. L., Henion, P. D., et al. (2005). Defective skeletogenesis with kidney stone formation in dwarf zebrafish mutant for *trpm7*. *Current Biology, 15*, 667–671.
Eom, D. S., Inoue, S., Patterson, L. B., Gordon, T. N., Slingwine, R., Kondo, S., et al. (2012). Melanophore migration and survival during zebrafish adult pigment stripe development require the immunoglobulin superfamily adhesion molecule Igsf11. *PLoS Genetics, 8*, e1002899.

Fernández-Díaz, C., Y´yfera, M., Cañavate, J., Moyano, F., Alarcón, F., & Díaz, M. (2001). Growth and physiological changes during metamorphosis of Senegal sole reared in the laboratory. *Journal of Fish Biology, 58*, 1086–1097.
Fukuhara, O. (1991). Size and age at transformation in red sea bream, *Pagrus major*, reared in the laboratory. *Aquaculture, 95*, 117–124.
Fukuhara, O., & Fushimi, T. (1988). Fin differentiation and squamation of artificially reared grouper, *Epinephelus akaara*. *Aquaculture, 69*, 379–386.
Galay-Burgos, M., Power, D. M., Llewellyn, L., & Sweeney, G. E. (2008). Thyroid hormone receptor expression during metamorphosis of Atlantic halibut (*Hippoglossus hippoglossus*). *Molecular and Cellular Endocrinology, 281*, 56–63.
Goldsmith, M., Iovine, M., O'Reilly-Pol, T., & Johnson, S. (2006). A developmental transition in growth control during zebrafish caudal fin development. *Developmental Biology, 296*, 450–457.
Grau, E. G., Specker, J. L., Nishioka, R. S., & Bern, H. A. (1982). Factors determining the occurrence of the surge in thyroid activity in salmon during smoltification. *Aquaculture, 28*, 49–57.
Harada, M., Yoshinaga, T., Ojima, D., & Iwata, M. (2008). cDNA cloning and expression analysis of thyroid hormone receptor in the coho salmon *Oncorhynchus kisutch* during smoltification. *General and Comparative Endocrinology, 155*, 658–667.
Harris, M. (2012). Comparative genetics of postembryonic development as a means to understand evolutionary change. *Journal of Applied Ichthyology, 28*, 306–315.
Harris, M. P., Rohner, N., Schwarz, H., Perathoner, S., Konstantinidis, P., & Nüsslein-Volhard, C. (2008). Zebrafish *eda* and *edar* mutants reveal conserved and ancestral roles of ectodysplasin signaling in vertebrates. *PLoS Genetics, 4*, e1000206.
Hashimoto, H., Mizuta, A., Okada, N., Suzuki, T., Tagawa, M., Tabata, K., et al. (2002). Isolation and characterization of a Japanese flounder clonal line, *reversed*, which exhibits reversal of metamorphic left-right asymmetry. *Mechanisms of Development, 111*, 17–24.
Hawkes, J. W. (1974). The structure of fish skin. *Cell and Tissue Research, 149*, 159–172.
Hildahl, J., Power, D. M., Björnsson, B. T., & Einarsdóttir, I. E. (2008). Involvement of growth hormone-insulin-like growth factor I system in cranial remodeling during halibut metamorphosis as indicated by tissue-and stage-specific receptor gene expression and the presence of growth hormone receptor protein. *Cell and Tissue Research, 332*, 211–225.
Hildahl, J., Sweeney, G., Galay-Burgos, M., Einarsdóttir, I. E., & Björnsson, B. T. (2007). Cloning of Atlantic halibut growth hormone receptor genes and quantitative gene expression during metamorphosis. *General and Comparative Endocrinology, 151*, 143–152.
Hoath, S. B., & Maibach, H. I. (2003). *Neonatal skin: Structure and function*. New York, NY: CRC Press.
Hoegg, S., Brinkmann, H., Taylor, J. S., & Meyer, A. (2004). Phylogenetic timing of the fish-specific genome duplication correlates with the diversification of teleost fish. *Journal of Molecular Evolution, 59*, 190–203.
Hultman, K. A., Budi, E. H., Teasley, D. C., Gottlieb, A. Y., Parichy, D. M., & Johnson, S. L. (2009). Defects in ErbB-dependent establishment of adult melanocyte stem cells reveal independent origins for embryonic and regeneration melanocytes. *PLoS Genetics, 5*, e1000544.
Hussain, N. A., & Higuchi, M. (1980). Larval rearing and development of the brown spotted grouper, *Epinephelus tauvina* (Forskål). *Aquaculture, 19*, 339–350.
Infante, C., Asensio, E., Cañavate, J. P., & Manchado, M. (2008). Molecular characterization and expression analysis of five different elongation factor 1 alpha genes in the flatfish Senegalese sole (*Solea senegalensis* Kaup): Differential gene expression and thyroid hormones dependence during metamorphosis. *BMC Molecular Biology, 9*, 19.
Infante, C., Manchado, M., Asensio, E., & Cañavate, J. P. (2007). Molecular characterization, gene expression and dependence on thyroid hormones of two type I keratin genes

(*sseKer1* and *sseKer2*) in the flatfish Senegalese sole (*Solea senegalensis* Kaup). *BMC Developmental Biology*, *7*, 118.
Inui, Y., & Miwa, S. (1985). Thyroid hormone induces metamorphosis of flounder larvae. *General and Comparative Endocrinology*, *60*, 450–454.
Inui, Y., Tagawa, M., Miwa, S., & Hirano, T. (1989). Effects of bovine TSH on the tissue thyroxine level and metamorphosis in prometamorphic flounder larvae. *General and Comparative Endocrinology*, *74*, 406–410.
Iovine, M. K. (2007). Conserved mechanisms regulate outgrowth in zebrafish fins. *Nature Chemical Biology*, *3*, 613–618.
Itoh, K., Watanabe, K., Wu, X., & Suzuki, T. (2010). Three members of the iodothyronine deiodinase family, *dio1, dio2* and *dio3*, are expressed in spatially and temporally specific patterns during metamorphosis of the flounder, *Paralichthys olivaceus*. *Zoological Science*, *27*, 574–580.
Iwamatsu, T. (1994). Stages of normal development in the medaka *Oryzias latipes*. *Zoological Science*, *11*, 825–839.
Iwashita, M., Watanabe, M., Ishii, M., Chen, T., Johnson, S. L., Kurachi, Y., et al. (2006). Pigment pattern in *jaguar/obelix* zebrafish is caused by a Kir7.1 mutation: Implications for the regulation of melanosome movement. *PLoS Genetics*, *2*, e197.
Jegstrup, I., & Rosenkilde, P. (2003). Regulation of post-larval development in the European eel: Thyroid hormone level, progress of pigmentation and changes in behaviour. *Journal of Fish Biology*, *63*, 168–175.
Johnson, S. L., Africa, D., Walker, C., & Weston, J. A. (1995). Genetic control of adult pigment stripe development in zebrafish. *Developmental Biology*, *167*, 27–33.
Johnston, C., & Saunders, R. (1981). Parr-smolt transformation of yearling Atlantic salmon (*Salmo salar*) at several rearing temperatures. *Canadian Journal of Fisheries and Aquatic Sciences*, *38*, 1189–1198.
Just, J. J., Kraus-Just, J., & Check, D. A. (1981). Survey of chordate metamorphosis. In L. I. Gilbert & E. Frieden (Eds.), *Metamorphosis: A problem in developmental biology*, Vol. 2, (pp. 265–326). New York: Plenum.
Kaji, T., Tanaka, M., Oka, M., Takeuchi, H., Ohsumi, S., Teruya, K., et al. (1999). Growth and morphological development of laboratory-reared yellowfin tuna *Thunnus albacares* larvae and early juveniles, with special emphasis on the digestive system. *Fisheries Science*, *65*, 700–707.
Kaji, T., Tanaka, M., Takahashi, Y., Oka, M., & Ishibashi, N. (1996). Preliminary observations on development of Pacific bluefin tuna *Thunnus thynnus* (Scombridae) larvae reared in the laboratory, with special reference to the digestive system. *Marine and Freshwater Research*, *47*, 261–269.
Kawakami, Y., Nozaki, J., Seoka, M., Kumai, H., & Ohta, H. (2008). Characterization of thyroid hormones and thyroid hormone receptors during the early development of Pacific bluefin tuna (*Thunnus orientalis*). *General and Comparative Endocrinology*, *155*, 597–606.
Kawakami, Y., Tanda, M., Adachi, S., & Yamauchi, K. (2003). Characterization of thyroid hormone receptor α and β in the metamorphosing Japanese conger eel, *Conger myriaster*. *General and Comparative Endocrinology*, *132*, 321–332.
Kawakami, Y., Yokoi, K., Kumai, H., & Ohta, H. (2008). The role of thyroid hormones during the development of eye pigmentation in the Pacific bluefin tuna (*Thunnus orientalis*). *Comparative Biochemistry and Physiology. Part B, Biochemistry & Molecular Biology*, *150*, 112–116.
Keefe, M., & Able, K. (1993). Patterns of metamorphosis in summer flounder, *Paralichthys dentatus*. *Journal of Fish Biology*, *42*, 713–728.
Keith, P., Hoareau, T., Lord, C., Ah-Yane, O., Gimonneau, G., Robinet, T., et al. (2008). Characterisation of post-larval to juvenile stages, metamorphosis and recruitment of an

amphidromous goby, *Sicyopterus lagocephalus* (Pallas) (*Teleostei: Gobiidae: Sicydiinae*). *Marine and Freshwater Research, 59*, 876–889.

Kelsh, R. N. (2004). Genetics and evolution of pigment patterns in fish. *Pigment Cell Research, 17*, 326–336.

Kelsh, R. N., Inoue, C., Momoi, A., Kondoh, H., Furutani-Seiki, M., Ozato, K., et al. (2004). The Tomita collection of medaka pigmentation mutants as a resource for understanding neural crest cell development. *Mechanisms of Development, 121*, 841–859.

Kitajima, C., Sato, T., & Kawanishi, M. (1967). On the effect of thyroxine to promote the metamorphosis of a conger eel-preliminary report. *Bulletin of the Japanese Society of Scientific Fisheries, 33*, 919–922.

Kitano, J., Lema, S. C., Luckenbach, J. A., Mori, S., Kawagishi, Y., Kusakabe, M., et al. (2010). Adaptive divergence in the thyroid hormone signaling pathway in the stickleback radiation. *Current Biology, 20*, 2124–2130.

Klaren, P. H. M., Wunderink, Y. S., Yúfera, M., Mancera, J. M., & Flik, G. (2008). The thyroid gland and thyroid hormones in Senegalese sole (*Solea senegalensis*) during early development and metamorphosis. *General and Comparative Endocrinology, 155*, 686–694.

Laforest, L., Brown, C. W., Poleo, G., Geraudie, J., Tada, M., Ekker, M., et al. (1998). Involvement of the *sonic hedgehog*, *patched 1* and *bmp2* genes in patterning of the zebrafish dermal fin rays. *Development (Cambridge, England), 125*, 4175–4184.

Lam, T., Sharma, R. (1985). Effects of salinity and thyroxine on larval survival, growth and development in the carp, *Cyprinus carpio*. *Aquaculture. 44,* 201–212.

Lang, M. R., Patterson, L. B., Gordon, T. N., Johnson, S. L., & Parichy, D. M. (2009). Basonuclin-2 requirements for zebrafish adult pigment pattern development and female fertility. *PLoS Genetics, 5*, e1000744.

Larson, T. A., Gordon, T. N., Lau, H. E., & Parichy, D. M. (2010). Defective adult oligodendrocyte and Schwann cell development, pigment pattern, and craniofacial morphology in puma mutant zebrafish having an alpha tubulin mutation. *Developmental Biology, 346*, 296–309.

Laudet, V. (2011). The origins and evolution of vertebrate metamorphosis. *Current Biology, 21*, R726–R737.

Le Guellec, D., Morvan-Dubois, G., & Sire, J. Y. (2004). Skin development in bony fish with particular emphasis on collagen deposition in the dermis of the zebrafish (*Danio rerio*). *International Journal of Developmental Biology, 48*, 217–232.

Ledent, V. (2002). Postembryonic development of the posterior lateral line in zebrafish. *Development (Cambridge, England), 129*, 597–604.

Leis, J. M., & McCormick, M. I. (2006). The biology, behavior and ecology of the pelagic, larval stage of coral reef fishes. In P. F. Sale (Ed.), *Coral reef fishes: Dynamics and diversity in a complex ecosystem* (pp. 171–200). Burlington, MA: Elsevier, Inc.

Lin, R. J., Cross, T. F., Mills, C. P. R., Nishioka, R. S., Grau, E. G., & Bern, H. A. (1988). Changes in plasma thyroxine levels during smoltification in hatchery-reared one-year and two-year Atlantic salmon (*Salmo salar*). *Aquaculture, 74*, 369–378.

Madsen, S. S., & Bern, H. A. (1992). Antagonism of prolactin and growth hormone—impact on seawater adaptation in two salmonids, *Salmotrutta* and *Oncorhynchus mykiss*. *Zoological Science, 9*, 775–784.

Manchado, M., Infante, C., Asensio, E., Crespo, A., Zuasti, E., & Cañavate, J. P. (2008). Molecular characterization and gene expression of six trypsinogens in the flatfish Senegalese sole (*Solea senegalensis* Kaup) during larval development and in tissues. *Comparative Biochemistry and Physiology. Part B, Biochemistry & Molecular Biology, 149*, 334–344.

Manchado, M., Infante, C., Asensio, E., Planas, J. V., & Cañavate, J. P. (2008). Thyroid hormones down-regulate thyrotropin β subunit and thyroglobulin during metamorphosis in the flatfish Senegalese sole (*Solea senegalensis* Kaup). *General and Comparative Endocrinology, 155*, 447–455.

Manchado, M., Infante, C., Rebordinos, L., & Cañavate, J. P. (2009). Molecular characterization, gene expression and transcriptional regulation of thyroid hormone receptors in Senegalese sole. *General and Comparative Endocrinology, 160*, 139–147.

Manchado, M., Salas-Leiton, E., Infante, C., Ponce, M., Asensio, E., Crespo, A., et al. (2008). Molecular characterization, gene expression and transcriptional regulation of cytosolic HSP90 genes in the flatfish Senegalese sole (*Solea senegalensis* Kaup). *Gene, 416*, 77–84.

Manzon, R. G. (2011). Thyroidal regulation of life history transitions in fish. In T. Flatt & A. Heyland (Eds.), *Mechanisms of life history evolution: The genetics and physiology of life history traits and trade-offs* (pp. 72–86). Oxford: OUP.

Marchand, O., Duffraisse, M., Triqueneaux, G., Safi, R., & Laudet, V. (2004). Molecular cloning and developmental expression patterns of thyroid hormone receptors and T3 target genes in the turbot (*Scophtalmus maximus*) during post-embryonic development. *General and Comparative Endocrinology, 135*, 345–357.

Marí-Beffa, M., & Murciano, C. (2010). Dermoskeleton morphogenesis in zebrafish fins. *Developmental Dynamics, 239*, 2779–2794.

Martinez, G. M., & Bolker, J. A. (2003). Embryonic and larval staging of summer flounder (*Paralichthys dentatus*). *Journal of Morphology, 255*, 162–176.

Martínez-Lagos, R., & Gracia-López, V. (2009). Morphological development and growth patterns of the leopard grouper *Mycteroperca rosacea* during larval development. *Aquaculture Research, 41*, 120–128.

Matsumoto, J., Lynch, T. J., Grabowski, S., Richards, C. M., Lo, S. L., Clark, C., et al. (1983). Fish tumor pigment cells: Differentiation and comparison to their normal counterparts. *American Zoologist, 23*, 569–580.

Matsumoto, S., & Tanaka, M. (1996). The influence of thyroid hormone on development and settlement of spottybelly greenling (*Hexagrammos agrammus*; Pisces). *Marine and Freshwater Behaviour and Physiology, 28*, 33–44.

Mayden, R. L., & Chen, W. J. (2010). The world's smallest vertebrate species of the genus *Paedocypris*: A new family of freshwater fishes and the sister group to the world's most diverse clade of freshwater fishes (Teleostei: Cypriniformes). *Molecular Phylogenetics and Evolution, 57*, 152–175.

McCormick, M., & Makey, L. (1997). Post-settlement transition in coral reef fishes: Overlooked complexity in niche shifts. *Marine Ecology Progress Series, 153*, 247–257.

McCormick, M., Makey, L., & Dufour, V. (2002). Comparative study of metamorphosis in tropical reef fishes. *Marine Biology, 141*, 841–853.

McCormick, S. D., Moriyama, S., & Björnsson, B. T. (2000). Low temperature limits photoperiod control of smolting in Atlantic salmon through endocrine mechanisms. *The American Journal of Physiology, 278*, R1352–R1361.

McCormick, S. D., & Saunders, R. L. (1987). Preparatory physiological adaptations for marine life of salmonids: Osmoregulation, growth, and metabolism. *American Fisheries Society Symposium, 1*, 211–229.

McCormick, S. D., Shrimpton, J. M., Moriyama, S., & Björnsson, B. T. (2002). Effects of an advanced temperature cycle on smolt development and endocrinology indicate that temperature is not a zeitgeber for smolting in Atlantic salmon. *The Journal of Experimental Biology, 205*, 3553–3560.

Miller, C. T., Beleza, S., Pollen, A. A., Schluter, D., Kittles, R. A., Shriver, M. D., et al. (2007). cis-Regulatory changes in *Kit ligand* expression and parallel evolution of pigmentation in sticklebacks and humans. *Cell, 131*, 1179–1189.

Miwa, S., & Inui, Y. (1991). Thyroid hormone stimulates the shift of erythrocyte populations during metamorphosis of the flounder. *The Journal of Experimental Zoology, 259*, 222–228.

Miwa, S., Tagawa, M., Inui, Y., & Hirano, T. (1988). Thyroxine surge in metamorphosing flounder larvae. *General and Comparative Endocrinology, 70*, 158–163.

Miwa, S., Yamano, K., & Inui, Y. (1992). Thyroid hormone stimulates gastric development in flounder larvae during metamorphosis. *The Journal of Experimental Zoology, 261*, 424–430.

Miyashita, S., Sawada, Y., Okada, T., Murata, O., & Kumai, H. (2001). Morphological development and growth of laboratory-reared larval and juvenile *Thunnus thynnus* (Pisces: Scombridae). *Fishery Bulletin, 99*, 601–616.

Moser, H. G. (1981). Morphological and functional aspects of marine fish larvae. In R. Lasker (Ed.), *Marine fish larvae, Washington Sea Grant Program* (pp. 89–131). Seattle: University Washington Press.

Nuñez, V. A., Sarrazin, A. F., Cubedo, N., Allende, M. L., Dambly-Chaudiére, C., & Ghysen, A. (2009). Postembryonic development of the posterior lateral line in the zebrafish. *Evolution & Development, 11*, 391–404.

Odenthal, J., Rossnagel, K., Haffter, P., Kelsh, R. N., Vogelsang, E., Brand, M., et al. (1996). Mutations affecting xanthophore pigmentation in the zebrafish, *Danio rerio*. *Development (Cambridge, England), 123*, 391–398.

Okimoto, K., Weber, M., & Grau, G. (1993). The effects of thyroxine and propylthiouracil treatment on changes in body form associated with a possible developmental thyroxine surge during post-hatching development of the tilapia, *Oreochromis mossambicus*. *Zoological Science, 10*, 803–811.

Orozco, A., & Valverde-R, C. (2005). Thyroid hormone deiodination in fish. *Thyroid, 15*, 799–813.

Page, R. B., Monaghan, J. R., Samuels, A. K., Smith, J. J., Beachy, C. K., & Voss, S. R. (2007). Microarray analysis identifies keratin loci as sensitive biomarkers for thyroid hormone disruption in the salamander *Ambystoma mexicanum*. *Comparative Biochemistry and Physiology Part C: Toxicology & Pharmacology, 145*, 15–27.

Parhar, I. S., & Iwata, M. (1996). Intracerebral expression of gonadotropin-releasing hormone and growth hormone-releasing hormone is delayed until smoltification in the salmon. *Neuroscience Research, 26*, 299–308.

Parichy, D. M. (2006). Evolution of danio pigment pattern development. *Heredity, 97*, 200–210.

Parichy, D. M., Elizondo, M. R., Mills, M. G., Gordon, T. N., & Engeszer, R. E. (2009). Normal table of postembryonic zebrafish development: Staging by externally visible anatomy of the living fish. *Developmental Dynamics, 238*, 2975–3015.

Parichy, D. M., Rawls, J. F., Pratt, S. J., Whitfield, T. T., & Johnson, S. L. (1999). Zebrafish *sparse* corresponds to an orthologue of c-kit and is required for the morphogenesis of a subpopulation of melanocytes, but is not essential for hematopoiesis or primordial germ cell development. *Development (Cambridge, England), 126*, 3425–3436.

Parichy, D. M., & Turner, J. M. (2003a). Temporal and cellular requirements for Fms signaling during zebrafish adult pigment pattern development. *Development (Cambridge, England), 130*, 817–833.

Parichy, D. M., & Turner, J. M. (2003b). Zebrafish *puma* mutant decouples pigment pattern and somatic metamorphosis. *Developmental Biology, 256*, 242–257.

Paris, M., Escriva, H., Schubert, M., Brunet, F., Brtko, J., Ciesielski, F., et al. (2008). Amphioxus postembryonic development reveals the homology of chordate metamorphosis. *Current Biology, 18*, 825–830.

Paris, M., & Laudet, V. (2008). The history of a developmental stage: Metamorphosis in chordates. *Genesis (New York, N.Y.: 2000), 46*, 657–672.

Patterson, S. E., Mook, L. B., & Devoto, S. H. (2008). Growth in the larval zebrafish pectoral fin and trunk musculature. *Developmental Dynamics, 237*, 307–315.

Policansky, D., & Sieswerda, P. (1979). Early life history of the starry flounder, *Platichthys stellatus*, reared through metamorphosis in the laboratory. *Transactions of the American Fisheries Society, 108*, 326–327.

Porazzi, P., Calebiro, D., Benato, F., Tiso, N., & Persani, L. (2009). Thyroid gland development and function in the zebrafish model. *Molecular and Cellular Endocrinology, 312*, 14–23.
Power, D., Silva, N., & Campinho, M. A. (2008). Metamorphosis. In R. N. Finn & B. G. Kapoor (Eds.), *Fish larval physiology* (pp. 607–638). Enfield, New Hampshire: Science Publishers.
Prunet, P., Boeuf, G., Bolton, J. P., & Young, G. (1989). Smoltification and seawater adaptation in Atlantic salmon (*Salmo salar*): Plasma prolactin, growth hormone, and thyroid hormones. *General and Comparative Endocrinology, 74*, 355–364.
Quigley, D. T. G., Harvey, M., Hayden, T., Dowling, C., & O'Keane, M. (2006). A comparative study of smoltification in sea trout (*Salmo trutta L.*) and Atlantic salmon (*Salmo salar L.*): Seawater tolerance and thyroid hormone titres. *Biology & Environment: Proceedings of the Royal Irish Academy, 106*, 35–47.
Quigley, I. K., Turner, J. M., Nuckels, R. J., Manuel, J. L., Budi, E. H., MacDonald, E. L., et al. (2004). Pigment pattern evolution by differential deployment of neural crest and post-embryonic melanophore lineages in *Danio* fishes. *Development (Cambridge, England), 131*, 6053–6069.
Raju, S. N. (1974). Distribution, growth and metamorphosis of leptocephali of the garden eels, *Taenioconger* sp. and *Gorgasia* sp. *Copeia, 2*, 494–500.
Rakers, S., Gebert, M., Uppalapati, S., Meyer, W., Maderson, P., Sell, A. F., et al. (2010). 'Fish matters': The relevance of fish skin biology to investigative dermatology. *Experimental Dermatology, 19*, 313–324.
Ramanathan, N., Natarajan, P., & Sukumran, N. (1985). Studies on the induced spawning and larval rearing of a freshwater catfish, *Mystus punctatus* (Jerdon). *Proceedings of the Indian Academy of Sciences, 94*, 389–398.
Ravi, V., & Venkatesh, B. (2008). Rapidly evolving fish genomes and teleost diversity. *Current Opinion in Genetics & Development, 18*, 544–550.
Richman, N. H., de Diaz, S. T., Nishioka, R. S., & Bern, H. A. (1985). Developmental study of coho gill functional morphology and the effects of cortisol. *Aquaculture, 45*, 386–387.
Roberts, R. B., Ser, J. R., & Kocher, T. D. (2009). Sexual conflict resolved by invasion of a novel sex determiner in Lake Malawi cichlid fishes. *Science, 326*, 998–1001.
Robertson, G., McGee, C., Dumbarton, T., Croll, R., & Smith, F. (2007). Development of the swimbladder and its innervation in the zebrafish, *Danio rerio*. *Journal of Morphology, 268*, 967–985.
Rousseau, K., Aroua, S., Schmitz, M., Elie, P., & Dufour, S. (2009). Silvering: Metamorphosis or puberty. In G. van den Thillart, S. Dufour & J. C. Rankin (Eds.), *Spawning migration of the European Eel* (pp. 39–63). Amsterdam: Springer.
Rüber, L., Kottelat, M., Tan, H., Ng, P., & Britz, R. (2007). Evolution of miniaturization and the phylogenetic position of *Paedocypris*, comprising the world's smallest vertebrate. *BMC Evolutionary Biology*, 7, 38.
Ryan, A. K., Blumberg, B., Rodriguez-Esteban, C., Yonei-Tamura, S., Tamura, K., Tsukui, T., et al. (1998). Pitx2 determines left-right asymmetry of internal organs in vertebrates. *Nature, 394*, 545–551.
Sæle, Ø., Solbakken, J., Watanabe, K., Hamre, K., Power, D., & Pittman, K. (2004). Staging of Atlantic halibut (*Hippoglossus hippoglossus* L.) from first feeding through metamorphosis, including cranial ossification independent of eye migration. *Aquaculture, 239*, 445–465.
Sarropoulou, E., Nousdili, D., Magoulas, A., & Kotoulas, G. (2008). Linking the genomes of nonmodel teleosts through comparative genomics. *Marine Biotechnology, 10*, 227–233.
Saunders, R., McCormick, S., Henderson, E., Eales, J., & Johnston, C. (1985). The effect of orally administered 3, 5, 3'-triiodo-L-thyronine on growth and salinity tolerance of atlantic salmon (*Salmo salar* L.). *Aquaculture, 45*, 143–156.

Schreiber, A. M. (2001). Metamorphosis and early larval development of the flatfishes (Pleuronectiformes): An osmoregulatory perspective. *Comparative Biochemistry and Physiology. Part B, Biochemistry & Molecular Biology, 129*, 587–595.
Schreiber, A. M., & Specker, J. L. (1998). Metamorphosis in the summer flounder (*Paralichthys dentatus*): Stage-specific developmental response to altered thyroid status. *General and Comparative Endocrinology, 111*, 156–166.
Seear, P. J., Carmichael, S. N., Talbot, R., Taggart, J. B., Bron, J. E., & Sweeney, G. E. (2010). Differential gene expression during smoltification of Atlantic salmon (*Salmo salar* L.): A first large-scale microarray study. *Marine Biotechnology, 12*, 126–140.
Shiao, J. C., Wu, S. M., Hwang, Y. P., Wu, D. P., & Hwang, P. P. (2008). Evaluation of thyroid-mediated otolith growth of larval and juvenile tilapia. *The Journal of Experimental Biology, 211*, 1919–1926.
Sims, K., Eble, D. M., & Iovine, M. K. (2009). Connexin43 regulates joint location in zebrafish fins. *Developmental Biology, 327*, 410–418.
Sire, J. Y., & Akimenko, M. A. (2004). Scale development in fish: A review, with description of sonic hedgehog (shh) expression in the zebrafish (*Danio rerio*). *The International Journal of Developmental Biology, 48*, 233–248.
Sire, J. Y., & Géraudie, J. (1983). Fine structure of the developing scale in the cichlid *Hemichromis bimaculatus* (Pisces, Teleostei, Perciformes). *Acta Zoologica, 64*, 1–8.
Solbakken, J. S., Norberg, B., Watanabe, K., & Pittman, K. (1999). Thyroxine as a mediator of metamorphosis of Atlantic halibut, *Hippoglossus hippoglossus. Environmental Biology of Fishes, 56*, 53–65.
Specker, J. L. (1982). Interrenal function and smoltification. *Aquaculture, 28*, 59–66.
Specker, J. L., & Schreck, C. B. (1982). Changes in plasma corticosteroids during smoltification of coho salmon, *Oncorhynchus kisutch. General and Comparative Endocrinology, 46*, 53–58.
Stefansson, S. O., Björnsson, B. T., Ebbesson, L. O., & McCormick, S. D. (2008). Smoltification. In R. N. Finn & B. G. Kapoor (Eds.), *Fish larval physiology*. Enfield, New Hampshire: Science Publishers.
Sullivan, C. V., Dickhoff, W. W., Mahnken, C. V. W., & Hersbberger, W. K. (1985). Changes in the hemoglobin system of the coho salmon *Oncorhynchus kisutch* during smoltification and triiodothyronine and propylthiouracil treatment. *Comparative Biochemistry and Physiology. Part A, Physiology, 81*, 807–813.
Suzuki, T., Washio, Y., Aritaki, M., Fujinami, Y., Shimizu, D., Uji, S., et al. (2009). Metamorphic *pitx2* expression in the left habenula correlated with lateralization of eye-sidedness in flounder. *Development, Growth & Differentiation, 51*, 797–808.
Sweeting, R., & McKeown, B. (1989). Changes in plasma growth hormone and various metabolic factors during smoltification of coho salmon, *Oncorhynchus kisutch. Aquaculture, 82*, 279–295.
Szisch, V., Papandroulakis, N., Fanouraki, E., & Pavlidis, M. (2005). Ontogeny of the thyroid hormones and cortisol in the gilthead sea bream, *Sparus aurata. General and Comparative Endocrinology, 142*, 186–192.
Tagawa, M., & Aritaki, M. (2005). Production of symmetrical flatfish by controlling the timing of thyroid hormone treatment in spotted halibut *Verasper variegatus. General and Comparative Endocrinology, 141*, 184–189.
Taillebois, L., Keith, P., Valade, P., Torres, P., Baloche, S., Dufour, S., et al. (2011). Involvement of thyroid hormones in the control of larval metamorphosis in *Sicyopterus lagocephalus* (Teleostei: Gobioidei) at the time of river recruitment. *General and Comparative Endocrinology, 173*, 281–288.
Tanaka, M., Kawai, S., Seikai, T., & Burke, J. (1996). Development of the digestive organ system in Japanese flounder in relation to metamorphosis and settlement. *Marine & Freshwater Behaviour & Physiology, 28*, 19–31.

Tanaka, Y., Satoh, K., Iwahashi, M., & Yamada, H. (2006). Growth-dependent recruitment of Pacific bluefin tuna *Thunnus orientalis* in the northwestern Pacific Ocean. *Marine Ecology Progress Series, 319*, 225–235.

Thorson, G. (1950). Reproductive and larval ecology of marine bottom invertebrates. *Biological Reviews, 25*, 1–45.

Tipsmark, C. K., Sørensen, K. J., & Madsen, S. (2010). Aquaporin expression dynamics in osmoregulatory tissues of Atlantic salmon during smoltification and seawater acclimation. *The Journal of Experimental Biology, 213*, 368–379.

Trijundo, D. D., Yoseda, K., Hirokawa, J., Tagawa, M., & Tanaka, M. (2002). Effects of thyroxine and thiourea on the metamorphosis of coral trout grouper *Plectropomus leopardus*. *Fisheries Science, 68*, 282–289.

Tsukamoto, K., Yamada, Y., Okamura, A., Kaneko, T., Tanaka, H., Miller, M. J., et al. (2009). Positive buoyancy in eel leptocephali: An adaptation for life in the ocean surface layer. *Marine Biology, 156*, 835–846.

Virtanen, E. (1987). Correlations between energy metabolism, osmotic balance and external smolt indices in smolting young salmon, *Salmo salar L. Annales Zoologici Fennici, 24*, 71–78.

Virtanen, E., & Soivio, A. (1985). The patterns of T3, T4, cortisol and Na+/K+-ATPase during smoltification of hatchery-reared *Salmo salar* and comparison with wild smolts. *Aquaculture, 45*, 97–109.

Volff, J.-N. (2005). Genome evolution and biodiversity in teleost fish. *Heredity, 94*, 280–294.

Wang, C., & Tzeng, W. (2000). The timing of metamorphosis and growth rates of American and European eel leptocephali: A mechanism of larval segregative migration. *Fisheries Research, 46*, 191–205.

Watanabe, M., Iwashita, M., Ishii, M., Kurachi, Y., Kawakami, A., Kondo, S., et al. (2006). Spot pattern of *leopard* Danio is caused by mutation in the zebrafish *connexin41. 8* gene. *EMBO Reports, 7*, 893–897.

Watanabe, K., Washio, Y., Fujinami, Y., Aritaki, M., Uji, S., & Suzuki, T. (2008). Adult-type pigment cells, which color the ocular sides of flounders at metamorphosis, localize as precursor cells at the proximal parts of the dorsal and anal fins in early larvae. *Development, Growth & Differentiation, 50*, 731–741.

Webb, J. (1999). Larvae in fish development and evolution. In B. K. Hall & M. H. Wake (Eds.), *The origin and evolution of larval forms*. San Diego: Academic Press.

Webb, J., & Shirey, J. (2003). Postembryonic development of the cranial lateral line canals and neuromasts in zebrafish. *Developmental Dynamics, 228*, 370–385.

Winans, G. A., & Nishioka, R. S. (1987). A multivariate description of change in body shape of coho salmon (*Oncorhynchus kisutch*) during smoltification. *Aquaculture, 66*, 235–245.

Wittenrich, M. L., Baldwin, C. C., & Turingen, R. G. (2010). Larval development of laboratory-reared Green Mandarinfish, *Synchiropus splendidus* (Teleostei: Callionymidae). *aqua, International Journal of Ichthyology, 16*, 7–18.

Woo, N., Bern, H., & Nishioka, R. (1978). Changes in body composition associated with smoltification and premature transfer to seawater in coho salmon (*Oncorhynchus kisutch*) and king salmon (*O. tschawytscha*). *Journal of Fish Biology, 13*, 421–428.

Yamada, T., Okauchi, M., & Araki, K. (2010). Origin of adult-type pigment cells forming the asymmetric pigment pattern, in Japanese flounder (*Paralichthys olivaceus*). *Developmental Dynamics, 239*, 3147–3162.

Yamano, K., & Miwa, S. (1998). Differential gene expression of thyroid hormone receptor α and β in fish development. *General and Comparative Endocrinology, 109*, 75–85.

Yamano, K., Miwa, S., Obinata, T., & Inui, Y. (1991). Thyroid hormone regulates developmental changes in muscle during flounder metamorphosis. *General and Comparative Endocrinology, 81*, 464–472.

Yamano, K., Nomura, K., & Tanaka, H. (2007). Development of thyroid gland and changes in thyroid hormone levels in Leptocephali of Japanese Eel (*Anguilla japonica*). *Aquaculture, 270*, 499–504.

Yamano, K., Tagawa, M., Jesus, E. G., Hirano, T., Miwa, S., & Inui, Y. (1991). Changes in whole body concentrations of thyroid hormones and cortisol in metamorphosing conger eel. *Journal of Comparative Physiology B: Biochemical, Systemic, and Environmental Physiology, 161*, 371–375.

Yen, P. M. (2001). Physiological and molecular basis of thyroid hormone action. *Physiological Reviews, 81*, 1097–1142.

Yoo, J. H., Takeuchi, T., Tagawa, M., & Seikai, T. (2000). Effect of thyroid hormones on the stage-specific pigmentation of the Japanese flounder *Paralichthys olivaceus*. *Zoological Science, 17*, 1101–1106.

Young, G., Bjornsson, B. T., Prunet, P., Lin, R. J., & Bern, H. A. (1989). Smoltification and seawater adaptation in coho salmon (*Oncorhynchus kisutch*): Plasma prolactin, growth hormone, thyroid hormones, and cortisol. *General and Comparative Endocrinology, 74*, 335–345.

Youngson, A., McLay, H., Wright, R., & Johnstone, R. (1988). Steroid hormone levels and patterns of growth in the early part of the reproductive cycle of adult Atlantic salmon (*Salmo salar* L.). *Aquaculture, 69*, 145–157.

Yousefian, M., & Shirzad, E. (2011). The review of the effect of growth hormone on immune system, metabolism and osmoregulation of fish. *Australian Journal of Basic and Applied Sciences, 5*, 467–475.

Youson, J. (1988). First metamorphosis. In: W. Hoar & D. Randall (Eds.), *Fish physiology*, Vol. 11B, (pp. 125–196). New York: Academic Press.

Youson, J. H. (1997). Is lamprey metamorphosis regulated by thyroid hormones? *American Zoologist, 37*, 441–460.

Youson, J. H. (2004). The impact of environmental and hormonal cues on the evolution of fish metamorphosis. In B. K. Hall, R. D. Pearson & G. B. Müller (Eds.), *Environment, development, and evolution: Toward a synthesis*. Vol. 1, (pp. 239-278). Cambridge: The MIT Press.

CHAPTER SIX

Flatfish: An Asymmetric Perspective on Metamorphosis

Alexander M. Schreiber[1]

Laboratory of Vertebrate Metamorphosis, St. Lawrence University, Biology Department, Canton, New York, USA

[1]Corresponding author: e-mail address: aschreiber@stlawu.edu

Contents

Abstract

The most asymmetrically shaped and behaviorally lateralized of all the vertebrates, the flatfishes are an endless source of fascination to all fortunate enough to study them. Although all vertebrates undergo left–right asymmetric internal organ placement during embryogenesis, flatfish are unusual in that they experience an additional period of postembryonic asymmetric remodeling during metamorphosis, and thus deviate from a bilaterally symmetrical body plan more than other vertebrates. As with amphibian metamorphosis, all the developmental programs of flatfish metamorphosis are ultimately under the control of thyroid hormone. At least one gene pathway involved in embryonic organ lateralization (*nodal-lefty-pitx2*) is re-expressed in the larval stage during flatfish metamorphosis. Aspects of modern flatfish ontogeny, such as the gradual translocation of one eye to the opposite side of the head and the appearance of key neurocranial elements during metamorphosis, seem to elegantly recapitulate flatfish phylogeny. This chapter highlights the current state of knowledge of the developmental biology of flatfish metamorphosis with emphases on the genetic, morphological, behavioral, and evolutionary origins of flatfish asymmetry.

Current Topics in Developmental Biology, Volume 103
ISSN 0070-2153
http://dx.doi.org/10.1016/B978-0-12-385979-2.00006-X

1. INTRODUCTION

Flatfishes (flounder, sole, halibut, turbot, plaice, tonguefish) are among the most easily recognizable of all groups of fishes, characterized by a distinct laterally compressed body morphology with both eyes located on the same side of the head. The flatfishes are a monophyletic group (Order Pleuronectiformes) within the Percomorpha, consisting of over 700 species with virtually global marine and estuarine distribution, including species that inhabit deep sea thermal vents and fresh water rivers (Friedman, 2012; Munroe, 2005; Tunnicliffe, Koop, Tyler, & So, 2010). Adults can range in size from 2 cm to over 2 m length, and they spend most of their lives as benthic (bottom-dwelling) creatures that lay on one side, and are the most left–right asymmetrically shaped and behaviorally lateralized vertebrates to ever exist. In contrast with adults, larval flatfish resemble typical bilaterally symmetrical fish that swim upright in the water column. Flatfish larvae metamorphose into their juvenile (immature adult) form, a process characterized in part by the migration of one eye to the opposite side of the head and the transition to a lateralized swim posture (Figs. 6.1 and 6.2A–C) (Schreiber, 2006). This transformation is accompanied by the development of various dramatic internal and external asymmetries, including extensive craniofacial remodeling (Fig. 6.2D–H) and development of pigmentation predominantly on one side (Fig. 6.2I and J). As with metamorphosis in amphibians, the developmental programs of flatfish metamorphosis are mediated by thyroid hormone (TH) (Inui & Miwa, 1985; Miwa, Tagawa, Inui, & Hirano, 1988; Power et al., 2008; Schreiber & Specker, 1998) and its nuclear-localized receptors (TRs) (Galay-Burgos, Power, Llewellyn, & Sweeney, 2008; Laudet, 2011; Manchado, Infante, Rebordinos, & Canavate, 2009; Marchand, Duffraisse, Triqueneaux, Safi, & Laudet, 2004; Yamano & Miwa, 1998) (Fig. 6.3). Deiodinases, enzymes that modulate TH activity in peripheral tissues, have also been recently characterized in flatfish (Isorna et al., 2009; Itoh, Watanabe, Wu, & Suzuki, 2010). Although all vertebrates undergo left–right asymmetric internal organ placement during embryogenesis (Hamada, Meno, Watanabe, & Saijoh, 2002; Levin, 2005), flatfish are unusual in that they experience an additional period of postembryonic asymmetric remodeling during metamorphosis, and thus deviate from a bilaterally symmetrical body plan more than other vertebrates. This chapter highlights the current state of knowledge of the developmental biology of flatfish metamorphosis with emphases on the genetic, morphological, behavioral, and evolutionary origins of flatfish asymmetry.

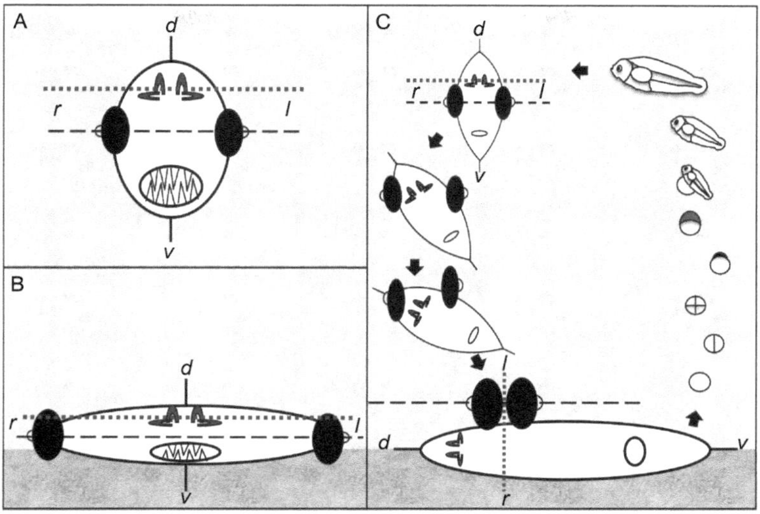

Figure 6.1 The flatfish interocular and interaural axes reorient perpendicular to each other during metamorphosis. The interocular and interaural axes are parallel to each other in typical larval and adult "round" fish (A) and also in fish with dorsoventrally compressed morphology (e.g., skates and rays, Superorder Batoidea; anglerfish, Family Lophiidae) (B). The interocular and interaural axes in laterally compressed premetamorphic larval flatfish (Order Pleuronectiformes) are initially parallel to each other but reorient during metamorphosis as one eye migrates to the opposite side of the head; by the postmetamorphic juvenile stage, the axes have become perpendicular to each other (C). Dashed line denotes interocular axis, and dotted line denotes the interaural axis. Dorsal, d; ventral, v; left, l; right, r. The dorsal-medial-localized curved lines denote the semicircular canals. (For color version of this figure, the reader is referred to the online version of this chapter.)

2. GENETIC BASIS OF FLATFISH ASYMMETRY

Within the currently recognized 15 flatfish families, most species are laterally monomorphic and, as adults, display either exclusively "dextral" (both eyes are located on the right side after metamorphosis) or "sinistral" (both eyes are located on the left side) morphology (Munroe, 2005). However, of the approximately 715 known flatfish species, 7 display varying proportions of dextral to sinistral polymorphism within geographically distinct populations (Bergstrom, 2007; Hubbs & Hubbs, 1944; Munroe, 2005; Policansky, 1982b). For example, in starry flounder (*Platichthys stellatus*), the proportion of sinistrality increases along a geographical gradient from 50% in California to 100% in Japan (Bergstrom, 2007; Hubbs & Kuronuma, 1942),

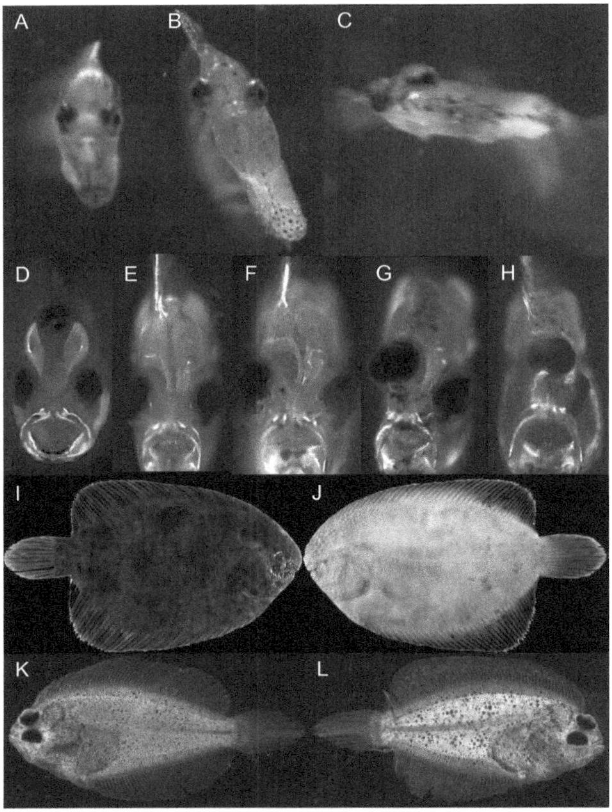

Figure 6.2 Development of lateralized swimming posture and morphological asymmetry during flatfish metamorphosis. Bilaterally symmetrical premetamorphic Southern flounder larvae (*Paralichthys lethostigma*) swim with an upright posture in the light (A). Migration of one eye to the opposite side of the head is accompanied by a transition to a tilted posture during metamorphic climax (B), and by the end of metamorphosis flounder swim on one side (C). Asymmetric remodeling of the skull (visualized *in vivo* by the fluorescent calcium-binding stain, calcein) accommodates eye translocation (D–H) in *P. lethostigma*. After metamorphosis, dextral flatfish, such as this hogchoker (*Trinectes maculatus*), have a pigmented "eyed" (right) side (I) and a nonpigmented "blind" (left) side (J). Some flatfish species, such as *P. lethostigma*, can metamorphose with either sinistral (K) or dextral (L) morphology and behavior. *Figure panels A–H are reproduced with permission from Schreiber (2006).* (See Color Insert.)

and distinct ecological gradients of lateral polymorphism have also been described for the congeneric European flounder (*P. flesus*) (Russo et al., 2012). Though breeding experiments suggest that the direction of asymmetry in populations of polymorphic flatfish is under genetic control (Boklage, 1984; Hashimoto et al., 2002; Policansky, 1982a, 1982b), environmental

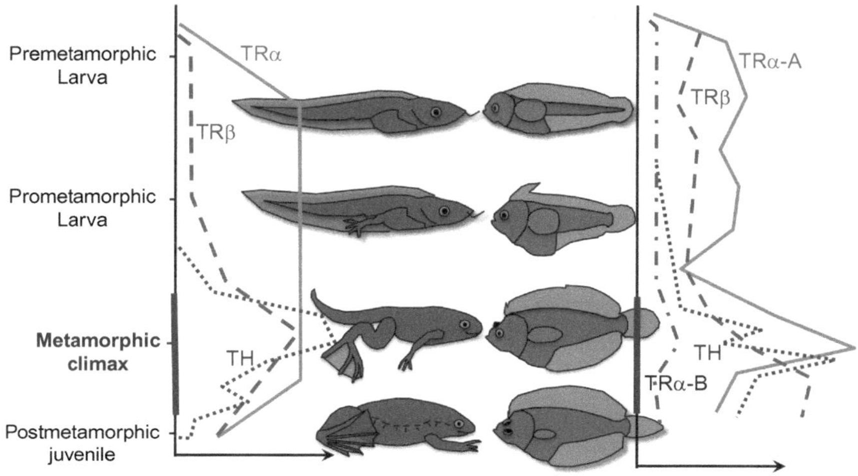

Figure 6.3 A comparison of developmental profiles for thyroid hormone and its receptors in the Japanese flounder (*Paralichthys olivaceus*) and in the African clawed frog (*Xenopus laevis*). Thyroid hormone (TH, dotted lines), thyroid hormone receptor α (TRα/TRα-A, solid lines), TRα-B (alternating dash-dot lines), and TRβ (dashed lines). *Data are derived from Shi (2000) (*Xenopus*) and Yamano and Miwa (1998) (flounder).* (For color version of this figure, the reader is referred to the online version of this chapter.)

variables may influence the relative distribution and selection of dextral and sinistral forms (Bergstrom, 2007; Bergstrom & Palmer, 2007; Russo et al., 2012). Interestingly, not all aspects of morphology and feeding behavior are found to be exact mirror images of each other with regards to dextral and sinistral morphs within the genus *Platichthys* (Bergstrom & Palmer, 2007; Russo et al., 2012), suggesting that for these species there is an adaptive value to lateral polymorphism. The precise selective advantage of sinistral versus dextral morphology, however, still remains unclear. In the case of flounders within the genus *Paralichthys*, the vast majority of wild-caught specimens are sinistral. However, when cultured in a laboratory or aquaculture environment, the incidence of offspring with "reversed" (dextral) asymmetry (see Fig. 6.2K and L) has been reported to be much higher than found in nature for Japanese flounder (*Paralichthys olivaceus*) (see Okada, Takagi, Tanaka, & Tagawa, 2003; Okada, Tanaka, & Tagawa, 2003), southern flounder (*Paralichthys lethostigma*) (Benetti et al., 2001; Schreiber, 2006), Brazilian flounder (*Paralichthys orbignyanus*) (López et al., 2009), and summer flounder (*Paralichthys dentatus*) (Bisbal & Bengtson, 1993). Even in the aforementioned *P. stellatus* populations reported to display virtually 100% sinistrality in Japanese waters, when adults were crossed in an aquaculture setting, 13–17% of the offspring were reported to have dextral morphology (Policansky, 1982b). The higher incidences of reversal in the above flounders in cultured versus wild populations may suggest that in at least some populations and ecological settings reversed flounder may be selected against in nature for reasons that still remain unknown.

During embryogenesis, all vertebrates display directional asymmetry in the placement of internal organs (Hamada et al., 2002; Levin, 2005), as well as morphological asymmetries in the paired habenulae, pineal, and parapineal glands of the brain's dorsal diencephalon (epithalamus) (Concha & Wilson, 2001). Asymmetric development of the epithalamus, heart, and gut during embryogenesis has been shown to be controlled by expression of the *nodal-lefty-pitx2* (NLP) pathway on the left side of their embryonic primordia in both the zebrafish (*Danio rerio*) (see reviews by Concha, Signore, & Colombo, 2009 and Snelson & Gamse, 2009) and the Japanese flounder (*P. olivaceus*) (Hashimoto et al., 2002, 2007; Suzuki et al., 2009). Interestingly, about 2 weeks following embryogenesis, *pitx2* has been found to be re-expressed asymmetrically on the left side of the habenulae of nearly metamorphic flounder larvae in both *P. olivaceus* (a sinistral flatfish) and the spotted halibut (*Verasper variegatus*, a dextral flatfish) (Suzuki et al., 2009). The same study observed that during metamorphosis flatfish habenulae undergo two distinct forms of asymmetric morphological development: (1) volumetric asymmetry, where the right habenula grows larger than the left in both sinistral and dextral flatfish species and (2) positional asymmetry, where in sinistral species the migration of the right eye is accompanied by a rightward movement of the habenulae along the ventral diencephalon, and in dextral species the migration of the left eye is accompanied by a leftward movement of the habenulae (refer to Fig. 6.4A–D). Whereas many fish species are known to display volumetric asymmetry of the habenulae (Concha & Wilson, 2001), the metamorphic development of such a dramatic positional asymmetry of the habenulae has to date only been reported for flatfish. It is not yet clear whether asymmetric positional development of the habenulae plays an active or a permissive role in facilitating metamorphic skull remodeling and eye migration.

Hashimoto et al. (2002) have isolated a Japanese flounder clonal line, *reversed* (*rev*), in which more than half of homozygous offspring derived from *rev* expressed the normally left-specific marker *pitx2* bilaterally in the lateral plate mesoderm (LPM) of embryos (Fig. 6.5). As reversal of gut coiling was observed at a high frequency in *rev* juveniles, and also considering that abnormal *pitx2* expression in the LPM has been shown to correspond with reversed visceral organ asymmetry in other vertebrates

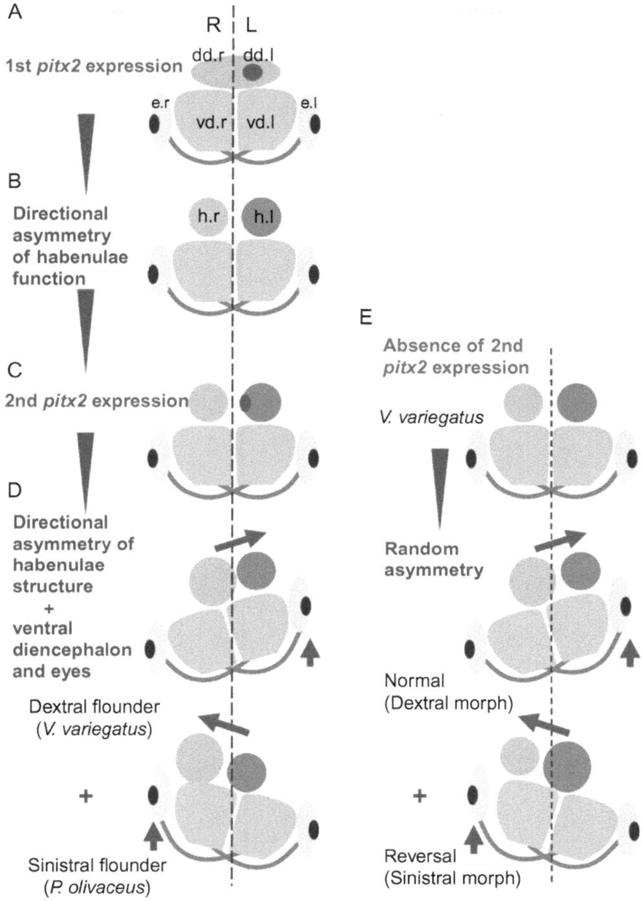

Figure 6.4 Hypothesized control system of flounder eye-sidedness by the *nodal-lefty-pitx2* (NLP) pathway. (a) First step: *pitx2* is expressed on the left side of the embryonic dorsal diencephalon with the involvement of nodal and lefty. (b) Second step: The paired habenulae develop asymmetrically from the dorsal diencephalon under the control of *pitx2*; functional asymmetry is apparent between the pair. Steps a and b refer to published descriptions of zebrafish (Concha et al., 2003; Gamse, Thisse, Thisse, & Halpern, 2003; Liang et al., 2000). (c) Third step: *pitx2* is re-expressed at the left habenula at the premetamorphic stage. (d) Fourth step: The right habenula becomes larger than the left habenula and the pair shifts the position along the ventral diencephalon under the control of the second *pitx2* expression, when the ventral diencephalon, tectum, and eyes undergo asymmetrical development in conjunction with habenulae, in the opposite direction in sinistral and dextral flounders. (e) When the second *pitx2* expression is suppressed, sinistral and dextral morphs appear in a population due to randomization of habenulae sidedness. e, eye; dd, dorsal diencephalon; h, habenula; l, left; r, right; vd, ventral diencephalon. *Reproduced with permission from Suzuki et al. (2009).* (For color version of this figure, the reader is referred to the online version of this chapter.)

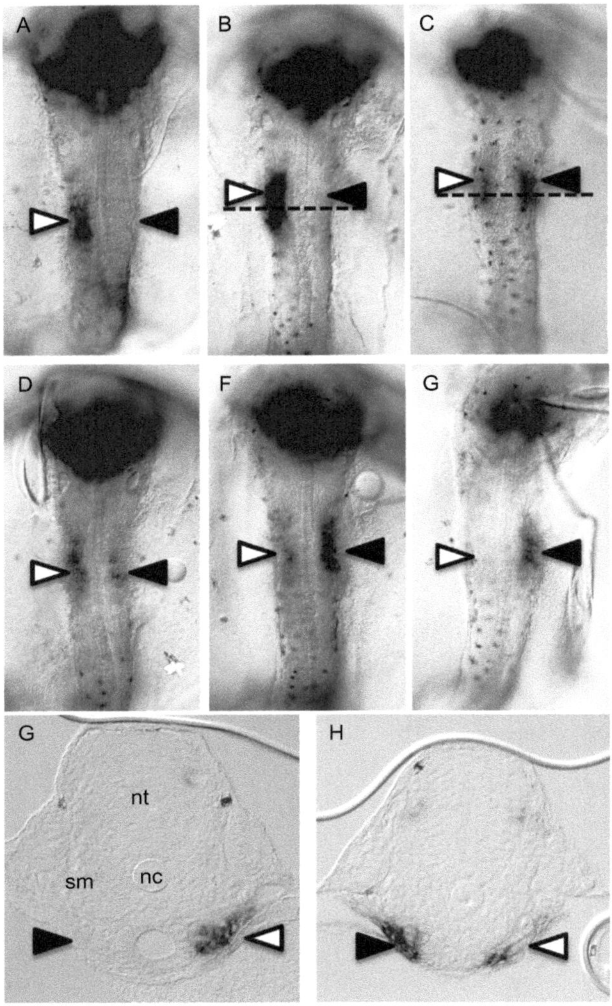

Figure 6.5 Expression of *pitx2* in the lateral plate mesoderm (LPM). The embryos at mid-somitogenesis (17–18 somite stage) were analyzed with *pitx2* antisense RNA probe. (A) Control F3 embryo. (B–F) *rev* F3 embryos. (G, H) Transverse section of *rev* embryos. Among these embryos, 43% exhibits *pitx2* expression in the left LPM (B) as in normal embryos (A). In another 43% of the embryos from *rev*, *pitx2* was expressed bilaterally without apparent L/R bias (C). In 4% and 6%, *pitx2* expression was observed bilaterally but in left- and right-biased fashions, respectively (D, E). Four percent exhibited *pitx2* expression only on the right side (F). Sections in (G, H) reveal that *pitx2* expression domain locates in the left LPM. Right-sided expression of *pitx2* is also detected in the LPM as shown in (H). Dashed lines in (B) and (C) denote regions of cross-section depicted in (G) and (H), respectively. Left LPM (open arrowhead), right LPM (closed arrowhead). (A–F) Dorsal view, anterior to the top. Bilaterally symmetrical *pitx2* signal in the anterior

(Logan, Pagan-Westphal, Smith, Paganessi, & Tabin, 1998; Piedra, Icardo, Albajar, Rodriguez-Rey, & Ros, 1998; Ryan et al., 1998), the findings suggest that the *rev* locus plays a role in patterning the embryonic left–right axis. Furthermore, *rev* offspring exhibit a high proportion (20–30%) of "reversed" (dextral) flounder, suggesting that *rev* also has a role in controlling the directionality of metamorphic eye migration (Hashimoto et al., 2002). Considering the previously discussed observation by Suzuki et al. (2009) that *pitx2* expression occurs twice in developing flatfish (once during embryogenesis and once prior to metamorphosis), it is interesting to note their additional observation that the establishment of eye-sidedness at metamorphosis does not correlate with the laterality of either *pitx2* expression or visceral organ placement during embryogenesis. However, the expression of *pitx2* in the left habenula at metamorphosis does indeed correspond with the establishment of eye-sidedness, as loss of *pitx2* expression in the habenula at this time (induced in wild-type flounder via treatment with ouabain to induce laterality defects in the NLP pathway) results in the randomization of eye-sidedness and the production of a population of juveniles containing both normal morphology (i.e., sinistral for *P. olivaceus* and dextral for *V. variegatus*) and "reversed" morphology (Suzuki et al., 2009). Interestingly, whereas in both dextral and sinistral flatfish species the right habenula always grows larger than the left, in flatfish with reversed metamorphic morphology the volumetric asymmetry becomes inverted and the left habenula grows larger than the right (Fig. 6.4E). In all vertebrates, axons projecting from the habenula are known to relay thalamic, telencephalic, and other inputs to the ventral midbrain (Butler & Hodos, 1996), and in the zebrafish, it has been shown that axons emanating from the left and right habenula nuclei innervate different dorsal and ventral regions of the midbrain in a pattern established by the NLP pathway during embryogenesis (see review by Snelson & Gamse, 2009). Although southern flounder (*P. lethostigma)* juveniles have also been shown to display a pattern of habenular–midbrain innervation similar to zebrafish, the innervation pattern is not different between siblings with sinistral and reversed (dextral) eye-sidedness (Kuan, Gamse, Schreiber, & Halpern, 2007), suggesting that the development of eye-sidedness during metamorphosis is independent of the differential

regions of the embryos localize to the cephalic mesoderm. (G, H) Posterior view, dorsal to the top. nt, neural tube; nc, notochord; sm, somite. *Reproduced with permission from Hashimoto et al. (2002).* (For color version of this figure, the reader is referred to the online version of this chapter.)

innervation of dorsal and ventral midbrain targets by left and right habenulae that is established during embryogenesis. Nonetheless, the observations of Suzuki et al. (2009) clearly suggest that in flatfish the expression of *pitx2* in the left habenula at metamorphosis corresponds with the development of eye-sidedness that is characteristic for the species in question (i.e., a dextral or sinistral species), and that the absence of this expression is accompanied by randomized sidedness that yields an abnormally high percentage of "reversed" individuals. The mechanism that specifies whether a monomorphic species metamorphoses with specifically sinistral or specifically dextral morphology, however, remains to be discovered.

In addition to eye-sidedness, another example of postembryonic asymmetry that is particularly evident in the largest of the flatfishes, the Atlantic halibut (*Hippoglossus hippoglossus*), is the presence of a much larger fast skeletal muscle mass on the eyed side of adults compared with the blind side (Hagen, Vieira, Solberg, & Johnston, 2008; Johnston, 2004). Interestingly, Galloway et al. (2006) reported that mRNA for the myogenic regulatory factor *myod2* is expressed with a left–right asymmetric pattern in fast muscle precursor cells during halibut somitogenesis, leading to an initial speculation that *myod2* may play a role in the development of the thicker muscle on the eyed side. Upon further analysis, the researchers discovered that embryonic *myod2* is expressed in a randomized left–right pattern (displaying left-sided, right-sided, or bilateral expression) (see Fig. 6.6), which did not correspond with the development of asymmetric muscle thickness in the majority of the halibut in the study that metamorphosed with dextral morphology, with very low incidence of reversal (Andersen et al., 2009). However, there still does remain the intriguing possibility that halibut embryos expressing *myod2* on their left sides are fated to metamorphose with reversed (sinistral) muscle morphology, but that for unknown reasons experience higher mortality such that only dextral morphs typically survive. In contrast to the asymmetric expression patterns of halibut *myod2*, its gene duplicate, *myod1*, is expressed in a bilaterally symmetrical fashion during embryogenesis (Andersen et al., 2009). Also, in spite of asymmetric *myod2* expression during embryogenesis, the somite pairs have been shown to form symmetrically along the halibut embryo midline, and that morphological muscle symmetry is maintained throughout metamorphosis and into the juvenile stage (Galloway, Kjørsvik, & Kryvi, 1999), suggesting that the development of muscle thickness asymmetry is a postmetamorphic event. The functional influence of asymmetric *myod2* expression during halibut embryogenesis on metamorphic asymmetry, if any, remains unresolved.

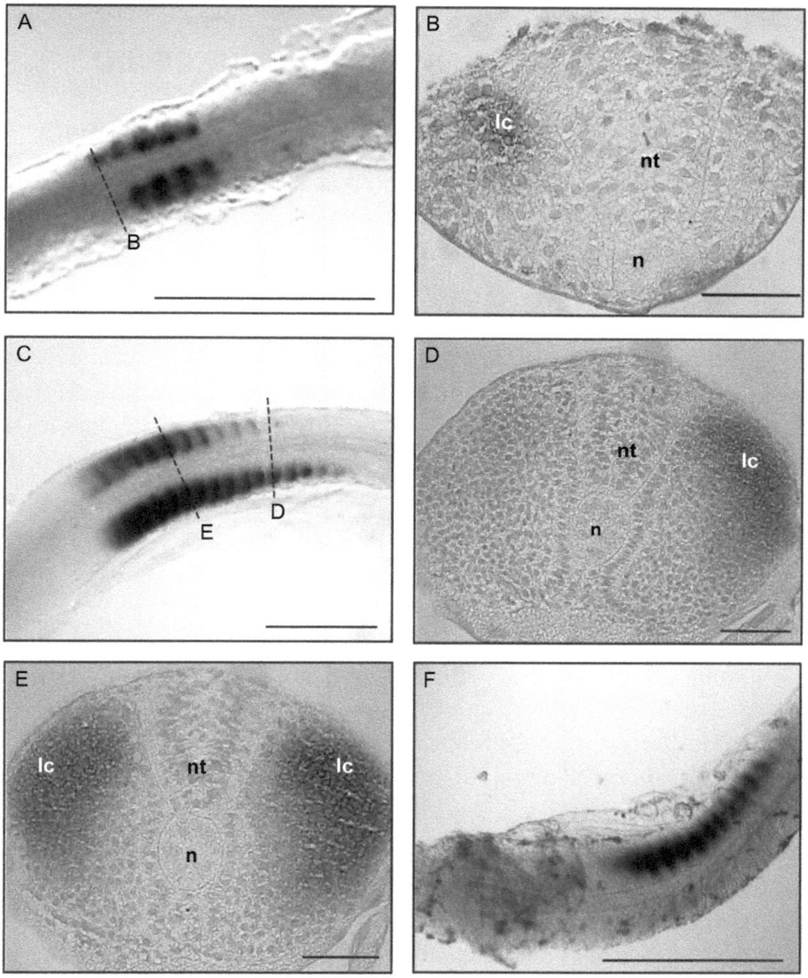

Figure 6.6 Left–right asymmetric expression of *myod2* in Atlantic halibut during somitogenesis analyzed by whole-mount in situ hybridization. (A) Asymmetric *myod2* signal shown as one additional positive somite at five somites (75% epiboly). Stippled line indicates site of transverse section. (B) Transverse section showing left-sided expression of *myod2* in lateral cells at five somites. (C) Right-sided asymmetric expression of *myod2* in embryo at 20 somites. Transverse sections shown in picture (D) and (E) are indicated by stippled lines. (D) Transverse section of the embryo in picture C demonstrating *myod2* expression in the lateral cells on the right side. (E) Transverse section of the embryo in picture C with stronger *myod2* signal on the right side. (F) Embryo at 45 somites with no *myod2* signal on the right side of the midline. All embryos are positioned with the head toward right and are seen dorsally. The sections are positioned with the dorsal side on the top of the picture. lc, lateral cells; n, notochord; nt, neural tube. Scale bars for the whole embryos and transverse sections equal 0.5 mm and 50 μm, respectively. *Reproduced with permission from Andersen et al. (2009).* (See Color Insert.)

3. METAMORPHIC ASYMMETRY OF THE BRAIN, SENSORY ORGANS, AND BEHAVIOR

In contrast with the eyes, the flatfish olfactory organs do not change location but do display dramatic morphological asymmetry, with the eyed-side olfactory organ, nerve, and telencephalon being significantly larger compared with that on the blind side (Brinon et al., 1993; Prasada Rao & Finger, 1984). The central olfactory projections have also been shown to be asymmetric in the winter flounder (*Pseudopleuronectes americanus*), with the blind side telencephalon receiving approximately equal olfactory input from the two sides, whereas the eyed-side telencephalon is measurably larger than and receives significantly more input from the eyed-side olfactory organ (Prasada Rao & Finger, 1984). The development of metamorphic telencephalic asymmetry in the Senegalese sole (*Solea senegalensis*) does not appear to result from asymmetric cell proliferation within the telencephalon (Pinuela, Rendon, Gonzalez de Canales, & Sarasquete, 2004). Interestingly, using *S. senegalensis*, Velez, Hubbard, Barata, and Canario (2005) demonstrated that these morphological olfactory asymmetries are accompanied by functional asymmetry in responsiveness to different odorants by each olfactory organ as measured by electro-olfactogram, the first report of an asymmetric olfactory response in a vertebrate.

Despite the dramatic migration of one eye to the opposite side of the head during flatfish metamorphosis, a documented length asymmetry between the right and left optic nerves (Murray, 1974), and an unusually nonrandom configuration of the dorsal–ventral crossing of the optic nerves at the optic chiasma in some flatfish species (the nerve of the migrating eye typically lies dorsal to the nonmigrating eye's optic nerve) (Ballard, Pickett, & Sivak, 1987; Parker, 1903; Policansky, 1982b), no evidence of asymmetric central organization of the retinal projection into the brain has yet been discovered (Luckenbill-Edds & Sharma, 1977; Medina, Reperant, Ward, Rio, & Lemire, 1993). Brinon et al. (1993) have, however, reported the existence of a transient volumetric asymmetry between the left and right sides of the optic tectum during turbot (*Scophthalmus maximus*) metamorphosis, with symmetry reestablished after metamorphosis. No morphological asymmetry has ever been reported to exist in the flatfish midbrain or hindbrain. Together with the previously described changes in metamorphic epithalamus morphology, these findings suggest that postmetamorphic flatfish morphological asymmetry of the neurocranium is confined to regions rostral to the optic tectum.

Unlike all other vertebrates, the axes of flatfish eyes and horizontal semicircular canals of the inner ear become oriented perpendicular to each other during metamorphosis (Graf & Baker, 1983), a process that is accompanied by the development of an increasingly tilted swimming posture (Finger, 1976; Neave, 1985; Schreiber, 2006) (see Figs. 6.1 and 6.2 A–C). The specific mechanisms responsible for inducing the tilted swimming posture and lateralized settling behaviors in flatfish remain essentially unknown, though in principle the abrupt transition to lateralized behavior could be influenced by asymmetric changes in eye position, inner ear morphology, and/or central changes in vestibular connectivity or activity. Although changes in swim posture have been shown to be accompanied by changes in central nervous pathways (Graf & Baker, 1983, 1985a, 1985b, 1990; Graf & Meyer, 1983; Graf, Spencer, Baker, & Baker, 2001; Graff, Simmons, Meyer, Martin-Morgan, & Kurtz, 1986; Platt, 1973a, 1973b), none of the aforementioned studies found evidence of an asymmetric remodeling of these pathways. Furthermore, no morphological evidence of asymmetry has yet been identified for the flatfish labyrinths (Graf et al., 2001; Jacob, 1928), the distribution of the vestibular nuclear complex in the hindbrain (Jansen & Enger, 1996a), or the peripheral or central oculomotor apparatus (Graf & Baker, 1985a), which all appear bilaterally symmetrical and similar to those of other teleosts. Interestingly, all flatfish except for the most primitive (i.e., *Psettodes*, which have been observed to swim upright on occasion; see De Groot, 1970) display nearly omnidirectional hair–cell polarization of the inner ear's saccular and lagenar otoliths (Jorgensen, 1976; Platt, 1973a, 1983). This pattern, which is unique among the vertebrates, is thought to accommodate, but not induce, the adult flatfish posture. Some evidence for asymmetric peripheral and postural control in flatfish adults (presumably established during metamorphosis) does exist, though. Meyer, von Seydlitz-Kurzbach, and Fiebig (1981) found that 2-deoxyglucose is taken up differentially by the bilateral vestibular nuclei suggesting that lateralized behavior may be caused by a "permanent imbalance in vestibular neuron activity." Left–right asymmetries for adult flatfish otolith morphology (Lychakov, 1996; Lychakov, Rebane, Lombarte, Demestre, & Fuiman, 2008; Sogard, 1991), mass (Helling, Scherer, Hausmann, & Clarke, 2005), and chemical composition (Loher, Wischniowski, & Martin, 2008) have also been described. Indeed, TH has been shown to induce growth and development of all three types of otoliths (saculus, utricle, and lagena) during flatfish metamorphosis (Schreiber et al., 2010; Wang et al., 2011), though it remains to be seen if otolith morphological asymmetries develop specifically during metamorphosis or if they develop after behavioral change has already been established.

The dorsal light reflex, which causes a typical upright swimming fish to tilt toward a lateral light source until equilibrium is established between visual and gravitational sensory input, may contribute to the development of flatfish lateralized swimming behavior as the migrating eye changes the perceived angle of light incidence (Finger, 1976; Graf & Baker, 1990; Neave, 1985). Indeed, while premetamorphic larval flatfish have been shown to exhibit a dorsal light response similar to that of a typical upright fish, after one eye begins to migrate the larvae tilt toward their future blind side when illuminated from above (Neave, 1985), a process that could be attributable to the angle of perceived light caused by eye migration. Additionally, Jansen and Enger (1996b) have shown that NADPH diaphorase activity is asymmetrically distributed in the optic tectum during metamorphosis, a time corresponding with the aforementioned transient morphological asymmetry of the tectum described by Brinon et al. (1993), though the functional significance of this asymmetry is not clear. Ultimately, studies with southern flounder (*P. lethostigma*) larvae have revealed three lines of evidence that suggest lateralized swimming and settling behaviors in flatfish can be completely dissociated from—and thus occur independently of—asymmetric eye positioning: (1) lateralized feeding and tilted swimming behaviors are apparent prior to the start of eye migration during premetamorphosis, (2) treatment of young premetamorphic larvae with TH induces acute tilted swimming and settling behavior prior to its induction of eye migration, and (3) symmetrically metamorphosed flounder variants also display normal lateralized swimming and settling behaviors despite the complete absence of eye migration (Schreiber, 2006). These findings suggest that both the transition to a lateralized swim posture and metamorphic craniofacial remodeling/eye migration are independent developmental events, each controlled autonomously by TH. The development of a lateralized swim posture occurs during late premetamorphosis (when TH levels are known to be relatively low; see Fig. 6.3) and precedes eye migration which takes place later during metamorphic climax (when TH levels are known to be highest; see Fig. 6.3). Assuming that the transition to a lateralized swim posture is ultimately mediated by central or peripheral changes in vestibular function, these regions might be expected to exhibit a higher sensitivity to low levels of TH (e.g., increased types 1 and 2 deiodinase activity or elevated TH receptor levels) compared with other regions, though this hypothesis remains untested.

4. ASYMMETRIC CRANIOFACIAL REMODELING AND THE BASIS OF EYE MIGRATION

The metamorphic remodeling of the flatfish skull is perhaps the most striking example of asymmetric postembryonic development in the vertebrate world, a process that ultimately accommodates eye migration and the transition from a pelagic to a benthic lifestyle. In adult winter flounder (*P. americanus*), the orbit of the migrated eye is formed entirely of bones whereas that of the nonmigrated eye consists partially of jaw muscles (Ballard et al., 1987). The orbital bones of the migrated eye induce a deflection in the path of the extraocular muscles prior to insertion, causing the lateral and medial rectus muscles from the migrated eye to be shorter than those of the opposite eye (Ballard et al., 1987). These extraocular muscle asymmetries are clearly necessary to accommodate the final eye position, though it is not known if changes in extraocular muscle length play an active role in actually promoting eye migration. The rearrangement of cranial elements to accommodate eye migration and transition to a benthic habitat is known to be accompanied by the development of asymmetric movements of the jaws during prey capture for some flatfish species (Francis & Turingan, 2008; Gibb, 1995, 1996, 1997, 2003).

Most of the metamorphic changes in cranial cartilage and bone appear to be conserved across flatfish species (Brewster, 1987; Chapleau, 1993; Saele, Silva, & Pittman, 2006), and the major events of asymmetric neurocranial remodeling that have been described include (1) asymmetric growth of the ethmoid plate and frontal bones that twist toward the future ocular direction (Brewster, 1987; Chapleau, 1993; Norman, 1934; Saele, Silva, et al., 2006; Saele, Smaradottir, & Pittman, 2006; Schreiber, 2006; Wagemans, Focant, & Vandewalle, 1998; Wagemans & Vandewalle, 2001) (also see Fig. 6.2D–H); (2) differential resorption of the left and right cartilaginous supraorbital bars (Okada, Takagi, Seikai, Tanaka, & Tagawa, 2001; Schreiber, 2006; Wagemans et al., 1998); (3) the proliferation of a dense cell population of dermal fibroblasts specifically under the migrating eye (Bao et al., 2011; Okada, Takagi, et al., 2003; Okada et al., 2001; Saele, Silva, et al., 2006; also see Fig. 6.7); (4) the deposition of osteoblasts within the dense cell layer to form the anterior flange of the blind side frontal (dermal bone), termed the "postlateral ethmoid" by Hoshino (2006) (and referred to previously as the "pseudomesial bar"—see Okada et al., 2001; Traquair, 1865), a structure unique to flatfishes that eventually covers the area vacated

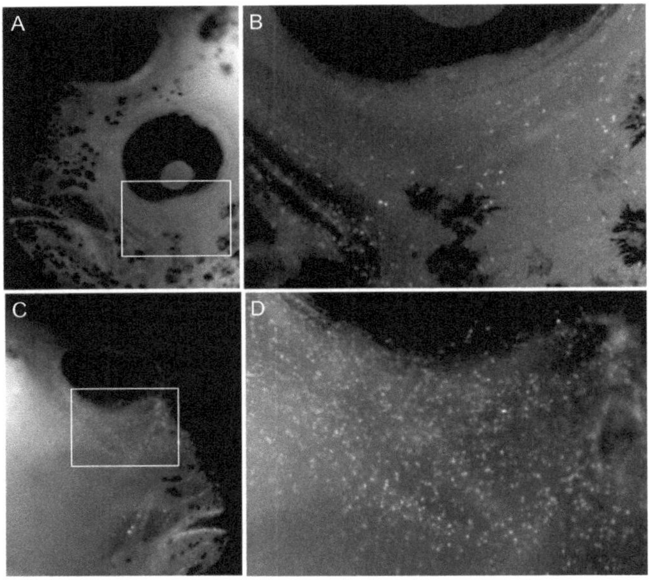

Figure 6.7 Asymmetric suborbital cell proliferation during eye translocation in the Southern flounder, *Paralichthys lethostigma*. Cell proliferation in the suborbital regions of the skin visualized using fluorescent-labeled antibodies against bromodeoxyuridine (BrdU) is relatively low under the left (nonmigrating) eye (A, B) and much higher under the right (migrating) eye (C, D) in a mid-metamorphic climax larva. Boxed regions in (A) and (C) are magnified in (B) and (D), respectively. Images by A.M. Schreiber.

by the migrating eye (Saele, Silva, et al., 2006); and (5) a differential increase in volume on the blind versus ocular sides of "retrorbital vesicles," sac-like structures of unknown function located under both eyes (Okada et al., 2001; Okada, Takagi, et al., 2003; Okada, Tanaka, et al., 2003; Saele, Smaradottir, et al., 2006). Each of the aforementioned changes in cranial asymmetry has been found to accompany normal eye migration and may be necessary for it to occur. For example, bilaterally symmetrical flatfish variants that lack eye migration have also been shown to lack the presence of a postlateral ethmoid and retain bilateral symmetry in all other cranial components (Okada et al., 2001; Okada, Tanaka, et al., 2003; Saele, Silva, et al., 2006; Schreiber, 2006). Furthermore, Bao et al. (2011) have shown that injections of colchicine (an inhibitor of mitosis) into the suborbital area of the blind side of larval flatfish result in an inhibition of eye migration in several different flatfish species. The study suggests that the proliferation of the suborbital tissues functions as the primary driving force that "pushes" the migrating eye in the dorsal direction (see Fig. 6.8), and that by inhibiting suborbital cell division

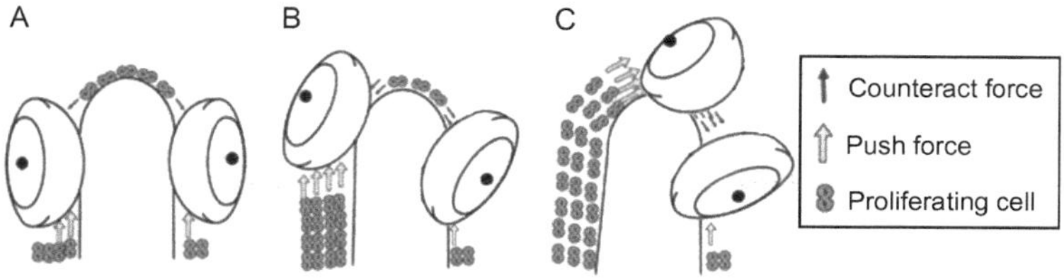

Figure 6.8 Schematic drawing to explain eye migration in flatfish. (A) Before eye migration has initiated, cells located in the suborbital area exhibit asymmetric proliferation between the left and right sides. (B) Once the migrating eye receives sufficient pushing force from proliferating cells in its suborbital tissue, it begins moving in the dorsal direction. (C) By the end of metamorphosis, both the migrating eye and the nonmigrating eye have experienced a 90° rotation from their initial orientations, with the migrating eye facing a dorsal trajectory, and the nonmigrating eye facing a ventral trajectory. *Reproduced in altered form with permission from Bao et al. (2011).* (For color version of this figure, the reader is referred to the online version of this chapter.)

the eye can no longer translocate. It is possible that inhibition of suborbital proliferation also prevents the development of the postlateral ethmoid (dermal bone produced by the suborbital tissue), indirectly inhibiting eye migration.

Endocrine signals clearly mediate the progression of metamorphic skull remodeling. Insulin-like growth factor I receptors have been shown to be expressed in fibroblasts, frontal bone osteocytes, and dorsal chondrocytes at the beginning of halibut metamorphosis (Hildahl, Power, Bjornsson, & Einarsdottir, 2008), as well as in the dense cell population of dermal fibroblasts under the migrating eye (Zhang, Shi, Cheng, & Chen, 2011). Ultimately, elevated levels of TH are necessary to induce eye migration and skull remodeling, and if endogenous levels of TH are chemically ablated, flatfish retain a bilaterally symmetrical skull (Okada, Tanaka, et al., 2003; Schreiber, 2006). Interestingly, Tagawa and Aritaki (2005) have shown that in the spotted halibut (*V. variegatus*) the timing of responsiveness to TH differs between the left and right sides of the larval body. Hypothyroid larvae (held in developmental stasis by treatment with the goitrogen thiourea) metamorphosed with two distinct bilaterally symmetrical phenotypes when treated with exogenous TH at different times: (1) an ambicolorate phenotype in which both sides possessed characteristics of the ocular side when treated with TH prior to 15 days after hatching (DAH) or after 60 DAH and (2) a symmetrical pseudoalbino phenotype where both sides possessed characteristics of the blind side when treated with TH beginning 25 DAH (Tagawa & Aritaki, 2005). These findings suggest that both sides of the larval body have the potential to metamorphose into either the juvenile ocular or the blind side, and the timing of TH release and differential sensitivity of each side to TH ultimately establishes flatfish metamorphic sidedness. Although TH receptors (Galay-Burgos et al., 2008; Laudet, 2011; Manchado et al., 2009; Marchand et al., 2004; Yamano & Miwa, 1998) and deiodinase enzymes (Isorna et al., 2009; Itoh et al., 2010) have been described for flatfish, differences in expression or activity between the two sides and among the various cranial bones and cartilages have not yet been reported. One last aspect of craniofacial remodeling that has been virtually ignored in the literature and hence deserves some mention are the changes that take place to the orientation and positioning of the nonmigrating eye during metamorphosis. The nonmigrating eye appears to move to a more ventral position to make room for the opposite migrating eye (Ballard et al., 1987; Norman, 1934), though this has not yet been established quantitatively. Furthermore, the nonmigrating eye also experiences a permanent change in orientation as it rotates by 90° and establishes a more ventral visual trajectory (see Figs. 6.1, 6.2A–C, and 6.8). The rotation of the nonmigrating eye is presumably mediated by permanent changes in its extraocular muscles, though this process remains uncharacterized.

The development of symmetrical juvenile flatfish due to abnormal metamorphic craniofacial remodeling ("arrested development" or failure of the eye to migrate) and symmetrical pigmentation (pseudoalbinism and ambicoloration, discussed above) is a common phenomenon in mass-cultured commercial flatfish species (reviewed in Power et al., 2008; also see Pittman, Jelmert, Naess, Harboe, & Watanabe, 1998). Pseudoalbinos possess the blind side pigment characteristics on both sides, whereas ambicolored juveniles express ocular side pigmentation on both sides and are considered malformations of asymmetry (Aritaki & Tagawa, 2012). In halibut, juveniles with arrested eye migration have been shown to have abnormal remodeling of the frontal bones with symmetrical osteoclast activity on either side, as well as the absence of the postlateral ethmoid (pseudomesial bar) dermal bone and abnormally symmetrical paraethmoids (Saele, Smaradottir, et al., 2006) (Fig. 6.9). Similar observations have been made for *P. olivaceus* (Okada, Takagi, et al., 2003; Okada, Tanaka, et al., 2003; Okada et al., 2001) and *P. lethostigma* (Schreiber, 2006) juveniles with arrested eye migration. Saele, Smaradottir, et al. (2006) conclude that "the lack of eye migration in flatfishes may not be a deformity like lordosis, short jaw, etc., but may result from lack of remodeling, leaving it with the neurocranial phenotype of its ancestors" (compare Fig. 6.9 with Fig. 6.10 in Section 5). These developmental abnormalities are known to be strongly

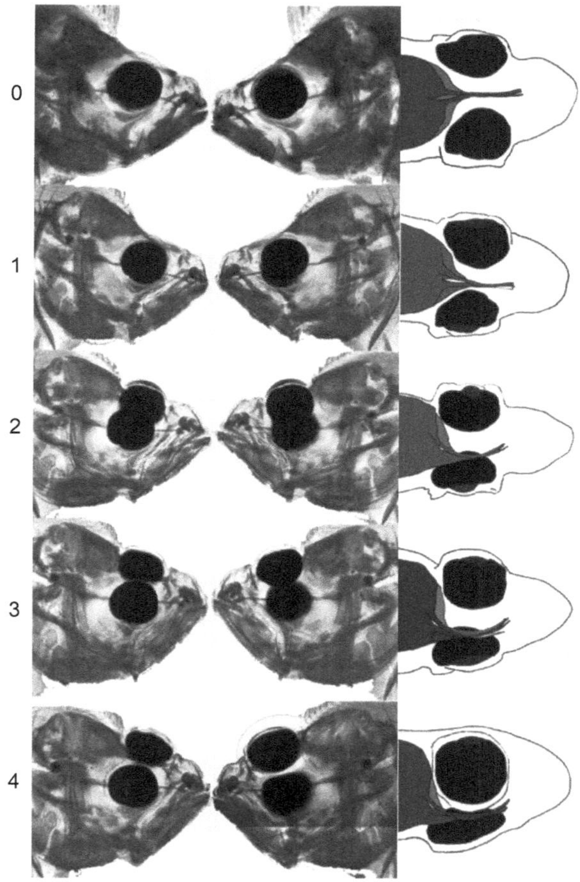

Figure 6.9 Postmetamorphic Atlantic halibut (*Hippoglossus hippoglossus*) with full ossification showing variation in cranial symmetry and degree of eye migration. Index of eye migration ranges from a scale of 0 (no eye migration) to 4 (complete eye migration). Profiles of the ocular side (left column) and blind side (middle column) are depicted. Each fish is presented with schematic drawings of cranial ossification in a top view showing torsion of the Frontal bones between the eyes (right column). Eye index was decided prior to photographing. Fish representing indices 0 and 1 are Artemia fed, whereas the remainder are zooplankton-fed. All fish are aged 40 days post start of feeding (dpsf) except the upper fish that is 46 dpsf. *Reproduced in modified form with permission from Saele, Silva, et al. (2006) and Saele, Smaradottir, et al. (2006).*

influenced by both larval nutritional status (see reviews by Fernandez & Gisbert, 2011; Hamre & Harboe, 2008; Hamre, Holen, & Moren, 2007; Hamre, Moren, Solbakken, Opstad, & Pittman, 2005) and environmental parameters such as photoperiod (Harboe, Mangor-Jensen, Moren, Hamre, & Rønnestad, 2009; Solbakken & Pittman, 2004) and temperature

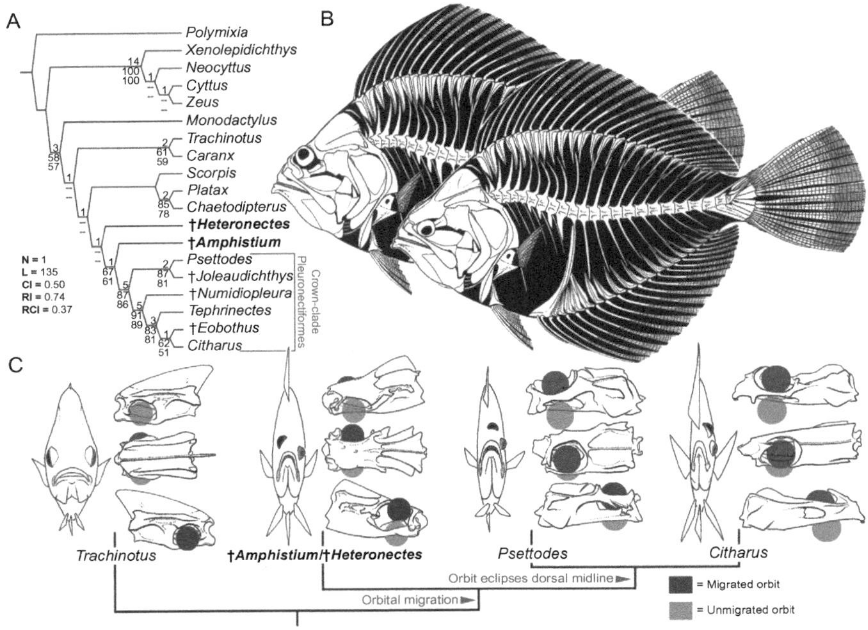

Figure 6.10 Phylogenetic placement of *Heteronectes* and *Amphistium* and implications for the origin of cranial asymmetry in flatfishes. (A) Topology arising from the analysis of a matrix comprising 19 taxa coded for 58 morphological characters (8 ordered) (number of cladograms = 1; cladogram length = 135; consistency index = 0.50; retention index = 0.74; rescaled consistency index = 0.37). *Heteronectes* and *Amphistium* are placed as successively more crownward plesions on the flatfish stem. Unordered analyses reconstruct these taxa at the same position. Numbers at nodes indicate Bremer decay index, bootstrap support, and jackknife support, from top to bottom, respectively. Extinct taxa are marked (†) and "–" indicates that bipartition occurs in fewer than half of cladograms arising from bootstrap or jackknife analysis. Previous placements of *Amphistium* outside Pleuronectiformes are rejected. (B) Reconstruction of *Amphistium*, showing sinistral (front) and dextral (back) individuals in the left lateral view (modified from Blot, 1969). (C) Simplified cladogram adapted from (A) showing the progression of orbital migration across flatfish phylogeny from adult specimens. Neurocrania are depicted in left lateral (top), dorsal (middle), and right lateral (bottom) views. *Reproduced with permission from Friedman (2008).*

(Aritaki & Seikai, 2004; Blanco-Vives et al., 2010). The developmental processes underlying abnormalities in metamorphic asymmetry likely originate during embryogenesis or early larval development (Power et al., 2008), and the identification of new molecular markers to differentiate normal from abnormal developing larvae will be needed to further study the cause of these abnormalities. Several molecular tools are now being developed and genome sequencing projects have been initiated for several flatfish species of commercial importance (e.g., Japanese flounder, Senegalese sole, Atlantic halibut) (see reviews by Cerda et al., 2008, Cerda et al., 2010, and Power et al., 2008 and studies by Bao et al., 2005, Fu et al., 2011, and Xie et al., 2011).

5. FLATFISH EVOLUTION: A DEVELOPMENTAL PERSPECTIVE

Flatfishes are thought to be descended from a bilaterally symmetrical, upright swimming common ancestor from the clade Percomorpha, though the specific sister group has not yet been identified (Chapleau, 1993; Friedman, 2012). The unique positioning of independently mobile eyes in juvenile and adult flatfish facilitates 360° vision, maximizing the ability to spot both prey and predators from multiple trajectories. Also, the extra-orbital positioning of the eyes allows these fish to burrow into the substrate leaving the eyes exposed (Gibson, 2005). These features, in combination with the ability to match their background color pattern through rapid adaptive camouflage, very effectively accommodate a predominantly benthic existence. From an ecological perspective, the retention of a bilaterally symmetric pelagic larval stage promotes dispersal of the larvae away from the benthic adults and toward future (often estuarine) nursery grounds (Bailey, Nakata, & Van der Veer, 2005).

Fossil records of the most ancestral flatfishes known (genus *Amphistium* and *Heteronectes*) from the Eocene epoch (53 million years ago) show incomplete eye migration, with the eyes remaining on opposite sides of the head in postmetamorphic fish (Friedman, 2008; Friedman, 2012; Fig. 6.10). The findings by Friedman also indicate that dextral and sinistral morphs within each ancestral species occurred in approximately equal frequency, in contrast with modern flatfishes, most of which are monomorphic (Munroe, 2005). An ancestral-like retention of dextral and sinistral polymorphism is still found in the most primitive of living flatfishes, *Psettodes* (Hubbs & Hubbs, 1944). Therefore, it appears that eye-sidedness was initially random in ancestral flatfishes, until developmental systems (such as the NLP pathway) evolved to control eye-sidedness (Suzuki et al., 2009) for reasons that remain unknown. Interestingly, these ancestral flatfish with incomplete eye migration were also shown to lack a postlateral ethmoid (pseudomesial bar) (Friedman, 2008, 2012), which appears to be required for normal eye migration in modern flatfishes. Friedman concludes "Thus, the evolutionary origin of flatfish asymmetry resembles its developmental origin, with increasing degrees of orbital migration transforming a symmetrical precursor into a fully asymmetric form" (Friedman, 2008) (Fig. 6.10). If modern flatfish ontogeny also recapitulates other aspects of their phylogeny, then studying the larval development and metamorphosis of these fascinating creatures may further our understanding of their evolution. For example, it has been shown for southern flounder that transitions to lateralized swimming postures and initiation of benthic settling and hiding behaviors are initiated prior to eye migration (Schreiber, 2006). These observations would support a hypothesis that the bilaterally symmetrical flatfish ancestor was a benthic fish with already established lateralized behavior (e.g., lying on one side and burrowing into the substrate for ambush or shelter); eye migration and craniofacial asymmetry may have evolved later to accommodate a preexisting lateralized behavior.

ACKNOWLEDGMENTS

The author is particularly grateful to the following for allowing reproduction of their previously published figures: Dr. Tohru Suzuki (Fig. 6.4), Dr. Hisashi Hashimoto (Fig. 6.5), Dr. Øivind Andersen (Fig. 6.6), Dr. Baolong Bao (Fig. 6.8), Dr. Øystein Sæle (Fig. 6.9), and Dr. Matt Friedman (Fig. 6.10).

REFERENCES

Andersen, O., Dahle, S. W., van Nes, S., Bardal, T., Tooming-Klunderud, A., Kjorsvik, E., et al. (2009). Differential spatio-temporal expression and functional diversification of the myogenic regulatory factors MyoD1 and MyoD2 in Atlantic halibut (*Hippoglossus hippoglossus*). *Comparative Biochemistry and Physiology. Part B, Biochemistry & Molecular Biology*, *154*, 93–101.

Aritaki, M., & Seikai, T. (2004). Temperature effects on early development and occurrence of metamorphosis-related morphological abnormalities in hatchery-reared brown sole *Pseudopleuronectes herzensteini*. *Aquaculture*, *240*, 517–530.

Aritaki, M., & Tagawa, M. (2012). Pseudoalbinism and ambicoloration in hatchery-reared pleuronectids as malformations of asymmetrical formation. *Fisheries Science*, *78*, 327–335.

Bailey, K. M., Nakata, H., & Van der Veer, H. W. (2005). The planktonic stages of flatfishes: Physical and biological interactions in transport processes. In R. N. Gibson (Ed.), *Flatfishes: Biology and exploitation* (pp. 94–119). Oxford: Blackwell Publishing.

Ballard, K. A., Pickett, R. L., & Sivak, J. G. (1987). Comparison of the musculoskeletal structure of the orbits of the migrating and non-migrating eyes in the winter flounder (*Pseudopleuronectes americanus*). *Experimental Biology*, *47*, 23–26.

Bao, B., Ke, Z., Xing, J., Peatman, E., Liu, Z., Xie, C., et al. (2011). Proliferating cells in suborbital tissue drive eye migration in flatfish. *Developmental Biology*, *351*, 200–207.

Bao, B., Yang, G., Liu, Z., Li, S., Wang, Z., & Ren, D. (2005). Isolation of SFRS3 gene and its differential expression during metamorphosis involving eye migration of Japanese flounder *Paralichthys olivaceus*. *Biochimica et Biophysica Acta*, *1725*, 64–70.

Benetti, D. D., Grabe, S. W., Feeley, M. W., Stevens, O. M., Powell, T. M., Leingang, A. J., et al. (2001). Development of aquaculture methods for southern flounder. *Paralichthys lethostigma*: I. Spawning and larval culture. *Journal of Applied Aquaculture*, *11*, 113–133.

Bergstrom, C. A. (2007). Morphological evidence of correlational selection and ecological segregation between dextral and sinistral forms in a polymorphic flatfish, *Platichthys stellatus*. *Journal of Evolutionary Biology*, *20*, 1104–1114.

Bergstrom, C. A., & Palmer, A. R. (2007). Which way to turn? Effect of direction of body asymmetry on turning and prey strike orientation in starry flounder *Platichthys stellatus* (Pallas) (Pleuronectidae). *Journal of Fish Biology*, *71*, 737–748.

Bisbal, G. A., & Bengtson, D. A. (1993). Reversed asymmetry in laboratory-reared summer flounder. *The progressive fish-culturist*, *55*, 106–108.

Blanco-Vives, B., Villamizar, N., Ramos, J., Bayarri, M. J., Chereguini, O., & Sánchez-Vázquez, F. J. (2010). Effect of daily thermo- and photo-cycles of different light spectrum on the development of Senegal sole (*Solea senegalensis*) larvae. *Aquaculture*, *306*, 137–145.

Blot, J. (1969). Les poissons fossiles du Monte Bolca. Classe´s jusqu'ici dans les familles des Carangidae, Menidae, Ephippidae, Scatophagidae. *Memorie del Museo Civico di Storia Naturale di Verona*, *1*, 1–525.

Boklage, C. E. (1984). On the inheritance of directional asymmetry (sidedness) in the starry flounder, *Platichthys stellatus*: Additional analyses of Policansky's data. *The Behavioral and Brain Sciences*, *7*, 725–730.

Brewster, B. (1987). Eye migration and cranial development during flatfish metamorphosis: A reappraisal (Teleostei: Pleuronectiformes). *Journal of Fish Biology*, *31*, 805–833.

Brinon, J. G., Medina, M., Arevalo, R., Alonso, J. R., Lara, J. M., & Aijon, J. (1993). Volumetric analysis of the telencephalon and tectum during metamorphosis in a flatfish, the turbot *Scophthalmus maximus*. *Brain, Behavior and Evolution*, *41*, 1–5.

Butler, A., & Hodos, W. (1996). *Comparative vertebrate euroanatomy: Evolution and adaptation*. New York: Wiley-Liss.

Cerda, J., Mercade, J., Lozano, J. J., Manchado, M., Tingaud-Sequeira, A., Astola, A., et al. (2008). Genomic resources for a commercial flatfish, the Senegalese sole (*Solea senegalensis*): EST sequencing, oligo microarray design, and development of the Soleamold bioinformatic platform. *BMC Genomics*, *9*, 508.

Cerda, J., Douglas, S., & Reith, M. (2010). Genomic resources for flatfish research and their applications. *Journal of Fish Biology*, 77, 1045–1070.

Chapleau, F. (1993). Pleuronectiform relationships: A cladistic reassessment. *Bulletin of Marine Science*, *52*, 516–540.

Concha, M. L., Russell, C., Regan, J. C., Tawk, M., Sidi, S., Gilmour, D. T., et al. (2003). Local tissue interactions across the dorsal midline of the forebrain establish CNS laterality. *Neuron*, *39*, 423–438.

Concha, M. L., Signore, I. A., & Colombo, A. (2009). Mechanisms of directional asymmetry in the zebrafish epithalamus. *Seminars in Cell & Developmental Biology*, *20*, 498–509.

Concha, M. L., & Wilson, S. W. (2001). Asymmetry in the epithalamus of vertebrates. *Journal of Anatomy*, *199*, 63–84.

De Groot, S. J. (1970). Some notes on an ambivalent behaviour of the Greenland halibut *Reinhardtius hippoglossoides* (Walb.) Pisces: Pleuronectiformes. *Journal of Fish Biology*, *2*, 275–279.

Fernandez, I., & Gisbert, E. (2011). The effect of vitamin A on flatfish development and skeletogenesis: A review. *Aquaculture and Fisheries Management*, *315*, 34–48.

Finger, T. E. (1976). An asymmetric optomotor response in developing flounder larvae (*Pseudopleuronectes americanus*). *Vision Research*, *16*, 941–943.

Francis, A. W., Jr., & Turingan, R. G. (2008). Morphological and biomechanical changes of the feeding apparatus in developing southern flounder, *Paralichthys lethostigma*. *Journal of Morphology*, *269*, 1169–1180.

Friedman, M. (2008). The evolutionary origin of flatfish asymmetry. *Nature*, *454*, 209–212.

Friedman, M. (2012). Osteology of †*Heteronectes chaneti* (Acanthomorpha, Pleuronectiformes), an Eocene stem flatfish, with a discussion of flatfish sister-group relationships. *Journal of Vertebrate Paleontology*, *32*, 735–756.

Fu, Y., Shi, Z., Wu, M., Zhang, J., Jia, L., & Chen, X. (2011). Identification and differential expression of microRNAs during metamorphosis of the Japanese flounder (*Paralichthys olivaceus*). *PLoS One*, *6*, e22957.

Galay-Burgos, M., Power, D. M., Llewellyn, L., & Sweeney, G. E. (2008). Thyroid hormone receptor expression during metamorphosis of Atlantic halibut (*Hippoglossus hippoglossus*). *Molecular and Cellular Endocrinology*, *281*, 56–63.

Galloway, T. F., Bardal, T., Kvam, S. N., Dahle, S. W., Nesse, G., Randol, M., et al. (2006). Somite formation and expression of MyoD, myogenin and myosin in Atlantic halibut (*Hippoglossus hippoglossus* L.) embryos incubated at different temperatures: Transient asymmetric expression of MyoD. *The Journal of Experimental Biology*, *209*, 2432–2441.

Galloway, T. F., Kjørsvik, E., & Kryvi, H. (1999). Muscle growth in yolk-sac larvae of the Atlantic halibut as influenced by temperature in the egg and yolk-sac stage. *Journal of Fish Biology*, *55*, 26–43.

Gamse, J. T., Thisse, C., Thisse, B., & Halpern, M. E. (2003). The parapineal mediates left-right asymmetry in the zebrafish diencephalon. *Development (Cambridge, England)*, *130*, 1059–1068.

Gibb, A. (1995). Kinematics of prey capture in a flatfish, *Pleuronichthys verticalis*. *The Journal of Experimental Biology*, *198*, 1173–1183.

Gibb, A. (1996). The kinematics of prey capture in *Xystreurys liolepis*: Do all flatfish feed asymmetrically? *The Journal of Experimental Biology*, *199*, 2269–2283.

Gibb, A. (1997). Do flatfish feed like other fishes? A comparative study of percomorph prey-capture kinematics. *The Journal of Experimental Biology*, *200*, 2841–2859.

Gibb, A. C. (2003). Modeling the jaw mechanism of *Pleuronichthys verticalis*: The morphological basis of asymmetrical jaw movements in a flatfish. *Journal of Morphology*, *256*, 1–12.

Gibson, R. N. (2005). The behaviour of flatfishes. In R. N. Gibson (Ed.), *Flatfishes: Biology and exploitation* (pp. 213–239). Oxford: Blackwell Publishing.

Graf, W., & Baker, R. (1983). Adaptive changes of the vestibulo-ocular reflex in flatfish are achieved by reorganization of central nervous pathways. *Science*, *221*, 777–779.

Graf, W., & Baker, R. (1985a). The vestibuloocular reflex of the adult flatfish. I. Oculomotor organization. *Journal of Neurophysiology*, *54*, 887–899.

Graf, W., & Baker, R. (1985b). The vestibuloocular reflex of the adult flatfish. II. Vestibulooculomotor connectivity. *Journal of Neurophysiology*, *54*, 900–916.

Graf, W., & Baker, R. (1990). Neuronal adaptation accompanying metamorphosis in the flatfish. *Journal of Neurobiology*, *21*, 1136–1152.

Graf, W., & Meyer, D. L. (1983). Central mechanisms counteract visually induced tonus asymmetries. A study on ocular responses to unilateral illumination. *Journal of Comparative Physiology*, *150*, 473–481.

Graf, W., Spencer, R., Baker, H., & Baker, R. (2001). Vestibuloocular reflex of the adult flatfish. III. A species-specific reciprocal pattern of excitation and inhibition. *Journal of Neurophysiology*, *86*, 1376–1388.

Graff, R. J., Simmons, D., Meyer, J., Martin-Morgan, D., & Kurtz, M. (1986). Abnormal bone production associated with mutant mouse genes pa and we. *The Journal of Heredity*, *77*, 109–113.

Hagen, Ø., Vieira, V. L. A., Solberg, C., & Johnston, I. A. (2008). Myotube production in fast myotomal muscle is switched-off at shorter body lengths in male than female Atlantic halibut *Hippoglossus hippoglossus* (L.) resulting in a lower final fibre number. *Journal of Fish Biology*, *73*, 139–152.

Hamada, H., Meno, C., Watanabe, D., & Saijoh, Y. (2002). Establishment of vertebrate left-right asymmetry. *Nature Reviews. Genetics*, *3*, 103–113.

Hamre, K., & Harboe, T. (2008). Critical levels of essential fatty acids for normal pigmentation in Atlantic halibut (*Hippoglossus hippoglossus* L.) larvae. *Aquaculture*, *277*, 101–108.

Hamre, K., Holen, E., & Moren, M. (2007). Pigmentation and eye migration in Atlantic halibut (*Hippoglossus hippoglossus* L.) larvae: New findings and hypotheses. *Aquaculture Nutrition, 13*, 65–80.
Hamre, K., Moren, M., Solbakken, J., Opstad, I., & Pittman, K. (2005). The impact of nutrition on metamorphosis in Atlantic halibut (*Hippoglossus hippoglossus* L.). *Aquaculture, 250*, 555–565.
Harboe, T., Mangor-Jensen, A., Moren, M., Hamre, K., & Rønnestad, I. (2009). Control of light condition affects the feeding regime and enables successful eye migration in Atlantic halibut juveniles. *Aquaculture, 290*, 250–255.
Hashimoto, H., Aritaki, M., Uozumi, K., Uji, S., Kurokawa, T., & Suzuki, T. (2007). Embryogenesis and expression profiles of charon and nodal-pathway genes in sinistral (*Paralichthys olivaceus*) and dextral (*Verasper variegatus*) flounders. *Zoological Science, 24*, 137–146.
Hashimoto, H., Mizuta, A., Okada, N., Suzuki, T., Tagawa, M., Tabata, K., et al. (2002). Isolation and characterization of a Japanese flounder clonal line, reversed, which exhibits reversal of metamorphic left-right asymmetry. *Mechanisms of Development, 111*, 17–24.
Helling, K., Scherer, H., Hausmann, S., & Clarke, A. H. (2005). Otolith mass asymmetries in the utricle and saccule of flatfish. *Journal of Vestibular Research, 15*, 59–64.
Hildahl, J., Power, D. M., Bjornsson, B. T., & Einarsdottir, I. E. (2008). Involvement of growth hormone-insulin-like growth factor I system in cranial remodeling during halibut metamorphosis as indicated by tissue- and stage-specific receptor gene expression and the presence of growth hormone receptor protein. *Cell and Tissue Research, 332*, 211–225.
Hoshino, K. (2006). Fixing the confused term "pseudomesial bar" and homologies of pleuronectiform cranial elements, with proposals of new terms. *Ichthyological Research, 53*, 435–440.
Hubbs, C. L., & Hubbs, L. C. (1944). Bilateral asymmetry and bilateral variation in fishes. Papers of the Michigan Academy of Science Arts and Letters (pp. 229–311).
Hubbs, C. L., & Kuronuma, K. (1942). Hybridization in nature between two genera of flounders in Japan. Papers of the Michigan Academy of Science, Arts and Letters, Vol. 27.
Inui, Y., & Miwa, S. (1985). Thyroid hormone induces metamorphosis of flounder larvae. *General and Comparative Endocrinology, 60*, 450–454.
Isorna, E., Obregon, M. J., Calvo, R. M., Vazquez, R., Pendon, C., Falcon, J., et al. (2009). Iodothyronine deiodinases and thyroid hormone receptors regulation during flatfish (*Solea senegalensis*) metamorphosis. *Journal of Experimental Zoology. Part B, Molecular and Developmental Evolution, 312B*, 231–246.
Itoh, K., Watanabe, K., Wu, X., & Suzuki, T. (2010). Three members of the iodothyronine deiodinase family, dio1, dio2 and dio3, are expressed in spatially and temporally specific patterns during metamorphosis of the flounder, *Paralichthys olivaceus*. *Zoological Science, 27*, 574–580.
Jacob, W. (1928). Uber das Labyrinth der Pleuronectiden. *Zoologische Jahrbucher-Abteilung Fur All Gemeine Zoologie Und Physiologie Der Tiere, 44*, 523–574.
Jansen, J. K., & Enger, P. S. (1996a). Pre- and postmetamorphic organization of the vestibular nuclear complex in the turbot examined by retrograde tracer substances. *The Journal of Comparative Neurology, 364*, 677–689.
Jansen, J. K. S., & Enger, P. S. (1996b). NADPH diaphorase activity is asymmetrically distributed in the optic tectum during the period of eye migration in turbots. *Acta Physiologica Scandinavica, 157*, 515–517.
Johnston, I. A. (2004). Development of molecular tools for investigating muscle growth and flesh quality in halibut. In I. A. Johnston, B. Norberg & C. Solberg (Eds.), *Halibut farming from A to Z* (pp. 259–265). Bodø, Norway: Bodø University College.

Jorgensen, I. M. (1976). Hair cell polarization in the flatfish inner ear. *Acta Zoologica, 57*, 37–39.
Kuan, Y. S., Gamse, J. T., Schreiber, A. M., & Halpern, M. E. (2007). Selective asymmetry in a conserved forebrain to midbrain projection. *Journal of Experimental Zoology. Part B, Molecular and Developmental Evolution, 308*, 669–678.
Laudet, V. (2011). The origins and evolution of vertebrate metamorphosis. *Current Biology, 21*, R726–R737.
Levin, M. (2005). Left-right asymmetry in embryonic development: A comprehensive review. *Mechanisms of Development, 122*, 3–25.
Liang, J. O., Etheridge, A., Hantsoo, L., Rubinstein, A. L., Nowak, S. J., Izpisua Belmonte, J. C., et al. (2000). Asymmetric nodal signaling in the zebrafish diencephalon positions the pineal organ. *Development (Cambridge, England), 127*, 5101–5112.
Logan, M., Pagan-Westphal, S. M., Smith, D. M., Paganessi, L., & Tabin, C. J. (1998). The transcription factor Pitx2 mediates situs-specific morphogenesis in response to left-right asymmetric signals. *Cell, 94*, 307–317.
Loher, T., Wischniowski, S., & Martin, G. B. (2008). Elemental chemistry of left and right sagittal otoliths in a marine fish *Hippoglossus stenolepis* displaying cranial asymmetry. *Journal of Fish Biology, 73*, 870–887.
López, A. V., Müller, M. I., Radonić, M., Bambill, G. A., Boccanfuso, J. J., & Bianca, F. A. (2009). Larval culture technique and quality control in juveniles of flounder. *Paralichthys orbignyanus* (Valenciennes, 1839) in Argentina. *Spanish Journal of Agricultural Research, 7*, 75–82.
Luckenbill-Edds, L., & Sharma, S. C. (1977). Retinotectal projection of the adult winter flounder (*Pseudopleuronectes americanus*). *The Journal of Comparative Neurology, 173*, 307–318.
Lychakov, D. V. (1996). Anomalous otoliths of Black Sea fishes. *Zhurnal Evoliutsionnoĭ Biokhimii i Fiziologii, 32*, 470–477.
Lychakov, D. V., Rebane, Y. T., Lombarte, A., Demestre, M., & Fuiman, L. A. (2008). Saccular otolith mass asymmetry in adult flatfishes. *Journal of Fish Biology, 72*, 2579–2594.
Manchado, M., Infante, C., Rebordinos, L., & Canavate, J. P. (2009). Molecular characterization, gene expression and transcriptional regulation of thyroid hormone receptors in Senegalese sole. *General and Comparative Endocrinology, 160*, 139–147.
Marchand, O., Duffraisse, M., Triqueneaux, G., Safi, R., & Laudet, V. (2004). Molecular cloning and developmental expression patterns of thyroid hormone receptors and T3 target genes in the turbot (*Scophtalmus maximus*) during post-embryonic development. *General and Comparative Endocrinology, 135*, 345–357.
Medina, M., Reperant, J., Ward, R., Rio, J. P., & Lemire, M. (1993). The primary visual system of flatfish: An evolutionary perspective. *Anatomy and Embryology, 187*, 167–191.
Meyer, D. L., von Seydlitz-Kurzbach, U., & Fiebig, E. (1981). Bilaterally asymmetrical uptake of [14C]2-deoxyglucose by the octavo-lateralis complexes in flatfish. *Cell and Tissue Research, 214*, 659–662.
Miwa, S., Tagawa, M., Inui, Y., & Hirano, T. (1988). Thyroxine surge in metamorphosing flounder larvae. *General and Comparative Endocrinology, 70*, 158–163.
Munroe, T. A. (2005). Systematic diversity of the Pleuronectiformes. In R. N. Gibson (Ed.), *Flatfishes: Biology and exploitation* (pp. 10–36). Oxford: Blackwell Publishing.
Murray, M. (1974). Axonal transport in the asymmetric optic axons of flatfish. *Experimental Neurology, 42*, 636–646.
Neave, D. A. (1985). The dorsal light reactions of larval and metamorphosing flatfish. *Journal of Fish Biology, 26*, 629–640.
Norman, J. R. (1934). A systematic monograph of the flatfishes (heterosomata). In *Psettodidae, Bothidae, Pleuronectidae*, Vol. 1. London: British Museum of Natural History.

Okada, N., Takagi, Y., Seikai, T., Tanaka, M., & Tagawa, M. (2001). Asymmetrical development of bones and soft tissues during eye migration of metamorphosing Japanese flounder, *Paralichthys olivaceus*. *Cell and Tissue Research*, *304*, 59–66.

Okada, N., Takagi, Y., Tanaka, M., & Tagawa, M. (2003). Fine structure of soft and hard tissues involved in eye migration in metamorphosing Japanese flounder (*Paralichthys olivaceus*). *The Anatomical Record*, *273A*, 663–668.

Okada, N., Tanaka, M., & Tagawa, M. (2003). Histological study of deformity in eye location in Japanese flounder *Paralichthys olivaceus*. *Fisheries Science (Japan)*, *69*, 777–784.

Parker, G. H. (1903). The optic chiasma in teleosts and its bearing on the asymmetry of the heterosomata (flatfishes). *Bulletin of the Museum of Comparative Zoology at Harvard College*, *40*, 221–242.

Piedra, M. E., Icardo, J. M., Albajar, M., Rodriguez-Rey, J. C., & Ros, M. A. (1998). Pitx2 participates in the late phase of the pathway controlling left-right asymmetry. *Cell*, *94*, 319–324.

Pinuela, C., Rendon, C., Gonzalez de Canales, M. L., & Sarasquete, C. (2004). Development of the Senegal sole, *Solea senegalensis* forebrain. *European Journal of Histochemistry*, *48*, 377–384.

Pittman, K., Jelmert, A., Naess, T., Harboe, T., & Watanabe, K. (1998). Plasticity of viable postmetamorphic forms of farmed Atlantic halibut, *Hippoglossus hippoglossus* L. *Aquaculture Research*, *29*, 949–954.

Platt, C. (1973a). Central control of postural orientation in flatfish. I. Postural change dependence on central neural changes. *The Journal of Experimental Biology*, *59*, 491–521.

Platt, C. (1973b). Central control of postural orientation in flatfish. II. Optic vestibular efferent modification of gravistatic input. *The Journal of Experimental Biology*, *59*, 523–541.

Platt, C. (1983). Retention of generalized hair cell patterns in the inner ear of the primitive flatfish Psettodes. *The Anatomical Record*, *207*, 503–508.

Policansky, D. (1982a). Flatfishes and the inheritance of asymmetries. *The Behavioral and Brain Sciences*, *5*, 262–266.

Policansky, D. (1982b). The asymmetry of flounders. *Scientific American*, *246*, 116–122.

Power, D., Einarsdottir, I., Pittman, K., Sweeney, G., Hildahl, J., Campinho, M., et al. (2008). The molecular and endocrine basis of flatfish metamorphosis. *Reviews in Fisheries Science*, *16*, 93–109.

Prasada Rao, P. D., & Finger, T. E. (1984). Asymmetry of the olfactory system in the brain of the winter flounder, *Pseudopleuronectes americanus*. *The Journal of Comparative Neurology*, *225*, 492–510.

Russo, T., Pulcini, D., Costantini, D., Pedreschi, D., Palamara, E., Boglione, C., et al. (2012). "Right" or "wrong"? insights into the ecology of sidedness in european flounder, *Platichthys flesus*. *Journal of Morphology*, *273*, 337–346.

Ryan, A. K., Blumberg, B., Rodriguez-Esteban, C., Yonei-Tamura, S., Tamura, K., Tsukui, T., et al. (1998). Pitx2 determines left-right asymmetry of internal organs in vertebrates. *Nature*, *394*, 545–551.

Saele, O., Silva, N., & Pittman, K. (2006). Post-embryonic remodelling of neurocranial elements: A comparative study of normal versus abnormal eye migration in a flatfish, the Atlantic halibut. *Journal of Anatomy*, *209*, 31–41.

Saele, O., Smaradottir, H., & Pittman, K. (2006). Twisted story of eye migration in flatfish. *Journal of Morphology*, *267*, 730–738.

Schreiber, A. M., & Specker, J. L. (1998). Metamorphosis in the summer flounder (*Paralichthys dentatus*): stage-specific developmental response to altered thyroid status. *General and Comparative Endocrinology*, *111*, 156–166.

Schreiber, A. M. (2006). Asymmetric craniofacial remodeling and lateralized behavior in larval flatfish. *The Journal of Experimental Biology*, *209*, 610–621.

Schreiber, A. M., Wang, X., Tan, Y., Sievers, Q., Sievers, B., Lee, M., et al. (2010). Thyroid hormone mediates otolith growth and development during flatfish metamorphosis. *General and Comparative Endocrinology, 169*, 130–137.

Shi, Y.-B. (2000). *Amphibian metamorphosis: From morphology to molecular biology*. New-York: Wiley-Liss.

Snelson, C. D., & Gamse, J. T. (2009). Building an asymmetric brain: Development of the zebrafish epithalamus. *Seminars in Cell & Developmental Biology, 20*, 491–497.

Sogard, S. M. (1991). Interpretation of otolith microstructure in juvenile winter flounder *Pseudopleuronectes americanus* ontogenetic development daily increment validation and somatic growth relationship. *Canadian Journal of Fisheries and Aquatic Sciences, 48*, 1862–1871.

Solbakken, J., & Pittman, K. (2004). Photoperiodic modulation of metamorphosis in Atlantic halibut (*Hippoglossus hippoglossus* L.). *Aquaculture, 232*, 613–625.

Suzuki, T., Washio, Y., Aritaki, M., Fujinami, Y., Shimizu, D., Uji, S., et al. (2009). Metamorphic pitx2 expression in the left habenula correlated with lateralization of eye-sidedness in flounder. *Development, Growth & Differentiation, 51*, 797–808.

Tagawa, M., & Aritaki, M. (2005). Production of symmetrical flatfish by controlling the timing of thyroid hormone treatment in spotted halibut *Verasper variegatus*. *General and Comparative Endocrinology, 141*, 184–189.

Traquair, R. H. (1865). On the asymmetry of the Pleuronectidae, as elucidated by an examination of the skeleton in the turbot, halibut and plaice. *Transactions of the Linnean Society of London, 25*, 263–296.

Tunnicliffe, V., Koop, B. F., Tyler, J., & So, S. (2010). Flatfish at seamount hydrothermal vents show strong genetic divergence between volcanic arcs. *Marine Ecology, 31*, 158–167.

Velez, Z., Hubbard, P. C., Barata, E. N., & Canario, A. V. (2005). Evidence for functional asymmetry in the olfactory system of the Senegalese sole (*Solea senegalensis*). *Physiological and Biochemical Zoology, 78*, 756–765.

Wagemans, F., Focant, B., & Vandewalle, P. (1998). Early development of the cephalic skeleton in the turbot. *Journal of Fish Biology, 52*, 166–204.

Wagemans, F., & Vandewalle, P. (2001). Development of the bony skull in common sole: Brief survey of morpho-functional aspects of ossification sequence. *Journal of Fish Biology, 59*, 1350–1369.

Wang, X., Tan, Y., Sievers, Q., Sievers, B., Lee, M., Burrall, K., et al. (2011). Thyroid hormone-responsive genes mediate otolith growth and development during flatfish metamorphosis. *Comparative Biochemistry and Physiology. Part A, Molecular & Integrative Physiology, 158*, 163–168.

Xie, C., Xu, S., Yang, L., Ke, Z., Xing, J., Gai, J., et al. (2011). mRNA/microRNA profile at the metamorphic stage of olive flounder (*Paralichthys olivaceus*). *Comparative and Functional Genomics, 2011*, 256038.

Yamano, K., & Miwa, S. (1998). Differential gene expression of thyroid hormone receptor alpha and beta in fish development. *General and Comparative Endocrinology, 109*, 75–85.

Zhang, J., Shi, Z., Cheng, Q., & Chen, X. (2011). Expression of insulin-like growth factor I receptors at mRNA and protein levels during metamorphosis of Japanese flounder (*Paralichthys olivaceus*). *General and Comparative Endocrinology, 173*, 78–85.

CHAPTER SEVEN

Neuroendocrinology of Amphibian Metamorphosis

Robert J. Denver[*,†,1]

[*]Department of Molecular, Cellular and Developmental Biology, The University of Michigan, Ann Arbor, Michigan, USA

[†]Department of Ecology and Evolutionary Biology, The University of Michigan, Ann Arbor, Michigan, USA

[1]Corresponding author: e-mail address: rdenver@umich.edu

Contents

Abstract

The timing of metamorphosis is a central amphibian life history trait and is controlled by the interplay of developmental progression, body size and condition, and environmental signals. These different processes and signals are integrated by the neuroendocrine system to regulate production of hormones by the thyroid gland. Thyroid hormone (TH) is the primary morphogen controlling metamorphosis, while corticosteroids (CSs) produced by the interrenal glands synergize with TH to promote metamorphic changes. The actions of TH are modulated by monodeiodinase enzymes expressed in TH target tissues. CSs act by sensitizing tissues to the actions of TH via the upregulation of TH receptors and monodeiodinases. The increase in thyroid gland activity during metamorphosis is controlled by the hypothalamus and pituitary gland. The hypothalamo–pituitary–thyroid and hypothalamo–pituitary–interrenal axes are regulated at multiple levels. Hypothalamic corticotropin-releasing factor (CRF) functions as a common, central regulator of pituitary thyroid-stimulating hormone (TSH) and adrenocorticotropic hormone (ACTH) secretion in tadpoles. CRF neurons transduce the signals of environmental change (e.g., pond drying, resource availability, etc.) on metamorphic timing by regulating TSH and ACTH secretion, and consequently the production of TH and CS.

Current Topics in Developmental Biology, Volume 103
ISSN 0070-2153
http://dx.doi.org/10.1016/B978-0-12-385979-2.00007-1

1. HORMONAL CONTROL OF METAMORPHOSIS

Metamorphosis is a stage of the amphibian life cycle that is characterized by dramatic morphological transformation accompanied by a transition in ecological niche and behavioral mode. Hormones orchestrate the diverse morphological and physiological changes that occur during metamorphosis, and also function as mediators of environmental effects on development. A striking characteristic of amphibian metamorphosis is that a single signaling molecule produced by the thyroid gland (thyroid hormone, TH) can orchestrate the entire suite of molecular, biochemical, and morphological changes. TH is required for amphibian metamorphosis (Brown & Cai, 2007); the hormone initiates gene expression programs in diverse tissues that lead to cell proliferation, death, differentiation, or migration (Brown & Cai, 2007). Hormones produced by the anterior pituitary gland and the interrenal glands (amphibian homologs of the mammalian adrenal cortex; corticosteroids, CSs) influence the rate of metamorphosis by controlling TH production and action on target tissues. Neurohormones produced in the hypothalamus control hormone biosynthesis and secretion by the pituitary gland, and the hypothalamus mediates the interaction between the external and internal environments, and the production of hormones that control metamorphosis.

1.1. Thyroid hormone

The thyroid gland develops early in the amphibian embryo and matures functionally at the time of hatching when it separates into two distinct lobes and is essentially completely developed by the onset of metamorphosis (Dodd & Dodd, 1976; Kaye, 1959, 1961; Nieuwkoop & Faber, 1956; Regard, Taurog, & Nakashima, 1978; Saxen, Saxen, Toivonen, & Salimaki, 1957a, 1957b). Thyroid activity increases markedly during prometamorphosis, peaks at metamorphic climax, and declines thereafter to reach an "adult" level of activity (Dodd & Dodd, 1976; Kaye, 1959, 1960; Kikuyama, Kawamura, Tanaka, & Yamamoto, 1993; Regard et al., 1978). The major product of the amphibian thyroid gland is 3,5,3′5′-tetraiodothyronine (thyroxine; T_4) with minor amounts of 3,5,3′-triiodothyronine (T_3) produced (Buscaglia, Leloup, & De Luze, 1985; Rosenkilde, 1978). Coincident with measures of thyroid activity, plasma concentration and whole-body content of T_3 and T_4 increase throughout prometamorphosis and peak at metamorphic climax (Denver,

2009a; Kikuyama et al., 1993). As in other vertebrates, T_3 has greater biological activity than T_4 in amphibia owing to the TH receptors (TRs) having 10–15 times greater affinity for T_3 than for T_4 (Frieden, 1981; Leonard & Visser, 1986; Lindsay, Buettner, Wimberly, & Pittman, 1967; Oppenheimer, Schwartz, & Strait, 1995; Rosenkilde, 1978; Wahlborg, Bright, & Frieden, 1964; White & Nicoll, 1981).

1.1.1 TH metabolism

An important point of control of TH bioactivity is at the target tissues, where monodeiodinase enzymes convert T_4 to T_3, or inactivate T_4 and T_3 (Fig. 7.1). The monodeiodinases catalyze two basic reactions: a 5′-monodeiodination (outer ring) that results in bioactivation; and a 5-monodeiodination (inner ring) that results in bioinactivation of the substrate, T_4 or T_3. There are three types of vertebrate deiodinases (types I, II, and III) that differ in their substrate specificity, kinetics, and sensitivity to inhibitors. Type I catalyzes both 5- and 5′-, type II 5′-, and type III 5-deiodination (St Germain, Galton, & Hernandez, 2009). Type II and type III, but not type I enzyme activities have been detected in tadpole tissues, and although frogs have a type I (*Dio1*) gene, little is known about its expression or function (Becker, Stephens, Davey, Schneider, & Galton, 1997; Dubois et al., 2006; Kuiper et al., 2006).

Three deiodinase genes have been isolated in amphibian species (Brown, 2005). The *Dio2* (type II) and *Dio3* (type III) genes exhibit tissue-specific and developmental stage-specific expression patterns (Becker et al., 1997; Brown, 2005; Cai & Brown, 2004). The expression patterns correlate with the asynchronous tissue morphogenesis, and the roles that the deiodinases play in modulating intracellular T_3 concentration during metamorphosis (Brown, 2005; St Germain et al., 2009). In many cells, both enzymes may be expressed, and the relative expression levels may establish a type of push–pull mechanism that regulates intracellular T_3 concentration (St Germain et al., 2009). Alternatively, in some tissues, the two genes show different temporal dynamics, leading to hormone inactivation or activation at different developmental stages.

For example, *Dio3* mRNA is expressed in several cell types in tadpole tail, but not in tail muscle cells (Berry, Schwartzman, & Brown, 1998); both *Dio3* mRNA and 5-deiodinase activity increase during late prometamorphosis (NF stage 59–61) but then decline sharply at metamorphic climax (Brown et al., 1996; St Germain et al., 1994). This pattern of *Dio3* expression may protect the tadpole tail, an essential locomotory organ, from premature resorption (Brown, 2005). By contrast, *Dio2* expression, which occurs mainly in tail

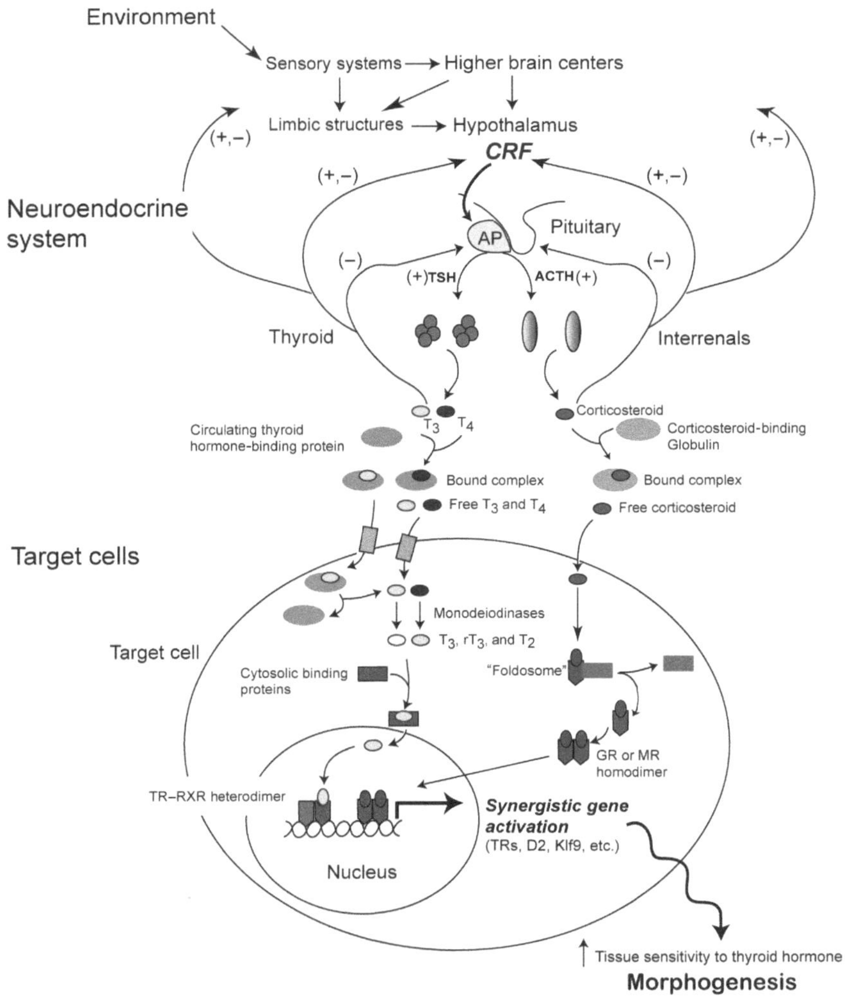

Figure 7.1 Central and peripheral organization of the thyroid and stress endocrine axes controlling amphibian metamorphosis. A schematic representation of the hypothalamo–pituitary–thyroid (HPT) and hypothalamo–pituitary–adrenal (HPA; stress) axes in amphibian tadpoles, their regulation by input from the external environment, transduction of this input by neural and neuroendocrine pathways, and synergistic interactions among thyroid hormones and corticosteroids in target cells leading to the promotion of metamorphosis. The two endocrine axes are controlled centrally by corticotropin-releasing factor (CRF) which acts on the anterior pituitary gland (AP) to stimulate the release of thyrotropin (TSH) and corticotropin (ACTH). TSH acts on the thyroid gland to stimulate release of thyroxine (T_4) and 3,5,3′-triiodothyronine (T_3). Thyroid hormones are transported in the blood bound by serum binding proteins (transthyretin, TBG, and albumin). ACTH acts on adrenal cortical cells in the interrenal glands to stimulate biosynthesis and

fibroblasts, is undetectable before late prometamorphosis at which time expression increases markedly and is maintained through the end of metamorphosis (Cai & Brown, 2004). This late expression of *Dio2* is hypothesized to generate bioactive T_3 at an appropriate developmental stage to accelerate tail resorption. Neither *Dio2* nor *Dio3* mRNAs are expressed in tail muscle cells (Berry et al., 1998; Bonett, Hoopfer, & Denver, 2010); nonmuscle tail cells may inactivate T_4 during pre- and prometamorphosis to protect tail muscle cells from apoptosis, and subsequently generate high local concentrations of T_3 to promote tail muscle cell apoptosis at metamorphic climax.

Tissue transformations during metamorphosis are asynchronous: some tissues respond early to low plasma concentrations of TH (e.g., hindlimb, brain), while other tissues respond later in development and require high TH concentration (e.g., intestine, tail—discussed above). The expression patterns of the monodiodinase genes may play a key role in establishing tissue competence to respond to the TH signal. For example, for tissues that respond early in metamorphosis to TH like the retina and hindlimb, *Dio2* expression was high during early prometamorphosis but declined at metamorphic climax. The importance of 5′-deiodinase activity for hindlimb development is supported by findings that T_4 has no effect on the hindlimb

---

release of glucocorticoids which are transported in the blood bound to corticosteroid-binding globulin. Cellular uptake of T_3 and T_4 is achieved by organic anion, monocarboxylate, and amino acid transporters; there is also evidence that thyroid hormones may enter cells bound to transthyretin via a receptor-mediated process. Glucocorticoids enter cells by passive diffusion across the plasma membrane. Upon entering the cell, thyroid hormone is bound by cytosolic-binding proteins, some of which (the monodeiodinases) convert the hormone to either active (T_3; deiodinases types I and II) or inactive forms (reverse T_3 [rT_3], diiodothyronine [T_2]; deiodinase type III). TRs form heterodimers with RXRs and are bound to DNA in the unliganded form where they actively repress gene transcription. Upon thyroid hormone binding to TR, gene transcription is derepressed and activated. Upon entering the cell, glucocorticoids bind to corticosteroid receptors (glucocorticoid receptor [GR] or mineralocorticoid receptor [MR]) that are located in the cytosol bound to heat shock proteins (the "foldosome"). Hormone binding causes a conformational change in the receptor, the release of heat shock proteins, and dimerization and translocation of receptors to the nucleus where they bind DNA to activate or repress target genes. When cells are exposed to low concentrations of thyroid hormone plus glucocorticoids, genes such as the TRs, deiodinase type 2, and the thyroid hormone-inducible transcription factor *Klf9* are activated in a synergistic manner. This leads to enhanced sensitivity of cells to the actions of thyroid hormone, which accelerates metamorphosis. "+" indicates an increase and "−" indicates a decrease in the regulated variable. (See Color Insert.)

in the presence of the deiodinase inhibitor iopanoic acid (Brown, 2005). *Dio2* mRNA expression showed a progressive decline in the brain throughout metamorphosis, while brain *Dio3* mRNA increased during late prometamorphosis and metamorphic climax (Hogan, Crump, Duarte, Lean, & Trudeau, 2007). TH induces cell proliferation in the early prometamorphic tadpole brain, but cells of the neurogenic zone become refractory to TH action on cell proliferation as metamorphic climax approaches, which may be explained by the temporal patterns of *Dio2* and *Dio3* expression and actions (Cai & Brown, 2004; Denver, Hu, Scanlan, & Furlow, 2009).

TH regulates the expression of the *Dio2* and *Dio3* genes. *Dio3* appears to be a direct T_3 response gene based on its T_3 response kinetics and the resistance of its upregulation to protein synthesis inhibition (Becker, Schneider, Davey, & Galton, 1995; Das, Heimeier, Buchholz, & Shi, 2009; Denver, Pavgi, & Shi, 1997; Hogan et al., 2007; Kawahara, Gohda, & Hikosaka, 1999; St Germain et al., 1994; Wang & Brown, 1993). TH positively regulates 5′-deiodinase activity and *Dio2* mRNA in tadpoles (Brown, 2005; Buscaglia et al., 1985; Hogan et al., 2007). However, unlike *Dio3*, which is an early T_3 response gene, *Dio2* exhibits delayed response kinetics; the gene was not isolated in screens for early response, direct TR target genes in *Xenopus* tadpole tissues (Bonett et al., 2010; Brown, 2005; Buchholz, Heimeier, Das, Washington, & Shi, 2007; Das et al., 2006). The dependence of *Dio2* expression on TH may vary among tissues. Tissues in which cell proliferation occurs as an early response to TH constitutively express relatively high levels of *Dio2* (e.g., neurogenic zones of the brain and spinal cord, limb buds); whereas, in tissues that transform later, *Dio2* is upregulated by TH (Cai & Brown, 2004). Treatment of early prometamorphic tadpoles with T_4 can induce cell proliferation in these tissues, which can be blocked by the deiodinase inhibitor iopanoic acid (Cai & Brown, 2004). *Dio2* mRNA in brain and spinal cord declines at metamorphic climax, as does TH-dependent cell proliferation (Cai & Brown, 2004; Denver et al., 2009). While the decline in *Dio2* mRNA may be permissive for the reduction in TH-dependent cell proliferation in the brain, it does not alone explain why the brain becomes refractory to TH action since treatment with T_3 could not increase cell proliferation (which normally declines) as the animals approach metamorphic climax (Denver et al., 2009). The decline in cell proliferation is likely due to processes, likely under TH control, that lead to a reduction in the stem cell/progenitor pool in the ventricular/subventricular zones of the tadpole brain. *Dio2* expression then appears in late-responding tissues such as the intestine, tail, and anterior pituitary and may be induced at this time by rising plasma titers of TH (Cai & Brown, 2004; Huang, Cai, Remo, & Brown, 2001; Manzon & Denver, 2004; discussed more below). Physiological roles for tissue monodeiodinases in the timing of metamorphosis are supported by experiments with iopanoic acid and transgenesis overexpression of *Dio3* (Becker et al., 1997; Buscaglia et al., 1985; Cai & Brown, 2004; Galton, 1989; Huang et al., 2001; Huang, Marsh-Armstrong, & Brown, 1999; Marsh-Armstrong, Huang, Remo, Liu, & Brown, 1999).

1.1.2 Plasma TH transport proteins

Thyroxine synthesized by thyroid follicular cells diffuses into the bloodstream where it is reversibly bound by plasma proteins that transport the hormone from the site of production to its target tissues (Fig. 7.1). Two plasma-binding proteins that bind T_4 and T_3 with moderate to high affinities have been identified in vertebrates. Thyroxine-binding globulin (TBG) binds T_4 with high affinity and low capacity but is found only in large, eutherian mammals (Power et al., 2000). Transthyretin (TTR; also known as prealbumin) is found in all vertebrates and it binds T_4 with moderate affinity and intermediate capacity. Both TBG and TTRs can also bind T_3, although in most cases with 10 times lower affinity than T_4 (Power et al., 2000); although, the situation in amphibia is the reverse, where TTR binds T_3 with greater affinity than T_4 (Yamauchi, Kasahara, Hayashi, & Horiuchi, 1993; Yamauchi, Nakajima, Hayashi, & Hara, 1999; Yamauchi, Prapunpoj, & Richardson, 2000; Yamauchi et al., 1998). The two primary sites for TTR expression in vertebrates are the liver and the choroid plexus (although it is expressed at other sites; Power et al., 2000). In amphibians, TTR is expressed primarily in the liver (Power et al., 2000). An essential function of TTR is its interaction with retinol-binding protein, which acts as a carrier for all-*trans*-retinol in the blood. The functional significance of this interaction is not known, but it is intriguing that T_3 and 9-*cis*-retinoic acid (which is a metabolite of all-*trans*-retinol) serve as ligands for the TR–retinoid X receptor (RXR) heterocomplex that regulates TH target genes. Serum albumin also binds T_3 and T_4 in many species with low affinity and high capacity, and Power et al. (2000) suggested that albumin might be the principal T_4-binding protein in amphibia.

Circulating TTR is present in tadpoles during premetamorphosis and prometamorphosis when thyroid activity is increasing, but declines at metamorphic climax (Prapunpoj, Yamauchi, Nishiyama, Richardson, & Schreiber, 2000; Yamauchi et al., 2000, 1998). The free hormone hypothesis (Ekins, 1990; Mendel, 1989) leads to the prediction that TTR during pre- and early prometamorphosis serves to reduce the free fraction of TH in blood thus limiting bioavailability. Conversely, hormone-binding proteins can serve as a reservoir for hormone in the blood; TTR could therefore help to sustain increasing plasma TH concentrations prior to the acceleration of thyroid gland activity induced by rising plasma thyroid-stimulating hormone (TSH) titers. The TTR concentration in the blood declines at metamorphic climax when plasma TH concentration is maximal. The continued rise in TH synthesis by the thyroid gland, paired with a decline in TTR, could result in an increase in the free hormone fraction in the blood. At the same time, the rate of clearance of T_3 from the circulation would likely increase. However, because the thyroid synthetic rate is high at metamorphic climax, total plasma T_3 concentration continues to rise. Thus, one would predict that, not only does the hormone production rate increases at metamorphic climax but also does the proportional availability of T_3 to the target tissues. To my knowledge, T_3 or T_4 clearance rates have not been calculated in tadpoles at different stages of development. Based on TTR expression profiles, one would predict that clearance rates would be lower during prometamorphosis compared with premetamorphosis or metamorphic climax. Further, given the lower affinity of TTR for T_4 compared with T_3, one would predict that the clearance rate for T_4 would be higher than T_3.

1.1.3 Membrane TH transporters and cytosolic thyroid hormone-binding proteins

Tadpole cells have the capacity to actively take up TH (see Krain & Denver, 2004), and this activity may be regulated during metamorphosis. Saturable, carrier-mediated uptake of THs has been demonstrated in tadpole RBCs (Galton, Stgermain, & Whittemore, 1986; Murata & Yamauchi, 2005; Yamauchi, Horiuchi, Koya, & Takikawa, 1989). The genes that encode TH transport proteins could be the important loci for the modulation of the timing of metamorphosis.

There are three general classes of proteins that allow for active uptake of TH by cells: the organic anion transporters (OATC), monocarboxylate transporters (MCT), and the L-amino acid permeases (LAT) (Friesema, Jansen, Milici, & Visser, 2005; Jansen, Friesema, Milici, & Visser, 2005; Ritchie, Peter, Shi, & Taylor, 1999; Ritchie et al., 2003; Visser, Frieserna, Jansen, & Visser, 2008). Orthologs of *oatc*, *mct* and *lat* genes have been isolated from frogs and patterns of expression throughout metamorphosis have been described (Connors, Korte, Anderson, & Degitz, 2010;

Liang, Sedgwick, & Shi, 1997; Shi & Brown, 1993). Only the amino acid permeases have so far received attention with regard to their function during tadpole metamorphosis. The T_3-inducible gene *iu12* from *Xenopus laevis* intestine (Liang et al., 1997; Shi & Brown, 1993) encodes a subunit of a heterodimeric amino acid permease complex (System L; Ritchie et al., 2003; Torrents et al., 1998). This permease complex efficiently transports T_3 and T_4 when expressed in the *Xenopus* oocyte expression system but is inhibited by reverse T_3 (Ritchie et al., 1999). Overexpression of System L in *Xenopus* oocytes increased cytoplasmic and nuclear delivery of THs from the external medium and enhanced transcriptional activation by TRs (Ritchie, Hayashi, Shi, & Taylor, 2002). By contrast, blocking endogenous System L activity in mammalian cells reduced both TH uptake and TR function (Ritchie et al., 2003). The fact that *iu12* is a T_3-inducible gene suggests that it could play a role in mediating T_3 uptake by cells during tadpole metamorphosis (Liang et al., 1997).

Upon entering cells, THs may bind to a series of intracellular-binding proteins, termed cytoplasmic TH-binding proteins (CTHBPs) that represent several classes of multifunctional proteins. These proteins may have a variety of enzymatic activities; for example, three genes isolated from *X. laevis* code for (1) a cytosolic aldehyde dehydrogenase which catalyzes the formation of retinoic acid (Yamauchi & Tata, 1994), (2) an M2 pyruvate kinase (Shi, Liang, Parkison, & Cheng, 1994), and (3) protein disulfide isomerase (PDI; induced by T_3 in tadpole brain; Denver et al., 1997), which catalyzes disulfide bond formation and human PDI has been shown to bind TH with high affinity (Cheng et al., 1987; Yamauchi et al., 1987). These CTHBPs may transport THs within the cytoplasm to the nucleus to bind to TRs, or alternatively, they may serve as chelators to limit the cellular free TH concentration (Shi, 2000). It is also possible that TH may regulate the enzymatic activity of these proteins (Ashizawa & Cheng, 1992).

1.2. Corticosteroids

1.2.1 CS production during metamorphosis

In addition to TH, CSs, the primary vertebrate stress hormones, play important roles in amphibian metamorphosis. Like TH, the production of CS changes with development and likely reflects the functional maturation of the hypothalamo–hypophyseal–interrenal axis. The major CS produced by the amphibian interrenal glands are corticosterone (CORT) and aldosterone (ALDO) (Carstensen, Burgers, & Li, 1961; Macchi & Phillips, 1966).

In most species studied, the plasma concentration and tissue content of CORT and ALDO increase during late prometamorphosis/metamorphic climax, more or less in parallel with the rise in TH production (Carr & Norris, 1988; Denver, 1998a; Glennemeier & Denver, 2002a; Jaffe, 1981; Jolivet-Jaudet & Leloup-Hatey, 1984; Kikuyama, Suzuki, & Iwamuro, 1986; Kloas, Reinecke, & Hanke, 1997; Krain & Denver, 2004; Krug, Honn, Battista, & Nicoll, 1983; Niinuma et al., 1989). The majority of these studies showed low to nondetectable CS during premetamorphosis and a marked increase at metamorphic climax. The only exception is in *X. laevis* where whole-body CS content may be highest during premetamorphosis (Kloas et al., 1997); although, there is a secondary, lower peak at metamorphic climax (Glennemeier & Denver, 2002a). CSs, being lipophilic, are transported in blood bound to plasma proteins. Recently, binding properties of a putative corticosteroid-binding globulin (CBG) were described in the serum of an amphibian (*Ambystoma tigrinum*) (Orchinik, Matthews, & Gasser, 2000). However, to my knowledge, the expression of CBG has not been studied in amphibians.

1.2.2 CS actions during metamorphosis

CSs exert complex effects on tadpole growth and development. Depending on the animal's developmental stage and TH status, CS can accelerate or decelerate metamorphosis. If elevated during premetamorphosis, CS typically inhibit tadpole growth and slow development (Belden, Moore, Wingfield, & Blaustein, 2005; Darras et al., 2002; Frieden & Naile, 1955; Glennemeier & Denver, 2002b; Gray & Janssens, 1990; Hayes, 1995; Hayes, Chan, & Licht, 1993; Hu, Crespi, & Denver, 2008; Kobayashi, 1958; Wright et al., 1994).

However, CS have been found to accelerate TH-induced and spontaneous metamorphosis (Darras et al., 2002; Frieden & Naile, 1955; Gray & Janssens, 1990; Hayes, 1995; Kikuyama et al., 1993, 1983; Kuhn, De Groef, Grommen, Van der Geyten, & Darras, 2004; Kuhn, De Groef, Van der Geyten, & Darras, 2005; Wright et al., 1994). Taken together, the findings suggest that elevated CS (e.g., in response to environmental stressors) during premetamorphosis retard growth and slow development, while increased CS during prometamorphosis accelerate metamorphosis. The mechanisms of CS inhibition of growth in tadpoles have not been investigated, but based on studies in mammals, these actions could manifest at multiple levels that likely include diverse catabolic actions (Sapolsky,

Romero, & Munck, 2000) and perhaps decreased anterior pituitary growth hormone (GH) biosynthesis (Harvey, Scanes, & Daughaday, 1995).

Recent findings suggest several molecular mechanisms for the positive interactions between CS and TH signaling in the acceleration of metamorphosis. CSs enhance TH bioactivity, increasing expression of TR and monodeiodinase genes. They increase maximal nuclear-binding capacity for T_3 (Kikuyama et al., 1993; Niki, Yoshizato, & Kikuyama, 1981; Suzuki & Kikuyama, 1983), which is paralleled by the upregulation of *tr* mRNAs in *X. laevis* tail and in frog cell lines; this occurs in a synergistic manner with low or subthreshold doses of TH plus CORT causing superinduction of TRs (Bonett et al., 2010). CORT also increased 5′-deiodinase activity and *Dio2* mRNA in tadpoles, thereby increasing T_3 in target tissues (Bonett et al., 2010; Darras et al., 2002; Galton, 1990; Kuhn et al., 2005). Notably, the action of CORT on *Dio2* was synergistic with T_3 in tadpole tail (Bonett et al., 2010).

Direct TH target genes may also be synergistically regulated by T_3 and CS through mechanisms that are not directly or immediately dependent on increased TRs or deiodinases (i.e., direct synergy between TRs and CS receptors at the target gene). For example, Krüppel-like factor 9 (*Klf9*; also known as *BTEB1*), a direct T_3 target gene, is induced by CORT (Bonett, Hu, Bagamasbad, & Denver, 2009) and is superinduced in tadpole tissues with rapid kinetics by combined treatment with T_3 plus CORT (P. Bagamasbad, R.M. Bonett, and R.J. Denver, unpublished data). Other genes are synergistically regulated by TH and CS, which could explain the mechanism by which these two hormones cooperate to accelerate metamorphosis (Kulkarni et al., 2012).

2. NEUROENDOCRINE CONTROL OF METAMORPHOSIS

The vertebrate neuroendocrine system is comprised of the hypothalamus and the pituitary gland (Fig. 7.1). The importance of hypothalamic control of metamorphosis has long been recognized (Denver, 1996; Kikuyama et al., 1993). The pituitary hormones that control TH and CS production, TSH and adrenocorticotropic hormone (ACTH), respectively, are primarily under stimulatory hypothalamic control in amphibians (Denver, 1996). The neuroendocrine system serves as an interface between the central nervous system and the endocrine system, and transduces signals derived from the external and internal environments into appropriate physiological/developmental responses.

2.1. Thyroid-stimulating hormone

Hypophysectomy of tadpoles arrests the development of the thyroid gland and leads to metamorphic stasis that is reversed by injecting TSH (Dodd & Dodd, 1976; Regard & Mauchamp, 1971, 1973). The rate of thyroid gland growth and TH production in the tadpole is coordinate with the development of the pituitary and the production of TSH (Buckbinder & Brown, 1993; Denver, 1996; Dodd & Dodd, 1976; Kaye, 1961; Korte et al., 2011; Manzon & Denver, 2004; Okada et al., 2009, 2004). The amphibian thyroid gland develops sensitivity to TSH just before hatching (Kaye, 1961).

Pituitary expression of *tshβ* mRNA and plasma TSH concentration increases throughout metamorphosis (Buckbinder & Brown, 1993; Manzon & Denver, 2004; Okada et al., 2000). Pituitary *tshβ* mRNA levels rise from premetamorphosis to peak values during late prometamorphosis/metamorphic climax (Buckbinder & Brown, 1993; Manzon & Denver, 2004; Okada et al., 2000). Korte et al. (2011) used a homologous radioimmunoassay to show that changes in plasma and pituitary TSH in *Xenopus* species throughout metamorphosis paralleled changes in pituitary *tshβ* mRNA. Thus, TSH biosynthesis is coordinate with thyroid gland development and hormone secretion, and the stimulatory action of pituitary TSH is necessary for thyroid gland growth and hormone biosynthesis.

The tripeptide amide thyrotropin-releasing hormone (TRH), which is the primary regulator of TSH release in adult mammals, is inactive on tadpole TSH secretion, although the *trh* gene is expressed in the brain and pituitary of amphibians (Denver, 1996; Denver & Licht, 1989; Kikuyama et al., 1993; Manzon & Denver, 2004; Norris & Dent, 1989; Okada et al., 2004), and TRH can stimulate TSH release in adult frogs (Denver, 2009a; Galas et al., 2009). Many studies now support that the secretion of TSH by the tadpole pituitary gland is under stimulatory control by corticotropin-releasing factor (CRF) and related peptides (e.g., sauvagine, urocortins; Denver, 2009b, 2009c). CRF-like peptides regulate neuroendocrine, autonomic, and behavioral responses to physical and emotional stress (Aguilera, 1998; Yao & Denver, 2007).

CRF was named for its role in inducing release of pituitary ACTH in mammals, a role that is shared in amphibia (Vale, Vaughan, & Perrin, 1997). Shortly after its discovery in 1981, CRF was discovered to be a potent stimulator of the thyroid axis in larval amphibians and other nonmammalian vertebrates (De Groef, Van der Geyten, Darras, & Kuhn, 2006; Denver, 1999, 2009b, 2009c). Treatment of amphibian pituitary explants or primary pituitary cells with CRF-like peptides stimulated TSH release (De Groef et al., 2006; Denver, 2009b, 2009c; Okada et al., 2009), and injections of CRF-like peptides elevated whole-body TH content in tadpoles of several species (Boorse & Denver, 2004; Denver, 1993, 1997; Gancedo et al., 1992).

Commensurate with their positive actions on tadpole thyroid activity, CRF-like peptides have been shown to accelerate tadpole metamorphosis (Boorse & Denver, 2002; Denver, 1993, 1997; Gancedo et al., 1992; Miranda, Affanni, & Paz, 2000). Conversely, blocking endogenous CRF by passive immunization with CRF antiserum, or by injection of the CRF receptor antagonist α-helical $CRF_{(9-41)}$ slowed spontaneous metamorphosis, or blocked simulated pond drying-induced metamorphosis (Denver, 1997). Further, hypothalamic *crf* mRNA and peptide content increased during spontaneous metamorphosis (Denver, 2009b), and hypothalamic CRF peptide content was increased in spadefoot toad tadpoles that accelerated metamorphosis in response to simulated pond drying (Denver, 1997). Kulkarni, Singamsetty, and Buchholz (2010) recently showed that CRF accelerates development of the direct developing frog *Eleutherodactylus coqui*. Because CRF is a stress neurohormone, endogenous CRF may participate in environmentally induced (stress-induced) metamorphosis (Boorse & Denver, 2004; Denver, 1997).

Work from Sakae Kikuyama's laboratory found that a majority of the TSH-releasing activity of tadpole and adult frog hypothalamic extracts on dispersed adult pituitary cells can be blocked by coincubation with the CRF receptor antagonist α-helical $CRF_{(9-41)}$ (Ito et al., 2004; Okada et al., 2009). These findings suggest that a significant proportion of TSH-releasing activity in the amphibian hypothalamus is contributed by CRF-like peptides. They also suggest that other factors could be involved in the regulation of TSH, or that α-helical $CRF_{(9-41)}$ may have only partial antagonist activity in amphibia as has been found in mammals (Rivier, Rivier, & Vale, 1984).

CRF actions are mediated by two G protein-coupled receptors (CRF_1 and CRF_2; Dautzenberg & Hauger, 2002) and are modulated by a secreted-binding protein (CRF-BP; Seasholtz, Valverde, & Denver, 2002). The action of CRF-like peptides on TSH release in the tadpole is mediated by the CRF_2 receptor expressed in thyrotropes (Okada et al., 2007, 2009); whereas, ACTH release may be controlled by the CRF_1 receptor in amphibians as it is in mammals (De Groef, Geris, et al., 2003; De Groef, Goris, Arckens, Kuhn, & Darras, 2003; Okada et al., 2009; Tonon et al., 1986; Van Pett et al., 2000). In *X. laevis*, crf_1 mRNA was expressed during premetamorphosis and its level increased during prometamorphosis, reaching a plateau through metamorphic climax (Manzon & Denver, 2004). In contrast, mRNA for the *crf*$_2$ was very low during pre- and early prometamorphosis, but increased dramatically during late prometamorphosis and metamorphic climax. The expression of the *crf*$_2$ in the tadpole pituitary paralleled the increase in sensitivity of the pituitary to CRF-like peptides during metamorphosis (Fig. 7.2; Kaneko et al., 2005). These findings support the hypothesis that the competence of tadpole pituitary thyrotropes to respond to hypothalamic CRF depends on the upregulation of the CRF$_2$ receptor during late prometamorphosis.

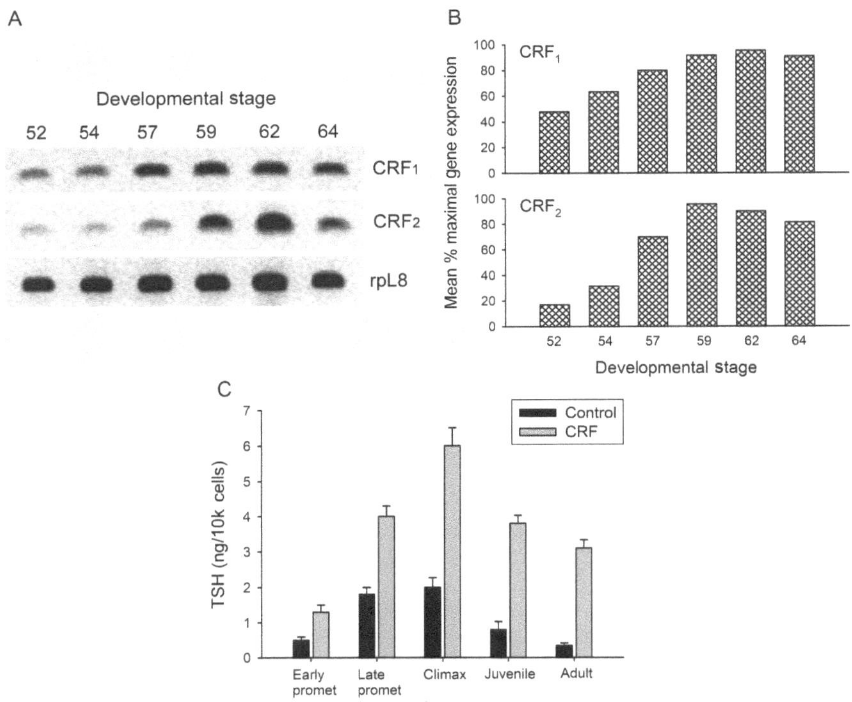

Figure 7.2 Pituitary expression of the corticotropin-releasing factor (CRF) receptor type 1 (CRF$_1$) and type 2 (CRF$_2$) show distinct patterns during tadpole metamorphosis. (A) Semiquantitative RT-PCR analysis of *crf*$_1$ and *crf*$_2$ mRNAs in *X. laevis* tadpole pituitary. The *rpL8* is a housekeeping gene whose expression did not change during development. (B) Quantitation of semiquantitative RT-PCR analysis of *crf*$_1$ and *crf*$_2$ mRNAs in *X. laevis* tadpole pituitary. *crf*$_1$ and *crf*$_2$ mRNA levels were normalized to *rpL8* mRNA (modified from Manzon & Denver, 2004). (C) Thyroid-stimulating hormone (TSH) secretion by dispersed tadpole or frog pituitary cells treated with vehicle or frog CRF (100 nM) for 24 h. *Data were derived from Kaneko et al. (2005), Fig. 4.*

2.1.1 Feedback regulation of TSH

Negative feedback by TH on the hypothalamus and pituitary plays a central role in thyroid homeostasis in all adult vertebrates that have been studied including frogs (Fig. 7.1; Jacobs & Kuhn, 1992; Kaneko et al., 2005). The discovery of a sustained rise in TSH production and thyroid activity during tadpole metamorphosis led Etkin (1968) to hypothesize that negative feedback on pituitary TSH does not develop until metamorphic climax. Huang et al. (2001) proposed that the onset of negative feedback at metamorphic climax was coincident with the expression of *Dio2* in the tadpole pituitary. However, many investigators have found that negative feedback by TH on TSH is active in the premetamorphic and early prometamorphic tadpole. For example, treatment of premetamorphic tadpoles with goitrogens caused enlargement of the thyroid gland and degranulation of pituitary thyrotropes, while replacement with T_4 reversed the effects, suggesting that negative feedback was functional in the premetamorphic tadpole (Dodd & Dodd, 1976; Goos, 1968, 1978; Goos, Deknecht, & Devries, 1968). Further, goitrogen treatment of premetamorphic tadpoles caused a dramatic elevation in *tshβ* mRNA (Buckbinder & Brown, 1993; Huang et al., 2001).

Physiological concentrations of T_4 or T_3 can act directly on pituitary explants of *X. laevis* tadpoles throughout metamorphosis to suppress *tshβ* mRNA expression and TSH secretion (Manzon & Denver, 2004; Sternberg et al., 2011). Pituitary sensitivity to negative feedback by TH may decline slightly during late prometamorphosis and metamorphic climax, perhaps due to the upregulation of pituitary *Dio3* at this time (Manzon & Denver, 2004; Sternberg et al., 2011). Kaneko et al. (2005) found that CRF-induced TSH release by bullfrog primary pituitary cells was suppressed by T_3 throughout metamorphosis. Taken together, these findings support that negative feedback at the level of the pituitary is active in the premetamorphic and early prometamorphic tadpole, which does not support Etkin's hypothesis and contradicts Huang et al. (2001).

Deiodinase type 2 plays an important role in TH-negative feedback on TSH in mammals (Schneider et al., 2001; St Germain, Hernandez, Schneider, & Galton, 2005). The *Dio2* gene is expressed in the tadpole from early prometamorphosis and shows a progressive increase during metamorphosis, reaching a maximum by NF stage 59 (Manzon & Denver, 2004). This supports the findings discussed above that T_4, likely through conversion to T_3, exerts negative feedback on TSH throughout tadpole metamorphosis. The downregulation of TSH expression by T_4 suggests that 5′-deiodinase activity is either present in the pituitary throughout prometamorphosis or the conversion of T_4 to T_3 is not required for negative feedback. TH receptor β is required for transcriptional repression of the *tshβ* and *trh* genes in mammals (Flamant & Samarut, 2003; Guissouma, Dupre, & Demeneix, 2005). *trβ* mRNA increased throughout metamorphosis in the tadpole pituitary (Manzon & Denver, 2004).

Despite the presence of functional negative feedback in the prometamorphic tadpole, TSH production shows a progressive increase throughout metamorphosis reaching a peak at metamorphic climax. Hypothalamic neurosecretory neurons and the median eminence, the structure necessary for the delivery of neurohormones to the pituitary, develop during prometamorphosis under the influence of TH (Denver, 1998b). The expression of neuropeptide receptors by anterior pituitary cells, and the responsiveness of these cells to secretagogues increases during metamorphosis (Kaneko et al., 2005; Manzon & Denver, 2004). Etkin (1968) first proposed that the maturation of the hypothalamus, median eminence, and pituitary under the influence of TH is responsible for the sustained rise in plasma TH concentration that drives metamorphosis. Thus, combined with a slight decrease in the sensitivity of the pituitary to negative feedback at metamorphic climax, the hypothalamic drive for TSH production may be sufficient to overcome negative feedback exerted by the elevated plasma TH concentration at this time. Negative feedback is likely to be physiologically important for limiting TSH secretion once the system has matured, and perhaps during maturation of the neuroendocrine system; that is, the coordination of morphogenesis may require the temperance of TSH expression by TH throughout metamorphosis. However, the sustained rise in thyroid activity during metamorphosis is likely to be due primarily to the maturational effects of TH on the CNS (and perhaps the pituitary) rather than the absence of negative feedback. The relatively lower levels of pituitary *Dio2* and *TRβ* expression during early prometamorphosis might be permissive for the sustained rise in TSH during prometamorphosis.

2.2. Adrenocorticotropic hormone

Expression of proopiomelanocortin mRNA in the anterior pituitary is low in premetamorphic tadpoles and increases during prometamorphosis peaking at metamorphic climax (Aida, Iwamuro, Miura, & Kikuyama, 1999). To my knowledge, there have been no direct measures of ACTH during tadpole development. Functional ACTH receptors are expressed by tadpole interrenal glands prior to the onset of metamorphosis (Glennemeier & Denver, 2002a). Premetamorphic tadpoles are capable of mounting a CORT response following exposure to a physical stressor (shaking/confinement stressor; Glennemeier & Denver, 2002a), which suggests that functional maturation of the hypothalamo–hypophyseal–interrenal axis occurs prior to metamorphosis (by contrast to the tadpole hypothalamo–hypophyseal–thyroid axis, which matures during prometamorphosis). The early functional maturation of the hypothalamo–hypophyseal–interrenal axis is reflected in the earlier expression of the CRF_1 receptor (expressed on corticotropes; expression at NF stage 52) compared with the CRF_2 receptor (expressed on thyrotropes; expression at NF stage 57) (Manzon & Denver, 2004; Okada et al., 2009; Fig. 7.2). The early maturation of the hypothalamo–hypophyseal–interrenal axis provides for environmental stressors to elevate endogenous CS production in premetamorphic tadpoles, which can have consequences for tadpole growth and development.

Compared with TSH, much less is known about how the hypothalamus controls ACTH secretion in amphibia. CRF and arginine vasopressin (AVT is the amphibian hormone) have been shown to stimulate ACTH secretion by cultured adult frog pituitaries (Tonon et al., 1986).

2.3. Growth hormone and prolactin

A central prediction of Etkin's model for the endocrine control of tadpole metamorphosis (Etkin, 1968) was that the metamorphic actions of TH were balanced by the inhibitory actions of the pituitary hormone prolactin (PRL). Etkin proposed that in the premetamorphic tadpole, PRL secretion was high, but declined at metamorphic climax. This prediction was based in large part on the inhibitory effects of injecting mammalian PRLs on metamorphosis (White & Nicoll, 1981), which led some investigators to suggest that PRL exerted a juvenilizing action in amphibian larvae similar to juvenile hormone in insects (Bern, Nicoll, & Strohman, 1967; Etkin & Gona, 1967).

The early studies that led to the development of the Etkin model have been extensively reviewed (Denver, 1996; Dodd & Dodd, 1976; Kaltenbach, 1996; Kikuyama et al., 1993; White & Nicoll, 1981). Work using mostly mammalian PRL or GH preparations suggested different roles for these hormones, with PRL enhancing larval growth and blocking the actions of TH on metamorphosis, and GH stimulating postmetamorphic growth as the hormone does in most vertebrates (Denver, 1996; Takada &

Kasai, 2003). A role for GH in regulating body growth in amphibia as in other vertebrates (Harvey et al., 1995) has been demonstrated by many studies in which GH was injected into tadpoles or frogs (Denver, 1996; Kikuyama et al., 1993; White & Nicoll, 1981) and more recently through transgenic overexpression of GH in *X. laevis* (Huang & Brown, 2000a).

In contrast to GH, a role for PRL in regulating tadpole growth and metamorphosis continues to be controversial (Huang & Brown, 2000b). Early work supporting that PRL inhibited metamorphosis and stimulated larval growth was conducted with mammalian PRL (and GH) preparations. These studies clearly showed that functional receptors for PRL/GH are expressed in amphibian tissues and their activation can both promote tadpole growth and block T_3-induced metamorphosis; the latter action likely through the prevention of *TRβ* autoinduction (Tata, Baker, Machuca, Rabelo, & Yamauchi, 1993). Further, injection of purified frog PRL had similar effects on tadpole growth and development as mammalian PRL (Kikuyama et al., 1993).

One can argue that the effects of exogenous hormones may represent pharmacological rather than physiological actions. However, it is noteworthy that blockade of endogenous PRL by passive immunization accelerated metamorphic changes, which supports a physiological role for the endogenous hormone (Denver, 1996; Kikuyama et al., 1993). Etkin (1968) proposed that larval growth and metamorphosis is controlled by a balance between TH and PRL, and that the two should show an inverse relationship in their blood concentrations at metamorphic climax. The rise in circulating concentrations of TH during prometamorphosis and climax have been confirmed (see above). However, circulating concentrations of PRL and levels of pituitary *prl* mRNA are low during premetamorphosis and also rise, more or less in parallel with TH, during late prometamorphosis and climax (Buckbinder & Brown, 1993; Clemons & Nicoll, 1977; Niinuma, Yamamoto, & Kikuyama, 1991; Takahashi et al., 1990; Yamamoto & Kikuyama, 1982), thus contradicting the earlier hypothesis of an inverse relationship of the two hormones (Etkin, 1968). The rise in PRL production tends to occur slightly later than the rise in TSH expression and circulating TH (Buckbinder & Brown, 1993). Similarly, [^{125}I]-PRL binding to kidney membrane fractions was low in premetamorphic bullfrog tadpoles and increased during metamorphic climax (White & Nicoll, 1979). Huang and Brown (2000b) measured PRL receptor mRNA in whole *X. laevis* tadpole and tail tissue and found increased expression at metamorphic climax. Hasunuma, Yamamoto, and Kikuyama (2004) found that PRL receptor mRNA increased in the tail fin and kidney of bullfrog tadpoles during metamorphic climax. Taken together, these findings argue against the hypothesis that PRL plays a juvenilizing role in amphibian metamorphosis (Buckbinder & Brown, 1993; Huang & Brown, 2000b). However, Kikuyama et al. (1993) have argued, based on their experiments with passive immunization with antiserum to bullfrog PRL, that low levels of PRL during the premetamorphic/early prometamorphic period might be sufficient to support larval growth and inhibit TH action.

Huang and Brown (2000a, 2000b) created transgenic *X. laevis* tadpoles that overexpressed *X. laevis* GH, *X. laevis* PRL, or ovine PRL. All tadpole tissues expressed the transgenes driven by the simian cytomegalovirus promoter; that is, expression was not restricted to the pituitary gland where the hormones are normally produced. They found that overexpression of GH did not affect the timing of metamorphosis but resulted in larger tadpoles and larger juvenile frogs. Overexpression of frog or ovine PRL had no effect on the timing of most metamorphic changes, but blocked tail resorption in some tadpoles. They concluded that their results disprove the hypothesis that PRL is a juvenile hormone in *X. laevis*. One note of caution in interpreting these findings is that PRL was overexpressed in all tissues throughout development, which could have led to compensatory changes (e.g., receptor desensitization) that masked the physiological roles of the hormone.

The elevation in PRL biosynthesis at metamorphic climax suggests that the hormone could modulate the actions of TH at a time when tissues are undergoing rapid and dramatic transformation (Denver, 1996). Shintani, Nohira, Hikosaka, and Kawahara (2002) showed that PRL and GH induced *Dio3* mRNA in tadpole tail, and they proposed that the effects of PRL and GH on metamorphosis may be mediated in part by the tissue-specific regulation of *Dio3*. PRL secretion is stimulated by TRH (Galas et al., 2009), and it has been hypothesized that, while TRH does not regulate TSH secretion in the tadpole, it plays a role in regulating the rise in PRL production at metamorphic climax (Buckbinder & Brown, 1993; Norris & Dent, 1989; White & Nicoll, 1981). The level of type 2 TRH receptor mRNA in the tadpole pituitary increased through late prometamorphosis and peaked at metamorphic climax (Manzon & Denver, 2004). In mammals, PRL secretion is induced by stressors (Cooke et al., 2004; Soares, Alam, Konno, Ho-Chen, & Ain, 2006). If a similar regulatory relationship exists in amphibia (e.g., see Lorenz, Opitz, Lutz, & Kloas, 2009), then it may be that the activation of neuroendocrine stress pathways during metamorphosis function in the late rise in PRL secretion.

3. ROLE OF THE NEUROENDOCRINE SYSTEM IN MEDIATING ENVIRONMENTAL INFLUENCES ON THE TIMING OF METAMORPHOSIS

The duration of the larval period varies considerably among and within amphibian species. The earliest time for the onset of metamorphosis is established by a genetically determined, species-specific minimum size for transformation. The time that it takes to reach the minimum size is determined in part by growth opportunity in the larval habitat (Werner, 1986; Wilbur & Collins, 1973). The better the resource supply, the earlier that a tadpole can reach its species-specific minimum size for metamorphosis. Variation in the proximate environment establishes trade-offs between growth opportunity and risk of mortality (environmental stress, predation risk, etc.) which ultimately determines the duration that the animal spends as a tadpole. Species that breed in permanent, predictable habitats can have relatively long larval periods (i.e., 3 years or greater); whereas those that breed in unpredictable, ephemeral ponds have short larval periods (as short as 10 days from hatching) (Denver, Boorse, & Glennemeier, 2002). The proximate mechanisms that govern the timing of metamorphosis involve the production, metabolism, and actions of hormones. Competence to respond to environmental signals depends on the development and activity of endocrine glands that produce the hormones that control metamorphosis.

Points of regulation by the environment include the neuroendocrine system, peripheral endocrine organs, hormone transport and metabolism, and hormone action. Thermal, osmotic, and effects related to the gaseous environment may be sensed directly by most or all tissues. Signals generated by other factors, such as photoperiod, resource availability, predator presence, and crowding are integrated by higher brain centers and transduced by the neuroendocrine system into changes in peripheral endocrine gland activity. The activity of the tadpole hypothalamo–pituitary–thyroid axis can be regulated at multiple levels, and thyroid activity determines when larvae enter metamorphosis, and the rate at which metamorphosis progresses. Because the stress hormonal axis is closely linked to the thyroid axis, central nervous stress pathways play a critical role in transducing environmental information and regulating metamorphic timing.

Work of Etkin (1968) suggested that the "clock" that determines the timing of metamorphosis is located in the hypothalamus. He showed that tadpoles in which the pituitary primordium was autotransplanted to the tail during embryogenesis grew more rapidly than controls, suggesting that pituitary GH is under inhibitory hypothalamic control; however, the tadpoles failed to metamorphose, supporting that a hypothalamic neurohormone was required to stimulate TSH secretion (Etkin, 1968). Destruction of the preoptic area or surgical removal of the primordium of the posterior hypothalamus (and thus isolation of the pituitary from the brain) prevented metamorphosis (reviewed by Denver, 1996). Investigations of the normal development of the neurosecretory centers of the hypothalamus and the median eminence further support Etkin's hypothesis (Etkin, 1968).

A striking example of the role of the hypothalamus in controlling metamorphosis, in particular, the role of hypothalamic CRF, comes from studies of desert toad species. The most important environmental variable for a tadpole is water availability, and duration of the aquatic habitat can profoundly influence the rate of metamorphosis in many species. This is especially true for desert amphibians that tend to breed in ephemeral habitats. As discussed earlier, CRF-like peptides control TSH secretion in tadpoles, acting via the CRF_2 receptor. Because the secretion of CRF is activated by stressors, CRF plays a central role in mediating a tadpole's developmental response to a deteriorating larval habitat (e.g., pond drying in the case of the Western spadefoot toad) (Boorse & Denver, 2004; Denver, 1997; Denver, 1998a; Denver, Mirhadi, & Phillips, 1998). The timing of the expression of receptors for neurohormones in the pituitary gland, particularly the CRF_2 receptor, could be important in establishing competence of pituitary thyrotropes to respond to stimulation by CRF-like peptides (Manzon & Denver, 2004; Okada et al., 2009).

Other environmental factors that are known to alter the timing of metamorphosis (e.g., food availability, crowding, predation) may also act through the neuroendocrine stress axis. For example, whole-body CORT content was elevated in tadpoles that were food restricted or subjected to high conspecific density, compared to their high resource, low density counterparts (Glennemeier & Denver, 2002b). Both low food and increased density resulted in slowed growth and development in premetamorphic tadpoles, which agrees with other studies showing growth- and development-inhibiting effects of these factors in premetamorphs (but contrast this with prometamorphic animals which accelerate development in response to food restriction or crowding). This slowed growth caused by crowding stress was reversed by treatment of tadpoles with the CORT synthesis inhibitor metyrapone, again suggesting a functional role for the hypothalamo– hypophyseal–interrenal axis in mediating the larval developmental response to environmental conditions (Glennemeier & Denver, 2002b). Hayes (1997) also reported an elevation in whole-body CORT content in tadpoles caused by crowding. By contrast, Belden, Rubbo, Wingfield, and Kiesecker (2007) did not find such a relationship in a mesocosm study. Predation, temperature, photoperiod, or other environmental factors could conceivably work through similar neuroendocrine pathways to exert their effects on larval development. If larvae have a means of detecting the state of environmental conditions, through visual, chemical, or other sensory systems, then the neuroendocrine system is a likely pathway through which developmental responses to the environment can operate.

While the hypothalamus and pituitary gland are required for metamorphosis through their control of thyroid and interrenal gland secretion, other processes occurring at target tissues may influence metamorphic timing. For example, the availability of biologically active TH is regulated within tissues by the monodeiodinases. Buchholz and Hayes (2005) showed that closely related species of spadefoot toads that differ in the duration of their larval periods show strong differences in the tissue content of T_3 and T_4, and the sensitivity of their tissues to TH. They speculated that these differences might be due to differences in TH uptake into cells and/or TH metabolism. The expression of monodeiodinases enzymes could be modified either directly or indirectly by environmental factors. CS increases 5′-deiodinase activity, with the result that more of the active hormone T_3 is generated. This regulatory relationship suggests that stress and stress hormones could accelerate metamorphosis by upregulating 5′-deiodinase activity.

Tissue expression of TRs influences sensitivity to the TH signal. TH receptor β is autoinduced in many tissues during metamorphosis, and evidence suggests that this is required to drive metamorphosis (Bagamasbad & Denver, 2011; Laudet, 2011). Hollar, Choi, Grimm, and Buchholz (2011) recently showed that TR expression level is negatively correlated with the duration of the larval period in different species of spadefoot toad; that is, higher TR equals shorter larval period.

Biosynthesis of TRs might be regulated directly or indirectly by environmental factors, which could influence metamorphic timing. Currently, relatively little is known about what factors, either physiological or environmental, regulate nuclear receptor expression in any species (Bagamasbad & Denver, 2011). As for monodeiodinases, evidence suggests that CS can enhance TH action by upregulating TR expression, and so TR biosynthesis is an additional site where stress and stress hormones may modulate metamorphic timing (Bonett et al., 2010).

ACKNOWLEDGMENT

The preparation of this chapter and the unpublished work reported herein was supported by NSF grant IOS 0922583 to R. J. D.

REFERENCES

Aguilera, G. (1998). Corticotropin releasing hormone, receptor regulation and the stress response. *Trends in Endocrinology and Metabolism, 9*, 329–336.

Aida, T., Iwamuro, S., Miura, S., & Kikuyama, S. (1999). Changes of pituitary proopiomelanocortin mRNA levels during metamorphosis of the bullfrog larvae. *Zoological Science, 16*, 255–260.

Ashizawa, K., & Cheng, S. Y. (1992). Regulation of thyroid hormone receptor mediated transcription by a cytosol protein. *Proceedings of the National Academy of Sciences of the United States of America, 89*, 9277–9281.

Bagamasbad, P., & Denver, R. J. (2011). Mechanisms and significance of nuclear receptor auto- and cross-regulation. *General and Comparative Endocrinology, 170*, 3–17.

Becker, K. B., Schneider, M. J., Davey, J. C., & Galton, V. A. (1995). The type III 5-deiodinase in *Rana catesbeiana* tadpoles is encoded by a thyroid hormone responsive gene. *Endocrinology, 136*, 4424–4431.

Becker, K. B., Stephens, K. C., Davey, J. C., Schneider, M. J., & Galton, V. A. (1997). The type 2 and type 3 iodothyronine deiodinases play important roles in coordinating development in Rana catesbeiana tadpoles. *Endocrinology, 138*, 2989–2997.

Belden, L. K., Moore, I. T., Wingfield, J. C., & Blaustein, A. R. (2005). Corticosterone and growth in Pacific Treefrog (Hyla regilla) tadpoles. *Copeia*, 424–430.

Belden, L. K., Rubbo, M. J., Wingfield, J. C., & Kiesecker, J. M. (2007). Searching for the physiological mechanism of density dependence: Does corticosterone regulate tadpole responses to density? *Physiological and Biochemical Zoology, 80*, 444–451.

Bern, H. A., Nicoll, C. S., & Strohman, R. C. (1967). Prolactin and tadpole growth. *Proceedings of the Society for Experimental Biology and Medicine, 126*, 518–520.

Berry, D. L., Schwartzman, R. A., & Brown, D. D. (1998). The expression pattern of thyroid hormone response genes in the tadpole tail identifies multiple resorption programs. *Developmental Biology, 203*, 12–23.

Bonett, R. M., Hoopfer, E. D., & Denver, R. J. (2010). Molecular mechanisms of corticosteroid synergy with thyroid hormone during tadpole metamorphosis. *General and Comparative Endocrinology, 168*, 209–219.

Bonett, R. M., Hu, F., Bagamasbad, P., & Denver, R. J. (2009). Stressor and glucocorticoid-dependent induction of the immediate early gene kruppel-like factor 9: Implications for neural development and plasticity. *Endocrinology, 150*, 1757–1765.

Boorse, G. C., & Denver, R. J. (2002). Acceleration of Ambystoma tigrinum metamorphosis by corticotropin-releasing hormone. *The Journal of Experimental Zoology, 293*, 94–98.

Boorse, G. C., & Denver, R. J. (2004). Endocrine mechanisms underlying plasticity in metamorphic timing in spadefoot toads. *Integrative and Comparative Biology, 43*, 646–657.

Brown, D. D. (2005). The role of deiodinases in amphibian metamorphosis. *Thyroid, 15*, 815–821.

Brown, D. D., & Cai, L. Q. (2007). Amphibian metamorphosis. *Developmental Biology, 306*, 20–33.

Brown, D. D., Wang, Z., Furlow, J. D., Kanamori, A., Schwartzman, R. A., Remo, B. F., et al. (1996). The thyroid hormone-induced tail resorption program during *Xenopus laevis* metamorphosis. *Proceedings of the National Academy of Sciences of the United States of America, 93,* 1924–1929.

Buchholz, D. R., & Hayes, T. B. (2005). Variation in thyroid hormone action and tissue content underlies species differences in the timing of metamorphosis in desert frogs. *Evolution & Development,* 7, 458–467.

Buchholz, D. R., Heimeier, R. A., Das, B., Washington, T., & Shi, Y. B. (2007). Pairing morphology with gene expression in thyroid hormone-induced intestinal remodeling and identification of a core set of TH-induced genes across tadpole tissues. *Developmental Biology, 303,* 576–590.

Buckbinder, L., & Brown, D. D. (1993). Expression of the *Xenopus laevis* prolactin and thyrotropin genes during metamorphosis. *Proceedings of the National Academy of Sciences of the United States of America, 90,* 3820–3824.

Buscaglia, M., Leloup, J., & De Luze, A. (1985). The role and regulation of monodeiodination of thyroxine to 3,5,3′-triiodothyronine during amphibian metamorphosis. In M. Balls & M. Bownes (Eds.), *Metamorphosis* (pp. 273–293). Oxford: Clarendon Press.

Cai, L. Q., & Brown, D. D. (2004). Expression of type II iodothyronine deiodinase marks the time that a tissue responds to thyroid hormone-induced metamorphosis in Xenopus laevis. *Developmental Biology, 266,* 87–95.

Carr, J. A., & Norris, D. O. (1988). Interrenal activity during metamorphosis of the tiger salamander, *Ambystoma tigrinum. General and Comparative Endocrinology, 71,* 63–69.

Carstensen, H., Burgers, A. C. J., & Li, C. H. (1961). Demonstration of aldosterone and corticosterone as the principle steroids formed in incubates of adrenals of the american bullfrog Rana catesbeiana and stimulation of their production by mammalian adrenocorticotropin. *General and Comparative Endocrinology, 1,* 37–50.

Cheng, S. Y., Gong, Q. H., Parkison, C., Robinson, E. A., Appella, E., Merlino, G. T., et al. (1987). The nucleotide sequence of a human cellular thyroid hormone binding protein present in endoplasmic reticulum. *The Journal of Biological Chemistry, 262,* 11221–11227.

Clemons, G. K., & Nicoll, C. S. (1977). Effects of antisera to bullfrog prolactin and growth hormone on metamorphosis of Rana catesbeiana tadpoles. *General and Comparative Endocrinology, 31,* 495–497.

Connors, K. A., Korte, J. J., Anderson, G. W., & Degitz, S. J. (2010). Characterization of thyroid hormone transporter expression during tissue-specific metamorphic events in Xenopus tropicalis. *General and Comparative Endocrinology, 168,* 149–159.

Cooke, P. S., Holsberger, D. R., Witorsch, R. J., Sylvester, P. W., Meredith, J. M., Treinen, K. A., et al. (2004). Thyroid hormone, glucocorticoids, and prolactin at the nexus of physiology, reproduction, and toxicology. *Toxicology and Applied Pharmacology, 194,* 309–335.

Darras, V. M., Van der Geyten, S., Cox, C., Segers, I. B., De Groef, B., & Kuhn, E. R. (2002). Effects of dexamethasone treatment on iodothyronine deiodinase activities and on metamorphosis-related morphological changes in the axolotl (*Ambystoma mexicanum*). *General and Comparative Endocrinology, 127,* 157–164.

Das, B., Cai, L. Q., Carter, M. G., Piao, Y. L., Sharov, A. A., Ko, M. S. H., et al. (2006). Gene expression changes at metamorphosis induced by thyroid hormone in Xenopus laevis tadpoles. *Developmental Biology, 291,* 342–355.

Das, B., Heimeier, R. A., Buchholz, D. R., & Shi, Y. B. (2009). Identification of direct thyroid hormone response genes reveals the earliest gene regulation programs during frog metamorphosis. *The Journal of Biological Chemistry, 284,* 34167–34178.

Dautzenberg, F. M., & Hauger, R. L. (2002). The CRF peptide family and their receptors: Yet more partners discovered. *Trends in Pharmacological Sciences, 23,* 71–77.

De Groef, B., Geris, K. L., Manzano, J., Bernal, J., Millar, R. P., Abou-Samra, A. B., et al. (2003). Involvement of thyrotropin-releasing hormone receptor, somatostatin receptor subtype 2 and corticotropin-releasing hormone receptor type 1 in the control of chicken thyrotropin secretion. *Molecular and Cellular Endocrinology, 203*, 33–39.

De Groef, B., Goris, N., Arckens, L., Kuhn, E. R., & Darras, V. M. (2003). Corticotropin-releasing hormone (CRH)-induced thyrotropin release is directly mediated through CRH receptor type 2 on thyrotropes. *Endocrinology, 144*, 5537–5544.

De Groef, B., Van der Geyten, S., Darras, V. M., & Kuhn, E. R. (2006). Role of corticotropin-releasing hormone as a thyrotropin-releasing factor in non-mammalian vertebrates. *General and Comparative Endocrinology, 146*, 62–68.

Denver, R. J. (1993). Acceleration of anuran amphibian metamorphosis by corticotropin-releasing hormone-like peptides. *General and Comparative Endocrinology, 91*, 38–51.

Denver, R. J. (1996). Neuroendocrine control of amphibian metamorphosis. In L. I. Gilbert, J. R. Tata & B. G. Atkinson (Eds.), *Metamorphosis: Post-embryonic reprogramming of gene expression in amphibian and insect cells* (pp. 433–464). San Diego, CA: Academic Press, Inc.

Denver, R. J. (1997). Environmental stress as a developmental cue: Corticotropin-releasing hormone is a proximate mediator of adaptive phenotypic plasticity in amphibian metamorphosis. *Hormones and Behavior, 31*, 169–179.

Denver, R. J. (1998a). Hormonal correlates of environmentally induced metamorphosis in the Western spadefoot toad, *Scaphiopus hammondii. General and Comparative Endocrinology, 110*, 326–336.

Denver, R. J. (1998b). The molecular basis of thyroid hormone-dependent central nervous system remodeling during amphibian metamorphosis. *Comparative Biochemistry and Physiology. Part C, Pharmacology, Toxicology & Endocrinology, 119*, 219–228.

Denver, R. J. (1999). Evolution of the corticotropin-releasing hormone signaling system and its role in stress-induced phenotypic plasticity. *Annals of the New York Academy of Sciences, 897*, 46–53.

Denver, R. J. (2009a). Endocrinology of complex life cycles: Amphibians. In D. W. Pfaff, A. P. Arnold, A. M. Etgen, R. T. Rubin & S. E. Fahrbach (Eds.), *Hormones, brain and behavior* (pp. 707–744). (2nd ed). San Diego, CA: Elsevier.

Denver, R. J. (2009b). Stress hormones mediate environment-genotype interactions during amphibian development. *General and Comparative Endocrinology, 164*, 20–31.

Denver, R. J. (2009c). Structural and functional evolution of vertebrate neuroendocrine stress systems. *Annals of the New York Academy of Sciences, 1163*, 1–16.

Denver, R. J., Boorse, G. C., & Glennemeier, K. A. (2002). Endocrinology of complex life cycles: Amphibians. In A. A. D. Pfaff, A. Etgen, S. Fahrbach, R. Moss & R. Rubin (Eds.), *Hormones, brain and behavior* (pp. 469–513). San Diego, CA: Academic Press, Inc.

Denver, R. J., Hu, F., Scanlan, T. S., & Furlow, J. D. (2009). Thyroid hormone receptor subtype specificity for hormone-dependent neurogenesis in *Xenopus laevis. Developmental Biology, 326*, 155–168.

Denver, R. J., & Licht, P. (1989). Neuropeptide stimulation of thyrotropin secretion in the larval bullfrog: Evidence for a common neuroregulator of thyroid and interrenal activity during metamorphosis. *The Journal of Experimental Zoology, 252*, 101–104.

Denver, R. J., Mirhadi, N., & Phillips, M. (1998). Adaptive plasticity in amphibian metamorphosis: Response of *Scaphiopus hammondii* tadpoles to habitat desiccation. *Ecology, 79*, 1859–1872.

Denver, R. J., Pavgi, S., & Shi, Y. B. (1997). Thyroid hormone-dependent gene expression program for Xenopus neural development. *The Journal of Biological Chemistry, 272*, 8179–8188.

Dodd, M. H. I., & Dodd, J. M. (1976). The biology of metamorphosis. In B. Lofts (Ed.), *Physiology of the amphibia* (pp. 467–599). New York: Academic Press.

Dubois, G. M., Sebillot, A., Kuiper, G., Verhoelst, C. H. J., Darras, V. M., Visser, T. J., et al. (2006). Deiodinase activity is present in Xenopus laevis during early embryogenesis. *Endocrinology*, *147*, 4941–4949.

Ekins, R. (1990). Measurement of free hormones in blood. *Endocrine Reviews*, *11*, 5–46.

Etkin, W. (1968). Hormonal control of amphibian metamorphosis. In W. Etkin & L. I. Gilbert (Eds.), *Metamorphosis: A problem in developmental biology* (pp. 313–348). New York: Appleton-Century-Crofts.

Etkin, W., & Gona, A. G. (1967). Antagonism between prolactin and thyroid hormone in amphibian development. *The Journal of Experimental Zoology*, *165*, 249–258.

Flamant, F., & Samarut, J. (2003). Thyroid hormone receptors: Lessons from knockout and knock-in mutant mice. *Trends in Endocrinology and Metabolism*, *14*, 85–90.

Frieden, E. (1981). The dual role of thyroid hormones in vertebrate development and calorigenesis. In L. Gilbert & E. Frieden (Eds.), *Metamorphosis: A problem in developmental biology* (pp. 545–564). New York: Plenum Press.

Frieden, E., & Naile, B. (1955). Biochemistry of amphibian metamorphosis. 1. Enhancement of induced metamorphosis by glucocorticoids. *Science*, *121*, 37–38.

Friesema, E. C. H., Jansen, J., Milici, C., & Visser, T. J. (2005). Thyroid hormone transporters. *Vitamins and Hormones*, *70*, 137–167.

Galas, L., Raoult, E., Tonon, M. C., Okada, R., Jenks, B. G., Castano, J. P., et al. (2009). TRH acts as a multifunctional hypophysiotropic factor in vertebrates. *General and Comparative Endocrinology*, *164*, 40–50.

Galton, V. A. (1989). The role of 3,5,3′-triiodothyronine in the physiological action of thyroxine in the premetamorphic tadpole. *Endocrinology*, *124*, 2427–2433.

Galton, V. (1990). Mechanisms underlying the acceleration of thyroid hormone-induced tadpole metamorphosis by corticosterone. *Endocrinology*, *127*, 2997–3002.

Galton, V. A., Stgermain, D. L., & Whittemore, S. (1986). Cellular uptake of 3,5,3′-triiodothyronine and thyroxine by red blood and thymus cells. *Endocrinology*, *118*, 1918–1923.

Gancedo, B., Corpas, I., Alonso-Gomez, A. L., Delgado, M. J., Morreale de Escobar, G., & Alonso-Bedate, M. (1992). Corticotropin-releasing factor stimulates metamorphosis and increases thyroid hormone concentration in prometamorphic *Rana perezi* larvae. *General and Comparative Endocrinology*, *87*, 6–13.

Glennemeier, K. A., & Denver, R. J. (2002a). Developmental changes in interrenal responsiveness in anuran amphibians. *Integrative and Comparative Biology*, *42*, 565–573.

Glennemeier, K. A., & Denver, R. J. (2002b). Role for corticoids in mediating the response of *Rana pipiens* tadpoles to intraspecific competition. *The Journal of Experimental Zoology*, *292*, 32–40.

Goos, H. J. T. (1968). Hypothalamic neurosecretion and metamorphosis in Xenopus laevis. 3. Effect of an interruption of thyroid hormone synthesis. *Zeitschrift Fur Zellforschung Und Mikroskopische Anatomie*, *92*, 583–587.

Goos, H. J. T. (1978). Hypophysiotropic centers in the brain of amphibians and fish. *American Zoologist*, *18*, 401–410.

Goos, H. J. T., Deknecht, A. M., & Devries, J. (1968). Hypothalamic neurosecretion and metamorphosis in Xenopus laevis. I. Effect of propylthiouracil. *Zeitschrift Fur Zellforschung Und Mikroskopische Anatomie*, *86*, 384–392.

Gray, K. M., & Janssens, P. A. (1990). Gonadal hormones inhibit the induction of metamorphosis by thyroid hormones in *Xenopus laevis* tadpoles *in vivo*, but not *in vitro*. *General and Comparative Endocrinology*, *77*, 202–211.

Guissouma, H., Dupre, S. M., & Demeneix, B. A. (2005). Transcriptional repression of the TRH gene. *Medecine Sciences*, *21*, 854–859.

Harvey, S., Scanes, C. G., & Daughaday, W. H. (1995). *Growth hormone*. Boca Raton, FL: CRC.

Hasunuma, I., Yamamoto, K., & Kikuyama, S. (2004). Molecular cloning of bullfrog prolactin receptor cDNA: Changes in prolactin receptor mRNA level during metamorphosis. *General and Comparative Endocrinology, 138*, 200–210.

Hayes, T. B. (1995). Interdependence of corticosterone hormones and thyroid hormones in larval toads (Bufo boreas). 1. Thyroid hormone dependent and hormone-independent effects of corticosterone on growth and development. *The Journal of Experimental Zoology, 271*, 95–102.

Hayes, T. B. (1997). Steroids as potential modulators of thyroid hormone activity in anuran metamorphosis. *American Zoologist, 37*, 185–194.

Hayes, T., Chan, R., & Licht, P. (1993). Interactions of temperature and steroids on larval growth, development and metamorphosis in a toad (*Bufo boreas*). *The Journal of Experimental Zoology, 266*, 206–215.

Hogan, N. S., Crump, K. L., Duarte, P., Lean, D. R. S., & Trudeau, V. L. (2007). Hormone cross-regulation in the tadpole brain: Developmental expression profiles and effect of T3 exposure on thyroid hormone-and estrogen-responsive genes in Rana pipiens. *General and Comparative Endocrinology, 154*, 5–15.

Hollar, A. R., Choi, J., Grimm, A. T., & Buchholz, D. R. (2011). Higher thyroid hormone receptor expression correlates with short larval periods in spadefoot toads and increases metamorphic rate. *General and Comparative Endocrinology, 173*, 190–198.

Hu, F., Crespi, E. J., & Denver, R. J. (2008). Programming neuroendocrine stress axis activity by exposure to glucocorticoids during postembryonic development of the frog *Xenopus laevis*. *Endocrinology, 149*, 5470–5481.

Huang, H. C., & Brown, D. D. (2000a). Overexpression of Xenopus laevis growth hormone stimulates growth of tadpoles and frogs. *Proceedings of the National Academy of Sciences of the United States of America, 97*, 190–194.

Huang, H. C., & Brown, D. D. (2000b). Prolactin is not a juvenile hormone in Xenopus laevis metamorphosis. *Proceedings of the National Academy of Sciences of the United States of America, 97*, 195–199.

Huang, H. C., Cai, L. Q., Remo, B. F., & Brown, D. D. (2001). Timing of metamorphosis and the onset of the negative feedback loop between the thyroid gland and the pituitary is controlled by type II iodothyronine deiodinase in Xenopus laevis. *Proceedings of the National Academy of Sciences of the United States of America, 98*, 7348–7353.

Huang, H., Marsh-Armstrong, N., & Brown, D. D. (1999). Metamorphosis is inhibited in transgenic *Xenopus laevis* tadpoles that overexpress type III deiodinase. *Proceedings of the National Academy of Sciences of the United States of America, 96*, 962–967.

Ito, Y., Okada, R., Mochida, H., Hayashi, H., Yamamoto, K., & Kikuyama, S. (2004). Molecular cloning of bullfrog corticotropin-releasing factor (CRF): Effect of homologous CRF on the release of TSH from pituitary cells *in vitro*. *General and Comparative Endocrinology, 138*, 218–227.

Jacobs, G. F. M., & Kuhn, E. R. (1992). Thyroid-hormone feedback-regulation of the secretion of bioactive thyrotropin in the frog. *General and Comparative Endocrinology, 88*, 415–423.

Jaffe, R. C. (1981). Plasma concentration of corticosterone during Rana catesbeiana tadpole metamorphosis. *General and Comparative Endocrinology, 44*, 314–318.

Jansen, J., Friesema, E. C. H., Milici, C., & Visser, T. J. (2005). Thyroid hormone transporters in health and disease. *Thyroid, 15*, 757–768.

Jolivet-Jaudet, G., & Leloup-Hatey, J. (1984). Interrenal function during amphibian metamorphosis - in vitro biosynthesis of radioactive corticosteroids from (4C-14) progesterone by interrenal in *Xenopus laevis* tadpoles. *Comparative Biochemistry and Physiology. Part B, Biochemistry & Molecular Biology, 79*, 239–244.

Kaltenbach, J. C. (1996). Endocrinology of amphibian metamorphosis. In L. I. Gilbert, J. R. Tata & B. G. Atkinson (Eds.), *Metamorphosis: Post-embryonic reprogramming of gene expression in amphibian and insect cells* (pp. 403–431). San Diego, CA: Academic Press, Inc.

Kaneko, M., Fujisawa, H., Okada, R., Yamamoto, K., Nakamura, M., & Kikuyama, S. (2005). Thyroid hormones inhibit frog corticotropin-releasing factor-induced thyrotropin release from the bullfrog pituitary in vitro. *General and Comparative Endocrinology, 144*, 122–127.

Kawahara, A., Gohda, Y., & Hikosaka, A. (1999). Role of type III iodothyronine 5-deiodinase gene expression in temporal regulation of *Xenopus* metamorphosis. *Development, Growth & Differentiation, 41*, 365–373.

Kaye, N. W. (1959). Evidence for early function of pituitary-thyroid axis in the frog tadpole. *The Anatomical Record, 134*, 591–592.

Kaye, N. W. (1960). The pituitary-thyroid axis in the premetamorphic tadpole. *The Anatomical Record, 137*, 370.

Kaye, N. W. (1961). Interrelationships of the thyroid and pituitary in embryonic and premetamorphic stages of the frog, *Rana pipiens*. *General and Comparative Endocrinology, 1*, 1–19.

Kikuyama, S., Kawamura, K., Tanaka, S., & Yamamoto, K. (1993). Aspects of amphibian metamorphosis: Hormonal control. *International Review of Cytology, 145*, 105–148.

Kikuyama, S., Niki, K., Mayumi, M., Shibayama, R., Nishikawa, M., & Shintake, N. (1983). Studies on corticoid action on the toad tadpole tail in vitro. *General and Comparative Endocrinology, 52*, 395–399.

Kikuyama, S., Suzuki, M. R., & Iwamuro, S. (1986). Elevation of plasma aldosterone levels of tadpoles at metamorphic climax. *General and Comparative Endocrinology, 63*, 186–190.

Kloas, W., Reinecke, M., & Hanke, W. (1997). Stage dependent changes in adrenal steroids and catecholamines during development in *Xenopus laevis*. *General and Comparative Endocrinology, 108*, 416–426.

Kobayashi, H. (1958). Effect of desoxycorticosterone acetate on metamorphosis induced by thyroxine in anuran tadpoles. *Endocrinology, 62*, 371–377.

Korte, J. J., Sternberg, R. M., Serrano, J. A., Thoemke, K. R., Moen, S. M., Lillegard, K. E., et al. (2011). Thyroid-stimulating hormone (TSH): Measurement of intracellular, secreted, and circulating hormone in Xenopus laevis and Xenopus tropicalis. *General and Comparative Endocrinology, 171*, 319–325.

Krain, L. P., & Denver, R. J. (2004). Developmental expression and hormonal regulation of glucocorticoid and thyroid hormone receptors during metamorphosis in Xenopus laevis. *The Journal of Endocrinology, 181*, 91–104.

Krug, E. C., Honn, K. V., Battista, J., & Nicoll, C. S. (1983). Corticosteroids in serum of *Rana catesbeiana* during development and metamorphosis. *General and Comparative Endocrinology, 52*, 232–241.

Kuhn, E. R., De Groef, B., Grommen, S. V. H., Van der Geyten, S., & Darras, V. M. (2004). Low submetamorphic doses of dexamethasone and thyroxine induce complete metamorphosis in the axolotl (Ambystoma mexicanum) when injected together. *General and Comparative Endocrinology, 137*, 141–147.

Kuhn, E. R., De Groef, B., Van der Geyten, S., & Darras, V. M. (2005). Corticotropin-releasing hormone-mediated metamorphosis in the neotenic axolotl *Ambystoma mexicanum*: Synergistic involvement of thyroxine and corticoids on brain type II deiodinase. *General and Comparative Endocrinology, 143*, 75–81.

Kuiper, G., Klootwijk, W., Dubois, G. M., Destree, O., Darras, V. M., Van der Geyten, S., et al. (2006). Characterization of recombinant Xenopus laevis type I iodothyronine deiodinase: Substitution of a proline residue in the catalytic center by serine (Pro132Ser) restores sensitivity to 6-propyl-2-thiouracil. *Endocrinology, 147*, 3519–3529.

Kulkarni, S. S., Singamsetty, S., & Buchholz, D. R. (2010). Corticotropin-releasing factor regulates the development in the direct developing frog, Eleutherodactylus coqui. *General and Comparative Endocrinology, 169*, 225–230.

Kulkarni, S. S., & Buchholz, D. R. (2012). Beyond Synergy: Corticosterone and Thyroid Hormone Have Numerous Interaction Effects on Gene Regulation in. *Xenopus tropicalis* Tadpoles. *Endocrinology, 153*, 5309–5324.

Laudet, V. (2011). The origins and evolution of vertebrate metamorphosis. *Current Biology, 21*, R726–R737.

Leonard, J. L., & Visser, T. J. (1986). Biochemistry of deiodination. In G. Hennemann (Ed.), *Thyroid hormone metabolism* (pp. 189–229). New York: Dekker.

Liang, V. C., Sedgwick, T., & Shi, Y. B. (1997). Characterization of the *Xenopus* homolog of an immediate early gene associated with cell activation: Sequence analysis and regulation of its expression by thyroid hormone during amphibian metamorphosis. *Cell Research, 7*, 179–193.

Lindsay, R. H., Buettner, L., Wimberly, N., & Pittman, J. A. (1967). Effects of thyroxine analogs on isolated tadpole tail tips. *General and Comparative Endocrinology, 9*, 416–421.

Lorenz, C., Opitz, R., Lutz, I., & Kloas, W. (2009). Corticosteroids disrupt amphibian metamorphosis by complex modes of action including increased prolactin expression. *Comparative Biochemistry and Physiology. Toxicology & Pharmacology, 150*, 314–321.

Macchi, I. A., & Phillips, J. G. (1966). In vitro effect of adrenocorticotropin on corticoid secretion in the turtle, snake, and bullfrog. *General and Comparative Endocrinology, 6*, 170–182.

Manzon, R. G., & Denver, R. J. (2004). Regulation of pituitary thyrotropin gene expression during Xenopus metamorphosis: Negative feedback is functional throughout metamorphosis. *The Journal of Endocrinology, 182*, 273–285.

Marsh-Armstrong, N., Huang, H. C., Remo, B. F., Liu, T. T., & Brown, D. D. (1999). Asymmetric growth and development of the Xenopus laevis retina during metamorphosis is controlled by type III deiodinase. *Neuron, 24*, 871–878.

Mendel, C. M. (1989). The free hormone hypothesis: A physiologically based mathematical model. *Endocrine Reviews, 10*, 232–274.

Miranda, L. A., Affanni, J. M., & Paz, D. A. (2000). Corticotropin-releasing factor accelerates metamorphosis in Bufo arenarum: Effect on pituitary ACTH and TSH cells. *The Journal of Experimental Zoology, 286*, 473–480.

Murata, T., & Yamauchi, K. (2005). Low-temperature arrest of the triiodothyronine-dependent transcription in Rana catesbeiana red blood cells. *Endocrinology, 146*, 256–264.

Nieuwkoop, P. D., & Faber, J. (1956). *Normal table of* Xenopus laevis *Daudin*. Amsterdam: North Holland Publishers.

Niinuma, K., Mamiya, N., Yamamoto, K., Iwamuro, S., Vaudry, H., & Kikuyama, S. (1989). Plasma concentrations of aldosterone and prolactin in Bufo japonicus tadpoles during metamorphosis. *Bulletin of Science and Engineering Research Laboratory, Waseda University, 122*, 17–21.

Niinuma, K., Yamamoto, K., & Kikuyama, S. (1991). Changes in plasma and pituitary prolactin levels in yoad (*Bufo japonicus*) larvae during metamorphosis. *Zoological Science, 8*, 97–101.

Niki, K., Yoshizato, K., & Kikuyama, S. (1981). Augmentation of nuclear binding capacity for triiodothyronine by aldosterone in tadpole tail. *Proceedings of the Japan Academy. Series B, Physical and Biological Sciences, 57*, 271–275.

Norris, D. O., & Dent, J. N. (1989). Neuroendocrine aspects of amphibian metamorphosis. In C. G. Scanes & M. P. Schreibman (Eds.), *Development, maturation and senescence of neuroendocrine systems: A comparative approach* (pp. 63–90). San Diego, CA: Academic Press.

Okada, R., Iwata, T., Kato, T., Kikuchi, M., Yamamoto, K., & Kikuyama, S. (2000). Cloning of bullfrog thyroid-stimulating hormone (TSH) beta subunit cDNA: Expression of TSH beta mRNA during metamorphosis. *General and Comparative Endocrinology, 119*, 224–231.

Okada, R., Kobayashi, T., Yamamoto, K., Nakakura, T., Tanaka, S., Vaudry, H., et al. (2009). Neuroendocrine regulation of thyroid-stimulating hormone secretion in amphibians. In H. Vaudry, E. W. Roubos, G. M. Coast & M. Vallarino (Eds.), *Annals of the New York Academy of Sciences* (pp. 262–270).

Okada, R., Miller, M. F., Yamamoto, K., De Groef, B., Denver, R. J., & Kikuyama, S. (2007). Involvement of the corticotropin-releasing factor (CRF) type 2 receptor in CRF-induced thyrotropin release by the amphibian pituitary gland. *General and Comparative Endocrinology*, *150*, 437–444.

Okada, R., Yamamoto, K., Koda, A., Ito, Y., Hayashi, H., Tanaka, S., et al. (2004). Development of radioimmunoassay for bullfrog thyroid-stimulating hormone (TSH): Effects of hypothalamic releasing hormones on the release of TSH from the pituitary in vitro. *General and Comparative Endocrinology*, *135*, 42–50.

Oppenheimer, J. H., Schwartz, H. L., & Strait, K. A. (1995). An integrated view of thyroid hormone actions *in vivo*. In B. D. Weintraub (Ed.), *Molecular endocrinology: Basic concepts and clinical correlations* (pp. 249–268). New York: Raven Press, Ltd.

Orchinik, M., Matthews, L., & Gasser, P. J. (2000). Distinct specificity for corticosteroid binding sites in amphibian cytosol, neuronal membranes, and plasma. *General and Comparative Endocrinology*, *118*, 284–301.

Power, D. M., Elias, N. P., Richardson, S. J., Mendes, J., Soares, C. M., & Santos, C. R. A. (2000). Evolution of the thyroid hormone-binding protein, transthyretin. *General and Comparative Endocrinology*, *119*, 241–255.

Prapunpoj, P., Yamauchi, K., Nishiyama, N., Richardson, S. J., & Schreiber, G. (2000). Evolution of structure, ontogeny of gene expression, and function of Xenopus laevis transthyretin. *American Journal of Physiology. Regulatory, Integrative and Comparative Physiology*, *279*, R2026–R2041.

Regard, E., & Mauchamp, J. (1971). Ultrastructure of thyroid gland of normal and hypophysectomized larval Xenopus—Correlation with biosynthesis of thyroglobulin. *Journal of Ultrastructure Research*, *37*, 664–678.

Regard, E., & Mauchamp, J. (1973). Peroxidase activity in Xenopus thyroid gland through larval development—Correlation with iodide organification and thyrotropic control. *Journal of Microscopy*, *18*, 291–306.

Regard, E., Taurog, A., & Nakashima, T. (1978). Plasma thyroxine and triiodothyronine levels in spontaneously metamorphosing *Rana catesbeiana* tadpoles and in adult anuran amphibia. *Endocrinology*, *102*, 674–684.

Ritchie, J. W. A., Hayashi, Y., Shi, Y. B., & Taylor, P. M. (2002). A role for amino acid transporters in cellular TH action. *The Journal of Physiology (London)*, *539*, 11P–12P.

Ritchie, J. W. A., Peter, G. J., Shi, Y. B., & Taylor, P. M. (1999). Thyroid hormone transport by 4F2hc-IU12 heterodimers expressed in Xenopus oocytes. *The Journal of Endocrinology*, *163*, R5–R9.

Ritchie, J. W. A., Shi, Y. B., Hayashi, Y., Baird, F. E., Muchekehu, R. W., Christie, G. R., et al. (2003). A role for thyroid hormone transporters in transcriptional regulation by thyroid hormone receptors. *Molecular Endocrinology*, *17*, 653–661.

Rivier, J., Rivier, C., & Vale, W. (1984). Synthetic competitive antagonists of corticotropin-releasing factor—Effect on ACTH secretion in the rat. *Science*, *224*, 889–891.

Rosenkilde, P. (1978). Thyroid hormone synthesis in metamorphosing and adult *Xenopus laevis*. *General and Comparative Endocrinology*, *34*, 95–96.

Sapolsky, R. M., Romero, L. M., & Munck, A. U. (2000). How do glucocorticoids influence stress responses? Integrating permissive, suppressive, stimulatory, and preparative actions. *Endocrine Reviews*, *21*, 55–89.

Saxen, L., Saxen, E., Toivonen, S., & Salimaki, K. (1957a). The anterior pituitary and the thyroid function during normal and abnormal development of the frog. *Annales Botanici Societatis Zoologicæ-Botanicæ Fennicæ Vanamo*, *18*, 1–44.

Saxen, L., Saxen, E., Toivonen, S., & Salimaki, K. (1957b). Quantitative investigation on the anterior pituitary-thyroid mechanism during frog metamorphosis. *Endocrinology, 61*, 35–44.
Schneider, M. J., Fiering, S. N., Pallud, S. E., Parlow, A. F., St Germain, D. L., & Galton, V. A. (2001). Targeted disruption of the type 2 selenodeiodinase gene (D102) results in a phenotype of pituitary resistance to T-4. *Molecular Endocrinology, 15*, 2137–2148.
Seasholtz, A. F., Valverde, R. A., & Denver, R. J. (2002). Corticotropin-releasing hormone-binding protein: Biochemistry and function from fishes to mammals. *The Journal of Endocrinology, 175*, 89–97.
Shi, Y. B. (2000). *Amphibian metamorphosis. From morphology to molecular biology*. New York: Wiley-Liss.
Shi, Y. B., & Brown, D. D. (1993). The earliest changes in gene expression in tadpole intestine induced by thyroid hormone. *The Journal of Biological Chemistry, 268*, 20312–20317.
Shi, Y. B., Liang, V. C., Parkison, C., & Cheng, S. Y. (1994). Tissue-dependent developmental expression of a cytosolic thyroid hormone protein gene in *Xenopus*: Its role in the regulation of amphibian metamorphosis. *FEBS Letters, 355*, 61–64.
Shintani, N., Nohira, T., Hikosaka, A., & Kawahara, A. (2002). Tissue-specific regulation of type III iodothyronine 5-deiodinase gene expression mediates the effects of prolactin and growth hormone in Xenopus metamorphosis. *Development, Growth & Differentiation, 44*, 327–335.
Soares, M. J., Alam, S. M. K., Konno, T., Ho-Chen, J. K., & Ain, R. (2006). The prolactin family and pregnancy-dependent adaptations. *Animal Science Journal, 77*, 1–9.
Sternberg, R. M., Thoemke, K. R., Korte, J. J., Moen, S. M., Olson, J. M., Korte, L., et al. (2011). Control of pituitary thyroid-stimulating hormone synthesis and secretion by thyroid hormones during Xenopus metamorphosis. *General and Comparative Endocrinology, 173*, 428–437.
St Germain, D. L., Galton, V. A., & Hernandez, A. (2009). Minireview: Defining the roles of the iodothyronine deiodinases: Current concepts and challenges. *Endocrinology, 150*, 1097–1107.
St Germain, D. L., Hernandez, A., Schneider, M. J., & Galton, V. A. (2005). Insights into the role of deiodinases from studies of genetically modified animals. *Thyroid, 15*, 905–916.
St Germain, D. L., Schwartzman, R. A., Croteau, W., Kanamori, A., Wang, Z., Brown, D. D., et al. (1994). A thyroid hormone-regulated gene in *Xenopus laevis* encodes a type III iodothyronine 5-deiodinase. *Proceedings of the National Academy of Sciences of the United States of America, 91*, 7767–7771 [published erratum appears in Proceedings of the National Academy of Sciences of the United States of America 1994 Nov 8;91 (23):11282].
Suzuki, R. M., & Kikuyama, S. (1983). Corticoids augment nuclear binding capacity for triiodothyronine in bullgrog tadpole tail fins. *General and Comparative Endocrinology, 52*, 272–278.
Takada, M., & Kasai, M. (2003). Growth hormone is a weaker candidate than prolactin for the hormone responsible for the development of a larval-type feature in cultured bullfrog skin. *The Journal of Experimental Biology, 206*, 1137–1142.
Takahashi, N., Yoshihama, K., Kikuyama, S., Yamamoto, K., Wakabayashi, K., & Kato, Y. (1990). Molecular cloning and nucleotide sequence analysis of complementary DNA for bullfrog prolactin. *Journal of Molecular Endocrinology, 5*, 281–287.
Tata, J. R., Baker, B. S., Machuca, I., Rabelo, E. M., & Yamauchi, K. (1993). Autoinduction of nuclear receptor genes and its significance. *The Journal of Steroid Biochemistry and Molecular Biology, 46*, 105–119.
Tonon, M. C., Cuet, P., Lamacz, M., Jegou, S., Cote, J., Gouteux, L., et al. (1986). Comparative effects of corticotropin-releasing factor, arginine vasopressin, and related

neuropeptides on the secretion of ACTH and alpha-MSH by frog anterior pituitary cells and neurointermediate lobes *in vitro*. *General and Comparative Endocrinology*, *61*, 438–445.
Torrents, D., Estevez, R., Pineda, M., Fernandez, E., Lloberas, J., Shi, Y. B., et al. (1998). Identification and characterization of a membrane protein (y(+)l amino acid transporter-1) that associates with 4f2hc to encode the amino acid transport activity y (+)l—A candidate gene for lysinuric protein intolerance. *The Journal of Biological Chemistry*, *273*, 32437–32445.
Vale, W., Vaughan, J., & Perrin, M. (1997). Corticotropin-releasing factor (CRF) family of ligands and their receptors. *The Endocrinologist*, 7(Suppl.), S3–S9.
Van Pett, K., Viau, V., Bittencourt, J. C., Chan, R. K. W., Li, H. Y., Arias, C., et al. (2000). Distribution of mRNAs encoding CRF receptors in brain and pituitary of rat and mouse. *The Journal of Comparative Neurology*, *428*, 191–212.
Visser, W. E., Friesema, E. C. H., Jansen, J., & Visser, T. J. (2008). Thyroid hormone transport in and out of cells. *Trends in Endocrinology and Metabolism*, *19*, 50–56.
Wahlborg, A., Bright, C., & Frieden, E. (1964). Activity of some new triiodothyronine analogs in tadpole. *Endocrinology*, *75*, 561–564.
Wang, Z., & Brown, D. D. (1993). Thyroid hormone-induced gene expression program for amphibian tail resorption. *The Journal of Biological Chemistry*, *268*, 16270–16278.
Werner, E. E. (1986). Amphibian metamorphosis: Growth rate, predation risk, and the optimal size at transformation. *The American Naturalist*, *128*, 319–341.
White, B. A., & Nicoll, C. S. (1979). Prolactin receptors in Rana catesbeiana during development and metamorphosis. *Science*, *204*, 851–853.
White, B. A., & Nicoll, C. S. (1981). Hormonal control of amphibian metamorphosis. In L. I. Gilbert & E. Frieden (Eds.), *Metamorphosis: A problem in developmental biology* (pp. 363–396). New York: Plenum.
Wilbur, H. M., & Collins, J. P. (1973). Ecological aspects of amphibian metamorphosis. *Science*, *182*, 1305–1314.
Wright, M. L., Cykowski, L. J., Lundrigan, L., Hemond, K. L., Kochan, D. M., Faszewski, E. E., et al. (1994). Anterior-pituitary and adrenal-cortical hormones accelerate or inhibit tadpole hindlimb growth and development depending on stage of spontaneous development or thyroxine concentration in induced metamorphosis. *The Journal of Experimental Zoology*, *270*, 175–188.
Yamamoto, K., & Kikuyama, S. (1982). Effect of prolactin antiserum on growth and resorption of tadpole tail. *Endocrinologia Japonica*, *29*, 81–85.
Yamauchi, K., Horiuchi, R., Koya, S., & Takikawa, H. (1989). Uptake of 3,5,3′-L-triiodothyronine into bullfrog red blood cells mediated by plasma membrane binding sites. *Zoological Science*, *6*, 749–755.
Yamauchi, K., Kasahara, T., Hayashi, H., & Horiuchi, R. (1993). Purification and characterization of a 3,5,3′-L-triiodothyronine specific binding-protein from bullfrog tadpole plasma—A homolog of mammalian transthyretin. *Endocrinology*, *132*, 2254–2261.
Yamauchi, K., Nakajima, J., Hayashi, H., & Hara, A. (1999). Purification and characterization of thyroid-hormone-binding protein from masu salmon serum—A homolog of higher-vertebrate transthyretin. *European Journal of Biochemistry*, *265*, 944–949.
Yamauchi, K., Prapunpoj, P., & Richardson, S. J. (2000). Effect of diethylstilbestrol on thyroid hormone binding to amphibian transthyretins. *General and Comparative Endocrinology*, *119*, 329–339.
Yamauchi, K., Takeuchi, H., Overall, M., Dziadek, M., Munro, S. L. A., & Schreiber, G. (1998). Structural characteristics of bullfrog (Rana catesbeiana) transthyretin and its cDNA—Comparison of its pattern of expression during metamorphosis with that of lipocalin. *European Journal of Biochemistry*, *256*, 287–296.

Yamauchi, K., & Tata, J. R. (1994). Purification and characterization of a cytosolic thyroid hormone binding protein (CTBP) in *Xenopus* liver. *European Journal of Biochemistry*, *225*, 1105–1112.

Yamauchi, K., Yamamoto, T., Hayashi, H., Koya, S., Takikawa, H., Toyoshima, K., et al. (1987). Sequence of membrane associated thyroid hormone binding protein from bovine liver—Its identity with protein disulfide isomerase. *Biochemical and Biophysical Research Communications*, *146*, 1485–1492.

Yao, M., & Denver, R. J. (2007). Regulation of vertebrate corticotropin-releasing factor genes. *General and Comparative Endocrinology*, *153*, 200–216.

CHAPTER EIGHT

Salamander Paedomorphosis: Linking Thyroid Hormone to Life History and Life Cycle Evolution

Carlena K. Johnson, S. Randal Voss[1]
Department of Biology, Spinal Cord and Brain Injury Research Center, University of Kentucky, Lexington, Kentucky, USA
[1]Corresponding author: e-mail address: srvoss@uky.edu

Contents

Abstract

Many salamanders have biphasic life cycles with aquatic larval and terrestrial adult phases. In these species, the transition between phases—metamorphosis—requires thyroid hormone (TH) activation of transcriptional programs that cause regression of larval traits and development of adult traits. During salamander evolution, TH signaling pathways have been altered in biphasic species to yield paedomorphic salamanders that retain larval traits and attain sexual maturity in larval aquatic habitats. We review literature concerning the ecology, evolution, and hormonal regulation of metamorphic, paedomorphic, and facultative salamander life histories. We then discuss recent microarray results that detail gene expression signatures of metamorphosis and paedomorphosis, and genetic results that establish TH responsiveness as a continuous trait with a quantitative trait locus (QTL) basis. TH-responsive QTL from ambystomatid salamanders explain variation in metamorphic timing, expression of metamorphosis versus paedomorphosis, and adult fitness traits. We propose a model for salamander life history evolution that links adaptation to aquatic habitats with TH-responsive loci that pleiotropically alter metamorphic timing and adult body size. Future studies that adopt

Current Topics in Developmental Biology, Volume 103
ISSN 0070-2153
http://dx.doi.org/10.1016/B978-0-12-385979-2.00008-3

genetic and genomic approaches will further establish salamanders as ideal models for investigating TH signaling mechanisms that regulate postembryonic development and the expression of alternate life histories.

Salamanders are one of three primary groups of amphibians, the other two being caecilians and anurans. Ancestrally, all three groups trace their origins to ancestors that present biphasic life cycles with an aquatic larval phase and a more terrestrial adult phase (Duellman & Trueb, 1986). However, alternate modes of development subsequently evolved within all three groups. Interestingly, ancestral vestiges of metamorphosis are observed during early development of direct developing anurans and salamanders (Callery & Elinson, 2000; Kerney, Blackburn, Muller, & Hanken, 2012), suggesting shared evolutionary potential for radical and early shifts in the timing of metamorphosis. Radical, later shifts in metamorphic timing are only observed in salamanders. In the most extreme cases, metamorphosis has been abolished completely, yielding bizarre larval-form adults with completely aquatic life cycles (Gould, 1977; Shaffer & Voss, 1996). These unique paedomorphic forms are ideal for investigating mechanisms of thyroid hormone (TH) regulation that are associated with adaptive delays in metamorphic timing and the evolution of novel life histories.

1. WHAT IS PAEDOMORPHOSIS AND WHY SALAMANDERS?

Paedomorphosis is a heterochronic term that describes a specific pattern of evolution, the retention of ancestral juvenile traits in the adult stage of a derived species (Gould, 1977). In this meaning, paedomorphosis references an evolutionary change in developmental timing between an ancestor and descendant species. However, paedomorphosis is often used in a more general sense to describe the retention of larval morphological traits in adult salamanders, irrespective of phylogeny. The evolution or expression of paedomorphosis is clearly associated with TH, the primary metamorphic hormone in anurans, salamanders, and some fish. In species that undergo a metamorphosis, TH induces the regression of larval traits and the development of traits typical of a more terrestrial adult (Fig. 8.1). As we discuss in more detail below, many paedomorphic salamanders can be induced to undergo partial or complete metamorphosis by simply placing them in a bath of TH. This suggests the possibility that paedomorphosis evolves "simply" by blocking the synthesis, secretion, or reception of TH in target cells (Page,

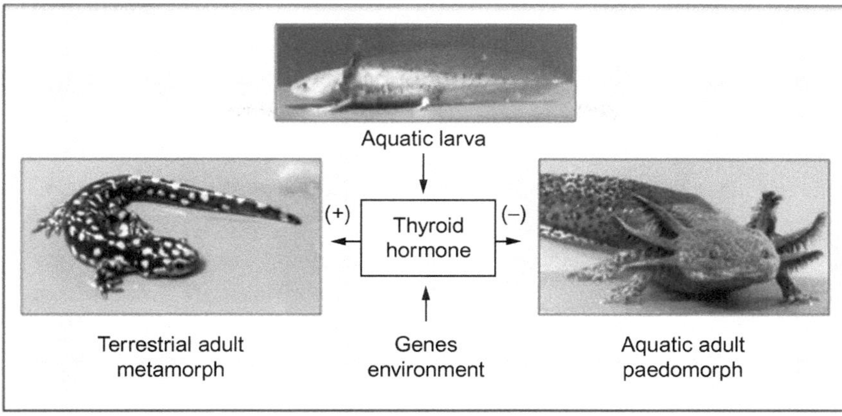

Figure 8.1 A representative salamander larva, adult metamorph (*Ambystoma tigrinum*), and adult paedomorph (*A. mexicanum*). Critical levels of thyroid hormone induce metamorphosis in salamanders. Environmental and genetic factors may cause thyroid hormone levels to be low, resulting in paedomorphic salamanders. *The pictures were taken by Jeramiah Smith.* (See Color Insert.)

Boley, Smith, Putta, & Voss, 2010). However, TH secretion is regulated by complex, neuroenodrine pathways that must first develop during the larval period before these systems become competent to signal TH release in response to intrinsic and extrinsic cues (Denver, Glennemeier, & Boorse, 2002). Thus, paedomorphosis could conceivably evolve by altering the development or function of a number of different cells, tissues, and organs that regulate the release or reception of TH. Given this complexity and the broad pleiotropic role that TH assumes in amphibian development and physiology, it seems likely that paedomorph evolution requires multiple changes across many genes.

The specific heterochronic process most commonly associated with salamander paedomorphosis is neoteny (Gould, 1977)—relative to the ancestor or metamorphic condition, somatic development is delayed but rate or timing of reproductive maturation is the same. However, paedomorphosis in some species is associated with an earlier time to first reproduction, achieved by accelerating the rate of gonadal development (progenesis in *Triturus alpestris*; Denoel & Joly, 2000) or by initiating reproductive maturation earlier in the life cycle (peramorphosis in *Ambystoma talpoideum*; Ryan & Semlitsch, 1998). Thus, very different patterns of growth and development may be associated with the expression of paedomorphosis within and between species (Whiteman, 1994). Among species that occur in stable aquatic habitats, paedomorphosis is a fixed trait with little or no incidence of metamorphosis. Within species that use less permanent bodies of water for breeding, individuals may reproduce initially as paedomorphs but undergo metamorphosis in subsequent years (Denoël et al., 2007). The different patterns of growth and development observed among paedomorphs yield different adult morphologies, and this suggests different mechanistic bases for the evolution of paedomorphosis within and among salamander families. However, exactly how patterns of development and morphology map onto mechanisms that block TH signaling and metamorphosis, or affect the rate of gonadal development, is essentially unknown. Without such mechanistic information, it is not possible to distinguish among examples of paedomorphosis that evolve convergently among lineages, which clearly has occurred during salamander phylogenesis (Shaffer & Voss, 1996; Weins & Hoverman, 2008). Thus, all terms that *describe* the evolutionary origin or developmental expression of paedomorphic salamanders are potentially useful metaphors, but if they are not based upon mechanistic insight and genetically based models, they are not likely to resolve outstanding questions concerning salamander life history variation and life cycle evolution.

It is interesting to consider why paedomorphosis is unique to salamanders and not other amphibians. In as much as paedomorphosis evolves as a prolongation of the larval period, it is notable that salamanders with biphasic life cycles typically have longer larval periods than caecilians or anurans (Duellman & Trueb, 1986). In biphasic salamander species, metamorphosis may occur within the same season or year that eggs are laid or larvae may overwinter in permanent habitats and metamorphose after one or more additional years of development (e.g., Beachy, 1995; Castanet, Francillon-Vieillot, & Bruce, 1996; Voss, Prudic, Oliver, & Shaffer, 2003). Although there are examples of anurans with multiyear larval periods, salamander larval periods are typically longer. So, there are fundamental differences between salamanders and other amphibians; salamanders have greater evolutionary potential to protract (and contract in cases of direct development) the length of the larval period. This greater evolutionary potential is associated with at least five aspects of salamander development and life history. First, salamanders exhibit slow rates of growth and development. Some anurans can complete metamorphosis in just a couple of weeks, while the fastest developing salamander larvae require more than a month, and typically several months (Duellman & Trueb, 1986). In general, salamander metamorphosis is developmentally less radical and more protracted than that of anurans. For example, salamander forelimb development does not

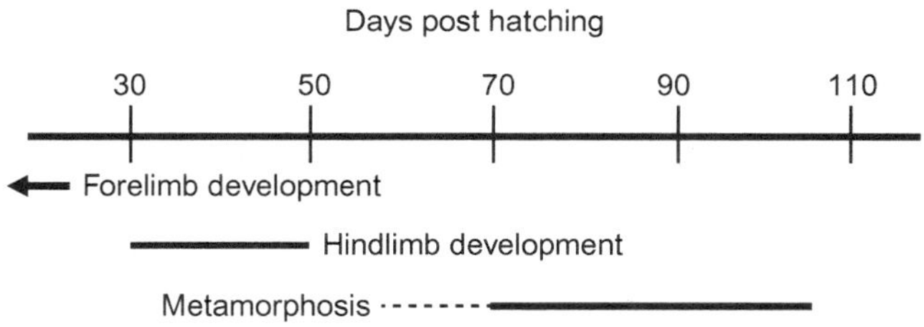

Figure 8.2 Timeline showing timing of fore and hindlimb development and the metamorphic period of laboratory reared *A. t. tigrinum*. The dashed line indicates variation among individuals in presenting early signatures of anatomical metamorphosis (bulging eyes, reduced tail fins). In the anuran *X. laevis*, hindlimb and forelimb development initiates 2 weeks after hatching. Hindlimbs complete development prior to metamorphosis, while forelimbs emerge during metamorphosis.

coincide with other morphological metamorphic changes, as it does in anurans (Fig. 8.2). Instead, it occurs shortly after hatching and before gills and tailfins are resorbed at metamorphosis. This suggests either a decoupling of forelimb development from metamorphosis or an incredibly gradual metamorphosis in salamanders that essentially spans the entirety of the larval period (Duellman & Trueb, 1986). Second, aspects of TH signaling may differ between salamanders and other amphibians. While it is clear that TH is required for salamander metamorphosis, *Ambystoma mexicanum* tolerates levels of TH that cause abnormal development and mortality in some anurans (Page et al., 2008). Also, several orthologous genes, including *thyroid hormone receptor beta* (*thrb*), do not show the same pattern of expression between TH-induced *A. mexicanum* and metamorphosing anurans (Page et al., 2007, 2008). Third, salamanders often utilize low temperature habitats, either because they occur at relatively high elevations or latitudes (Bizer, 1978; Sexton & Bizer, 1978). Low temperature slows growth and development, depresses activity of neurons and glands that regulate metamorphosis, and reduces the responsiveness of tissues to TH (Moriya, 1983a, 1983b; Uhlenhuth, 1919). Fourth, salamander larvae present the same body design as adults. Thus, unlike anuran tadpoles that must reorganize the body cavity at metamorphosis to allow space for gonads to develop, salamanders can potentially allocate resources to developing gonads and associated fat bodies during the larval period. Fifth, salamander larvae are carnivorous like adults, and thus unlike herbivorous anuran tadpoles, do not rely upon seasonal, primary productivity of ponds for food. Many anuran species have tadpoles that are highly specialized for rapid development, utilizing seasonally abundant resources in ephemeral ponds that select for finite larval periods (Wilbur, 1980). Multiyear larval periods are simply not possible within the context of anuran life history strategies that depend upon seasonally limiting resources. In contrast, salamander larvae present greater developmental flexibility to utilize more permanent aquatic habitats by extending the length of the larval period, indefinitely in the case of some paedomorphic species.

2. PAEDOMORPHOSIS AND SALAMANDER LIFE HISTORY VARIATION

Salamander life histories can be broadly classified into three categories: biphasic, paedomorphic, and facultative (Table 8.1). These are simply categories of convenience and do not adequately treat the continuum of life histories represented by each class, and the possibility of different genetic, developmental, and physiological bases within classes. Species with biphasic life histories almost invariably undergo a metamorphosis. This life history strategy is viewed as an adaptation to exploit transient opportunities for larval

Table 8.1 General characteristics of metamorphic, paedomorphic, and facultative salamander life histories

Metamorphic	Paedomorphic	Facultative
Associated with ephemeral aquatic habitats[a]	Associated with stable aquatic habitats[b]	Associated with landscapes with ephemeral and stable aquatic habitats[c]
Individuals almost invariably metamorphose in nature[a]	Individuals rarely/never metamorphose in nature[b]	Populations often vary in the proportion of individuals that metamorphose or exhibit paedomorphosis[c]
Generally reach sexual maturity after metamorphosis[c,d]	Reach sexual maturity while remaining in larval aquatic habitat[d,e]	Timing of sexual maturity may vary between metamorphic and paedomorphic individuals of the same population (but not always)[f–h]
Thyroid hormones cause morphological changes (e.g., gill reduction, larval to adult skin changes, tail fin reduction, etc.)[i]	Individuals may or may not respond to thyroid hormone with morphological signs of metamorphosis[i]	Thyroid hormones cause morphological changes[j,k]

Table 8.1 General characteristics of metamorphic, paedomorphic, and facultative salamander life histories—cont'd

Metamorphic	Paedomorphic	Facultative
Species may respond to stressful environmental factors with precocious metamorphosis[l]	In rare cases, stressful environmental factors may cause some species to metamorphose (e.g., *A. mexicanum*)[m,n]	Environmental factors affect the probability of an individual to express metamorphosis or paedomorphosis[b,o,p]

[a]Wilbur (1980).
[b]Sprules (1974).
[c]Denoel, Joly, and Whiteman (2005).
[d]Duellman and Trueb (1986).
[e]Armstrong, Duhon, and Malacinski (1989).
[f]Scott (1993).
[g]Ryan and Semlitsch (1998).
[h]Denoel and Joly (2000).
[i]Bruce (2003).
[j]Snyder (1956).
[k]Voss, Kump, Walker, Shaffer, and Voss (2012).
[l]Wilbur and Collins (1973).
[m]Smith (1969).
[n]Voss (unpublished observation).
[o]Eagleson and McKeown (1980).
[p]Semlitsch (1987).

growth and is often found where aquatic habitats are ephemeral (Wilbur, 1980; Wilbur & Collins, 1973). At the opposite extreme are salamanders that almost invariably remain paedomorphic until they die as aquatic adults. Paedomorphic salamanders are associated with isolated and stable aquatic habitats, such as large closed-basin lakes, spring-fed lakes, caldera lakes, and river systems (Shaffer, 1984; Sprules, 1974). In facultative species, metamorphs and paedomorphs are observed at varying frequencies within the same population. In such species, it is thought that the ability to express paedomorphosis and metamorphosis is advantageous when a landscape contains a mixture of ephemeral and stable aquatic habitats (Denoel et al., 2005; Semlitsch, Harris, & Wilbur, 1990; Sexton & Bizer, 1978; Whiteman, 1994; Wilbur, 1980; Wilbur & Collins, 1973). Facultative life histories are sometimes associated with aquatic habitats that may only be permanent for a single year of reproduction; the fitness advantage must be considerable because it is not uncommon to observe a paedomorphic brood in a cattle-watering tank on an arid landscape with no permanent ponds (Randal Voss, personal observation). When facultative species lay eggs in habitats that decline in quality during the larval period, metamorphosis is initiated in all or some proportion of individuals in the population (Semlitsch, 1987; Semlitsch et al., 1990).

Whereas metamorphic and facultative taxa are capable of dispersal among aquatic habitats, obligatorily paedomorphic taxa are confined to isolated bodies of water. This suggests that biphasic or facultative ancestors colonized aquatic habitats and ecological conditions (e.g., permanent aquatic habitat and/or arid terrestrial conditions) selected for paedomorphic life histories. The expression of paedomorphic life histories is known to affect population structure and may affect the probability of speciation (Shaffer, 1984; Shaffer & Breeden, 1989). The fitness benefits of paedomorphosis are an earlier time to first reproduction, more than one breeding event per year, larger clutch size, and a higher probability of mating success (Krenz & Server, 1995; Ryan & Semlitsch, 1998; Scott, 1993). It is important to note that paedomorphosis is not a pathology or default life history strategy—a failure to undergo metamorphosis. Paedomorphic species are highly adapted to their aquatic habitats. For example, *Ambystoma dumerilii* has secondarily evolved webbed-feet and extra gill filaments that develop at the time metamorphosis occurs in related metamorphic species (Fig. 8.3). Thus, evolution of paedomorphosis may release tissues from the constraint of TH remodeling, allowing new functions to evolve that better adapt adults to aquatic habitats. The combination of selection and genetic drift acting on isolated paedomorphic lineages may explain rapid divergence of morphology and gene expression within some species groups (Page et al., 2010; Shaffer & Voss, 1996).

Paedomorphic life history strategies are widespread among salamanders, occurring in nine out of 10 families with a total of 57 species. Studies show that paedomorphic salamanders have evolved multiple times in different lineages (Denoel et al., 2005; Shaffer & Voss, 1996; Weins & Hoverman, 2008). Four salamander families (Amphiumidae, Sirenidae, Proteidae, and Cryptobranchidae) are comprised entirely of paedomorphic species, having no extant biphasic species. Some salamanders, such as species in the families Cryptobranchidae and Amphiumidae, are said to undergo an incomplete metamorphosis—meaning they undergo some anatomical changes, but not others, and remain aquatic throughout their lives. For example, adult Hellbenders (*Cryptobranchus alleganiensis*) lose their larval external gills, but do not develop eyelids and retain a single gill-slit that appears as a circular opening on the neck (Larson, Weisrock, & Kozak, 2003). The most species-rich salamander family, Plethodontidae, includes many biphasic and direct developing species, with the Tribe Hemidactyliini (Subfamily Plethodontinae) containing numerous paedomorphic species that are cave dwelling and exhibit reduced pigmentation and vision (Larson et al., 2003). Perhaps the best-studied salamander families with respect to life history variation are

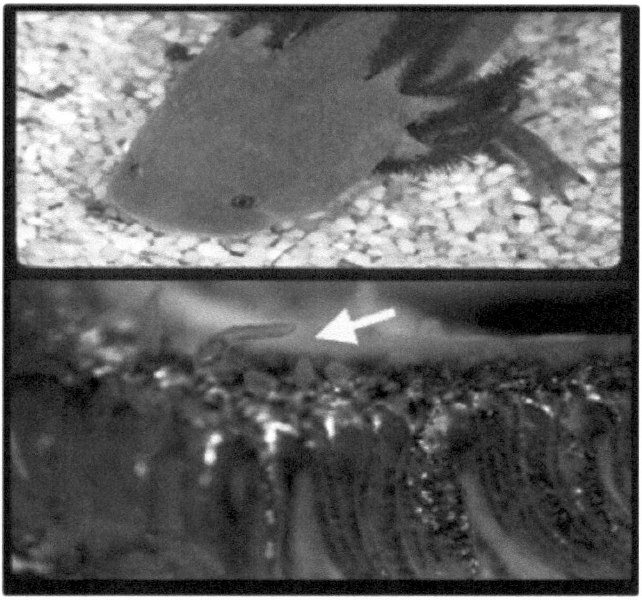

Figure 8.3 *Ambystoma dumerilii* is a paedomorph that has secondarily evolved webbed-feet and extra filaments that develop on the dorsal surfaces of each gill. The bottom panel is a magnified image of the dorsal gill surface. The arrow indicates dorsal filaments (<1.0 mm in length) that are sprouting on the gill surface. These extra fibers form late in the larval period, at the time metamorphosis normally occurs in related metamorphic species. *The A. dumerilii picture was taken by Brad Shaffer.* (See Color Insert.)

Salamandridae and Ambystomatidae. Studies of paedomorphic and facultative species within these families have better elucidated environmental and broad-sense genetic correlates of metamorph and paedomorph expression, and these species continue to provide exceptional models for investigating ecological aspects of life history evolution (Denoel, Ivanovic, Dzukic, & Kalezic, 2009; Denoel et al., 2005; Doyle & Whitemann, 2008; Takahashi & Parris, 2008; Whiteman et al., 2012).

3. HORMONAL BASIS OF SALAMANDER METAMORPHOSIS

Biphasic, paedomorphic, and facultative life histories begin the same way, with embryonic and larval development occurring in an aquatic habitat. The primary difference is when and if metamorphosis occurs. Although not as dramatic as metamorphosis in anurans, multiple tissues undergo hormonally regulated changes in cellular composition, biochemistry, and morphology during salamander metamorphosis (Dodd & Dodd, 1976). These changes include degeneration of external gills and tailfins; remodeling of skeletal elements and integument; shortening of the intestine; modification of pigmentation; muscle degeneration and differentiation; and changes in nitrogen excretion, urogenital function, and oxygen transport (Duellman & Trueb, 1986). Also, new structures arise *de novo*, including eyelids and skin glands. All of these changes adapt aquatic larvae for survival and reproduction as adults in more terrestrial environments.

Although salamanders have a more protracted and gradual metamorphosis than anurans, TH is the primary metamorphic hormone in both amphibian groups. Serum TH in the form of tetraiodothyroine (T4) is low during embryonic and early larval development, but as larvae approach metamorphosis, T4 levels increase (Norman, Carr, & Norris, 1987). This pattern of hormone secretion is not unlike that observed for other examples of hormonally regulated, postembryonic development (e.g., insect and fish metamorphosis; Gilbert, Tata, & Atkinson, 1996; Laudet, 2011; Power et al., 2001). The rise in T4 during amphibian larval development is associated with maturation and stimulation of brain regions that regulate thyroid activity, in response to extrinsic and intrinsic cues. These brain regions include the hypothalamus and pituitary gland. Increasing levels of T4 during larval development stimulate corticotropin-releasing hormone (CRH) release from modified nerve endings within the median eminence. CRH travels through capillaries to the anterior pituitary where it stimulates thyrotropes to release thyrotropin (TSH) into the blood stream. TSH in turn stimulates the thyroid gland to release T4. The TSH-releasing factor in amphibians is CRH, which also stimulates the release of corticotropin (ACTH) and subsequently, glucocorticoids (Denver & Licht, 1989). The potent, synergistic action of corticoids and TH in promoting anuran and salamander metamorphosis clearly implicates the hypothalamic-pituitary-thyroid (HPT) and HP-interrenal (HPI) axes in the physiological regulation of metamorphosis (Buscaglia, Leloup, & De Luze, 1985; Dodd & Dodd, 1976; Galton, 1991, 1992; Gancedo et al., 1992; Hayes, Chan, & Licht, 1993; Kuhn, De Groef, Van der Geyten, & Darras, 2005). Exactly how T4 levels initially increase at appropriate times during amphibian development to trigger the onset of metamorphosis, and sustain T4 at high levels during metamorphosis, is not fully understood (Manzon & Denver, 2004).

Peripherally, the availability and activity of TH is precisely regulated within cells. In mammals, TH is transported into target cells via TH transporters, which actively and preferentially import the hormone across the plasma membrane (Hennemann et al., 2001). TH transporters have been studied in *X. laevis* (Connors, Korte, Anderson, & Degitz, 2010; Ritchie et al., 2003), but there have been no studies using salamanders. Deiodinases have received considerable attention because they directly activate or inactivate intracellular forms of TH in amphibians (Buscaglia et al., 1985; Galton, 1991; Kuhn et al., 2005). The low activity T4 form of TH that is delivered to cells is converted intracellularly by deiodinase-type 2 (DIO2) into highly active 3,5,3′-triiodothyronine (T3). A second deiodinase, DIO3, inactivates intracellular T3. Although T4 may participate in cellular signaling (Caria, Dratman, Kow, Mameli, & Pavlides, 2008), T3 is the primary TH signaling molecule in anurans (Denver et al., 2002) and presumably this is also true for salamanders, acting through nuclear receptors (*thra*, *thrb*) that function as transcription factors. The higher activity of T3 is associated with a higher affinity for TH receptor (TR) binding. The binding of T3 activates transcriptional regulatory networks in cell-specific manners (Shi, 2000). TH is also known to cause cellular changes via nongenomic mechanisms, including membrane receptors, transporter molecules, and cytoplasmic receptors that elicit changes in cells through signaling pathways (Caria et al., 2008; Cheng, Leonard, & Davis, 2010; Davis, Leonard, & Davis, 2008; Furuya, Lu, Guigon, & Cheng, 2009). The effects of these nongenomic mechanisms of TH action and their role in amphibian metamorphosis have yet to be studied in detail, however, Buchholz, Tomita, Fu, Paul, and Shi (2004) used transgenic analysis to show that TR is sufficient to mediate TH signaling during metamorphosis in *X. laevis*. Regulation of deiodinases, T4/T3 concentrations, and developmental expression of TRs help explain how a single hormone can coordinate responses among different cell types, and more generally regulate the temporal sequence of remodeling events during amphibian metamorphosis (Brown & Cai, 2007).

4. HORMONAL BASIS OF PAEDOMORPHOSIS

Only a year after, Gudernatsch (1912) established TH as the anuran metamorphic hormone, TH was shown by Laufberger to induce metamorphosis in *A. mexicanum* (Mexican axolotl) (reviewed by Huxley & Hogben, 1922). Over the next few decades, the responsiveness of many different paedomorphic salamanders to TH was investigated. Studies showed that morphologically similar paedomorphic salamanders (with external gills or gill slits, no eyelids, larval pigmentation, etc.) responded differently to TH. Members of the family Proteidae, such as the American mudpuppy (*Necturus maculosus*), are known to be refractory to TH, such that treating them with exogenous or injected TH does not induce morphological metamorphosis (Swingle, 1922). Larsen (1968) proposed that peripheral tissues of *N. maculosus* had lost sensitivity to TH, given that it exhibits a morphologically normal thyroid gland that had been tested for functionality via transplant experiments (Charipper, 1929; Grant, 1930; Swingle, 1922). This prompted investigations into the functionality of the animal's deiodinase activity and TR expression. Galton (1985) found that deiodinase activity capable of converting T4 to T3 was present within *N. maculosus*, and Safi et al. (2006) showed that *N. maculosus* expresses functional TRs that bind to DNA and activate transcription in response to TH treatment, including transcriptional changes of classical TR target genes like *stromelysin 3* (*mmp11*) and *thrb* (Safi et al., 2006; Vlaeminck-Guillem et al., 2006). Additionally, *in situ* hybridization has been used to show expression of *thra* and *thrb* in central nervous system, epithelia of the digestive tract, myocardium, and skeletal muscle (Vlaeminck-Guillem et al., 2006). These studies suggest that the paedomorphic phenotype exhibited by *N. maculosus* may be due to the loss of key TH response genes needed to induce morphological changes, even though juveniles experience a postembryonic period of high TH sensitivity. Apparently, individuals undergo natural transcriptional and biochemical changes during a "cryptic metamorphosis" that does not result in gross morphological alterations (Laudet, 2011; Vlaeminck-Guillem et al., 2006). Species of the families Cryptobranchidae and Amphiumidae are responsive to THs in some respects, but not others. They often reabsorb external gills and develop adult characteristics of some features, but do not develop eyelids and retain gill slits as mature adults, indicating that some metamorphic processes may have been decoupled from TH regulation (Larson et al., 2003). When species of these families are treated with high doses of TH, often the only result is skin shedding, a trait that is regulated by TH in metamorphic species, though otherwise the individual continues to exhibit paedomorphic characteristics (Dent, 1968; Fox, 1984). It was originally thought that the Texas blind salamander (*Eurycea rathbuni*, formerly assigned to the genus *Typhlomolge*) was the only known vertebrate to lack a thyroid gland (Emerson, 1905). However, Gorbman (1957) later showed the presence of a thyroid gland in *E. rathbuni* and Dundee (1957) showed that an individual of this species underwent some morphological changes (gill and tail reabsorption) in response to T4, but other larval features were resistant (histological sections of skin revealed only larval characteristics). However, related paedomorphic species (*E. tynerensis* and *E. neotenes*) complete metamorphosis after T4 treatment in less than 2 weeks (Kezer, 1952). Paedomorphic species of *Ambystoma* also complete metamorphosis when treated with T4 but show species-specific variation in metamorphic timing (Voss et al., 2012). Collectively, these studies establish that paedomorphosis is associated with TH regulation and tissue responsiveness, and these characteristics vary among closely and distantly related species.

Based on results of early TH induction experiments, it was proposed that paedomorphic species evolve via several mechanisms, including disruption of peripheral and central mechanisms of TH regulation (Dent, 1968; Kezer, 1952). Species that were shown to be refractory to TH but had active thyroid glands were presumed to be deficient in peripheral mechanisms that receive and transduce the T4 signal, such as TRs. Species that responded to high levels of TH but did not apparently metamorphose in nature were assumed to be deficient in some aspect of central regulation within the hypothalamus or pituitary. Such a deficiency would normally prevent T4 release but could artificially be over-ridden in the lab by T4 treatment. It also seems likely that central regulatory mechanisms have been altered in the evolution of facultative species from biphasic ancestors. This is because the brain and pituitary regulate appropriate physiological and developmental responses in response to environmental cues. In the case of facultative paedomorphosis, the HPT axis is not sufficiently activated if cues favor a paedomorphic life history, or environmental conditions suppress hormone activity, or if larval development and maturation is delayed (Whiteman, 1994). Several environmental conditions that are associated with habitat quality affect the probability that an individual will express paedomorphosis, including low density of conspecifics (Harris, 1987; Semlitsch, 1987), food availability (Denoel & Poncin, 2001; Ryan & Semlitsch, 2003; Semlitsch, 1987), pond permanence (Semlitsch, 1987; Semlitsch et al., 1990), and low temperature (Eagleson & McKeown, 1980; Snyder, 1956; Sprules, 1974).

5. PAEDOMORPHOSIS IN THE MEXICAN AXOLOTL

The exemplar of all paedomorphic salamanders is the Mexican axolotl (*A. mexicanum*). The Mexican axolotl is native to the Xochimilco lake system, near present-day Mexico City. Whereas historical populations were sufficiently large to provide an important food source to Indians for thousands of years (Smith, 1989), the current *A. mexicanum* population at Xochimilco is on the verge of extinction (Graue, 1998; Zambrano, Vega, Herrera, Prado, & Reynoso, 2007). Smith (1989) provides an enjoyable historical account of the original collection of *A. mexicanum* from Xochimilco in 1863. Six *A. mexicanum* were transferred to Paris where Dumeril (1870) subsequently reported that individuals reproduced in an aquatic, larval state and that some of the resulting offspring underwent a metamorphosis. Smith (1989) suggests that some of the original stock that arrived in Paris included closely related members of the Tiger salamander species complex, that are capable of expressing a paedomorphic or metamorphic life history. However, it seems likely that domestication altered the penetrance for expressing paedomorphosis and the original axolotl stock was pure *A. mexicanum* that maintained a higher propensity to express metamorphosis in nature (Voss & Shaffer, 2000). Even though metamorphic forms have been culled from the Ambystoma Genetic Stock Center axolotl collection for decades, the frequency of spontaneous metamorphosis is ~1–2%, with a 10% frequency observed if *A. mexicanum* experience stressful conditions (Randal Voss, unpublished data).

The simplest explanation for paedomorphosis in *A. mexicanum* is a faulty thyroid gland. However, the thyroid gland is functional; it can synthesize T4 and release T4 after TSH stimulation (Prahlad, 1968; Taurog, Oliver, Eskay, Porter, & Mckenzie, 1974). Numerous studies, including our own (Page et al., 2007, 2008), have demonstrated that individuals can be stimulated to undergo morphological metamorphosis by administering T4. This suggests that intracellular receptors and deiodinase enzymes are functional. Putative T3 receptors have been identified in red blood cells (Galton, 1991) and TR function has been shown *in vitro* using mammalian cells (Safi et al., 2004). Also, the activity of deiodinases has been demonstrated (Darras et al., 2002; Galton, 1991). Blount (1950) showed that metamorphosis is inhibited if the pituitary of *A. mexicanum* is transplanted into the metamorph *A. tigrinum*, while the reciprocal transplantation experiment resulted in metamorphosis. Tassava (1969) pointed out that these grafts were probably not precise enough to rule out a hypothalamic contribution. TSH is present in the *A. mexicanum* pituitary (Taurog et al., 1974) at a quantity sufficient for inducing metamorphosis (Darras & Kuhn, 1983). Thus, the pituitary is capable of producing sufficient functional TSH for metamorphosis, but for some reason, it is not released. There are at least four explanations for this observation in *A. mexicanum*: (1) The pituitary does not receive appropriate stimulation from the hypothalamus, (2) the pituitary is insensitive to hypothalamic stimulation, (3) the pituitary is defective in releasing TSH, and (4) the pituitary is functional but stimulated to release TSH at the wrong time during ontogeny (Rosenkilde & Ussing, 1996). Galton (1992) suggested that T4 levels might never eclipse a threshold during development to stimulate brain regions to increase activity of the thyroid gland. Kuhn et al. (2005) similarly proposed that brain T3 availability is the limiting factor for metamorphic onset. They found that DIO2 activity in the brain, which is required to generate T3, requires appropriate doses of corticoids and T4. Thus, stimulation of both the HPA and HPI axes is required for metamorphosis, and in fact, it is possible to induce metamorphosis in *A. mexicanum* if submetamorphic levels of T4 and corticoids are administered together. Because injection of amniote CRH stimulates corticotropin but not thyrotropin release, Kuhn et al. (2005) implicated the thyrotroph CRH receptor as a likely candidate gene for paedomorphosis in *A. mexicanum*. However, researchers have demonstrated T4 increases after stimulating hypothalamic regions in neotenic tiger salamanders (*A. tigrinum*) and *A. mexicanum*; this could seemingly only occur if downstream thyrotrophin signaling pathways are functional (Norris & Gern, 1976; Rosenkilde & Ussing, 1996).

It is also possible that a recently discovered TSH signaling pathway may be implicated in salamander paedomorphosis. Studies of neuroendocrine regulation of amniote reproductive cycles showed that TSH from the *pars tuberalis* (PT) of the anterior pituitary gland is released to signal DIO2 transcription in the hypothalamus (Hanon et al., 2008, Nakao et al., 2008). This increases local T3 concentrations and thus stimulates canonical TH signaling pathways. Surprisingly, these studies show that the flow of information may be bidirectional between the pituitary and hypothalamus. If a similar PT signaling pathway is necessary for metamorphosis, this pathway would not be activated in *A. mexicanum* because of a block in TSH release. This might also explain why *A. mexicanum* does not show strong seasonality in breeding and can breed multiple times a year. Whether paedomorphosis involves a PT signaling mechanism or a classical TSH signaling mechanism, the metamorphic block in *A. mexicanum* is likely associated with hypothalamus and pituitary function (Kuhn et al., 2005; Laudet, 2011; Norris et al., 1973; Prahlad & DeLanney, 1965; Rosenkilde & Ussing, 1996).

6. GENOMIC ANALYSIS OF SALAMANDER METAMORPHOSIS AND PAEDOMORPHOSIS

Most mechanistic studies of metamorphosis and paedomorphosis have focused on physiology and to a lesser extent biochemistry. Recently, microarrays developed for *A. mexicanum* have been used to characterize global transcriptional responses of metamorphic and paedomorphic salamanders.

These studies have focused on the skin and brain, the former because it is a peripheral target tissue of TH and much is known about anatomical changes that occur during metamorphosis (Fahrmann, 1971a, 1971b, 1971c; Fox, 1983; Page, Monaghan, Walker, & Voss, 2009). The brain has also been targeted because it is where HPT and HPI axes are centrally regulated. We review three important findings from these studies.

T4 precisely activates metamorphic transcriptional programs in a concentration-dependent manner. Page et al. (2008) varied T4 concentration by an order of magnitude and examined transcriptional patterns in *A. mexicanum* skin after 0, 2, 12, and 28 days of T4 treatment. Individuals exposed to 50 nM T4 initiated metamorphosis approximately 1 week prior to those in 5 nM, however, the sequence of transcriptional and morphological changes during metamorphosis, and the length of the metamorphic period did not differ (Fig. 8.4A). Over 1000 genes were identified as differently expressed in both treatments, showing the same directional trends, but the patterns were temporally shifted by T4 concentration. The study showed that the timing of metamorphic onset is T4 concentration dependent, and surprisingly, high T4 concentration did not disassociate subsequent transcriptional programs. It remains to be determined if the T4 concentration effect in *A. mexicanum* is manifested at the level of target cells, the HPT axis, or possibly both (Rosenkilde & Ussing, 1996).

Patterns of gene expression are complex, reflecting tissue remodeling, changes in cellular metabolism, and loss and gain of larval and adult cell types. One of the advantages in working with *A. mexicanum* is that they present a natural, hypothyroid condition throughout life (Huggins et al., 2012). Thus, transcriptional programs can be activated precisely at juvenile or adult stages of the life cycle. When T4 is administered, thousands of genes show significant changes in transcript abundance. In a very general sense, these genes show four temporal gene expression patterns: they increase or decrease in abundance, or they transiently increase or decrease during metamorphosis (Fig. 8.4B). Linear, quadratic, and piece-wise regression models have been used to identify groups of genes that similarly increase or decrease in abundance as a function of days of T4 treatment. The earliest gene expression changes precede tissue-remodeling events and are associated with cells that are specific to larvae or cells that differentiate to form adult tissues. For example, after 2 days of T4 treatment, genes that are specifically expressed in apical and Leydig cells of the larval skin show decreases in transcript abundances (Page et al., 2007, 2008, 2009). At this time, apical and Leydig cell numbers are normal for larval skin and there is no histological evidence of remodeling. However, after 12 days of T4

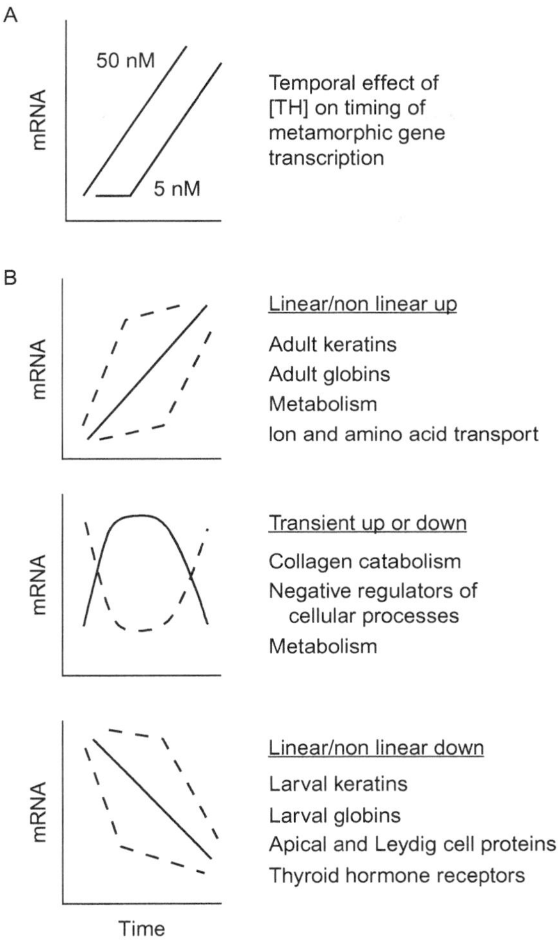

Figure 8.4 (A) Representative expression profile for a gene that shows a linear increase in mRNA abundance after TH induction of *A. mexicanum* using 5 or 50 nM T4. (B) Four general types of gene expression response observed after TH induction of metamorphosis in *A. mexicanum*. See text for details.

treatment, apical cells are no longer present in histological sections and Leydig cells are significantly reduced in number (Page et al., 2009); presumably, both cell types undergo TH-induced, programmed cell death, as has been demonstrated for apical cells in metamorphosing anurans (Schreiber & Brown, 2002). As larval cells and their gene expression products decrease in abundance during metamorphosis, TH induces the proliferation and differentiation of adult progenitor cells. As a result, transcripts associated with adult cells increase in abundance. The most dramatic changes are seen for basal keratinocytes that proliferate and differentiate during metamorphosis and transcribe adult specific keratins that generate a more cornified epithelial layer to limit desiccation during the adult phase. Genes that transiently increase in abundance code for matrix-remodeling proteins (e.g., MMPs) and cellular processes like metabolism, transcription, and translation, while genes that decrease in abundance include negative regulators of cellular processes (Page et al., 2008). These gene expression patterns illuminate tissue remodeling and macromolecular biosynthetic processes that occur during metamorphic climax.

Although some genes are expressed similarly between metamorphic and paedomorphic individuals during larval development, the majority of the differentially expressed genes are unique. For example, brain transcription has been compared between *A. mexicanum* and *A. t. tigrinum* individuals during larval development and the period spanning the metamorphic period of the later species (Page et al., 2010). As *A. t. tigrinum* individuals near the onset of anatomical metamorphosis, more gene expression differences are observed between larvae of these species. Presumably, many of these expression differences trace to increasing levels of glucocorticoids and THs in *A. t. tigrinum* individuals as they initiate early metamorphic changes. In support of this idea, a glucocorticoid receptor (*nr3c2*) showed significantly higher expression in *A. t. tigrinum*. Also, some genes that were more highly expressed during natural metamorphosis in *A. t. tigrinum* were also similarly upregulated in *A. mexicanum* brains when individuals were induced with T4 (Huggins et al., 2012). However, hundreds of genes were expressed more highly in *A. t. tigrinum* than *A. mexicanum* throughout the larval period. At this point, it is not clear if all of these differences are associated with metamorphic and paedomorphic life histories; however, the observed systematic bias in transcription suggests more than species-specific divergence of gene expression.

7. GENETIC ANALYSIS OF PAEDOMORPHOSIS

The idea that paedomorphic salamanders simply need a dose of TH to induce the second half of the ancestral ontogeny influenced the way paedomorphic life histories were thought to originate and evolve. For example, Goldschmidt (1940) cited *A. mexicanum* as the "hopeful monster" (Page et al., 2010), an example of evolution waiting around for the right macromutation—simply block a single physiological step in TH regulation and instantly a novel form is originated. The idea that paedomorphosis results from a simple mechanistic change gained further support when genetic studies showed that metamorphic and paedomorphic forms could be segregated according to Mendelian expectations. In particular, Humphrey (1967) showed that when *A. mexicanum*/*A. t. tigrinum* hybrids were backcrossed to *A. mexicanum*, there is 1:1 segregation of phenotypes. Later, Tompkins (1978) and Gould (1981) argued from Humphrey's single cross that *A. mexicanum* was an example of single gene, macromutational evolution. To further test the single gene hypothesis, Voss (1995) crossed domesticated *A. mexicanum* and *A. t. tigrinum* to create F_1 hybrids and like Humphrey (1967), backcrossed these to *A. mexicanum*. The ratios of metamorphs and paedomorphs arising from two relatively large backcrosses were consistent with single locus control, thus supporting the classical idea of a single mutation underlying the evolution of paeodomorphosis. In these crosses, the *A. t. tigrinum* and *A. mexicanum* alleles were dominant and recessive in effect, respectively. Subsequently, DNA was isolated from individuals of these crosses, amplified fragment length polymorphisms (AFPLs) were typed, and a genetic linkage map was constructed. This map was used to roughly locate the position of the major gene (*met1*) for paedomorphosis (Voss & Shaffer, 1996). To determine if *met1* arose independently in the domesticated stock, the backcross design was repeated using *A. mexicanum* collected from Lake Xochimilco (Voss & Shaffer, 2000). The segregation of phenotypes did not fit a single gene model, and the segregation of AFLPs marking the *met1* region exhibited segregation distortion, suggesting an epistatic effect between *met1* and other loci on viability. Still, *met1* genotypes did associate with metamorphic and paedomorphic phenotypes, indicating a difference in genetic background between wild-caught *A. mexicanum* and the domesticated stock.

Voss and Smith (2005) made additional *A. mexicanum*/*A. t. tigrinum* backcrosses and reared over 500 offspring to rigorously quantify the genetic effect of *met1* from wild-caught *A. mexicanum*. These crosses showed that *met1* not only explained discrete variation in the expression of metamorphosis and paedomorphosis but also continuous variation in metamorphic timing. Approximately 20% of the individuals that inherited two *met1* alleles from wild *A. mexicanum* were paedomorphic. The remainder of these homozygotes delayed metamorphic timing by an average of 36 days, relative to heterozygotes. These results showed that *met1* determines the expression of metamorphosis and paedomorphosis by altering metamorphic timing. Within the context of hybrid crosses, alleles from metamorphic and paedomorphic species decrease and delay the time to metamorphosis, respectively. Variation in the number of paedomorphic individuals generated from wild-caught versus domesticated *A. mexicanum* crosses reflect differences in genetic background. Here, we define genetic background as the summative effect of loci whose independent effects are too small to identify and quantify statistically. Within the genetic background of the domesticated stock, which has undergone strong artificial selection for paedomorphosis, the penetrance of *met1* is greater, and this results in a higher proportion of paedomorphic individuals in hybrid crosses (Voss & Smith, 2005).

8. THE LINK BETWEEN *met1* AND TH REGULATION OF METAMORPHOSIS

The link between *met1* and TH was recently established by combining quantitative trait locus (QTL) analysis with TH induction. Recognizing that paedomorphic species show variation in time to complete metamorphosis when treated with TH, Voss et al. (2012) crossed two paedomorphic species (*A. mexicanum* and *A. andersoni*) and then performed a backcross to segregate TH response alleles among second generation offspring. At 120 days postfertilization (dpf), backcross offspring were treated with 2.5 nM T4 and then scored for time to complete metamorphosis. Essentially, all offspring initiated and completed metamorphosis over an ~160 day interval of time. A QTL screen identified the genomic position of *met1* and two additional QTL (*met2*, *met3*), each explaining approximately 10% of the variation in metamorphic timing; the effects of all alleles were additive with *A. andersoni* alleles decreasing the time to metamorphosis. On average, individuals that inherited *met1-3* alleles from *A. andersoni* metamorphosed earlier, just as did hybrid *A. mexicanum* that inherited *met1* alleles from metamorphic *A. t. tigrinum* (Voss & Smith, 2005). The decrease in metamorphic timing was approximately 19 days for *A. andersoni met1* versus 36 days for *A. t. tigrinum met1*. These results demonstrate that variation in metamorphic timing is determined in part by alleles that segregate at TH-responsive QTL, and alleles from paedomorphic species may vary in effect.

In natural populations of amphibians, variation in metamorphic timing can affect the values of other life history traits. Life history theories predict that individuals should metamorphose early if conditions for larval growth are poor, but delay metamorphosis and attain large larval body sizes if growth conditions are optimal (Gould, 1977; Wilbur & Collins, 1973). This is because larger body sizes at metamorphosis translate into increased reproductive success in the adult phase (Semlitsch et al., 1988). From a physiological perspective, metamorphic timing and body size are expected to covary as a function of resource allocation. From an evolutionary perspective, metamorphic timing and body size are expected to covary and coevolve if they share a common genetic basis. This prediction holds in the case of *met2*. In the *A. andersoni*/*A. mexicanum* backcross study described above, *met2* not only explained significant variation in metamorphic timing but it also explained significant variation in adult body size (Voss et al., 2012). Individuals that delayed metamorphosis significantly increased body length at 300 dpf and weight at 400 dpf, and the effect was approximately equivalent for sexually dimorphic males and females. Overall, the genetic results reviewed here show that TH-responsive QTL accounts for variation in metamorphic timing, expression of metamorphosis versus paedomorphosis, and adult fitness traits.

9. WHAT GENES MAP TO METAMORPHIC TIMING QTL?

The *A. mexicanum* genome is in early stages of genome sequencing and thus comparative mapping has been used to more finely resolve the positions of *met1-3* and identify candidate genes. The most promising candidate that has been identified is *pou1f1*, a transcription factor associated with combined pituitary hormone deficiency (CPHD) in humans. *pou1f1* is predicted to map to the position of *met3* on linkage group (LG) 7. CPHD is associated with deficiencies in the secretion of pituitary hormones that regulate growth and development, but not hormones associated with reproductive physiology. Thus, *pou1f1* is a candidate gene for evolutionary developmental changes that are independent of reproductive maturation (neoteny), as is seen in the example of salamander paedomorphosis.

In contrast to *met3*, identifying candidates for *met1 and met2* has proven more difficult. For example, the 2 cm interval defining the location of *met1* on LG2 marks a unique synteny disruption in *Ambystoma* and perhaps other salamanders (Voss et al., 2011). This is a bit surprising and unlucky because the *Ambystoma* genome has undergone relatively few chromosomal rearrangements relative to other vertebrates and especially amniotes (Smith & Voss, 2006). Genes (e.g., *rai1*, *shmt1*, *drg2*, *med9*) from the Smith–Magenis disease syndrome region in the human genome flank one side of *met1*, while several genes associated with neural development and function (e.g., *setd2*, *ngfr*, *ccm2*) flank the other side. In no other vertebrate genome are these groups of flanking markers observed in synteny. Recently, we used microarray analysis to identify genes that are expressed differently during larval development as a function of *met1* genotype (Robert Page and Randal Voss, unpublished data). We reasoned that this combined genetic-genomic approach would identify differentially expressed genes that map to the position of *met1*, potentially identifying candidate genes. As predicted, the majority of differentially expressed genes from LG2 mapped within 10 cm of *met1*, and six of the 14 genes that located within 2 cm of *met1* were differentially expressed as a function of genotype during larval development and/or metamorphosis. At this point, it is not clear if the *met1* genomic region is more enriched for differentially expressed genes than other regions of the *Ambystoma* genome, and perhaps genes that are TH responsive and associated with brain development. Regardless, it is clear from microarray studies that many genes are differentially expressed between paedomorphic *A. mexicanum* and metamorphic *A. t. tigrinum*, and *met1* locates to a uniquely structured genomic region with several candidate genes for paedomorphosis.

10. SYNTHESIS: EVOLUTION OF SALAMANDER PAEDOMORPHOSIS

Many times during salamander evolution, mechanisms that regulate TH induction of metamorphosis were altered, resulting in paedomorphic salamanders that retain larval traits into the adult, breeding phase. Over the past 100 years, researchers have treated paedomorphic salamanders with TH to investigate the physiological and evolutionary basis of this unique salamander trait. Studies show that tissues within some species are entirely resistant to TH, while others show partial or complete metamorphosis. These varied responses are not explained by phylogeny or morphology. Salamanders are a morphologically conservative group and paedomorphosis has evolved independently during salamander evolution. It is not likely that investigations of convergent morphological differences among deeply divergent paedomorphic species will yield a synthetic model for the origin of salamander paedomorphosis. Instead, an understanding of the origins of paedomorphosis requires knowledge of physiological and genetic mechanism, environmental context, and life history traits associated with fitness (Fig. 8.5). In the case of ambystomatid salamanders, paedomorph expression is associated with TH-responsive loci that pleiotropically affect metamorphic timing and adult fitness traits. According to our model, paedomorphosis and metamorphosis are mechanistically connected by loci that segregate allelic variation for responsiveness to T4. Phenotypes arising from hybrid ambystomatid crosses clearly show that it is possible to artificially phenocopy the complete spectrum of metamorphic timing variation observed among species in nature. Given variation in habitat stability, selection should favor

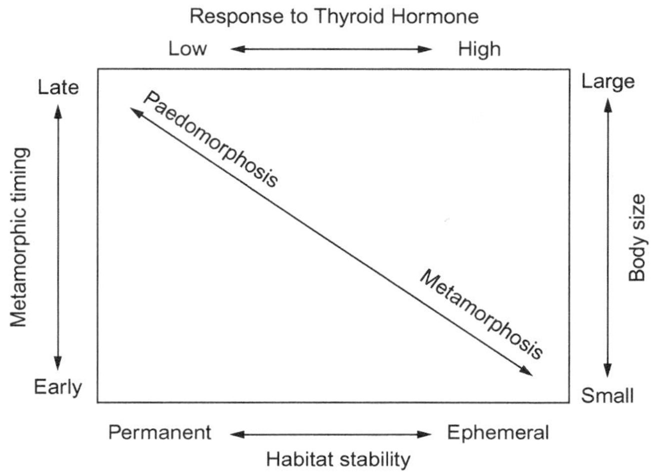

Figure 8.5 Model showing relationships among salamander life history variation, physiology, and habitat stability. Paedomorphosis is associated with low thyroid hormone responsiveness, permanent aquatic habitats, delayed metamorphic timing, and large body size. Metamorphosis is associated with contrary values for these variables.

low TH response alleles in biphasic populations when habitats allow for extensions of the larval period, as these alleles will delay metamorphic timing and increase adult body size. Under directional selection for mutations that increase the length of the larval period, central mechanisms of TH-release become progressively desensitized to intrinsic and extrinsic signals that provide reliable cues for initiating metamorphosis within the context of a biphasic life history. We hypothesize that central mechanisms are desensitized because TH-response loci become fixed for alleles that are less responsive to positive TH feedback. As genetic background changes, the likelihood of paedomorph expression increases. In support of this model, studies of facultative paedomorphic populations (*A. talpoideum*) have demonstrated the efficacy of selecting for higher paedomorph expression in relatively few generations (Semlitsch & Wilbur, 1989). If directional selection occurs over many generations and within the context of large and stable aquatic habitats, as has happened independently in the case of Tiger salamanders species in Mexico (Shaffer, 1984; Shaffer & McKnight, 1996; Shaffer & Voss, 1996), further changes in genetic background are predicted to yield individuals that are no longer capable of initiating metamorphosis in nature.

Many studies have shown that environmental factors, and especially low temperature, are often associated with paedomorphic populations (Eagleson & McKeown, 1980; Snyder, 1956; Sprules, 1974). It is possible that some examples of paedomorphosis may be pathological, reflecting extreme physiological conditions that may be exacerbated by inbreeding depression within small, isolated populations. However, Matsuda (1982; 1987) suggested that low temperature is pivotal in the adaptive evolution of paedomorphic salamanders; low temperature initially causes the expression of paedomorphosis in a population, and then through genetic assimilation (Waddington, 1953) the paedomorphic phenotype becomes fixed. The model that we proposed above is consistent with genetic assimilation because we also assume directional selection of TH-response alleles that increase an individual's likelihood of expressing a paedomorphic phenotype. The selection process and allelic effects depend upon environmental context.

The experimental approaches and evolutionary model outlined above provide a possible framework from which to initiate studies of the genetic basis of metamorphic timing in natural amphibian populations. At this point, it is not clear how many TH-response genes contribute to adaptive differences in metamorphic timing within natural populations and among species, or whether the same loci and physiological mechanisms are implicated in independent examples of paedomorph evolution. Also, much remains to be learned about facultative paedomorphosis at a mechanistic level. For example, does facultative expression of paedomorphosis mechanistically present an intermediated condition between metamorphosis and paedomorphosis, and how is it regulated? It is exciting to think that these and other questions can now be resolved using genetic and genomic methods that are applicable to nongenetic model organisms. The future looks bright for investigating evolutionary, developmental, and physiological aspects of salamander life history and life cycle evolution.

REFERENCES

Armstrong, J. B., Duhon, S. T., & Malacinski, G. M. (1989). Raising the axolotl in captivity. In J. B. Armstrong & G. M. Malacinski (Eds.), *Developmental biology of the axolotl* (pp. 220–227). New York: Oxford University Press.

Beachy, C. K. (1995). Age at maturation, body size, and life-history evolution in the salamander family Plethodontidae. *Herpetological Review*, *26*, 179–181.

Bizer, J. R. (1978). Growth rates and size at metamorphosis of high elevation populations of *Ambystoma tigrinum*. *Oecologia*, *34*, 175–184.

Blount, R. F. (1950). The effects of heteroplastic hypophyseal grafts upon the axolotl, *Ambystoma mexicanum*. *The Journal of Experimental Zoology*, *113*, 717–739.

Brown, D. D., & Cai, L. (2007). Amphibian metamorphosis. *Developmental Biology*, *306*, 20–33.

Bruce, R. C. (2003). Life histories. In D. M. Sever (Ed.), *Reproductive biology and phylogeny of Urodela* (pp. 477–525). Enfield, NH: Science Publishers, Inc.

Buchholz, D. R., Tomita, A., Fu, L., Paul, B. D., & Shi, Y. B. (2004). Transgenic analysis reveals that thyroid hormone receptor is sufficient to mediate the thyroid hormone signal in frog metamorphosis. *Molecular and Cellular Biology*, *24*, 9026–9037.

Buscaglia, M., Leloup, J., & De Luze, A. (1985). The role and regulation of monodeiodination of thyroxine to 3,5,3'-triiodothyronine during amphibian metamorphosis. In M. Balls & M. Bownes (Eds.), *Metamorposis* (pp. 273–293). Oxford, UK: Oxford University Press.

Callery, E. M., & Elinson, R. P. (2000). Thyroid hormone-dependent metamorphosis in a direct developing frog. *Proceedings of the National Academy of Sciences of the United States of America*, *97*, 2615–2620.

Caria, M. A., Dratman, M. B., Kow, L. M., Mameli, O., & Pavlides, C. J. (2008). Thyroid hormone action: Non-genomic modulation of neuronal excitability in the hippocampus. *Journal of Neuroendocrinology*, *21*, 98–107.

Castanet, J., Francillon-Vieillot, H., & Bruce, R. C. (1996). Age estimation in Desmognathine salamanders assessed by skeletochronology. *Herpetologica*, *52*, 160–171.

Charipper, H. A. (1929). Studies on amphibian endocrines. V. Accelerated metamorphosis of *Rana clamitans* larvae by means of implants of fresh anterior pituitary from adult *Necturus maculosus*. *The Anatomical Record*, *44*, 117–141.

Cheng, S. Y., Leonard, J. L., & Davis, P. J. (2010). Molecular aspects of thyroid hormone actions. *Endocrine Reviews*, *31*, 139–170.

Connors, K. A., Korte, J. J., Anderson, G. W., & Degitz, S. J. (2010). Characterization of thyroid hormone transporter expression during tissue-specific metamorphic events in *Xenopus tropicalis*. *General and Comparative Endocrinology*, *168*, 149–159.

Darras, V. M., & Kuhn, E. R. (1983). Effects of TRH, bovine TSH, and pituitary extracts on thyroidal T4 release in *Ambystoma mexicanum*. *General and Comparative Endocrinology*, *51*, 286–291.

Darras, V. M., Van der Geyten, S., Cox, C., Segers, I. B., De Groef, B., & Kühn, E. R. (2002). Effects of dexamethasone treatment on iodothyronine deiodinase activities and on metamorphosis-related morphological changes in the axolotl (*Ambystoma mexicanum*). *General and Comparative Endocrinology*, *127*, 157–164.

Davis, P. J., Leonard, J. L., & Davis, F. B. (2008). Mechanisms of nongenomic actions of thyroid hormone. *Frontiers in Neuroendocrinology*, *29*, 211–218.

Denoel, M., Ivanovic, A., Dzukic, G., & Kalezic, M. L. (2009). Sexual size dimorphism in the evolutionary context of facultative paedomorphosis: Insights from European newts. *BMC Evolutionary Biology*, *9*, 278.

Denoel, M., Lena, J. P., & Joly, P. (2007). Morph switching in a dimorphic population of *Triturus alpestris* (Amphibia, Caudata). *Evolutionary Ecology*, *21*, 325–335.

Denoel, M., & Joly, P. (2000). Neoteny and progenesis as two heterochronic processes involved in paedomorphosis in *Triturus alpestris* (Amphibia: Caudata). *Proceedings of the Royal Society of London, Series B: Biological Sciences*, *267*, 1481–1485.

Denoel, M., Joly, P., & Whiteman, H. H. (2005). Evolutionary ecology of facultative paedomorphosis in newts and salamanders. *Biological Reviews*, *80*, 663–671.

Denoel, M., & Poncin, P. (2001). The effect of food on growth and metamorphosis of paedomorphs in *Triturus alpestris apuanus*. *Archiv für Hydrobiologie*, *152*, 661–670.

Dent, J. N. (1968). Survey of amphibian metamorphosis. In W. Etkin & L. I. Gilbert (Eds.), *Metamorphosis: A problem in developmental biology* (pp. 271–311). New York: Appleton-Century-Crofts.

Denver, R. J., Glennemeier, K. A., & Boorse, G. C. (2002). Endocrinology of complex life cycles: Amphibians. In D. W. Pfaff, A. P. Arnld, A. M. Etgen, S. E. Fahrbach & R. T. Rubin (Eds.), *Hormones, brain and behavior* (pp. 469–513). San Diego: Elsevier.

Denver, R. J., & Licht, P. (1989). Neuropeptide stimulation of thyrotropin secretion in the larval bullfrog: Evidence for a common neuroregulator of thyroid and interrenal activity in metamorphosis. *The Journal of Experimental Zoology*, *252*, 101–104.

Dodd, M. H. I., & Dodd, J. M. (1976). The biology of metamorphosis. In B. Lofts (Ed.), *Physiology of amphibia*, Vol. 3, (pp. 467–599). New York: Academic Press.
Doyle, J. M., & Whitemann, H. H. (2008). Paedomorphosis in *Ambystoma talpoideum*: Effects of initial body size variation and density. *Oecologia, 156*, 87–94.
Duellman, W. E., & Trueb, L. (1986). *Biology of amphibians*. New York: McGraw-Hill.
Dumeril, A. (1870). Quatrieme notice sur la menagerie des reptiles du Museum d'Histoire Naturelle. *Nouvelles Archives du Museum D'Histoire Naturelle De Paris, 5*, 47–60.
Dundee, H. A. (1957). Partial metamorphosis induced in *Typhlomolge rathbuni*. *Copeia, 1957*, 52–53.
Eagleson, G. W., & McKeown, B. A. (1980). T4-induced metamorphosis in *Ambystoma gracile* (Baird) from two populations: The effects of aging and temperature. *Biochemistry and Experimental Biology, 16*, 111–126.
Emerson, E. T. (1905). General anatomy of *Typhlomolge rathbuni*. *Proceedings of the Boston Society of Natural History, 32*, 43–76.
Fahrmann, W. (1971a). Morphodynamics of the epidermis of the axolotl (*Siredon mexicanum* Shaw) under the influence of exogenously administered thyroxine. I. The epidermis of neotenic axolotls. *Zeitschrift für Mikroskopisch-Anatomische Forschung, 83*, 472–506.
Fahrmann, W. (1971b). Morphodynamics of the epidermis of the axolotl (*Siredon mexicanum* Shaw) under the influence of exogenously administered thyroxine. II. The epidermis during metamorphosis. *Zeitschrift für Mikroskopisch-Anatomische Forschung, 83*, 535–568.
Fahrmann, W. (1971c). Morphodynamics of the epidermis of the axolotl (*Siredon mexicanum* Shaw) under the influence of exogenously administered thyroxine. III. The epidermis of the metamorphosed axolotl. *Zeitschrift für Mikroskopisch-Anatomische Forschung, 84*, 1–25.
Fox, H. (1983). Changes in amphibian skin during larval development and metamorphosis. In M. Balls & M. Bownes (Eds.), *Metamorphosis. British society for developmental biology symposium*, Vol. 8, (pp. 59–87). Oxford: Clarendon Press.
Fox, H. (1984). *Amphibian morphogenesis*. Clifton, NJ: The Humana Press, Inc.
Furuya, F., Lu, C., Guigon, C. J., & Cheng, S. Y. (2009). Nongenomic activation of phosphatidylinositol 3-kinase signaling by thyroid hormone receptors. *Steroids, 74*, 628–634.
Galton, V. A. (1985). 3,5,3′-triiodothyronine receptors and thyroxine 5′-monodeiodinating activity in thyroid hormone-insensitive Amphibia. *General and Comparative Endocrinology, 57*, 465–471.
Galton, V. A. (1991). Thyroid hormone receptors and iodothyronine deiodinases in the developing Mexican axolotl, *Ambystoma mexicanum*. *General and Comparative Endocrinology, 85*, 62–70.
Galton, V. A. (1992). The role of thyroid hormone in amphibian metamorphosis. *Trends in Endocrinology and Metabolism, 3*, 96–100.
Gancedo, B., Corpas, I., Alonso-Gomez, A. L., Delgado, M. J., Morreale de Excobar, G., & Alonso-Bedate, M. (1992). Corticotropin-releasing factor stimulates metamorphosis and increases thyroid hormone concentration in prometamorphic *Rana perezi* larvae. *General and Comparative Endocrinology, 87*, 6–13.
Gilbert, L. I., Tata, J. R., & Atkinson, B. G. (Eds.), (1996). *Metamorphosis: Postembryonic reprogramming of gene expression in amphibians and insect cells*. San Diego: Academic Press.
Goldschmidt, R. (1940). *The material basis of evolution*. New Haven, CT: Yale University Press.
Gorbman, A. (1957). The thyroid gland of *Typhlomolge rathbuni*. *Copeia, 1957*, 41–43.
Gould, S. J. (1977). *Ontogeny and phylogeny*. Cambridge, UK: Belkap Press.
Gould, S. J. (1981). Change in developmental timing as a mechanism of macroevolution. In J. T. Bonner (Ed.), *Evolution and development* (pp. 333–346). Berlin, Germany: Springer-Verlag.
Grant, M. P. (1930). The release of follicular colloid from the thyroid of *Necturus maculosus* following heteroplastic anterior-pituitary implants. *The Anatomical Record, 46*, 205–221.

Graue, W. V. (1998). *Estudio genético y demográfico de la poblacion del amfibio* Ambystoma mexicanum *(Caudata: Ambystomaidae) del Lago de Xochimilco.* Mexico: Instituto de Ciencias del Mar y Limnologia, UNAM.
Gudernatsch, J. F. (1912). Feeding experiments in tadpoles: I. The influence of specific organs given as food on growth and differentiations: A contribution to the knowledge of organs with internal secretion. *Archiv für Entwicklungsmechanik der Organismen, 35,* 457–483.
Hanon, E. A., Lincoln, G. A., Fustin, J.-M., Dardente, H., Masson-Pevet, M., Morgan, P. J., et al. (2008). Ancestral TSH mechanism signals summer in a photoperiodic mammal. *Current Biology, 18,* 1147–1152.
Harris, R. N. (1987). Density-dependent paedomorphosis in the salamander *Notophthalamus viridescens dorsalis. Ecology, 68,* 705–712.
Hayes, T., Chan, R., & Licht, P. (1993). Interactions of temperature and steroids on larval growth, development, and metamorphosis in a toad (*Bufo boreas*). *The Journal of Experimental Zoology, 266,* 206–215.
Hennemann, G., Docter, R., Friesema, E. C., De Jong, M., Krenning, E. P., & Visser, T. J. (2001). Plasma membrane transport of thyroid hormones and its role in thyroid hormone metabolism and bioavailability. *Endocrine Reviews, 22,* 451–476.
Huggins, P., Johnson, C. K., Schoergendorfer, A., Putta, S., Bathke, A. C., Stromberg, A. J., et al. (2012). Identification of differentially expressed thyroid hormone responsive genes from the brain of the Mexican Axolotl (*Ambystoma mexicanum*). *Comparative Biochemistry and Physiology Part C: Toxicology, 155,* 128–135.
Humphrey, R. R. (1967). Albino axolotls from an albino tiger salamander through hybridization. *The Journal of Heredity, 58,* 95–101.
Huxley, J. S., & Hogben, L. T. (1922). Experiments on amphibian metamorphosis and pigment response in relation to internal secretions. *Proceedings of the Royal Society B, 93,* 36–53.
Kerney, R. R., Blackburn, D. C., Muller, H., & Hanken, J. (2012). Do larval traits re-evolve? Evidence from the embryogenesis of a direct-developing salamander, *Plethodon cinereus. Evolution, 66,* 252–262.
Kezer, J. (1952). Thyroxine-induced metamorphosis of the neotenic salamanders *Eurycea tynerensis* and *Eurycea neotenes. Copeia, 1952,* 234–237.
Krenz, J. D., & Server, D. M. (1995). Mating and ovoposition in paedomorphic *Ambystoma talpoideum* preceeds the arrival of terrestrial males. *Herpetologica, 51,* 387–393.
Kuhn, E. R., De Groef, B., Van der Geyten, S., & Darras, V. M. (2005). Corticotropin-releasing hormone-mediated metamorphosis in the neotenic axolotl *Ambystoma mexicanum*: Synergistic involvement of thyroxine and corticoids on brain type II deiodinase. *General and Comparative Endocrinology, 143,* 75–81.
Larsen, J. H. (1968). Ultrastructure of thyroid follicle cells of three salamanders (*Ambystoma, Amphiuma,* and *Necturus*) exhibiting varying degrees of neoteny. *Journal of Ultrastructure Research, 24,* 190–209.
Larson, A., Weisrock, D. W., & Kozak, K. H. (2003). Phylogenetic systematics of salamanders (Amphibia:Urodela), a review. In D. M. Sever (Ed.), *Reproductive biology and phylogeny of Urodela* (pp. 31–108). Enfield, NH: Science Publishers, Inc.
Laudet, V. (2011). The origins and evolution of vertebrate metamorphosis. *Current Biology, 21,* R726–R737.
Manzon, R. G., & Denver, R. J. (2004). Regulation of pituitary thyrotropin gene expression during *Xenopus* metamorphosis: Negative feedback is functional throughout metamorphosis. *The Journal of Endocrinology, 182,* 273–285.
Matsuda, R. (1982). Evolutionary process of talitrid amphipods and salamanders in changing environments, with a discussion of genetic assimilation and some other evolutionary concepts. *Canadian Journal of Zoology, 60,* 733–749.

Matsuda, R. (1987). *Animal evolution in changing environments: With special reference to abnormal metamorphosis*. New York: John Wiley and Sons.

Moriya, T. (1983a). The effect of temperature on the action of thyroid hormone and prolactin in larvae of the salamander *Hynobius retardatus*. *General and Comparative Endocrinology*, *49*, 1–7.

Moriya, T. (1983b). Cytological changes induced by low temperature in the thyroid glands of larvae of the salamander *Hynobius retardatus*. *General and Comparative Endocrinology*, *49*, 8–14.

Nakao, N., Ono, H., Yamamura, T., Anraku, T., Takagi, T., Higashi, K., et al. (2008). Thyrotrophin in the pars tuberalis triggers photoperiodic response. *Nature*, *452*, 317–322.

Norman, M. F., Carr, J. A., & Norris, D. O. (1987). Adenohypophysial-thyroid activity of the tiger salamander, *Ambystoma tigrinum*, as a function of metamorphosis and captivity. *The Journal of Experimental Zoology*, *242*, 55–66.

Norris, D. O., & Gern, W. A. (1976). Thyroxine-induced activation of hypothalamo-hypophysial axis in neotenic salamander larvae. *Science*, *194*, 525–527.

Norris, D. O., & Platt, J. E. (1973). Effects of pituitary hormones, melatonin, and thyroidal inhibitors on radioiodide uptake by the thyroid glands of larval and adult tiger salamanders, *Ambystoma tigrinum* (Amphibia: Caudata). *General and Comparative Endocrinology*, *21*, 368–376.

Page, R. B., Boley, M. A., Smith, J. J., Putta, S., & Voss, S. R. (2010). Microarray analysis of a salamander hopeful monster reveals transcriptional signatures of paedomorphic brain development. *BMC Evolutionary Biology*, *10*, 199.

Page, R. B., Monaghan, J. R., Samuels, A. K., Smith, J. J., Beachy, C. K., & Voss, S. R. (2007). Microarray analysis identifies keratin loci as sensitive biomarkers for thyroid hormone disruption in salamanders (*Ambystoma*). *Comparative Biochemistry and Physiology Part C: Toxicology and Pharmacology*, *145*, 15–27.

Page, R. B., Monaghan, J. R., Walker, J. A., & Voss, S. R. (2009). A model of transcriptional and morphological changes during thyroid hormone-induced metamorphosis of the axolotl. *General and Comparative Endocrinology*, *162*, 219–232.

Page, R. B., Voss, S. R., Samuels, A. K., Smith, J. J., Putta, S., & Beachy, C. K. (2008). Effect of thyroid hormone concentration on the transcriptional response underlying induced metamorphosis in the Mexican axolotl (*Ambystoma*). *BMC Genomics*, *9*, 78.

Power, D. M., Llewellyn, L., Faustino, M., Nowell, M. A., Bjornsson, B. T., Einarsdottir, I. E., et al. (2001). Thyroid hormones in growth and development of fish. *Comparative Biochemistry and Physiology. Part C: Toxicology and Pharmacology*, *130*, 447–459.

Prahlad, K. V. (1968). Induced metamorphosis: Rectification of a genetic disability by thyroid hormone in the Mexican axolotl, *Siredon mexicanum*. *General and Comparative Endocrinology*, *11*, 21–30.

Prahlad, K. V., & DeLanney, L. E. (1965). A study of induced metamorphosis in the axolotl. *The Journal of Experimental Zoology*, *160*, 137–145.

Ritchie, J. W., Shi, Y. B., Hayashi, Y., Baird, F. E., Muchekehu, R. W., Christie, G. R., et al. (2003). A role for thyroid hormone transporters in transcriptional regulation by thyroid hormone receptors. *Molecular Endocrinology*, *17*, 653–661.

Rosenkilde, P., & Ussing, A. P. (1996). What mechanisms regulate neoteny and induced metamorphosis in urodeles? *The International Journal of Developmental Biology*, *40*, 665–673.

Ryan, T. J., & Semlitsch, R. D. (1998). Intraspecific heterochrony and life history evolution: Decoupling somatic and sexual development in a facultatively paedomorphic salamander. *Proceedings of the National Academy of Sciences of the United States of America*, *95*, 5643–5648.

Ryan, T. J., & Semlitsch, R. D. (2003). Growth and the expression of alternative life cycles in the salamander *Ambystoma talpoideum* (Caudata: Ambystomatidae). *Biological Journal of the Linnean Society (London), 80*, 639–646.
Safi, R., Bertrand, S., Marchand, O., Duffraisse, M., de Luze, A., Vanacker, J. M., et al. (2004). The axolotl (*Ambystoma mexicanum*), a neotenic amphibian, expresses functional thyroid hormone receptors. *Endocrinology, 145*, 760–772.
Safi, R., Vlaeminck-Guillem, V., Duffraisse, M., Seugnet, I., Plateroti, M., Margotat, A., et al. (2006). Paedomorphosis revisited: Thyroid hormone receptors are functional in *Necturus maculosus. Evolution and Development, 8*, 284–292.
Schreiber, A. M., & Brown, D. D. (2002). Tadpole skin dies autonomously in response to thyroid hormone at metamorphosis. *Proceedings of the National Academy of Sciences of the United States of America, 100*, 1769–1774.
Scott, D. E. (1993). Timing of reproduction in paedomorphic and metamorphic *Ambystoma talpoideum. The American Midland Naturalist, 129*, 397–402.
Semlitsch, R. D. (1987). Paedomorphosis in *Ambystoma talpoideum*: Effects of density, food, and pond drying. *Ecology, 68*, 994–1002.
Semlitsch, R. D., Harris, R. D., & Wilbur, H. M. (1990). Paedomorphosis in *Ambystoma talpoideum*: Maintenance of population variation and alternative life history pathways. *Evolution, 44*, 1604–1613.
Semlitsch, R. D., Scott, D. E., & Pechmann, J. H. (1988). Time and size at metamorphosis related to adult fitness in Ambystoma talpoideum. *Ecology, 69*, 184–192.
Semlitsch, R. D., & Wilbur, H. M. (1989). Artificial selection for paedomorphosis in the salamander *Ambystoma talpoideum. Evolution, 43*, 105–112.
Sexton, O. J., & Bizer, J. R. (1978). Life history patterns of *Ambystoma tigrinum* in montane Colorado. *The American Midland Naturalist, 99*, 101–118.
Shaffer, H. B. (1984). Evolution in a paedomorphic lineage. I. An electrophoretic analysis of the Mexican ambystomatid salamanders. *Evolution, 38*, 1194–1206.
Shaffer, H. B., & Breeden, F. (1989). The relationship between allozyme variation and life history: Non-transforming salamanders are less variable. *Copeia, 1989*, 1016–1023.
Shaffer, H. B., & McKnight, M. L. (1996). The polytypic species revisited: Genetic differentiation and molecular phylogenetics of the tiger salamander *Ambystoma tigrinum* (Amphibia: Caudata) complex. *Evolution, 50*, 417–433.
Shaffer, H. B., & Voss, S. R. (1996). Phylogenetic and mechanistic analysis of a developmentally integrated character complex: Alternate life history modes in ambystomatid salamanders. *American Zoologist, 36*, 24–35.
Shi, Y. B. (2000). *Amphibian metamorphosis: From morphology to molecular biology*. New York: Wiley.
Smith, H. M. (1969). The Mexican axolotl: Some misconceptions and problems. *BioScience, 19*, 593–597.
Smith, H. M. (1989). Discovery of the axolotl and its early history in biological research. In G. M. Malacinski & J. B. Armstrong (Eds.), *Developmental biology of the axolotl*. New York: Oxford University Press.
Smith, J. J., & Voss, S. R. (2006). Gene order data from a model amphibian: New perspectives on vertebrate genome structure and evolution. *BMC Genomics*, 7, 219.
Snyder, R. C. (1956). Comparative features of the life histories of *Ambystoma gracile* (Baird) from populations at low and high altitudes. *Copeia, 1956*, 41–50.
Sprules, W. G. (1974). The adaptive significance of paedogenesis in North American species of *Ambystoma* (Amphibia: Caudata): An hypothesis. *Canadian Journal of Zoology, 52*, 393–400.
Swingle, W. W. (1922). Experiments on the metamorphosis of neotenous amphibians. *The Journal of Experimental Biology, 36*, 397–421.
Takahashi, M. K., & Parris, M. J. (2008). Life cycle polyphenism as a factor affecting ecological divergence within *Notophthalmus viridescens. Oecologia, 158*, 23–34.

Tassava, R. A. (1969). Survival and limb regeneration of hypophysectomized newts with pituitary xenografts from larval axolotls, *Ambystoma mexicanum*. *The Journal of Experimental Zoology*, *171*, 451–458.
Taurog, A. C., Oliver, R., Eskay, L., Porter, J. C., & Mckenzie, J. M. (1974). The role of TRH in the neoteny of the Mexican axolotl (*Ambystoma mexicanum*). *General and Comparative Endocrinology*, *24*, 267–279.
Tompkins, R. (1978). Genic control of axolotl metamorphosis. *American Zoologist*, *18*, 313–319.
Uhlenhuth, E. (1919). Relation between metamorphosis and other developmental phenomena in amphibians. *Journal of General Physiology*, *1*, 525–544.
Vlaeminck-Guillem, V., Safi, R., Guillem, P., Leteurtre, E., Duterque-Coquillaud, M., & Laudet, V. (2006). Thyroid hormone receptor expression in the obligatory paedomrophic salamander *Necturus maculosus*. *The International Journal of Developmental Biology*, *50*, 553–560.
Voss, S. R. (1995). Genetic basis of paedomorphosis in the axolotl, *Ambystoma mexicanum*: A test of the single gene hypothesis. *The Journal of Heredity*, *86*, 441–447.
Voss, S. R., Kump, D. K., Putta, S., Pauly, N., Reynolds, A., Henry, R. J., et al. (2011). Origin of amphibian and avian chromosomes by fission, fusion, and retention of ancestral chromosomes. *Genome Research*, *21*, 1306–1312.
Voss, S. R., Kump, K. D., Walker, J. A., Shaffer, H. B., & Voss, G. J. (2012). Thyroid hormone responsive QTL and the evolution of paedomorphic salamanders. *Heredity*, *109*, 293–298.
Voss, S. R., Prudic, K., Oliver, J., & Shaffer, H. B. (2003). Candidate gene analysis of metamorphic timing in ambystomatid salamanders. *Molecular Ecology*, *12*, 1217–1223.
Voss, S. R., & Shaffer, H. B. (1996). What insights into the developmental traits of urodeles does the study of interspecific hybrids provide? *The International Journal of Developmental Biology*, *40*, 885–893.
Voss, S. R., & Shaffer, H. B. (2000). Evolutionary genetics of metamorphic failure using wild-caught versus laboratory axolotls (*Ambystoma mexicanum*). *Molecular Ecology*, *9*, 1401–1408.
Voss, S. R., & Smith, J. J. (2005). Evolution of salamander life cycles: A major effect QTL contributes to both continuous and discrete variation for metamorphic timing. *Genetics*, *170*, 275–281.
Waddington, C. H. (1953). Genetic assimilation of an acquired character. *Evolution*, 7, 118–126.
Weins, J. J., & Hoverman, J. T. (2008). Digit reduction, body size, and paedomorphosis in salamanders. *Evolution and Development*, *10*, 449–463.
Whiteman, H. H. (1994). Evolution of facultative paedomorphosis. *The Quarterly Review of Biology*, *69*, 205–221.
Whiteman, H. H., Wissinger, S. A., Denoel, M., Mecklin, C. J., Gerlanc, N. M., & Gutrich, J. J. (2012). Larval growth in polyphenic salamanders: Making the best of a bad lot. *Oecologia*, *168*, 109–118.
Wilbur, H. M. (1980). Complex life cycles. *Annual Review of Ecology and Systematics*, *11*, 67–93.
Wilbur, H. M., & Collins, J. P. (1973). Ecological aspects of amphibian metamorphosis. *Science*, *182*, 1305–1314.
Zambrano, L., Vega, E. E., Herrera, M., Prado, E., & Reynoso, V. H. (2007). A population matrix model and population viability analysis to predict the fate of endangered species in highly managed water systems. *Animal Conservation*, *10*, 297–303.

CHAPTER NINE

Metamorphosis in a Frog That Does Not Have a Tadpole

Richard P. Elinson[1]

Department of Biological Sciences, Duquesne University, Pittsburgh, Pennsylvania, USA

[1]Corresponding author: e-mail address: elinson@duq.edu

Contents

Abstract

The evolutionary removal of the tadpole from the frog life history is a very successful strategy, particularly in the tropics. These direct developers form limbs and a frog-like head early in embryogenesis, and they have reduced or lost tadpole-specific structures, like gills, a long, coiled intestine, and tadpole teeth and jaws. Despite the apparently continuous development to the frog morphology, the direct developer, *Eleutherodactylus coqui*, undergoes a cryptic metamorphosis requiring thyroid hormone. As in *Xenopus laevis*, there is a stimulation by corticotrophin-releasing factor (CRF) and an upregulation of thyroid hormone receptor β (thrb). In addition to changes in skin and muscle, thyroid hormone stimulates yolk utilization for froglet growth from a novel tissue, the nutritional endoderm. The activities of CRF and corticosterone (CORT) in metamorphosis may provide the basis for the multiple evolutionary origins of direct development in anuran amphibians. Potential roles for maternally supplied thyroid hormone and its receptor and for deiodinases in regulating tissue sensitivity to thyroid hormone should be the subjects of future investigations.

Current Topics in Developmental Biology, Volume 103
ISSN 0070-2153
http://dx.doi.org/10.1016/B978-0-12-385979-2.00009-5

1. INTRODUCTION

Most people, at least those living in temperate zones, are familiar with a typical tadpole frog life history. Frog eggs are laid in water, where they develop into tadpoles. Tadpoles feed for weeks or months, and then metamorphose to frogs. At metamorphosis, the animal's body and its lifestyle change completely. The tadpole's tail is resorbed, and limbs develop. Rather than swimming in water and eating plant material or detritus, the frog jumps around on land and eats live insects and other animals. This metamorphosis is controlled by thyroid hormone.

It often comes as a surprise to learn that this typical life history is not found in many parts of the world, particularly in tropical areas. Frogs have evolved several strategies for keeping their young out of the dangerous water environment. One strategy is to keep the embryos and tadpoles hidden somewhere in the body of the father or the mother. Body sites include the vocal pouch of the male, a pouch on the back of the female, or the uterine-like ovisac of the female (Duellman & Trueb, 1986; Elinson & del Pino, 2012). A second strategy is to delete the tadpole from the life history and to have the embryos develop to frogs on land. This strategy, known as direct development, is quite common. It has evolved independently at least a dozen times (Duellman & Trueb, 1986; Hanken, 1999), and it is a remarkably successful strategy. One large group of direct developers is the Terrarana, mainly in Central and South America and the Caribbean. It contains about 900 species in four families (Hedges, Duellman, & Heinicke, 2008; Heinicke, Duellman, & Hedges, 2007; Schmid et al., 2010). Another group is from Sri Lanka, which contains more than 100 species so far (Meegaskumbura et al., 2002; Pethiyagoda & Manamendra-Arachchi, 1998).

The lack of a tadpole in these direct developing frogs raises the question as to whether they undergo a thyroid hormone-dependent metamorphosis. This question has been investigated in only one species of direct developer, the Puerto Rican tree frog *Eleutherodactylus coqui*, a member of the Terrarana. These frogs, indeed, have a cryptic thyroid-dependent metamorphosis (Callery & Elinson, 2000a; Elinson & del Pino, 2012), and these results form the basis of this chapter.

2. LOSS OF TADPOLE STRUCTURES IN FROG DIRECT DEVELOPMENT

Despite the dozen independent origins of frog direct development, the different groups tend to converge on a common embryo appearance. This is due in large part to the lack of a feeding tadpole, leading to the requirement for a greatly increased yolk content of the egg. Most direct developers have eggs that are greater than 3 mm in diameter, as compared to 1.3–1.8 mm diameter eggs, common among frogs with tadpoles (Callery, Fang, & Elinson, 2001; Thibaudeau & Altig, 1999). A notable exception is the direct developing *Sooglossus gardineri*, found in the Seychelles, with an egg diameter of only 1.8 mm (Thibaudeau & Altig, 1999).

With respect to maximal size, frog egg size is constrained, probably because of processes of cleavage and gastrulation (Buchholz et al., 2007; Elinson, 2009; Elinson & Beckham, 2002). Although there are exceptions, most species have eggs less than 5 mm in diameter (Duellman & Trueb, 1986; Elinson & del Pino, 2012; Thibaudeau & Altig, 1999). *E. coqui* is fairly typical for a direct developer, with a 3.5 mm egg and a hatched froglet with a snout-vent length of 6 mm. Direct developers have the correlated suite of characters of small numbers per clutch of large eggs which develop into small froglets.

As a result of this large egg, the appearance of the early embryo is a set of dorsal axial structures (neural tube, notochord, somites) resting on a ball of yolk (Fig. 9.1). In addition, different direct developers tend to have the early

Figure 9.1 Direct development of *Eleutherodactylus coqui*. *E. coqui* develops on land in jelly capsules brooded by the father. These photographs, taken at the same magnification, show Townsend–Stewart (TS) stages of embryos within their capsules. Limb buds are apparent as early as TS4, about 4 days after fertilization. After 3 weeks, the TS15 froglet is ready to hatch. (See Color Insert.)

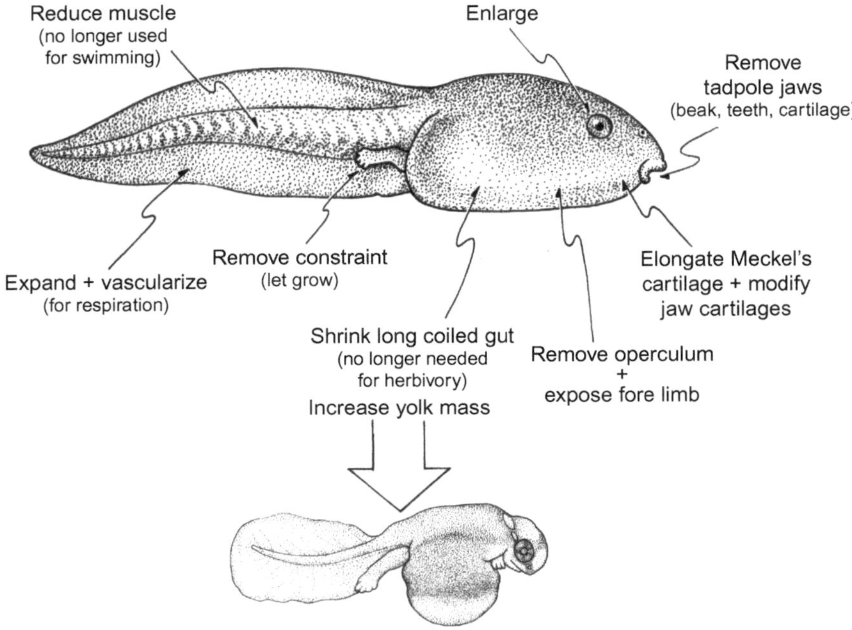

Figure 9.2 Comparison between a typical tadpole and an *E. coqui* TS10 embryo. A number of changes occurred as direct development was derived from the ancestral biphasic life history. *From Elinson (2001).*

formation of large limb buds in contrast to the delayed origin and slow development of limbs in tadpoles. They also have a frog-like head with large eyes and a gaping jaw in contrast to small eyes and a tadpole jaw designed for rasping or filter feeding (Figs. 9.1 and 9.2).

Many structures, characteristic of aquatic embryos and of tadpoles, are rudimentary or do not form in direct developers (Elinson, 2001). Using *E. coqui* as the exemplar, their embryos lack cement glands (Fang & Elinson, 1996, 1999), used by aquatic embryos to attach themselves to plants or other substrates while they develop swimming capability. Gills are greatly reduced in *E. coqui*, and absent in many direct developers. Development of the operculum is incomplete (Callery & Elinson, 2000b). *E. coqui* embryos lack a hatching gland, as they use an egg tooth for hatching from the jelly capsule (Fig. 9.3). Their egg tooth is modified from the keratinous teeth and beak, characteristic of tadpole jaws. They do not develop lateral line organs (Schlosser, Kintner, & Northcutt, 1999), used by aquatic organisms to sense water currents, and their tail musculature is greatly reduced, since the tail is no longer used for swimming (Fig. 9.2).

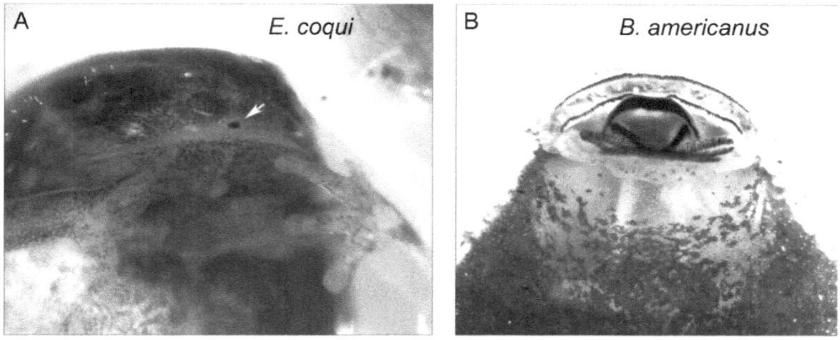

Figure 9.3 Comparison between jaws of an *E. coqui* TS13 embryo and a tadpole. (A) The *E. coqui* embryo has a froglike mouth, with a single keratinous egg tooth (arrow) in the middle of the upper jaw. The egg tooth is used to hatch from the jelly capsule. (B) A tadpole, here of the toad *Bufo americanus*, has rows of keratinous teeth and a keratinous beak. The teeth and beak are used for scraping surfaces to collect decayed plant material and other detritus. Note the very different shape of the tadpole jaw, compared to that of *E. coqui*. These photographs are at the same magnification. (See Color Insert.)

As yolk amounts in the egg increase to provide more and more of the nutritional needs, structures and tissues, used by tadpoles for feeding, are also lost. These include the tadpole-specific jaw cartilages, the suprarostral and infrarostral (Hanken, Klymkowsky, Summers, Seufert, & Ingebrigtsen, 1992; Kerney, Gross, & Hanken, 2010), the larval pattern of the olfactory system (Jermakowicz et al., 2004), and the long coiled gut, required by tadpoles for their plant diet (Buchholz et al., 2007). Mutations affecting the development of these structures can accumulate with no deleterious effect on the animal. Indeed, we see the loss of these structures to a greater or lesser degree among direct developing species (e.g., Kerney, Meegaskumbura, Manamendra-Arachchi, & Hanken, 2007).

The reduction or loss of all structures, used by tadpoles for feeding and swimming, eliminates one of the regulatory functions of thyroid hormone, namely, to cause destruction and resorption of tadpole-specific tissues. That leaves the thyroid hormone role of promoting development of adult-specific structures.

3. THYROID HORMONE DEPENDENCY IN FROG DIRECT DEVELOPMENT

Development from the embryo to the froglet in a direct developer looks seamless, without the radical transition from a tadpole's morphology to that of a frog (Fig. 9.1). Nonetheless, development to the froglet requires thyroid hormone in *E. coqui* as in indirect developers. This critical period was called "cryptic metamorphosis" (Callery & Elinson, 2000a).

Embryos are staged according to Townsend and Stewart (1985). Townsend and Stewart divided the development into 15 stages from uncleaved egg to the froglet, which hatches from its jelly capsule (Fig. 9.1). Each stage takes about 1.5 days at 23 °C. When thyroid hormone synthesis is inhibited with methimazole, the embryos arrest at TS12. They have a frog-like appearance of head, eyes, legs, and general body morphology, but development is incomplete of skin, body musculature, jaw cartilage, and digestive tract (Callery & Elinson, 2000a; Singamsetty & Elinson, 2010). The embryos do not hatch and remain in the aqueous environment, contained within their fertilization envelope and jelly capsules. When removed from their jelly capsules, these methimazole-inhibited embryos stay in water and do not climb onto land.

Two further features that require thyroid hormone late in development are tail and nutritional endoderm. The modified tail is used for respiration, rather than swimming, and thyroid hormone is required for its regression, around the time of hatching at TS15 (Callery & Elinson, 2000a). Nutritional endoderm is a specialized, yolk-rich, cellularized tissue which provides nutrition in late embryogenesis and for the first few days after hatching (Buchholz et al., 2007). Utilization of yolk and subsequent disappearance of the nutritional endoderm requires thyroid hormone (Singamsetty & Elinson, 2010). All of the effects of methimazole can be reversed by adding thyroid hormone (Callery & Elinson, 2000a; Singamsetty & Elinson, 2010).

Nath, Fisher, & Elinson. (2013) recently showed that the cardiac myosin gene, *myh6*, is expressed in legs and other muscle-containing tissues prior to thyroid hormone synthesis not only in *E. coqui* embryos but also in *Xenopus tropicalis* tadpoles. The microRNA, *miR-208*, is contained in an *myh6* intron, and its co-expression in mouse mediates thyroid hormone dependent heart muscle growth (van Rooij et al., 2007). This raises the possibility that *myh6* and its intronic *miR-208* is involved in responses of different muscles to thyroid hormone in frog embryos and tadpoles.

The mechanisms of thyroid hormone production and use in *E. coqui* parallel those found in metamorphosis of tadpoles. The thyroid gland differentiates around TS10 (Jennings & Hanken, 1998), and activation of the thyroid gland depends on hypothalamic stimulation of the pituitary by corticotrophin-releasing factor (CRF), as in tadpoles (Kulkarni, Singamsetty, & Buchholz, 2010). The expression of one of the thyroid hormone receptors, *thrb*, is induced by thyroid hormone (Callery, 1999; Singamsetty & Elinson, 2010) as in *Xenopus laevis*, and *thrb* is upregulated at TS10 (Callery & Elinson, 2000a), indicating the activity of thyroid hormone at this stage. CRF also upregulates *thrb* expression (Kulkarni et al., 2010). This expression is likely due to stimulation of the central neuroendocrine axis leading to increased circulating thyroid hormone, but an alternative exists at the peripheral tissue level. In *X. laevis* tadpoles, corticosterone (CORT), derived from interrenal glands in response to CRF, can synergize with thyroid hormone to increase the expression of *thrb* (Bonett, Hoopfer, & Denver, 2010).

In addition to the upregulation of *thrb* associated with cryptic metamorphosis, another receptor, *thra*, is constitutively expressed through development, and mRNAs for both *thra* and *thrb* are present in unfertilized oocytes (Callery & Elinson, 2000a). Potential roles for these maternally expressed receptors will be considered later.

4. MECHANISMS UNDERLYING TADPOLE DELETION

Two hypotheses have been presented to account for the evolutionary loss of the tadpole from the life history of the direct developing frog: precocious thyroid activation versus larval reduction. The hypothesis of precocious thyroid activation is that development and activation of the thyroid gland occur earlier in development (Hanken, Jennings, & Olsson, 1997; Jennings & Hanken, 1998). This early activation causes early metamorphosis. The hypothesis of larval reduction is that the animal spends less time as a larva, leading eventually to the joining of metamorphic events with embryonic ones (Fig. 9.4) (Callery & Elinson, 2000b; Callery et al., 2001). Essentially, precocious thyroid activation states that the earlier activation of the thyroid gland is the cause of larval reduction, while larval reduction states that the shorter larval period is the cause of earlier activation of the thyroid gland.

When the larval period is reduced, there is less time for the tadpole to feed and to grow. Consequently, reduction of the larval period requires both an increase in yolk content of the egg and the reduction of froglet size at metamorphosis (Callery et al., 2001). The Costa Rican golden toad, *Bufo periglenes*, likely recently extinct, illustrates a potential intermediate in tadpole deletion. It has a facultatively nonfeeding larva (Crump, 1989). Large eggs develop into tadpoles, which in the presence of food, will eat, grow, and metamorphose. If there is little food, the tadpoles have a sufficient amount of yolk to metamorphose at a small size without feeding and to try their luck on land.

A difficulty with the precocious thyroid activation hypothesis is that exposure of tadpoles to thyroid hormone leads to metamorphic changes but generally does not lead to normal metamorphosis. For example, treatment of young *X. laevis* tadpoles with thyroid hormone causes elongation of

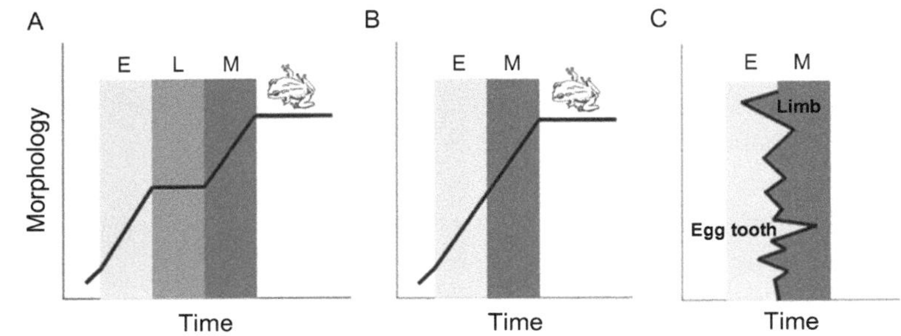

Figure 9.4 Larval deletion in the evolution of direct development. (A) In the typical biphasic frog life history with a tadpole, there is a rapid change of morphology in the development from egg to embryo (E). Once the tadpole larva (L) is produced, its morphology remains relatively constant while it eats and grows. During metamorphosis (M) induced by thyroid hormone, the morphology of the tadpole undergoes tremendous change to produce the frog. (B) In the direct developer, the larval period is reduced until it is essentially gone. That loss fuses the morphological changes of the embryonic period (E) to those of metamorphosis (M). The fusion line is around TS10 of *E. coqui* development (Callery & Elinson, 2000b). (C) The fusion between the embryonic and metamorphic periods is not completely smooth. For example, limbs are prominent in the embryonic period, well before thyroid hormone induced changes. Conversely, the egg tooth appears after TS10, while the homologous tadpole teeth and beak develop in the embryonic period. *From Callery et al. (2001).* (For color version of this figure, the reader is referred to the online version of this chapter.)

Meckel's cartilage, loss of branchial arch cartilages, enlargement of limb buds, thickening of the intestinal lining, and a decrease of tail musculature (Huang, Marsh-Armstrong, & Brown, 1999; Schreiber, Das, Huang, Marsh-Armstrong, & Brown, 2001). They end up as abnormal tadpoles, however, and do not acquire the morphology of a frog.

The disorganized response is due to differences in tissue sensitivity to thyroid hormone, and this sensitivity changes over time (Buchholz, Moskalik, Kulkarni, Hollar, & Ng, 2011). Only by the careful treatment with increasing amounts of thyroid hormone has a semblance of normal metamorphosis been achieved precociously (Etkin, 1935; Kollros, 1961). An evolutionary movement of thyroid gland activation earlier in development would require concomitant changes in tissue sensitivity in order to elicit a proper tissue response.

Perhaps these parallel changes can be coordinated by CRF and CORT. Hypothalamic CRF stimulates both the thyroid and the interrenal glands, which in turn synthesize thyroid hormone and CORT (Denver, 2009). CORT affects peripheral tissues in ways that enhance their response to thyroid hormone (Bonett et al., 2010; Kikuyama, Kawamura, Tanaka, & Yamamoto, 1993). Acting synergistically with thyroid hormone, CORT upregulates the expression of both *thrb* and *type II deiodinase* (*D2*) (Bonett et al., 2010).

D2 catalyzes the conversion of the thyroxine (T_4) to the more active 3,5,3′-triiodothyronine (T_3). Both effects of CORT increase tissue sensitivity to thyroid hormone, so in addition to CRF stimulation of the thyroid gland, CORT changes target tissue sensitivity.

The environmental pressure for direct development is to obtain a safer place for the embryos to develop. Ponds are unsafe, so frogs generally lay thousands of eggs of which no more than a couple survive to adulthood. In tropical areas, not only does the high humidity provide moist opportunities for development on land, but also ponds may be even more dangerous, increasing stress levels in tadpoles. The interplays at several levels in tadpoles between the thyroid hormone system and the CORT stress system could be the nexus that has facilitated the multiple evolutionary origins of anuran direct development.

5. PRECOCIOUS APPEARANCE OF FROG FEATURES IN DIRECT DEVELOPERS

Along with the deletion of most tadpole-specific structures in direct developers, some frog-like structures develop precociously. Prominent eyes and limbs are two features which distinguish most frogs from tadpoles, and both of these develop in indirect developers in response to thyroid hormone at metamorphosis (Brown & Cai, 2007; Hoskins, 1990; Mann & Holt, 2001; Marsh-Armstrong, Huang, Remo, Liu, & Brown, 1999; Rot-Nikcevic & Wassersug, 2004; Schreiber et al., 2001; Shi, 2000). Large eyes and limbs develop early in direct developers (Fig. 9.1), and this timing is before the thyroid gland becomes functional in *E. coqui* (Jennings & Hanken, 1998).

5.1. Eyes

The eyes of tadpoles are generally smaller relative to the head compared to frogs. The tadpole optic nerve, connecting the retina to the tectum and thalamus in the brain, crosses over completely at the optic chiasm (Mann & Holt, 2001). In response to thyroid hormone in *X. laevis*, cells on the ventral side of the retina proliferate and generate neural connections to the thalamus on the same side as the eye (Hoskins, 1986, 1990; Marsh-Armstrong et al., 1999). This new ipsilateral projection contributes to bilateral vision for the frog. The preferential action of thyroid hormone on the ventral retina is due to the expression of *type III deiodinase* (*D3*) in the dorsal retina (Marsh-Armstrong et al., 1999). D3 inactivates both T_4 and T_3, preventing interaction with their receptor.

It should be noted that most of this information on metamorphic changes of the tadpole eye is derived from analysis of *X. laevis*. Adult *X. laevis* remains aquatic and relies more on smell than sight as compared to big-eyed terrestrial frogs of fairy tales. It may be expected that these eye changes found in *X. laevis* are more extreme in terrestrial frogs, since they must see moving prey in order to eat.

In *E. coqui*, there is early cell proliferation of the eye, so that the eye undergoes three-fourth of its growth before the thyroid gland develops (Schlosser, 2008; Schlosser & Roth, 1997). Unlike the ventral retinal growth of metamorphosing *X. laevis*, cell division occurs throughout the *E. coqui* retina. Despite the rapid retinal growth, the major ipsilateral projections do not occur until several weeks after hatching. This delay is odd, since the froglets begin to eat live prey about a week after hatching. There is presently no information on the role of thyroid hormone in the development of the *E. coqui* eye.

5.2. Limbs

The other frog structures that arise early in direct developers are the limbs. In species with tadpoles, limb buds arise late, near the time that the tadpole begins feeding, and they develop slowly. Hindlimb development accelerates with increased thyroid hormone levels, and the start of metamorphic climax is indicated by the emergence of the forelimb through a hole in the operculum. When the primordium for the thyroid or pituitary gland was removed or otherwise did not develop, there was a variable and limited degree of hindlimb development, sometimes including a knee joint and digits as well as some cartilage (Allen, 1916, 1918, 1925; Hanaoka, 1967; Hoskins & Hoskins, 1919; Kerney, Wassersug, & Hall, 2009; Kollros, 1961; Rot-Nikcevic & Wassersug, 2004; Terry, 1918). On the other hand, blockage of thyroid hormone synthesis with methimazole arrests limb development in *X. laevis* early, without muscle or cartilage development (Brown et al., 2005). Although there may be species differences, this methimazole result raises the possibility that retention of a small amount of thyroid or pituitary tissue in the extirpation experiments allowed limb development to a limited degree while not allowing metamorphosis.

In contrast, four large limb buds are present in *E. coqui* within a day after closure of the neural folds (Fig. 9.1). The buds develop continuously over the next 3 weeks to become the limbs of the free-living froglet (Elinson, 1994; Townsend & Stewart, 1985). There is a late requirement for stimulation by thyroid hormone, as the limbs do not elongate fully in its absence (Callery & Elinson, 2000a). Nonetheless, the limbs develop cartilage and muscle without thyroid hormone synthesis by the thyroid gland (Callery & Elinson, 2000a). Attempts to accelerate limb development in *E. coqui* with thyroid hormone, either *in vivo* or *in vitro*, were not successful (Elinson, 1994), but CRF *in vivo* accelerated development, including hindlimb length (Kulkarni et al., 2010). *E. coqui* limb buds, when transplanted to embryos of the frog *Rana pipiens*, continue development including cartilage elements, while the host's limb buds remain small and undifferentiated (Elinson, 1994). These results indicate that most of limb development in *E. coqui* does not require embryo-derived thyroid hormone.

6. A POSSIBLE ROLE FOR MATERNAL THYROID HORMONE AND THYROID HORMONE RECEPTOR

Development of eyes and limbs in species with tadpoles, but not in direct developers, requires thyroid hormone synthesis. There are two possibilities for the derived, direct developer condition. The first is that the eyes and limbs have lost the requirement for regulation by thyroid hormone. Alternatively, thyroid hormone and its receptor are supplied maternally in the egg cytoplasm.

One change of early eye and limb development in *E. coqui* that is probably independent of thyroid hormone is their initial large sizes. Both eyes and limbs are proportionately larger when they first appear in direct developers compared to tadpoles. These large initial sizes are likely due to large embryonic fields, specified by inductive interactions involving growth factors and other ligands. Beyond that the activity of maternal stores of thyroid hormone and receptors needs to be explored.

The possibility that thyroid hormone is used in the early development is strongly suggested by the presence of thyroid hormone in the eggs of shark (McComb, Gelsleichter, Manire, Brinn, & Brown, 2005), teleosts (Power et al., 2001), amphibians (Morvan-Dubois et al., 2006; Weber, Farrar, Tom, & Grau, 1994), and birds (Flamant & Samarut, 1998; Ho, Reed, & Burggren, 2011; McNabb & Wilson, 1997; Prati, Calvo, Morreale, & Morreale de Escobar, 1992). In addition, mRNA or protein for thyroid hormone receptors is present in the eggs of zebrafish (Essner, Breuer, Esner, Fahrenkrug, & Hackett, 1997; Liu, Lo, & Chan, 2000), chicken

(Flamant & Samarut, 1998), the frogs *E. coqui* (Callery & Elinson, 2000a), *X. laevis* (Eliceiri & Brown, 1994; Kawahara, Baker, & Tata, 1991), and *Silurana tropicalis* (Duarte-Guterman, Langlois, Pauli, & Trudeau, 2010). RNAs for deiodinases are also present in oocytes with D1 detected in *X. laevis* (Morvan-Dubois et al., 2006) and D1, D2, and D3 detected in *S. tropicalis* (Duarte-Guterman et al., 2010; Tindall, Morris, Pownall, & Isaacs, 2007). As mentioned earlier, D2 activates thyroid hormone signaling by converting T_4 to the more active T_3, and D3 inactivates both T_4 and T_3. D1 has both activating and inactivating activities.

Experiments in *X. laevis* indicate that thyroid hormone receptor plays an inhibitory role in the early development. Unliganded thyroid hormone receptor represses the expression of genes activated by thyroid hormone, and this repressive activity functions to prevent precocious metamorphosis of tadpoles (Buchholz et al., 2011; Sachs et al., 2000; Sato, Buchholz, Paul, & Shi, 2007; Tomita, Buchholz, & Shi, 2004). In embryos, overexpression of a thyroid hormone receptor without the repression domain led to underdeveloped eyes and other head abnormalities (Havis et al., 2006; Morvan-Dubois, Demeneix, & Sachs, 2008), suggesting a role for repression in the early development. Head development is also abnormal when embryos, overexpressing thyroid hormone receptor, are given exogenous T_3 (Old, Jones, Sweeney, & Smith, 1992).

These experiments leave open the possibility that maternal thyroid hormone and thyroid hormone receptor have an activity in the early embryo. The most straightforward test is to interfere with hormone–receptor interactions in the early embryo. In one approach, transgenic *X. laevis* was produced expressing a dominant negative thyroid receptor (DN-TR) (Schreiber et al., 2001). Embryogenesis was not affected, but it was not determined whether sufficient DN-TR protein from the transgene was synthesized early enough to interfere with embryogenesis. A way to circumvent this problem would be to inject DN-TR RNA into fertilized eggs, or better into full-grown oocytes (Mir & Heasman, 2008). Another approach is to treat early embryos with a receptor antagonist, such as NH-3 (Havis et al., 2006; Grover et al., 2007). Treated embryos of *X. laevis* had reduced eyes and abnormal tails and intestines, but yolk retention was not examined. Based on the thyroid hormone dependency of yolk utilization in the *E. coqui* nutritional endoderm (Singamsetty & Elinson, 2010), it would be particularly interesting to know whether NH-3 inhibited yolk utilization in *X. laevis*.

In *E. coqui*, RNAs for both thyroid hormone receptors α and β (*thra*, *thrb*) (Callery & Elinson, 2000a) are present in full-grown oocytes. Based on the data discussed above from other animals, interactions between maternal thyroid hormone and its receptor could play roles in early embryos. If so, *E. coqui* may utilize those interactions to promote the early development of eyes, limbs, and other structures.

Alternatively, if activation roles for thyroid hormone are not found in embryos of other animals, very simple evolutionary changes could have generated activation roles in embryos of *E. coqui*. For example, the thyroid hormone inactivating enzyme D3 is expressed in many places in the *S. tropicalis* early embryo. A reduction of D3 expression would allow maternal thyroid hormone to interact with its receptor and change gene repression by unliganded receptor to gene activation by liganded receptor. A more thorough examination of the components of the thyroid hormone regulation system in *E. coqui* may reveal significant differences compared to tadpole species. These differences could positively regulate the development of eyes, limbs, and other tissues.

7. QUESTIONS FOR THE FUTURE

Direct developing frogs provide many questions for future investigation of metamorphosis and its evolution. Prime among these is whether thyroid hormone regulates events of early development in *E. coqui* as well as in embryos of fish, frogs with tadpoles, and birds. An important corollary to this question is whether thyroid hormone is involved in utilization of yolk in embryonic tissues of many animals and even in the yolk sac of birds and other amniotes. With regard to this latter possibility, thyroid hormone receptor β is upregulated in chicken yolk sac at day 9 of incubation, and expression remains elevated through the major period of yolk utilization (Forrest, Sjöberg, & Vennström, 1990).

Another question is how the patterns of expression of deiodinases compare between *E. coqui* and *X. laevis* or *S. tropicalis*. The activities of deiodinases affect tissue sensitivity to thyroid hormone, so differences in their expression in *E. coqui* may contribute to precocious development of specific tissues and structures. With respect to tissue sensitivity, there is the question of whether CORT played a role in the repeated evolution of frog direct development.

Finally, many of the novel features of frog direct development, such as cryptic metamorphosis and nutritional endoderm, are known from only one species, *E. coqui*. The multiple origins will allow exploration of variation in the evolution of this derived life history.

REFERENCES

Allen, B. M. (1916). The results of extirpation of the anterior lobe of the hyphophysis and of the thyroid of *Rana pipiens* larvae. *Science, 44*, 755–758.

Allen, B. M. (1918). The results of thyroid removal in the larvae of *Rana pipiens*. *The Journal of Experimental Zoology, 24*, 499–519.

Allen, B. M. (1925). The effects of extirpation of the thyroid and pituitary glands upon limb development of anurans. *The Journal of Experimental Zoology, 42*, 13–30.

Bonett, R. M., Hoopfer, E. D., & Denver, R. J. (2010). Molecular mechanisms of corticosteroid synergy with thyroid hormone during tadpole metamorphosis. *General and Comparative Endocrinology, 168*, 209–219.

Brown, D. D., & Cai, L. (2007). Amphibian metamorphosis. *Developmental Biology, 306*, 20–33.

Brown, D. D., Cai, L., Das, B., Marsh-Armstrong, N., Schreiber, A. M., & Juste, R. (2005). Thyroid hormone controls multiple independent programs required for limb development in *Xenopus laevis* metamorphosis. *Proceedings of the National Academy of Sciences of the United States of America, 102*, 12455–12458.

Buchholz, D. R., Moskalik, C. L., Kulkarni, S. S., Hollar, A. R., & Ng, A. (2011). Hormone regulation and evolution of frog metamorphic diversity. In T. Flatt & A. Heyland (Eds.), *Mechanisms of life history evolution* (pp. 87–97). Oxford: Oxford University Press.

Buchholz, D. R., Singamsetty, S., Karadge, U., Williamson, S., Langer, C. E., & Elinson, R. P. (2007). Nutritional endoderm in a direct developing frog: A potential parallel to the evolution of the amniote egg. *Developmental Dynamics, 236*, 1259–1272.

Callery, E. M. (1999). *Cryptic metamorphosis in the evolution of anuran direct development*. Ph.D. Dissertation. University of Toronto, 126 pp.

Callery, E. M., & Elinson, R. P. (2000a). Thyroid hormone-dependent metamorphosis in a direct developing frog. *Proceedings of the National Academy of Sciences of the United States of America, 97*, 2615–2620.

Callery, E. M., & Elinson, R. P. (2000b). Opercular development and ontogenetic reorganization in a direct-developing frog. *Development Genes and Evolution, 210*, 377–381.

Callery, E. M., Fang, H., & Elinson, R. P. (2001). Frogs without polliwogs: Evolution of anuran direct development. *BioEssays, 23*, 233–241.

Crump, M. L. (1989). Life history consequences of feeding versus non-feeding in a facultatively non-feeding toad larva . *Oecologia, 78*, 486–489.

Denver, R. J. (2009). Stress hormones mediate environment-genotype interactions during amphibian development. *General and Comparative Endocrinology, 164*, 20–31.

Duarte-Guterman, P., Langlois, V. S., Pauli, B. D., & Trudeau, V. L. (2010). Expression and T3 regulation of thyroid hormone- and sex steroid-related genes during *Silurana (Xenopus) tropicalis* early development. *General and Comparative Endocrinology, 166*, 428–435.

Duellman, W. E., & Trueb, L. (1986). *Biology of amphibians*. New York: McGraw-Hill Book Co, 670 pp.

Eliceiri, B. P., & Brown, D. D. (1994). Quantitation of endogenous thyroid hormone receptors alpha and beta during embryogenesis and metamorphosis in *Xenopus laevis*. *The Journal of Biological Chemistry, 269*, 24459–24465.

Elinson, R. P. (1994). Leg development in a frog without a tadpole (*Eleutherodactylus coqui*). *The Journal of Experimental Zoology, 270*, 202–210.

Elinson, R. P. (2001). Direct development: An alternative way to make a frog. *Genesis, 29*, 91–95.

Elinson, R. P. (2009). Nutritional endoderm: A way to breach the holoblastic-meroblastic barrier in tetrapods. *Journal of Experimental Zoology. Part B, Molecular and Developmental Evolution, 312*, 526–532.

Elinson, R. P., & Beckham, Y. (2002). Development in frogs with large eggs and the origin of amniotes. *Zoology (Jena, Germany), 105*, 105–117.

Elinson, R. P., & del Pino, E. M. (2012). Developmental diversity of amphibians. *WIREs Developmental Biology*, *1*, 345–369. http://dx.doi.org/10.1002/wdev.23.

Essner, J. J., Breuer, J. J., Esner, R. D., Fahrenkrug, S. C., & Hackett, P. B., Jr. (1997). The zebrafish thyroid hormone receptor alpha 1 is expressed during early embryogenesis and can function in transcriptional repression. *Differentiation*, *62*, 107–117.

Etkin, W. (1935). The mechanisms of anuran metamorphosis. I. Thyroxine concentrations and the metamorphic pattern. *Journal of Experimental Zoology*, *71*, 317–340.

Fang, H., & Elinson, R. P. (1996). Patterns of distal-less gene expression and inductive interactions in the head of the direct developing frog *Eleutherodactylus coqui*. *Developmental Biology*, *179*, 160–172.

Fang, H., & Elinson, R. P. (1999). Evolutionary alteration in anterior patterning: *otx2* expression in the direct developing frog *Eleutherodactylus coqui*. *Developmental Biology*, *205*, 233–239.

Flamant, F., & Samarut, J. (1998). Involvement of thyroid hormone and its alpha receptor in avian neurulation. *Developmental Biology*, *197*, 1–11.

Forrest, D., Sjöberg, M., & Vennström, B. (1990). Contrasting developmental and tissue-specific expression of α and β thyroid hormone receptor genes. *The EMBO Journal*, *9*, 1519–1528.

Grover, G. J., Dunn, C., Nguyen, N.-H., Boulet, J., Dong, G., Domogauer, J., et al. (2007). Pharmacological profile of the thyroid hormone receptor antagonist NH3 in rats. *Journal of Pharmacology and Experimental Therapeutics*, *322*, 385–390.

Hanaoka, Y. (1967). The effects of posterior hypothalectomy upon the growth and metamorphosis of the tadpole of *Rana pipiens*. *General and Comparative Endocrinology*, *8*, 417–431.

Hanken, J. (1999). Larvae in amphibian development and evolution. In B. K. Hall & M. H. Wake (Eds.), *The origin and evolution of larval forms* (pp. 61–108). San Diego: Academic Press.

Hanken, J., Jennings, D. H., & Olsson, L. (1997). Mechanistic basis of life-history evolution in anuran amphibians: Direct development. *American Zoologist*, *37*, 160–171.

Hanken, J., Klymkowsky, M. W., Summers, C. H., Seufert, D. W., & Ingebrigtsen, N. (1992). Cranial ontogeny in the direct-developing frog, *Eleutherodactylus coqui* (Anura: Leptodactylidae), analyzed by using whole-mount immunohistochemistry. *Journal of Morphology*, *211*, 95–118.

Havis, E., Le Mevel, S., Morvan Dubois, G., Shi, D.-L., Scanlan, T. S., Demeneix, B. A., et al. (2006). Unliganded thyroid hormone receptor is essential for *Xenopus laevis* eye development. *The EMBO Journal*, *25*, 4943–4951.

Hedges, S. B., Duellman, W. E., & Heinicke, M. P. (2008). New World direct-developing frogs (Anura:Terrarana): Molecular phylogeny, classification, biogeography, and conservation. *Zootaxa*, *1737*, 1–182.

Heinicke, M. P., Duellman, W. E., & Hedges, S. B. (2007). Major Caribbean and Central American frog faunas originated by ancient oceanic dispersal. *Proceedings of the National Academy of Sciences of the United States of America*, *104*, 10092–10097.

Ho, D. H., Reed, W. L., & Burggren, W. W. (2011). Egg yolk environment differentially influences physiological and morphological development of broiler and layer chicken embryos. *The Journal of Experimental Biology*, *214*, 619–628.

Hoskins, S. G. (1986). Control of the development of the ipsilateral retinothalamic projection in *Xenopus laevis* by thyroxine: Results and speculation. *Journal of Neurobiology*, *17*, 203–229.

Hoskins, S. G. (1990). Metamorphosis of the amphibian eye. *Journal of Neurobiology*, *21*, 970–989.

Hoskins, E. R., & Hoskins, M. M. (1919). Growth and development of amphibia as affected by thyroidectomy. *The Journal of Experimental Zoology*, *29*, 1–69.

Huang, H., Marsh-Armstrong, N., & Brown, D. D. (1999). Metamorphosis is inhibited in transgenic *Xenopus laevis* tadpoles that overexpress type III deiodinase. *Proceedings of the National Academy of Sciences of the United States of America, 96*, 962–967.

Jennings, D. H., & Hanken, J. (1998). Mechanistic basis of life history evolution in anuran amphibians: Thyroid gland development in the direct-developing frog, *Eleutherodactylus coqui. General and Comparative Endocrinology, 111*, 225–232.

Jermakowicz, W. J., III, Dorsey, D. A., Brown, A. L., Wojciechowski, K., Giscombe, C. L., Graves, B. M., et al. (2004). Development of the nasal chemosensory organs in two terrestrial anurans: The direct developing frog *Eleutherodactylus coqui* (Anura: Leptodactylidae), and the metamorphosing toad, *Bufo americanus* (Anura: Bufonidae). *Journal of Morphology, 261*, 225–248.

Kawahara, A., Baker, B. S., & Tata, J. R. (1991). Developmental and regional expression of thyroid hormone receptor genes during *Xenopus* metamorphosis. *Development, 112*, 933–943.

Kerney, R., Gross, J. B., & Hanken, J. (2010). Early cranial patterning in the direct-developing frog *Eleutherodactylus coqui* revealed through gene expression. *Evolution & Development, 12*, 373–382.

Kerney, R., Meegaskumbura, M., Manamendra-Arachchi, K., & Hanken, J. (2007). Cranial ontogeny in *Philautus silus* (Anura: Ranidae: Rhacophorinae) reveals few similarities with other direct-developing anurans. *Journal of Morphology, 268*, 715–725.

Kerney, R., Wassersug, R., & Hall, B. K. (2009). Skeletal advance and arrest in giant non-metamorphosing African clawed frog tadpoles (*Xenopus laevis*: Daudin). *Journal of Anatomy, 216*, 132–143.

Kikuyama, S., Kawamura, K., Tanaka, S., & Yamamoto, K. (1993). Aspects of amphibian metamorphosis: Hormonal control. *International Review of Cytology, 145*, 105–148.

Kollros, J. J. (1961). Mechanisms of amphibian metamorphosis: Hormones. *American Zoologist, 1*, 107–114.

Kulkarni, S. S., Singamsetty, S., & Buchholz, D. R. (2010). Corticotropin-releasing factor regulates development in the indirect developing frog, *Eleutherodactylus coqui. General and Comparative Endocrinology, 169*, 225–230.

Liu, Y.-W., Lo, L.-J., & Chan, W.-K. (2000). Temporal expression and T3 induction of thyroid hormone receptors α1 and β1 during early embryonic and larval development in zebrafish, *Danio rerio. Molecular and Cellular Endocrinology, 159*, 187–195.

Mann, F., & Holt, C. E. (2001). Control of retinal growth and axon divergence at the chiasm: Lessons from *Xenopus. BioEssays, 23*, 319–326.

Marsh-Armstrong, N., Huang, H., Remo, B. F., Liu, T. T., & Brown, D. D. (1999). Asymmetric growth and development of the *Xenopus laevis* retina during metamorphosis is controlled by Type III deiodinase. *Neuron, 24*, 871–878.

McComb, D. M., Gelsleichter, J., Manire, C. A., Brinn, R., & Brown, C. L. (2005). Comparative thyroid hormone concentration in maternal serum and yolk of the bonnethead shark (*Sphyrna tiburo*) from two sites along the coast of Florida. *General and Comparative Endocrinology, 144*, 167–173.

McNabb, F. M. A., & Wilson, C. M. (1997). Thyroid hormone deposition in avian eggs and effects on embryonic development. *Integrative and Comparative Biology, 37*, 553–560.

Meegaskumbura, M., Bossuyt, F., Pethiyagoda, R., Manamendra-Arachchi, K., Bahir, M., Milinkovitch, M. C., et al. (2002). Sri Lanka: An amphibian hot spot. *Science, 298*, 379.

Mir, A., & Heasman, J. (2008). How the mother can help: Studying maternal Wnt signaling by anti-sense-mediated depletion of maternal mRNAs and the host transfer technique. *Methods in Molecular Biology, 469*, 417–429.

Morvan-Dubois, G., Demeneix, B. A., & Sachs, L. M. (2008). *Xenopus laevis* as a model for studying thyroid hormone signalling: From development to metamorphosis. *Molecular and Cellular Endocrinology, 293*, 71–79.

Morvan-Dubois, G., Sebillot, A., Kuiper, G. G., Verhoelst, C. H., Darras, V. M., Visser, T. J., et al. (2006). Deiodinase activity is present in *Xenopus laevis* during early embryogenesis. *Endocrinology, 147*, 4941–4949.

Nath, K., Fisher, C., & Elinson, R. P. (2013). Expression of a cardiac myosin gene in non-heart tissues of developing frogs. *Development Genes and Evolution*. (in press).

Old, R., Jones, E. A., Sweeney, G., & Smith, D. P. (1992). Precocious synthesis of a thyroid hormone receptor in *Xenopus* embryos causes hormone-dependent developmental abnormalities. *Roux's Archives of Developmental Biology, 201*, 312–321.

Pethiyagoda, R., & Manamendra-Arachchi, K. (1998). Evaluating Sri Lanka's amphibian diversity. *Occasional Papers of Wildlife Heritage Trust, 2*, 1–12.

Power, D. M., Llewellyn, L., Faustino, M., Nowell, M. A., Bjornsson, B. T., Einarsdottir, I. E., et al. (2001). Thyroid hormones in growth and development of fish. *Comparative Biochemistry and Physiology. Part C: Toxicology & Pharmacology, 130*, 447–459.

Prati, M., Calvo, R., Morreale, G., & Morreale de Escobar, G. (1992). L-thyroxine and 3,5,3'-triiodothyronine concentrations in the chicken egg and in the embryo before and after the onset of thyroid function. *Endocrinology, 130*, 2651–2659.

Rot-Nikcevic, I., & Wassersug, R. J. (2004). Arrested development in *Xenopus laevis* tadpoles: How size constrains metamorphosis. *The Journal of Experimental Biology, 207*, 2133–2145.

Sachs, L. M., Damjanovski, S., Jones, P. L., Li, Q., Amano, T., Ueda, S., et al. (2000). Dual functions of thyroid hormone receptors during *Xenopus* development. *Comparative Biochemistry and Physiology. B, 126*, 199–211.

Sato, Y., Buchholz, D. R., Paul, B. D., & Shi, Y.-B. (2007). A role for unliganded thyroid hormone receptor in postembryonic development in *Xenopus laevis*. *Mechanisms of Development, 124*, 476–488.

Schlosser, G. (2008). Development of the retinotectal system in the direct-developing frog *Eleutherodactylus coqui* in comparison with other anurans. *Frontiers in Zoology, 5*, 9.

Schlosser, G., Kintner, C., & Northcutt, R. G. (1999). Loss of ectodermal competence for lateral line placode formation in the direct developing frog *Eleutherodactylus coqui*. *Developmental Biology, 213*, 354–369.

Schlosser, G., & Roth, G. (1997). Development of the retina is altered in the directly developing frog *Eleutherodactylus coqui* (Leptodactylidae). *Neuroscience Letters, 224*, 153–156.

Schmid, M., Steinlein, C., Bogart, J. P., Feichtinger, W., Leon, P., La Marca, E., et al. (2010). The chromosomes of terraranan frogs. Insights into vertebrate cytogenetics. *Cytogenetic and Genome Research, 130–131*, 1–568.

Schreiber, A. M., Das, B., Huang, H., Marsh-Armstrong, N., & Brown, D. D. (2001). Diverse developmental programs of *Xenopus laevis* metamorphosis are inhibited by a dominant negative thyroid hormone receptor. *Proceedings of the National Academy of Sciences of the United States of America, 98*, 10739–10744.

Shi, Y.-B. (2000). *Amphibian metamorphosis*. New York: John Wiley & Sons 288 pp.

Singamsetty, S., & Elinson, R. P. (2010). Novel regulation of yolk utilization by thyroid hormone in embryos of the direct developing frog *Eleutherodactylus coqui*. *Evolution & Development, 12*, 437–448.

Terry, G. S. (1918). Effects of the extirpation of the thyroid gland upon ossification in *Rana pipiens*. *The Journal of Experimental Zoology, 24*, 567–587.

Thibaudeau, G., & Altig, R. (1999). Endotrophic anurans, development and evolution. In R. W. McDiarmid & R. Altig (Eds.), *Tadpoles, the biology of anuran larvae* (pp. 170–188). Chicago: University of Chicago Press.

Tindall, A. J., Morris, I. D., Pownall, M. E., & Isaacs, H. V. (2007). Expression of enzymes involved in thyroid hormone metabolism during the early development of *Xenopus tropicalis*. *Biology of the Cell, 99*, 151–163.

Tomita, A., Buchholz, D. R., & Shi, Y.-B. (2004). Recruitment of N-CoR/SMRT-TBLR1 corepressor complex by unliganded thyroid hormone receptor for gene repression during frog development. *Molecular and Cellular Biology*, *24*, 3337–3346.
Townsend, D. S., & Stewart, M. M. (1985). Direct development in *Eleutherodactylus coqui* (Anura: Leptodactylidae): A staging table. *Copeia*, *1985*, 423–436.
Van Rooij, E., Sutherland, L. B., Qi, X., Richardson, J. A., Hill, J., & Olsen, E. N. (2007). Control of stress-dependent cardiac growth and gene expression by a microRNA. *Science*, *316*, 575–579.
Weber, G. M., Farrar, E. S., Tom, C. K., & Grau, E. G. (1994). Changes in whole-body thyroxine and triiodothyronine concentrations and total content during early development and metamorphosis of the toad *Bufo marinus*. *General and Comparative Endocrinology*, *94*, 62–71.

CHAPTER TEN

High-Throughput Sequencing Will Metamorphose the Analysis of Thyroid Hormone Receptor Function During Amphibian Development

Alexis G. Grimaldi[*], Nicolas Buisine[*], Patrice Bilesimo[*,1], Laurent M. Sachs[*,2]

[*]Laboratoire d'Evolution des Regulations Endocrinienne, Muséum National d'Histoire Naturelle, Département Régulation, Développement et Diversité Moléculaire, UMR 7221 CNRS, Paris, France
[1]Present address: Translational Research Center, Perelman School of Medicine, University of Pennsylvania, Philadelphia, Pennsylvania, USA
[2]Corresponding author: e-mail address: sachs@mnhn.fr

Contents

Abstract

Amphibian metamorphosis is marked by dramatic thyroid hormone (T_3)-induced changes including *de novo* morphogenesis, tissue remodeling, and organ resorption through programmed cell death. These changes involve cascades of gene regulation initiated by thyroid hormone (TH). TH functions by regulating gene expression through TH receptors (TR). TR are DNA-binding transcription factors that belong to the steroid hormone receptor superfamily. In the absence of ligand, TR can repress gene expression by recruiting a corepressor complex, whereas liganded TR recruits a coactivator

Current Topics in Developmental Biology, Volume 103
ISSN 0070-2153
http://dx.doi.org/10.1016/B978-0-12-385979-2.00010-1

complex for gene activation. Earlier studies have led us to propose a dual function model for TR during development. In premetamorphic tadpoles, unliganded TR represses transcription involving corepressors. During metamorphosis, endogenous T_3 allows TR to activate gene expression. To fully understand the diversity of T_3 effects during metamorphosis, whole genome analysis of transcriptome and mechanism of TR action should be carried out. To this end, the new sequencing technologies have dramatically changed how fundamental questions in biology are being addressed and is now making the transition from technology development to being a standard for genomic and functional genomic analysis. This review focuses on the applications of high-throughput technologies to the field of amphibian metamorphosis.

1. INTRODUCTION

Amphibian development is marked by an indirect development. Embryonic development and juvenile life style are separated by a larval period (tadpole) that ends with metamorphosis. Metamorphosis is a switch that results in the reprogramming of the morphological and biochemical characteristics of nearly all tadpole organs (Shi, 1999). Transformations involve apoptosis of larval cells and concurrent proliferation and differentiation of adult cell types. Although complex, metamorphosis is directly initiated by thyroid hormone (TH) with 3,5,3′-triiodothyronine (T_3) as the most biologically active agent. Interestingly, this hormone-controlled developmental phase shows strong similarities with the perinatal period in mammals, bird hatching, and of course other kind of metamorphosis such as in fish and insects (Laudet, 2011). Amphibian metamorphosis thus provides a powerful model to analyze the essential roles for TH, notably in maturation of the nervous system, the metabolism, the respiratory system, the intestine, and body shape modification (Shi, 1999; Tata, 1993).

TH acts through thyroid hormone receptor (TR) that belongs to the family of nuclear receptors (Gronemeyer & Laudet, 1995). TR is a transcription factor that regulates gene expression by binding to DNA at specific sites known as T_3 response elements (T_3REs), mostly composed of two direct repeats of the consensus sequence "AGGTCA" separated by four bases (DR4). TH is a versatile player. It can not only upregulate, but also downregulate gene expression in target tissues or cells (Flamant, Gauthier, & Samarut, 2007). Moreover, in the absence of TH, TR can also repress or increase the transcription. As in mammals, there are two types of TR (α and β) in amphibian (Yaoita, Shi, & Brown, 1990). Their amino acid sequences are well conserved in evolution and amphibian TR behaves similarly to the mammalian or avian TR. Unfortunately, there are only a very limited number of studies on gene repression by liganded TR (Santos, Fairall, & Schwabe, 2011). Thus, most discussions have been focused on the mechanism for T_3-response genes induced by T_3 (Grimaldi, Buisine, Miller, Shi, & Sachs, 2012). For these genes, in absence of hormone, TR binds T_3RE and recruits corepressor complexes carrying chromatin modification activity that leads to a closed chromatin conformation, inaccessible for the transcriptional machinery. T_3-binding induces a conformational change of TR that will lead first, to the release of the corepressor complexes and second, to the recruitment of coactivator complexes carrying a different set of chromatin modification enzymes. Their actions lead to chromatin opening favorable to transcription activation.

The expression profiles (Yaoita & Brown, 1990), together with the ability of TR to both repress and activate T_3-inducible genes in the absence and presence of TH (Sachs & Shi, 2000), respectively, suggest dual functions for TRs during development (Sachs et al., 2000). TRα and TRβ mRNA are present at very low levels during embryogenesis where their function has started to be addressed (Havis et al., 2006). Next, high levels of TRα mRNA are present after larval hatching and increase to reach a maximum during metamorphosis. TRβ mRNA levels are barely detectable before metamorphosis, when they will increase in parallel with endogenous T_3 level, again reaching a peak at the climax. After metamorphosis, the levels of both TR isoforms decrease to stay low in the juvenile and the adult. The same profiles were observed analyzing protein levels (Eliceiri & Brown, 1994). The tissue distribution of TR has shown to be correlated with organ transformation (Eliceiri & Brown, 1994; Fairclough & Tata, 1997; Kawahara, Baker, & Tata, 1991). Thus, in premetamorphic tadpoles, TRs, mostly TRα, act to repress T_3-response genes. As many of these genes are likely to participate in metamorphosis, their repression by unliganded TR will help prevent premature metamorphosis and ensure a proper period of tadpole growth. TH synthesis and secretion into the plasma transforms TRs from repressors to activators, which will induce the expression of TH response genes leading to metamorphosis.

Analysis of TH action in amphibian metamorphosis will provide a powerful tool to understand the diversity of biological activities known for TH and the mechanisms that lead to this diversity. Studies in the past two decades using subtractive hybridization or DNA array have identified many T_3-response genes during amphibian metamorphosis (Brown et al., 1996; Buchholz, Heimeier, Das, Washington, & Shi, 2007; Buckbinder &

Brown, 1992; Cai, Das, & Brown, 2007; Das et al., 2006; Das, Heimeier, Buchholz, & Shi, 2009; Shi & Brown, 1993; Wang & Brown, 1991, 1993). Among them, only a small fraction has been characterized as direct response gene (Das et al., 2009; Furlow & Brown, 1999; Ranjan, Wong, & Shi, 1994). The identification of the TH-induced transcriptional regulatory programs needs now to be addressed in a physiological context at the level of the whole genome to understand how TH regulates metamorphosis. Such ambitious studies depend on the exploitation of a sequenced amphibian genome. While *Xenopus leavis (X. laevis)* is the most used model, *Xenopus tropicalis (X. tropicalis)* is by far a more appropriate model because its genome sequence is in a large part already accessible (Hellsten et al., 2010). Now next-generation sequencing (NGS)-based transcriptome sequencing, transcription factor binding, and epigenetic analyses (chromatin modification) have dramatically changed how fundamental questions in biology are addressed (Hawkins, Hon, & Ren, 2010; Metzker, 2010; Sboner, Mu, Greenbaum, Auerbach, & Gerstein, 2011). NGS is well suited to provide all the primary data required for functional genomic analyses with an unprecedented resolution enabling the detection of even subtle differences. This review emphasizes the particular contribution of NGS-based technologies to functional genomics research with a special focus on gene regulation by TH.

2. *XENOPUS TROPICALIS* GENOMIC SEQUENCE AND ANNOTATION: CURRENT STATE

2.1. A fragmented genome assembly

The sequencing of *X. tropicalis* genome has been an ongoing project for nearly 10 years. It was initiated by the JGI in late 2003, and the first public version (3.0) was released in October 2004. The draft assembly version 4.1 released in 2005 had been in use for over 5 years before its publication (Hellsten et al., 2010). A new assembly (version 7.1) quickly followed and was made available to download on the XenBase Web site in 2011(Bowes et al., 2008). These assemblies have been an invaluable resource for the scientific community. In fact, as opposed to *X. laevis*, which is pseudotetraploid and has a genome of about 3.1 Gbp, *X. tropicalis* has a relatively small genome (1.7 Gbp) and much less genetic redundancy. This is expected since it is a diploid species, making it a very appropriate model to probe, for example, the TH-induced gene regulatory networks. This general purpose genomic resource has been instrumental in key recent findings of *X. tropicalis* biology (Bowes et al., 2008; Hellsten et al., 2010).

Unlike mouse and human genomes, *X. tropicalis* genome project did not benefit from the inputs and curations of a large and extensive scientific community. Indeed, this is the case for almost all eukaryotes genomes published so far, except for a few model organisms. As a result, the draft genome assembly v4.1 of *X. tropicalis* is composed of 19,501 scaffolds (see Fig. 10.1A; Buisine & Sachs, 2009) representing unordered fragments of the 10 chromosomes (Tymowska & Fischberg, 1973). In the version 4.1, the largest scaffold is 7.8 Mb long and the N50 = 1.6 Mb. The version 7.1 is quite less fragmented, with only 7730 scaffolds, an N50 of 130 Mb, with the longest scaffold being 216 Mb long. Interestingly, a small number of the largest scaffolds capture a majority of the genes: 80% of the Ensembl genes in the 834 top scaffolds (Fig. 10.1A). In addition, many assembly gaps (stretches of "N"s corresponding to unsequenced regions) are found in scaffolds and add an extra level of fragmentation (Fig. 10.1B and C). Given that many multiexon genes are several tens of kilobases long, such gaps will frequently be found in gene loci, resulting in alteration of apparent distances (Fig. 10.1B), as well as missing exons or gene fragments. Thus, although most of the genome sequence is reflected by a small subset of scaffolds (80% of the assembly in the top 2000 scaffolds), fragmentation remains high (Fig. 10.1C). BAC, cosmid and fosmid libraries have been end-sequenced to resolve the physical continuity over large size range (Hellsten et al., 2010). Nonetheless, end-sequenced scaffolds are still present in the assembly, artificially increasing the size of the assembled genome by ~85 Mbp. They cluster as dense clumps of points on Fig. 10.1A (highlighted by the three leftmost ellipses). In addition, some regions (end-sequences) may be represented multiple times in the genome sequence. While the assembled genome is 1.5 Gbp long, one has to take into account that overall, 20% of the released assembly is actually not sequenced. However, unassembled regions of the genome assembly result from low complexity regions, or repeated sequences, which makes both base detection—in the case of long stretches of similar nucleotides—and assembly—because of the ambiguity introduced—more difficult. In addition, assembly gaps could also be due to the presence of allelic polymorphism, which could induce hard to resolve ambiguities during the assembly process. Intraspecies polymorphism can be ruled out in the case of *X. tropicalis* since only one inbred individual frog was used for genome sequencing (Hellsten et al., 2010).

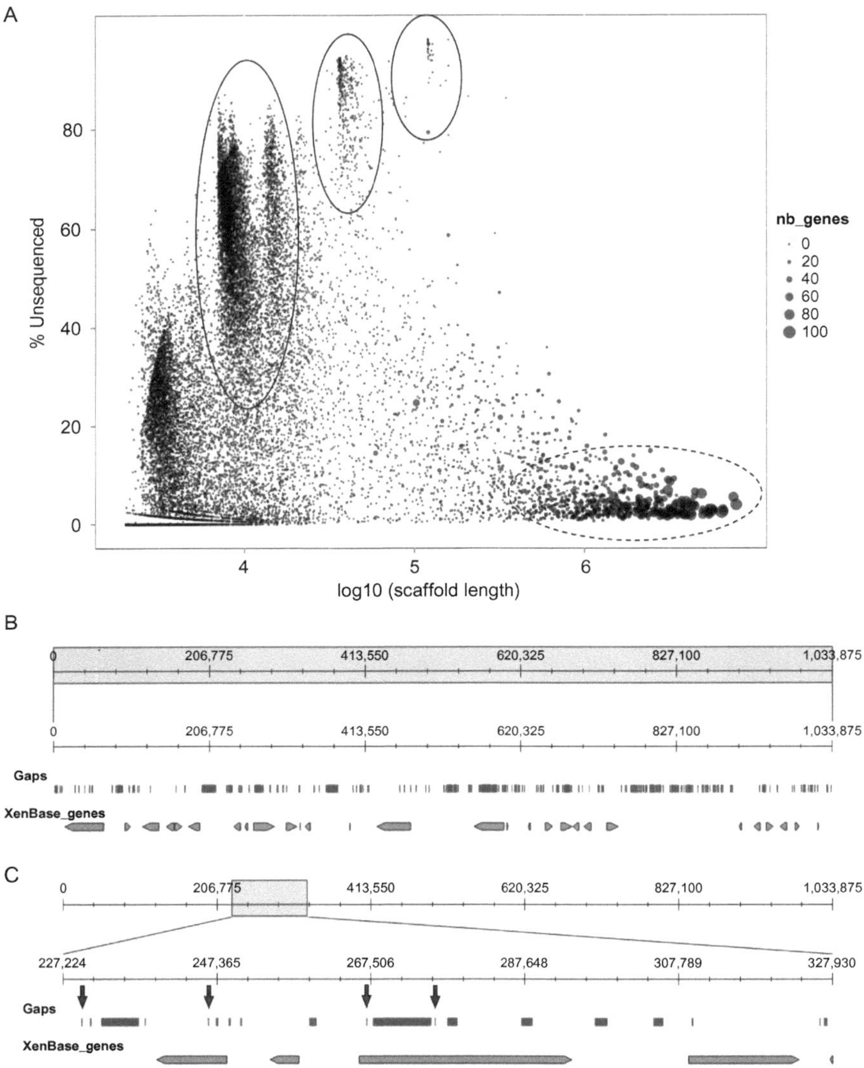

Figure 10.1 The *X. tropicalis* genome assembly 4.1 is highly fragmented. (A) Scatterplot representing the size of each scaffold (in $\log_{10}$ scale) compared to the proportion of assembly gaps (% Unsequenced). The gene number per scaffold (ranging from 0 to 100) is indicated by gray circles of various sizes. The three ellipses in the upper half of the plot correspond to a fraction of the cosmid, fosmid, and BAC-end libraries used to help the assembly process. The gene rich and almost fully sequenced scaffolds capturing ~80% of the assembly are indicated at the bottom-right of the plot (dashed-circle). (B) Illustrative example of the assembly gaps found in a scaffold. The first track shows the assembly gaps (stretches of 'N's). The second track represents XenBase gene models. The orientation of each gene is indicated by the direction of the arrow. The scaffold shown is scaffold_448. (C) Close-up of a 100 kb region of the scaffold_448. Assembly gaps that are exactly 50 bp long ("gaps of 50," see text) are indicated by arrows.

A specific subset of assembly gaps find their origin in a technical choice made during the assembly process. Indeed, nearly half of the ~170,000 assembly gaps are exactly 50 bp long. They were in fact arbitrarily introduced when the length of an unsequenced region could not be determined. They are heterogeneously distributed across the genome and can be found both inside of genes as well as in intergenic regions (Fig. 10.1C). The use of the gPET technology (Fullwood, Wei, Liu, & Ruan, 2009) suggest that the size of these "gaps of 50" is in fact underestimated by at least an order of magnitude, although the individual size of each gap is subject to large variations (unpublished data). Such inconsistencies can have a dramatic impact in the exploitation of the genome sequence as a general purpose resource by a biologist unaware of these pitfalls, since regions apparently close together can actually be located several kilobases apart from each other. This is not much of a problem when studying a small region of the genome which can be visually inspected (in the range of a few hundred base pairs up to a few kilobases), but this can become a real issue when large genomic regions are being analyzed. Furthermore, annotated features as short as 50 bp do not show up well in genome browsers when displaying genomic regions larger than 10 kbs, which deprives the biologist of visual clues for regions containing "gaps of 50." Of note, this point has been partially corrected in the version 7.1, although 42% 47,505 assembly gaps are exactly 100 bp and 10 kb.

2.2. An incomplete annotation

Despite the fragmentation level of the genome assembly, the majority of genes are included in the longest scaffolds (Fig. 10.1A, bottom-right circled cluster), which supports the idea that the current assembly contains most of the gene rich (euchromatic) regions. Many different gene annotation datasets are available for *X. tropicalis:* Ensembl (Flicek et al., 2012), XenBase (Bowes et al., 2008), and RefSeq (Pruitt, Tatusova, Klimke, & Maglott, 2009). They are more or less comprehensive, redundant, and complementing each other. For example, while Ensembl has more gene models than XenBase (18,025 vs. 15,929 out of more than 22,000 expected), it also has more models with no associated name ("unknown," 6293 vs. 1074). In fact, as shown in Fig. 10.2A, the three databases provide gene models that are quite comparable qualitatively, but clearly lack consensus in the definition of their boundaries. Indeed, the operational value of a genome sequence is (in part) dictated by the quality of its structural and functional annotations.

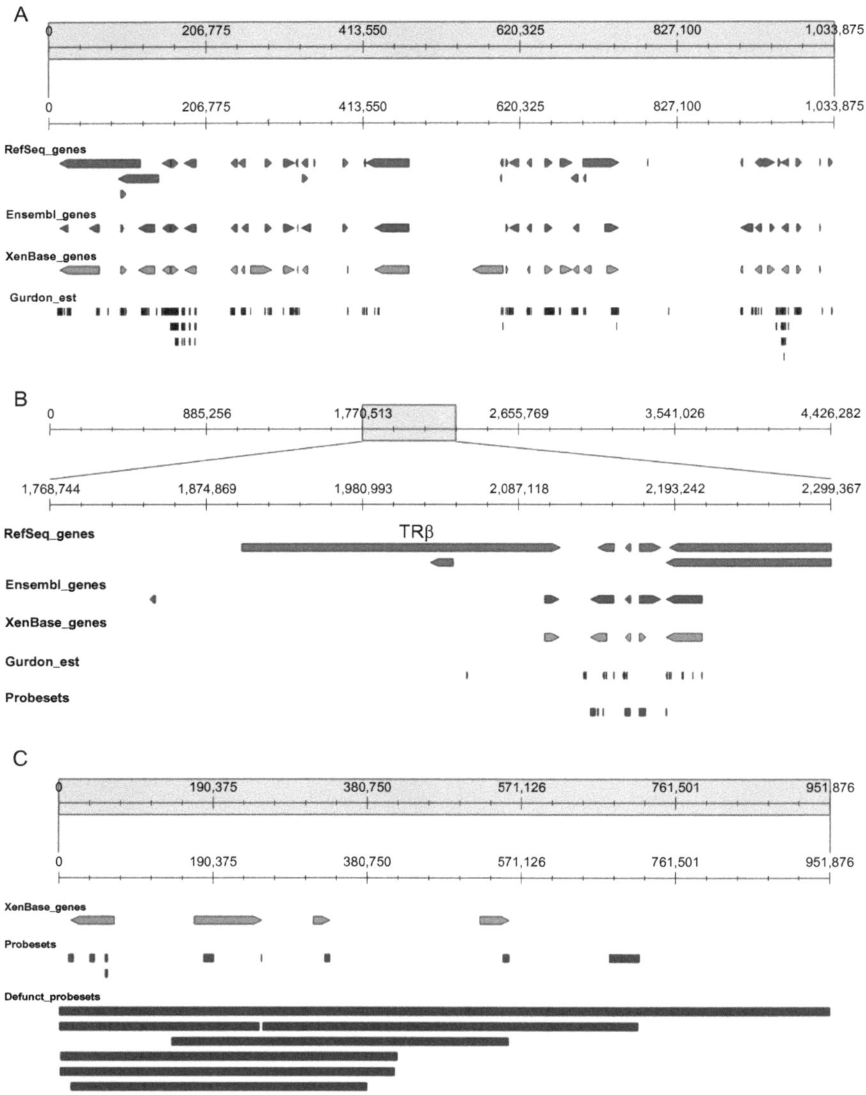

Figure 10.2 Limits of the current gene annotation and microarray design. (A) Comparison gene annotations from different sources. The first three tracks represent gene models described in RefSeq, Ensembl, and XenBase. The orientation of each gene model is indicated by arrows. The fourth track corresponds to the EST clusters computed at the Gurdon institute (available on the XenBase website). The scaffold shown is scaffold_448. (B) Affymetrix probesets located around the TRβ locus (scaffold_26). The four top tracks correspond to the same datasets as in A. The last track represents the genomic region spanned by the Affymetrix *X. tropicalis* transcriptome microarrays probesets, curated from defunct probesets. Note that there are no probeset associated with TRβ. Thus,

Precise definition of gene boundaries is of utmost importance when trying to link coordinated variation of gene expression with the specific recruitment of transcription factors or modification of the chromatin architecture. In particular, most of the current gene models capture only the coding sequence (CDS), without taking into account the untranslated regions (UTRs). The upstream boundary of the 5′UTR defines the actual transcription start site (TSS) and proximal promoter region, with which many regulatory elements are associated and which are required for transcriptional regulation. The same is true for 3′UTRs, which are underrepresented and often contain numerous microRNA-binding sites, which may be important posttranscriptional regulators.

Traditionally, the definition of gene models relies on the use of a collection of softwares (*ab initio* gene prediction, comparison to databases, etc.) and functional data (ESTs, full-length cDNA, etc.). Between 2000 and 2007, the number of EST for *X. tropicalis* has increased tremendously, reaching more than 1.2 million ESTs (grouped by unigene library entry, ~5 million altogether), making it one of the model with the highest number of ESTs. The ESTs were clustered and aligned to the genome sequence in order to define potential exons (Nagaraj, Gasser, & Ranganathan, 2007). Although this strategy can be quite effective when large collections of EST are used, the process remains error prone (depending on the quality of the collections) and often lacks the resolution required to describe alternative transcripts (Nagaraj et al., 2007). As can be seen in Fig. 10.2B, the gene TRβ is associated to a single EST cluster far upstream of the annotated TSS in Ensembl or XenBase, while only the RefSeq annotation reflects boundaries that actually encompass the EST cluster.

Gene models can be inferred from the genomic sequence using computational methods. Popular gene prediction softwares include genewise (Birney, Clamp, & Durbin, 2004), geneMark, FGENESH (Salamov & Solovyev, 2000), and EUGENE (Schiex, Moisan, & Rouzé, 2001), to name a few. This process suffers several pitfalls and always benefits from multiple lines of evidence. In addition, assembled genome must be of relatively good

TRβ transcription cannot be measured with Affymetrix arrays. (C) Illustrative example of defunct Affymetrix probesets. The defunct probesets are composed of probes poorly specific of a given gene/transcript. The first track corresponds to the XenBase gene models as in (A). The second track corresponds to genomic region spanned by the curated probesets that can be used for analysis, while the third track (defunct_probesets) corresponds to the genomic region spanned by defunct probesets. The region shown is located on scaffold_464.

quality, as unsequenced portions as well as assembly errors can lead to missing or truncated gene models. Also, the information content along the genome can be extremely variable, in such a way that pseudogenes, duplications, short tandem repeats, and the overall complexity level of a region can mislead the software. Many misannotated gaps inside the gene body, as shown in Fig. 10.1C, can further reduce the efficiency of gene prediction. Furthermore, predicted gene models are often limited to the CDS, and miss UTRs which, as previously stated, often contain important regulatory elements. In *X. tropicalis* genome, some key genes involved in metamorphosis or TH signaling are poorly annotated, which can hamper downstream analysis of functional data. For example, based on the current annotation, it would be impossible to find the T_3RE located at the TSS of TRβ by looking for it near the 5′ end of the gene (a very common practice, although many binding sites may be missed as they are located away from the TSS, see below). In fact, the 5′ end of the TRβ gene was determined by 5′ RACE PCR and is correctly annotated only in the manually curated RefSeq database. It is located 210 kb upstream of the XenBase and Ensembl models (Fig. 10.2B).

Gene models can be determined by full cDNA sequencing (Klein et al., 2002; Voigt, Chen, Gilchrist, Amaya, & Papalopulu, 2005). Although this method is quite expensive and slow, it is also very accurate. Individual cDNA clones are assembled from multiple single-pass sequenced reads, ensuring the whole coverage of the cloned molecules. When combined, these approaches can yield accurate models, although extensive curation is still required. While there is an expected ~22,000 protein coding genes, only half are actually linked to accession numbers pointing to cDNA in the RefSeq database, and only one-third are linked to full-length cDNA (Gilchrist, 2012).

The lack of clear definition of gene boundaries renders the analysis of coordinated variation of gene expression quite tricky. In fact, genome-wide measure of differential gene expression has traditionally been carried out using microarrays, whose design relies heavily on a good annotation (Altmann et al., 2001). In this technology, fluorescent labeled cDNA is hybridized to probes fixed on a solid surface. The readout of the signal for a given spot then depends on the relative amount of cDNA present in the sample. On the chip, transcript-specific probes define probesets, with the redundancy of the individual probes allowing the assay to be more specific and sensitive. The design of microarrays thus requires a comprehensive annotation, and the quality and exhaustiveness of the chip will greatly depend on it. In model organisms (mouse, drosophila, etc.), microarrays have been constantly refined, and inconsistent probes were detected in the numerous datasets available in GEO and Array Express and then excluded from analyses. In *X. tropicalis*, the number of microarray datasets available in the databases has not reached the critical value needed to do such curation, and large portion of the probes present on available microarrays are not usable, as they are located outside of exons/genes, or in misannotated regions. As a result, probesets (i.e., a collection of probes aimed at measuring the expression level of a single transcript or gene) often span over large genomic regions, irrespective of the underlying gene definition provided by the annotation (Fig. 10.2C). In addition, after filtering of the defunct probesets, the remaining probesets do not interrogate the whole transcriptome. For example, some genes critically involved in metamorphosis, like TRβ (Fig. 10.2B), are missing in the probesets.

3. TH AND TR REGULATING NETWORK USING NGS-BASED TRANSCRIPTOME ANALYSIS

3.1. Toward a better annotation

Traditional resources used to annotate genomes are either low-throughput (in the case of EST and full-length cDNAs) or error prone (in the case of *de novo, in silico* predictions) and they are combined to take the best of both and increase specificity and sensitivity of gene prediction methods (see above). The recent rise of NGS technologies and their applications to RNA sequencing (RNA-Seq) opens new perspectives in term of genome annotation, given their very high throughput, scalability, and robustness (Wang, Gerstein, & Snyder, 2009). RNA-Seq is reminiscent of EST sequencing, except that the sequencing step is carried out by an NGS system rather than a conventional capillary sequencing system. To this end, cDNAs are synthesized by reverse transcription from oligo-dT purified messenger RNAs, and sheared in smaller fragments suitable for sequencing. With regular RNA-Seq, only one end of each fragment of sheared cDNAs is sequenced over a short region, typically $\leq$50 bp. Sequence reads are then mapped onto the genome with one of the many short reads aligners. This simple protocol is best suited for quantitative measure of gene expression, as the normalized number of reads per gene is proportional to its expression level, but can also provide a quick description of transcription units (Garber, Grabherr, Guttman, & Trapnell, 2011). This is well illustrated in the case of the TRβ gene (Fig. 10.3A), in which the RNA-Seq signal can be detected

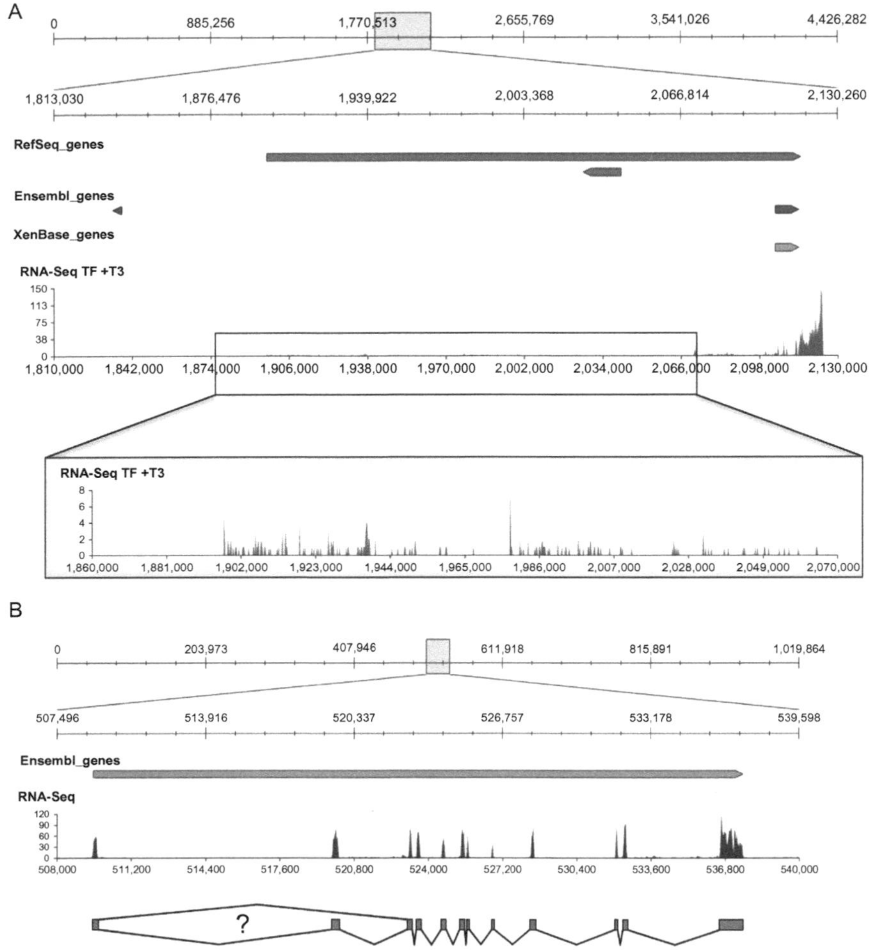

Figure 10.3 RNA-Seq profiling provides a quick overview of transcription units. (A) RNA-Seq profile at the TRβ locus. The first three tracks represent the gene models from the same databases as in Fig. 10.2. The fourth track represents the RNA-Seq profile obtained from the tail fin of premetamorphic tadpoles treated with T_3. The bottom part is a close-up of the TRβ transcription unit which is occulted by the very strong signal at the 3′ end of the gene. (B) Conventional RNA-Seq does not provide information on alternative splicing: Example of the wntless homolog (wls) on scaffold_431. The first track represents Ensembl gene models. The second track corresponds to the RNA-Seq profile. The third track corresponds to annotated exons and indicates two known alternative transcript are generated by exon skipping (connecting lines).

up to 210 kb upstream of the TSS annotated in XenBase or Ensembl databases. A limiting factor of conventional RNA-Seq is related to the length of the sequence reads ($\leq$50 bp), that can be difficult to map when they overlap over two exons or when exons are small. Because of this, conventional RNA-Seq does not allow precise definition of exon boundaries and is not well suited for the definition of genes structure. This technical limitation is overcome with longer reads of 75–100 bp, and pair-end sequencing, which allows a better resolution when defining exons boundaries (Fig. 10.3B) (see below). This technology has been successfully used to refine current gene models in *X. tropicalis* (Akkers et al., 2009).

Several recent protocols allow longer sequence reads ($\geq$100 bp) and pair-end sequencing, in which the two ends of each cDNA fragment are sequenced. Although the analysis is more complex, these protocols are well suited for *de novo* transcript reconstruction, exon definition, detection, and quantitative measure of alternative splicing. Two strategies can be used to process the sequenced reads (Garber et al., 2011): The reads can be assembled into transcripts, including splicing variants, which can be mapped to a reference genome, thus providing exon annotation. Alternatively, the reads can be mapped directly to the reference genome, without *de novo* assembly, and alternative splicing can be inferred from reads split in two and mapping over the ends of two distinct exons (also known as "spliced reads"). These methods provide a single base pair resolution of exon boundaries.

The 5′ end of transcripts are often missed and are poorly annotated. To this regard, the RNA-PET technology holds many promises. Originally developed to detect fusion transcripts in cancer cells, the RNA-PET specifically captures the 5′ and the 3′ ends of transcripts (Ruan & Ruan, 2011). The principle of this method is depicted in Fig. 10.4A. Full-length cDNA is synthesized from mRNA transcripts with CAP-Trapper and GsuI-dT(16) purification. A linker containing the binding site of the Mme1 (or *Eco*P15I) endonuclease is then ligated to the full-length cDNA, allowing it to be circularized. The circular DNA molecule is then restricted with *Mme*1, which cuts 27 bp downstream of its binding site, thus resulting in 27 bp tags of each end of the original transcript, spaced by the linker. Sequencing adaptors are then ligated for high-throughput sequencing. The resulting paired-end tags (PETs) are then mapped to the genome, thus providing experimental evidence for the 5′ and 3′ ends of transcripts, including potential alternative TSS and transcription termination sites (TTSs), but provides no information relative to the their internal structure (Fig. 10.4B). Thus, RNA-PET provides complementary information about the boundaries of the transcriptional

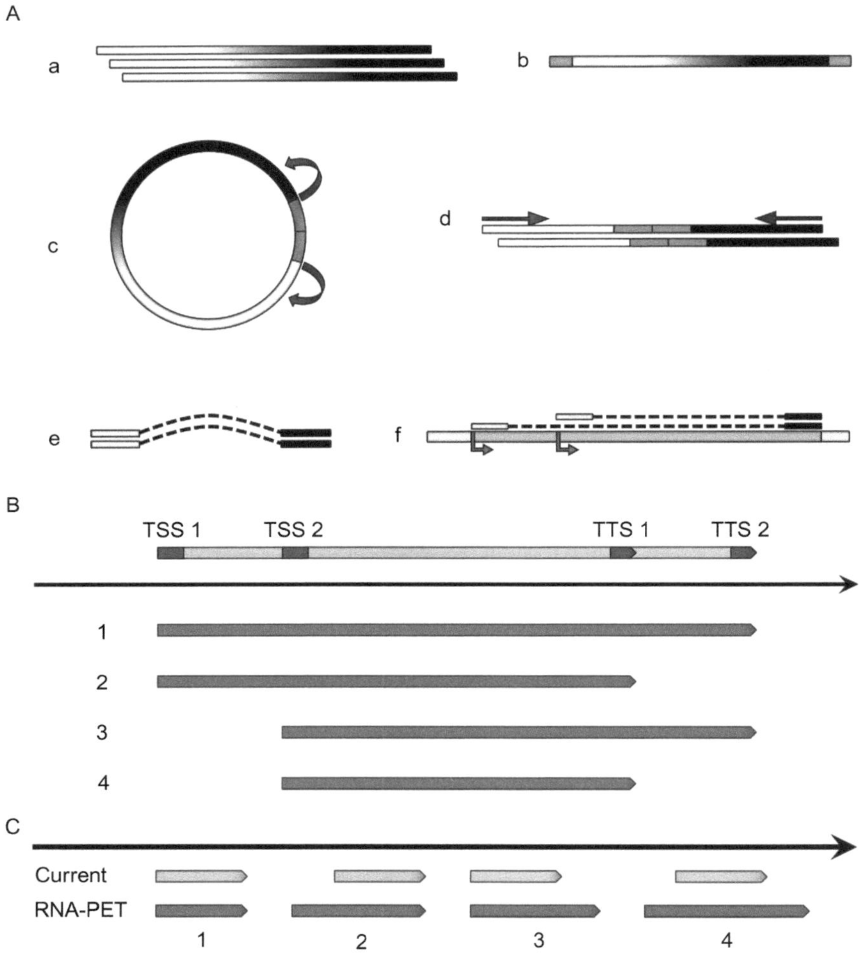

Figure 10.4 Principle of RNA-PET. (A) The starting point of RNA-PET is a library of full-length cDNAs made from poly-A RNA (a), on which linkers containing *Mme*1 (or *Eco*P15I) restriction sites are ligated (b). The resulting molecules are circularized and restricted with *Mme*1 (c), which cuts DNA 27 bp downstream of its binding site and thus produces a linear fragment containing the two ends of the original transcript separated by the linker (d) prior to deep sequencing (from each end). The resulting paired-end tags (PETs) contain information about the association of one transcript extremity to the other—respectively, white and black rectangles—(e). After mapping on the reference genome, this provides experimental evidence for the location of the 5′ and 3′ ends of transcripts (f). (B) Since RNA-PET captures the 5′ and 3′ ends of transcripts, it can be used to measure the relative abundance of transcripts that use alternative TSS or TTS (dark segments on the full gene model). (C) RNA-PET can improve the annotation of transcription units and thus gene boundaries. From left to right, RNA-PET determined gene boundaries (dark gray) can concord (1) with the current annotation (light gray), or extend in 5′ (2), 3′ (3), or both (4) extremities.

units, by confirming current models or redefining them at their 5′, 3′ (or both) ends (Fig. 10.4C). The potential of RNA-Seq coupled to RNA-PET in term of genome annotation is therefore tremendous.

3.2. NGS provides functional data to analyze TH-induced differential expression

Amphibian metamorphosis involves extensive modifications/remodeling of the organism as a whole. Since this biological process is strictly controlled by TH, documenting expression variation of TH-responsive genes is central to understand the molecular mechanisms controlling metamorphosis, decrypt the gene network controlled by TH, and its spatial and temporal dynamics. In the pre-NGS era, the simultaneous measure of the expression level of thousand of transcripts was carried out by microarrays. Yet, as discussed in a previous section, this high-throughput technology is not "naive" *per se*, in that the design of the microarrays is based on a prior knowledge of the transcripts to be detected. Indeed, RNA-Seq can benefit from the evolution and ongoing curation of a genome annotation, whereas detecting novel transcripts with microarrays would require a new design. In fact, important genes involved in amphibian metamorphosis might not even be represented on the chip (e.g., TRβ gene on *X. tropicalis* Affymetrix array, see above). Nowadays, in the NGS era, RNA-Seq represents a more naive, genome-wide alternative for measuring gene expression, in that the strength of the method is not based on a prior knowledge of transcripts, or gene annotation (Fig. 10.3). For example, in the case of TRβ, even if the gene is not well annotated, the RNA-Seq signal spreads over 210 kb and can be assigned to it *a posteriori*, whereas this is not possible with microarrays. RNA-Seq also has the benefit of having a much larger dynamic range than microarrays (Łabaj et al., 2011). In fact, the hybridization step in RNA-Seq is carried out *in silico*, which provides full control over it, and does not saturate. This is in contrast with microarrays, with which the hybridization signal can quickly reach the saturation limit of the photomultiplier. Once RNA-Seq data have been mapped to either a reference genome or transcriptome, several layers of analysis can be carried out. The absolute level of gene expression can be measured by computing the reads per kilobase of exon model per million of mapped reads (RPKM) value. Additionally, if several samples have been subject to the same analysis, one can perform differential analysis in a manner similar to that of microarray data, but using read count instead of photons. As shown in Fig. 10.5A, the expression of the TH regulated gene THb/ZIP is strongly induced upon T_3 treatment. Although the THb/ZIP gene model and the Affymetrix microarray probesets partially overlap with the RNA-Seq signal (indicating that they can be used to assess the expression levels of the gene), a better measure of gene expression is achieved when counting the RNA-Seq reads along the entire gene. This is clearly shown in Fig. 10.5A, where the RNA-Seq signal extends up to 5 kb upstream of the annotated TSS for THb/ZIP, or even ~200 kb for TRβ (Figs. 10.3A and 10.5B). If read count was restricted to the annotated region, a lot of signal could actually be missed, potentially introducing biases comparable to those found in microarray analysis. In this context, using RNA-PET to define precise gene boundaries would prove particularly helpful in achieving better differential analysis of RNA-Seq data.

In addition, when using RNA-Seq derivatives (Wang et al., 2009), fragments spanning multiple exons can be easily mapped to the genome or transcriptome, giving precise data about the expression levels of the various alternative transcripts (Fig. 10.3B). Furthermore, beyond the strict definition of exon boundaries, RNA-PET can provide important data about TH effect on alternative TSS and TTS usage (Fig. 10.4B). The best-case scenario for accurately measuring gene expression with RNA-Seq requires high quality gene models.

4. TR INTERACTOME

4.1. Whole genome TR mapping

Recent advances provide the technologies to go beyond the simple measurement of transcripts abundance, making it possible to investigate the detailed molecular mechanisms involved in the transcriptional dynamics of TR mediated regulatory cascades.

Upon binding to T_3RE, TR recruits cofactors that will modify the transcriptional permissiveness of chromatin (Wong, Shi, & Wolffe, 1995). Although T_3RE are located near TH regulated genes (Ranjan et al., 1994), recent experimental evidence on the regulation of gene expression by transcription factors involves loop-mediated physical interactions between regulatory factors bound to DNA at distant enhancers and the core transcriptional machinery located at the TSS of regulated genes (Fullwood, Liu, et al., 2009). In fact, in a manner similar to other transcription factors, TR could regulate the expression of target genes over long genomic distances (up to several tens of kilobases, unpublished data), although some TR-binding site are located nearby TH-responsive genes.

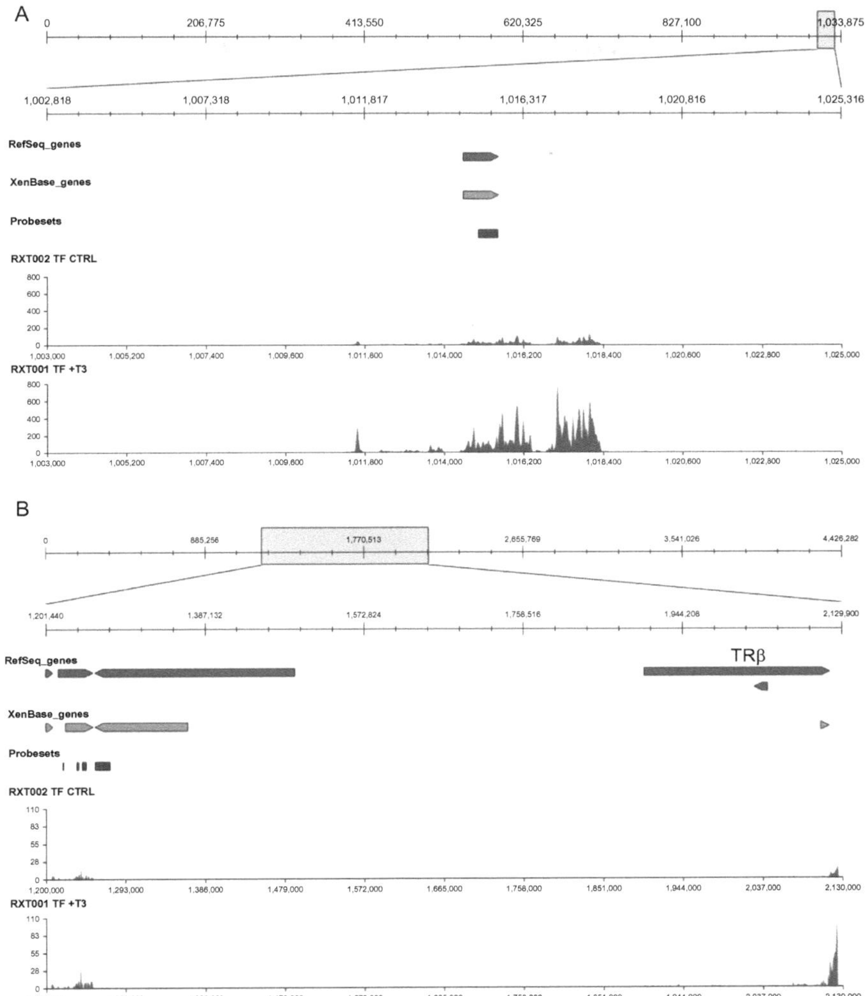

Figure 10.5 RNA-Seq is less limited than microarrays for quantitative measures of gene expression. (A) Genome view of the THb/ZIP locus. The first two tracks represent the RefSeq and XenBase gene models. The second track represents the (curated) probesets from the commercially available *X. tropicalis* Affymetrix microarray, with which the expression level of THb/ZIP appears induced 2.8 fold. The last two tracks correspond to the RNA-Seq profiles before and after T_3 treatment, respectively. In this case, the induction level is ~fivefolds, which fits better with RT-qPCR data. (B) Genome view of the TRβ locus. The tracks are the same as in A. No probeset is associated to the TRβ gene (on the right-hand side of the panel), as opposed to other genes (left-hand side). Thus, the Affymetrix microarray cannot monitor the expression of a key regulator of metamorphosis.

The canonical T_3RE sequence is well defined and it could be tempting to perform an exhaustive search of this motif genome wide with tools such as NHRscan (Sandelin & Wasserman, 2005). Unfortunately, given its small size and degeneracy, search for this motifs will be flooded by strong statistical noise (Fig. 10.6A). For example, more than 200,000 DR4 could be found across the genome. However, it is improbable that there are this many binding sites across the *X. tropicalis* genome, given that so far, the number of nuclear receptor binding site in a genome has been estimated between 2000 and 10,000 (Hamza et al., 2009; Nielsen et al., 2008; Welboren et al., 2009). In addition, identifying canonical DR4 motifs does not guaranty finding functional TR-binding sites. Indeed, TR was shown to bind other motifs (e.g., ER6 and DR1, see Desvergne, 1994). Nuclear receptors can also bind to motifs that depart significantly from the canonical motif (Zirngibl, Chan, & Aubin, 2008). Thus, it is imperative to use methods such as chromatin immunoprecipitation (ChIP) to provide experimental evidence of TR binding at specific genomic locations. To this end, protein–protein and protein–DNA interactions are stabilized by cross-linking, and chromatin is fragmented by sonication to a size range of 200–400 bp. The DNA fragments bound to the protein of interest are captured by immunoprecipitation with a specific antibody against the protein. After de-cross-linking and purification, the resulting DNA sample is enriched with the fragments originally associated with the protein. These samples can be subject to different detection techniques. With conventional ChIP–qPCR, the abundance of a known DNA fragment is measured by real-time quantitative PCR (qPCR). This method, which requires to know *a priori* the genomic region being monitored, has proved to be remarkably helpful in the context of amphibian metamorphosis to measure the differential binding patterns of TR in the vicinity of the TRβ gene, as well as the dynamics of histone modifications (Bilesimo et al., 2011). Although it is a very efficient technique, its throughput is very low and labor intense, since every location has to be tested individually by qPCR, and it requires to know *a priori* where to look for binding. With the ChIP-Seq technology, the detection method involves sequencing the purified DNA fragments. In this case, the number of sequence reads per genomic location provides a quantitative measure of DNA binding. This approach is naive and does not rely on prior knowledge of the putative binding sites, and rather provides genome-wide binding profiles (Fig. 10.6B).

The profiling of a few chromatin modification marks as well as RNA-Polymerase II occupancy has been carried out on *X. tropicalis* early

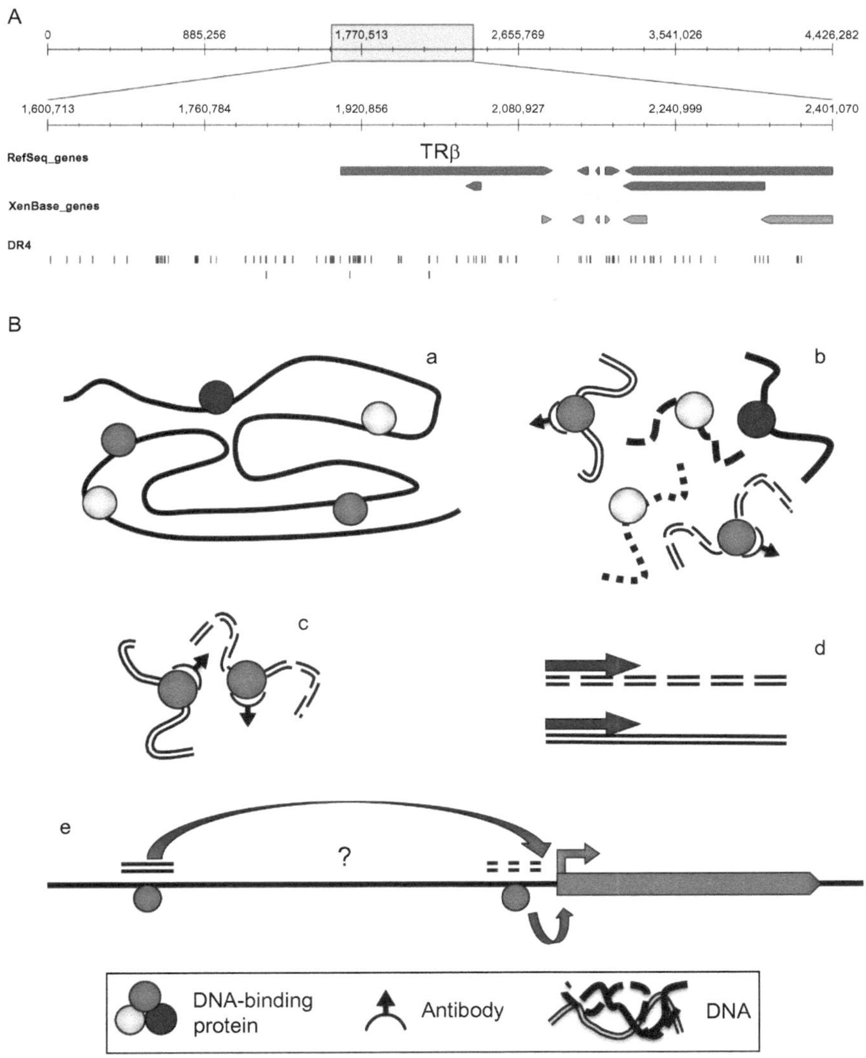

Figure 10.6 ChIP-Seq is a naive alternative to *in silico* predictions for mapping DNA-binding sites. (A) Genome view of the TRβ locus. The first two tracks represent the XenBase and Ensembl gene models, respectively. The third track represents 21,424 putative DR4 found in this region with a modified version of NHRscan (Sandelin & Wasserman, 2005). (B) Principle and limitations of ChIP-Seq. Physical interactions between chromatin components (protein, DNA) are stabilized by cross-linking (a). Chromatin is then fragmented by sonication in fragments of 200–400 bp (b). The DNA-binding sites are captured by immunoprecipitation with a protein-specific antibody (c). After cross-link reversal, the DNA sample is purified and sequenced (d). Binding density profiles are derived from sequenced reads after mapping to the reference genome, and provide an accurate description of the binding sites (e). Transcription factors (TFs) regulate the transcription of target genes after binding to DNA at enhancers, which can be located far from genes. Unfortunately, ChIP-Seq cannot connect DNA-binding sites of TFs to their target gene and thus cannot be used to describe the topology of gene regulatory networks.

development (Akkers et al., 2009), but not on amphibian metamorphosis. It would be of great interest to derive genome-wide TR-binding profiles to help understand the underlying mechanistic events of amphibian metamorphosis.

However, as seducing as it may sound, ChIP-Seq suffers from a major limitation as it provides no experimental evidence of the target gene(s) whose transcription is regulated by the DNA-bound transcription factor, if any. At best, one assumes that the target gene is located nearby the DNA-bound transcription factor, and infers such functional connection. Although this might be true for a few genes, this postulate is clearly not true genome wide (see below). It is clear that DNA-bound transcriptional regulators can act over large genomic regions and physically interact with the basal transcription machinery through DNA looping (Kim, Bresnick, & Bultman, 2009). Also, transcriptional regulators can interact with one another, thus adding more complexity (Fullwood, Liu, et al., 2009).

4.2. A need for 3D analysis

Recent technological breakthrough bypass the ChIP-Seq limitations and proceed a step further in the analysis of functional interactions between DNA-bound proteins. Ten years ago, the development of the chromosome conformation capture (3C) technology allowed to assess physical interactions between chromosome loci (Dekker, Rippe, Dekker, & Kleckner, 2002; de Wit & de Laat, 2012).The throughput was quite low in that 3C was used to detect the physical proximity of two DNA fragments in the nucleus in a "one to one" manner. It also required a prior knowledge of the genomic locations to probe. The first improvements of the 3C technology resulted in the circularized chromosome conformation capture (4C) technique, which allows to probe physical interactions in a "one to many" manner (Simonis et al., 2006; Zhao et al., 2006). In 2009, the Genome Institute of Singapore developed the chromatin interaction analysis by paired-end tags sequencing (ChIA-PET) (Fullwood, Liu, et al., 2009; Fullwood & Ruan, 2009), which combines the principles of ChIP-Seq and 4C, and produces two lines of evidence from a single experiment: a genome-wide binding profile together with a map of the physical interactions between protein-bound DNA regions. It is important to note that this method only reveals if two regions of DNA are close to each other but does not tell which or if any proteins is involved in the interaction. It does not either reveal any physical interactions among proteins, especially if no other proteins are analyzed. The experimental procedure is quite complex and is aimed at capturing DNA-bound proteins engaging physical contacts with other protein-bound DNA complexes. The procedure can be summarized as follows (Fig. 10.7): DNA-bound protein complexes are captured by immunoprecipitation with a specific antibody, following chromatin fragmentation (this step is similar to that of a regular ChIP). However, as opposed to conventional ChIP, cross-linking is not reversed after this step. Instead, a double-stranded DNA linker containing the binding site of the restriction enzyme *Mme*1, which cuts DNA 27 bp downstream of its binding site is ligated to each end of the DNA fragment. The DNA fragments are then ligated. Two kinds of ligation events can take place: *cis*-ligation (self-ligation of a DNA fragment, providing ChIP-Seq like data but no information on long range interactions) and *trans*-ligation (ligation between two fragments far from each other on the genome but brought close to each other within the large protein–DNA complex). After ligation, the complexes are de-cross-linked, and the ligated DNA fragments are restricted with *Mme*1 to produce fragments containing pairs of ends ligated to the linker. After amplification and deep sequencing, they are mapped to the reference genome. One of the strengths of ChIA-PET relative to 3C and 4C is that it can probe "many to many" interactions of DNA regions associated with the protein, be it via interactions of the protein with other transcription factors (Fullwood, Liu, et al., 2009) or core components of the chromatin such as histones or other factors such as insulator-binding proteins (CCCTC-binding protein) (Chepelev, Wei, Wangsa, Tang, & Zhao, 2012; Handoko et al., 2011). As a consequence, in addition to profiling DNA-bound transcription factors, ChIA-PET also provides experimental evidence for physical interactions between DNA regions far away from each other, and thus can unambiguously identify direct target genes.

Using ChIA-PET on TR would help document with unprecedented details the topology and dynamics of regulatory networks of TH regulated genes. Importantly, this would not only provide a genome-wide map of TR binding but also identify direct TR target genes. Perhaps most of all, ChIA-PET will provide us with the opportunity to go beyond the local chromatin organization and probe higher orders of chromatin architecture. Our preliminary data show that indeed, TR can act over large genomic distances and that the local chromatin context might be marginally relevant to understand the mechanisms of TH-induced transcription cascades (unpublished data).

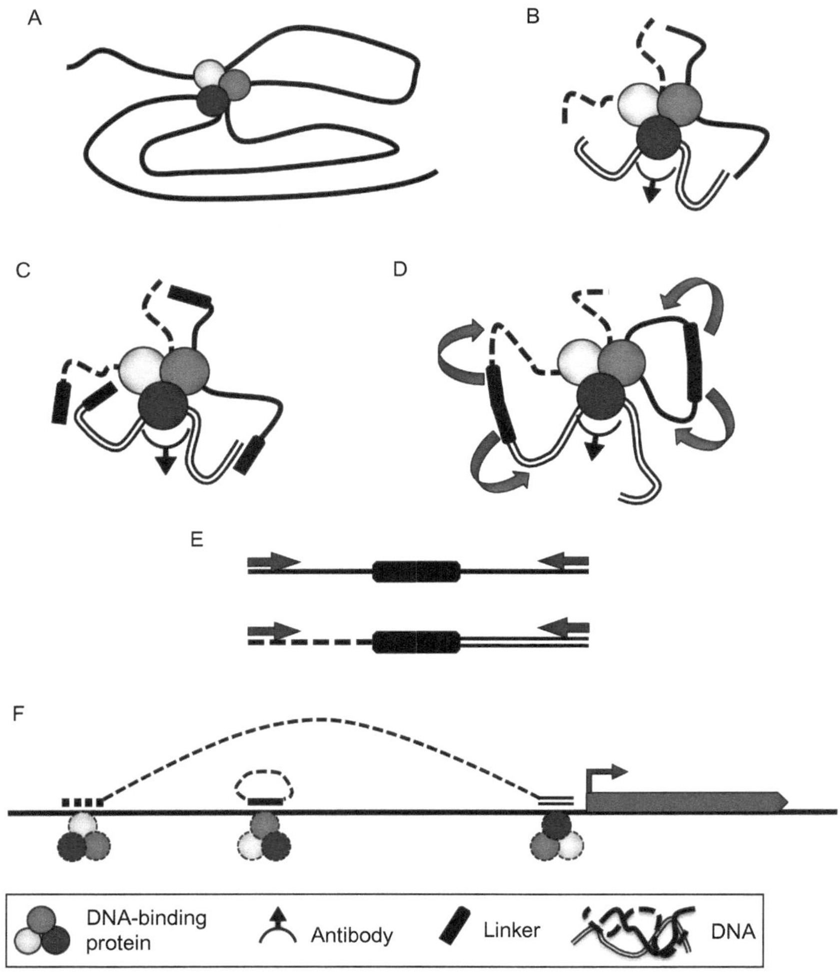

Figure 10.7 ChIA-PET describes DNA-binding sites and their physical interactions through DNA looping. The first two steps of ChIA-PET are similar to those of ChIP-Seq. Briefly, chromatin is cross-linked to stabilize the interactions between its proteins and DNA components and sonicated (A). Complexes are then purified with a specific antibody (B). Double stranded linkers containing the binding site of the *Mme*1 endonuclease are ligated to the ends of the fragmented DNA (C). Ligation can occur in *cis*, when a DNA fragment is circularized, or in *trans*, when two different DNA fragments are linked together (D). After restriction with *Mme*1, which cuts DNA 27 bp downstream of its binding site in the genomic fragment (D), the resulting linear fragments are sequenced, providing paired-end tags (PETs) reflecting the ligation products (E). After mapping to the reference genome, *cis*-ligation products are essentially similar to ChIP-Seq end products, and serve to derive binding profiles and identify DNA-binding sites. Given that the two tags of *trans*-ligation products map at distinct genomic locations, they do describe the network of physical interactions between the DNA-bound proteins (F). All the possible ligation products are not shown for the sake of simplicity.

5. CONCLUSION

Metamorphosis is a very common process which marks the transition between embryonic and adult life stages. Not only does it correspond to an ecological transition, in that the ecological niche often differs between the larva and the young adult, it also sets deep anatomical, physiological, cellular, and molecular changes. As such, metamorphosis is a deep, fundamental, and ancient biological process.

The problem of studying amphibian metamorphosis is that amphibian genomes are often polyploids, which adds extra levels of genetic redundancy and complexity, and limits to the tools which can be used to characterize the molecular mechanisms of metamorphosis. In this regard, the diploid *X. tropicalis* presents many advantages. Not surprisingly, at the beginning of genome sequencing, the scientific community quickly committed itself to sequence *X. tropicalis* genome and produced functional genomic resources. These were instrumental for the development of our current perception of the molecular mechanisms of *X. tropicalis* metamorphosis.

The recent breakthroughs in sequencing technologies push once again the scale and the precision of our molecular toolkits a step further. It is now possible to describe with exquisite details the genome-wide remodeling of transcriptional programs. Even better, such projects that used to require the combined workforce of one to several research institutes can now be carried out by a single lab. The promises of the molecular dissection of amphibian metamorphosis are enormous. The downside of it is that the current state of the *X. tropicalis* genome assembly and annotation does not reach the standard needed to make use of the full power of these technologies. It is thus important to revisit and update the *X. tropicalis* genomic resources.

The sequencing throughput has increased by three orders of magnitude in the past few years, posing many computing constraints: storage disks space is quickly saturated by the flood of sequencing data (because the advances of hard drive technologies is much slower than the progresses of sequencing throughput), large-scale analysis and processing often require large computing clusters (with their own constraints on dedicated facilities, power consumption, etc.), and data processing through large and complex bioinformatic pipeline. Theses weaknesses are counterbalanced (in part) by the progress made in software engineering with new and efficient data types, optimized database systems, and computing algorithms.

But perhaps the most challenging is our capacity, as scientists, to embrace the different layers of the technological fields deployed in such projects, which can go from breeding tadpoles in a tank to run metagenomic analysis on a distributed computing system. The success surely lies in the hands of next-generation scientists (Wickware, 2000).

ACKNOWLEDGMENTS

The work was supported by the "Centre National de la Recherche Scientifique," the "Muséum National d'Histoire Naturelle," "CRESCENDO," a European Integrated Project funding from FP6 (contract n°LSHM-CT-2005-018652), "IDEAL" a European Large Integrated Project funding from FP7 (n°259679), and "TRIGGER" (ANR-08-JCJC-0100-01). Part of this work was carried out in collaboration with the Genome Institute of Singapore, to whom we are grateful for introducing us to their technologies.

REFERENCES

Akkers, R. C., van Heeringen, S. J., Jacobi, U. G., Janssen-Megens, E. M., FranCoijs, K.-J., Stunnenberg, H. G., et al. (2009). A hierarchy of H3K4me3 and H3K27me3 acquisition in spatial gene regulation in Xenopus embryos. *Developmental Cell, 17*, 425–434.

Altmann, C. R., Bell, E., Sczyrba, A., Pun, J., Bekiranov, S., Gaasterland, T., et al. (2001). Microarray-based analysis of early development in Xenopus laevis. *Developmental Biology, 236*, 64–75.

Bilesimo, P., Jolivet, P., Alfama, G., Buisine, N., Le Mevel, S., Havis, E., et al. (2011). Specific histone lysine 4 methylation patterns define TR-binding capacity and differentiate direct T3 responses. *Molecular Endocrinology, 25*, 225–237.

Birney, E. E., Clamp, M. M., & Durbin, R. R. (2004). GeneWise and genomewise. *Genome Research, 14*, 988–995.

Bowes, J. B., Snyder, K. A., Segerdell, E., Gibb, R., Jarabek, C., Noumen, E., et al. (2008). Xenbase: A Xenopus biology and genomics resource. *Nucleic Acids Research, 36*, D761–D767.

Brown, D. D., Wang, Z., Furlow, J. D., Kanamori, A., Schwartzman, R., Remo, B., et al. (1996). The thyroid hormone-induced tail resorption program during Xenopus laevis metamorphosis. *Proceedings of the National Academy of Sciences of the United States of America, 93*, 1924–1929.

Buchholz, D. R., Heimeier, R. A., Das, B., Washington, T., & Shi, Y.-B. (2007). Pairing morphology with gene expression in thyroid hormone-induced intestinal remodeling and identification of a core set of TH-induced genes across tadpole tissues. *Developmental Biology, 303*, 576–590.

Buckbinder, L., & Brown, D. D. (1992). Thyroid hormone-induced gene-expression changes in the developing frog limb. *The Journal of Biological Chemistry, 267*, 25786–25791.

Buisine, N., & Sachs, L. (2009). Impact of genome assembly status on ChIP-Seq and ChIP-PET data mapping. *BMC Research Notes, 2*, 257.

Cai, L., Das, B., & Brown, D. D. (2007). Changing a limb muscle growth program into a resorption program. *Developmental Biology, 304*, 260–271.

Chepelev, I., Wei, G., Wangsa, D., Tang, Q., & Zhao, K. (2012). Characterization of genome-wide enhancer-promoter interactions reveals co-expression of interacting genes and modes of higher order chromatin organization. *Cell Research, 22*, 490–503.

Das, B., Cai, L., Carter, M. G., Piao, Y.-L., Sharov, A. A., Ko, M. S. H., et al. (2006). Gene expression changes at metamorphosis induced by thyroid hormone in Xenopus laevis tadpoles. *Developmental Biology, 291*, 342–355.

Das, B., Heimeier, R. A., Buchholz, D. R., & Shi, Y.-B. (2009). Identification of direct thyroid hormone response genes reveals the earliest gene regulation programs during frog metamorphosis. *The Journal of Biological Chemistry, 284*, 34167–34178.

Dekker, J., Rippe, K., Dekker, M., & Kleckner, N. (2002). Capturing chromosome conformation. *Science, 295*, 1306–1311.

de Wit, E., & de Laat, W. (2012). A decade of 3C technologies: Insights into nuclear organization. *Genes & Development, 26*, 11–24.

Desvergne, B. (1994). How do thyroid hormone receptors bind to structurally diverse response elements? *Molecular and Cellular Endocrinology, 100*, 125–131.

Eliceiri, B., & Brown, D. D. (1994). Quantitation of endogenous thyroid-hormone receptor-alpha and receptor-beta during embryogenesis and metamorphosis in Xenopus-laevis. *The Journal of Biological Chemistry, 269*, 24459–24465.

Fairclough, L., & Tata, J. R. (1997). An immunocytochemical analysis of the expression of thyroid hormone receptor alpha and beta proteins during natural and thyroid hormone-induced metamorphosis in Xenopus. *Development, Growth & Differentiation, 39*, 273–283.

Flamant, F., Gauthier, K., & Samarut, J. (2007). Thyroid hormones signaling is getting more complex: STORMs are coming. *Molecular Endocrinology, 21*, 321–333.

Flicek, P., Amode, M. R., Barrell, D., Beal, K., Brent, S., Carvalho-Silva, D., et al. (2012). Ensembl 2012. *Nucleic Acids Research, 40*, D84–D90.

Fullwood, M. J., & Ruan, Y. (2009). ChIP-based methods for the identification of long-range chromatin interactions. *Journal of Cellular Biochemistry, 107*, 30–39.

Fullwood, M. J., Liu, M. H., Pan, Y. F., Liu, J., Xu, H., Mohamed, Y. B., et al. (2009). An oestrogen-receptor-alpha-bound human chromatin interactome. *Nature, 462*, 58–64.

Fullwood, M. J., Wei, C.-L. C., Liu, E. T. E., & Ruan, Y. (2009). Next-generation DNA sequencing of paired-end tags (PET) for transcriptome and genome analyses. *Genome Research, 19*, 521–532.

Furlow, J. D., & Brown, D. D. (1999). In vitro and in vivo analysis of the regulation of a transcription factor gene by thyroid hormone during Xenopus laevis metamorphosis. *Molecular Endocrinology, 13*, 2076–2089.

Garber, M., Grabherr, M. G., Guttman, M., & Trapnell, C. (2011). Computational methods for transcriptome annotation and quantification using RNA-seq. *Nature Methods, 8*, 469–477.

Gilchrist, M. J. (2012). From expression cloning to gene modeling: The development of Xenopus gene sequence resources. *Genesis, 50*, 143–154.

Grimaldi, A., Buisine, N., Miller, T., Shi, Y.-B., & Sachs, L. M. (2012). Mechanisms of thyroid hormone receptor action during development: Lessons from amphibian studies. *Biochimica et Biophysica Acta*. http://dx.doi.org/10.1016/j.bbagen.2012.04.020.

Gronemeyer, H., & Laudet, V. (1995). Transcription factors 3: Nuclear receptors. *Protein Profile, 2*, 1173–1308.

Hamza, M. S., Pott, S., Vega, V. B., Thomsen, J. S., Kandhadayar, G. S., Ng, P. W. P., et al. (2009). De-novo identification of PPARgamma/RXR binding sites and direct targets during adipogenesis. *PloS One, 4*, e4907.

Handoko, L., Xu, H., Li, G., Ngan, C. Y., Chew, E., Schnapp, M., et al. (2011). CTCF-mediated functional chromatin interactome in pluripotent cells. *Nature Genetics, 43*, 630–638.

Havis, E., Le Mevel, S., Morvan Dubois, G., Shi, D.-L., Scanlan, T. S., Demeneix, B. A., et al. (2006). Unliganded thyroid hormone receptor is essential for Xenopus laevis eye development. *The EMBO Journal, 25*, 4943–4951.

Hawkins, R. D., Hon, G. C., & Ren, B. (2010). Next-generation genomics: An integrative approach. *Nature Reviews. Genetics*, *11*, 476–486.
Hellsten, U. U., Harland, R. M., Gilchrist, M. J., Hendrix, D., Jurka, J., Kapitonov, V., et al. (2010). The genome of the Western clawed frog Xenopus tropicalis. *Science*, *328*, 633–636.
Kawahara, A., Baker, B. S., & Tata, J. R. (1991). Developmental and regional expression of thyroid hormone receptor genes during Xenopus metamorphosis. *Development*, *112*, 933–943.
Kim, S.-I., Bresnick, E. H., & Bultman, S. J. (2009). BRG1 directly regulates nucleosome structure and chromatin looping of the alpha globin locus to activate transcription. *Nucleic Acids Research*, *37*, 6019–6027.
Klein, S., Strausberg, R., Wagner, L., Pontius, J., Clifton, S., & Richardson, P. (2002). Genetic and genomic tools for Xenopus research: The NIH Xenopus initiative. *Developmental Dynamics*, *225*, 384–391.
Łabaj, P. P., Leparc, G. G., Linggi, B. E., Markillie, L. M., Wiley, H. S., & Kreil, D. P. (2011). Characterization and improvement of RNA-Seq precision in quantitative transcript expression profiling. *Bioinformatics*, *27*, i383–i391.
Laudet, V. (2011). The origins and evolution of vertebrate metamorphosis. *Current Biology*, *21*, R726–R737.
Metzker, M. L. (2010). Sequencing technologies—The next generation. *Nature Reviews. Genetics*, *11*, 31–46.
Nagaraj, S. H., Gasser, R. B., & Ranganathan, S. (2007). A hitchhiker's guide to expressed sequence tag (EST) analysis. *Briefings in Bioinformatics*, *8*, 6–21.
Nielsen, R., Pedersen, T. A., Hagenbeek, D., Moulos, P., Siersbaek, R., Megens, E., et al. (2008). Genome-wide profiling of PPARgamma: RXR and RNA polymerase II occupancy reveals temporal activation of distinct metabolic pathways and changes in RXR dimer composition during adipogenesis. *Genes & Development*, *22*, 2953–2967.
Pruitt, K. D., Tatusova, T., Klimke, W., & Maglott, D. R. (2009). NCBI reference sequences: Current status, policy and new initiatives. *Nucleic Acids Research*, *37*, D32–D36.
Ranjan, M., Wong, J., & Shi, Y. B. (1994). Transcriptional repression of Xenopus TR beta gene is mediated by a thyroid hormone response element located near the start site. *Journal of Biological Chemistry*, *269*, 24699–24705.
Ruan, X., & Ruan, Y. (2011). Genome wide full-length transcript analysis using 5" and 3" paired-end-tag next generation sequencing (RNA-PET). *Methods in Molecular Biology*, *809*, 535–562.
Sachs, L., Damjanovski, S., Jones, P., Li, Q., Amano, T., Ueda, S., et al. (2000). Dual functions of thyroid hormone receptors during Xenopus development. *Comparative Biochemistry and Physiology. Part B, Biochemistry & Molecular Biology*, *126*, 199–211.
Sachs, L. M., & Shi, Y. B. (2000). Targeted chromatin binding and histone acetylation in vivo by thyroid hormone receptor during amphibian development. *Proceedings of the National Academy of Sciences of the United States of America*, *97*, 13138–13143.
Salamov, A. A., & Solovyev, V. V. (2000). Ab initio gene finding in Drosophila genomic DNA. *Genome Research*, *10*, 516–522.
Sandelin, A., & Wasserman, W. W. (2005). Prediction of nuclear hormone receptor response elements. *Molecular Endocrinology*, *19*, 595–606.
Santos, G. M. G., Fairall, L., & Schwabe, J. W. R. J. (2011). Negative regulation by nuclear receptors: A plethora of mechanisms. *Trends in Endocrinology and Metabolism*, *22*, 87–93.
Sboner, A., Mu, X. J., Greenbaum, D., Auerbach, R. K., & Gerstein, M. B. (2011). The real cost of sequencing: Higher than you think!. *Genome Biology*, *12*, 125.

Schiex, T., Moisan, A., & Rouzé, P. (2001). Lecture notes in computer science. In O. Gascuel & M.-F. Sagot (Eds.), Berlin, Heidelberg: Springer-Verlag Berlin Heidelberg.

Shi, Y.-B. (1999). *Amphibian metamorphosis: From morphology to molecular biology* (1st ed). Wiley-Liss, New York.

Shi, Y. B., & Brown, D. D. (1993). The earliest changes in gene expression in tadpole intestine induced by thyroid hormone. *The Journal of Biological Chemistry, 268*, 20312–20317.

Simonis, M., Klous, P., Splinter, E., Moshkin, Y., Willemsen, R., de Wit, E., et al. (2006). Nuclear organization of active and inactive chromatin domains uncovered by chromosome conformation capture-on-chip (4C). *Nature Genetics, 38*, 1348–1354.

Tata, J. R. (1993). Gene expression during metamorphosis: An ideal model for post-embryonic development. *BioEssays, 15*, 239–248.

Tymowska, J., & Fischberg, M. (1973). Chromosome complements of the genus Xenopus. *Chromosoma, 44*, 335–342.

Voigt, J., Chen, J., Gilchrist, M., Amaya, E., & Papalopulu, N. (2005). Expression cloning screening of a unique and full-length set of cDNA clones is an efficient method for identifying genes involved in Xenopus neurogenesis. *Mechanisms of Development, 122*, 289–306.

Wang, Z., & Brown, D. D. (1991). A gene-expression screen. *Proceedings of the National Academy of Sciences of the United States of America, 88*, 11505–11509.

Wang, Z., & Brown, D. D. (1993). Thyroid hormone-induced gene expression program for amphibian tail resorption. *The Journal of Biological Chemistry, 268*, 16270–16278.

Wang, Z., Gerstein, M., & Snyder, M. (2009). RNA-Seq: A revolutionary tool for transcriptomics. *Nature Reviews. Genetics, 10*, 57–63.

Welboren, W.-J., van Driel, M. A., Janssen-Megens, E. M., van Heeringen, S. J., Sweep, F. C., Span, P. N., et al. (2009). ChIP-Seq of ERalpha and RNA polymerase II defines genes differentially responding to ligands. *The EMBO Journal, 28*, 1418–1428.

Wickware, P. (2000). Next-generation biologists must straddle computation and biology. *Nature, 404*, 683–684.

Wong, J., Shi, Y. B., & Wolffe, A. P. (1995). A role for nucleosome assembly in both silencing and activation of the Xenopus TR beta A gene by the thyroid hormone receptor. *Genes & Development, 9*, 2696–2711.

Yaoita, Y., & Brown, D. D. (1990). A correlation of thyroid hormone receptor gene expression with amphibian metamorphosis. *Genes & Development, 4*, 1917–1924.

Yaoita, Y., Shi, Y., & Brown, D. D. (1990). Xenopus laevis alpha and beta thyroid hormone receptors. *Proceedings of the National Academy of Sciences of the United States of America, 87*, 7090–7094.

Zhao, Z., Tavoosidana, G., Sjölinder, M., Göndör, A., Mariano, P., Wang, S., et al. (2006). Circular chromosome conformation capture (4C) uncovers extensive networks of epigenetically regulated intra- and interchromosomal interactions. *Nature Genetics, 38*, 1341–1347.

Zirngibl, R. A., Chan, J. S. M., & Aubin, J. E. (2008). Estrogen receptor-related receptor alpha (ERRalpha) regulates osteopontin expression through a non-canonical ERRalpha response element in a cell context-dependent manner. *Journal of Molecular Endocrinology, 40*, 61–73.

CHAPTER ELEVEN

Establishment of Intestinal Stem Cell Niche During Amphibian Metamorphosis

Atsuko Ishizuya-Oka[1], Takashi Hasebe
Department of Biology, Nippon Medical School, Kawasaki, Kanagawa, Japan
[1]Corresponding author: e-mail address: a-oka@nms.ac.jp

Contents

Abstract

In the amphibian intestine during metamorphosis, most of the larval epithelial cells undergo apoptosis, whereas a small number of them survive. These cells dedifferentiate into stem cells through interactions with the microenvironment referred to as "stem cell niche" and generate the adult epithelium analogous to the mammalian counterpart. Since all processes of the larval-to-adult intestinal remodeling can be experimentally induced by thyroid hormone (TH) both *in vivo* and *in vitro*, the amphibian intestine provides us a valuable opportunity to study how adult stem cells and their niche are formed during postembryonic development. To address this issue, a number of expression and functional analyses of TH response genes have been intensely performed in the *Xenopus laevis* over the past two decades, by using organ culture and transgenic techniques. We here review recent progress in this field, focusing on key signaling pathways involved in establishment of the stem cell niche and discuss their evolutionarily conserved roles in the vertebrate intestine.

Current Topics in Developmental Biology, Volume 103
ISSN 0070-2153
http://dx.doi.org/10.1016/B978-0-12-385979-2.00011-3

1. INTRODUCTION

During amphibian metamorphosis, the digestive organs undergo rapid and extensive remodeling to adapt from the aquatic herbivorous to terrestrial carnivorous life. The major digestive organ, the small intestine dramatically shortens and develops multiple intestinal folds (Ishizuya-Oka & Shimozawa, 1987a; McAvoy & Dixon, 1977). The adult intestinal epithelium after metamorphosis acquires a cell renewal system along the trough–crest axis of the intestinal folds, similar to that along the crypt–villus axis in the adult mammalian intestine (Bjerknes & Cheng, 1981; Cheng & Bjerknes, 1985; Madara & Trier, 1994). This suggests that stem cells analogous to those in the adult mammalian intestine develop in the amphibian intestine during metamorphosis. In fact, in the *Xenopus laevis* intestine, undifferentiated cells expressing markers for mammalian intestinal stem cells become detectable at the onset of metamorphosis (Ishizuya-Oka & Shi, 2007). In the adult mammalian intestine, the microenvironment around stem cells called as the niche is generally known to play important roles in the maintenance of the stem cells throughout adulthood and are thus clinically important for tissue regeneration and/or cancer therapies. However, it still remains unknown how the adult stem cells and their niche are formed during normal development in any vertebrate intestine.

Amphibian metamorphosis can be easily manipulated by simply adding a single hormone, thyroid hormone (TH) to the tadpoles (Dodd & Dodd, 1976; Kikuyama, Kawamura, Tanaka, & Yamamoto, 1993; Shi, 1999). Our previous studies have shown that TH can organ-autonomously reproduce the whole process of larval-to-adult remodeling including development of the adult stem cells and their niche in the *X. laevis* intestine *in vitro* (Ishizuya-Oka, Shimizu, Sakakibara, Okano, & Ueda, 2003; Ishizuya-Oka & Shimozawa, 1992). Thus, in this animal model, TH response genes provide important clues to clarify molecular mechanisms regulating the adult stem cells. Until now, a number of TH response genes have been identified in the *X. laevis* intestine by subtractive differential screening (Amano & Yoshizato, 1998; Shi & Brown, 1993) and, more recently, cDNA microarrays (Buchholz, Heimeier, Das, Washington, & Shi, 2007; Heimeier, Das, Buchholz, Fiorentino, & Shi, 2010). Their functions can be now easily analyzed by using culture and transgenic (Tg) techniques. In this review, we will first summarize findings obtained by previous morphological observations and culture experiments on adult epithelial development in the *X. laevis* intestine during metamorphosis and then survey recent progress in expression and functional analyses of TH response genes, focusing on potential signaling pathways involved in establishment of the stem cell niche.

2. ADULT EPITHELIAL DEVELOPMENT DURING INTESTINAL REMODELING

The larval-to-adult remodeling has been well characterized in the *X. laevis* intestine at the cellular level. Throughout pre- and prometamorphosis, the long small intestine has a simple tubular structure with only a single longitudinal fold, the typhlosole, which is localized in the anterior part of the small intestine (Marshall & Dixon, 1978b). Histologically, the intestine mainly consists of a single layer of the primary (larval) epithelium, the immature connective tissue which is very thin except in the typhlosole, and thin layers of inner circular and outer longitudinal muscles (Fig. 11.1).

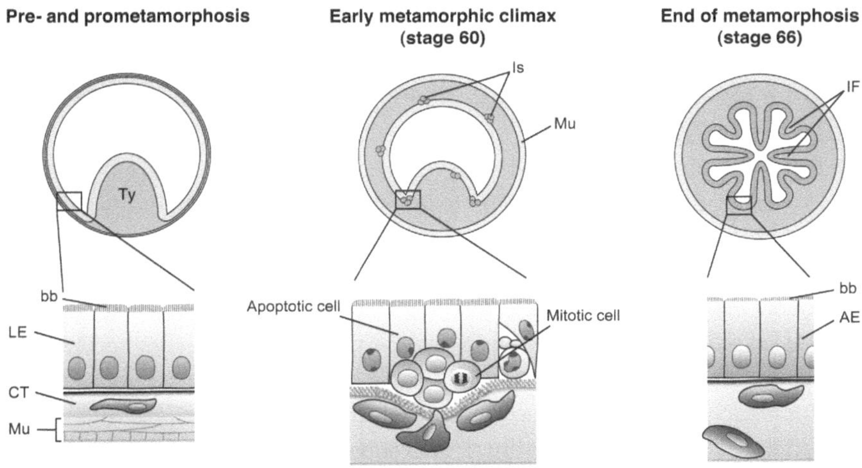

Figure 11.1 Larval-to-adult intestinal remodeling during *Xenopus laevis* metamorphosis. During pre- and prometamorphosis, the small intestine has a single fold, typhlosole (Ty), and consists of the larval epithelium (LE) possessing the brush border (bb), the immature connective tissue (CT), and thin layers of inner and outer muscles (Mu). At the early metamorphic climax (stage 60), most of the larval epithelial cells begin to undergo apoptosis, whereas adult progenitor/stem cells strongly stained red with pyronin Y appear as islets (Is) between the degenerating larval epithelium and the developing connective tissue. They actively proliferate and gradually replace the larval epithelial cells. Then, with the progress of intestinal fold formation, they differentiate into a single layer of the adult epithelium (AE) possessing the shorter brush border. After the end of metamorphosis (stage 66), the adult epithelium acquires a cell renewal system along the trough–crest axis of the intestinal folds (IF). (For interpretation of the references to color in this figure legend, the reader is referred to the online version of this chapter.)

The major epithelial cells are intestinal absorptive cells possessing the brush border that is much longer than that of the adult intestine (Bonneville, 1963; Fox, Bailey, & Mahoney, 1972). Although a small number of proliferating cells are randomly distributed in the larval epithelium (Ishizuya-Oka & Ueda, 1996; Marshall & Dixon, 1978a), no undifferentiated cells are morphologically identified by light and electron microscopy.

During metamorphosis, the small intestine rapidly shortens and remodels from the larval to adult form (Ishizuya-Oka & Shimozawa, 1987a; Marshall & Dixon, 1978b). At the early metamorphic climax (stage 60) (Nieuwkoop & Faber, 1967), when the plasma level of TH becomes high (Leloup & Buscaglia, 1977), most of the larval epithelial cells (larval proper cells) begin to undergo apoptosis (Ishizuya-Oka & Ueda, 1996). At the same time, a small number of undifferentiated cells become morphologically detectable as small islets between the larval proper cells and the connective tissue in the entire small intestine (Hourdry & Dauca, 1977; McAvoy & Dixon, 1977). These cells are strongly stained red with pyronin Y (PY) (Ishizuya-Oka & Shi, 2007), actively proliferate, and gradually replace the larval proper cells undergoing apoptosis (Ishizuya-Oka et al., 1997). Then, with the progress of intestinal fold formation, they differentiate into a single layer of the secondary (adult) epithelium (Hourdry & Dauca, 1977; McAvoy & Dixon, 1977). After the completion of metamorphosis (stage 66), the adult epithelium continuously undergoes cell renewal along the trough–crest axis of multiple intestinal folds (Shi & Ishizuya-Oka, 1996). That is, the epithelial cells proliferate in the trough region of the folds and, as they migrate upward, gradually differentiate into major absorptive cells, which possess the shorter brush border and express intestinal fatty acid binding protein (IFABP), goblet cells, and enteroendocrine cells (Ishizuya-Oka et al., 1997; McAvoy & Dixon, 1978). Finally, they undergo apoptosis at the tip of the folds (Ishizuya-Oka & Ueda, 1996). These chronological observations imply that the undifferentiated cells stained red with PY at stage 60 are adult progenitor cells that include multipotent stem cells. Consistent with this, previous immunohistochemical and *in situ* hybridization analyses have shown that the adult progenitor cells express sonic hedgehog (Shh) (Hasebe, Kajita, Shi, & Ishizuya-Oka, 2008; Ishizuya-Oka et al., 2001), Musashi-1 (Msi1) (Ishizuya-Oka et al., 2003), phosphorylated form of phosphatase and tensin homolog (P-PTEN), Akt (Ishizuya-Oka & Shi, 2007), and protein arginine methyltransferase 1 (Matsuda & Shi, 2010; Shi, Hasebe, Fu, Fujimoto, & Ishizuya-Oka, 2011), all of which are also expressed in adult stem cells and their descendants of the mammalian intestine (de Santa Barbara, van den Brink, & Roberts, 2003; He et al., 2007; Kayahara et al., 2003; Potten et al., 2003; van Den Brink, de Santa Barbara, & Roberts, 2001). More definitively, the orphan leucine-rich repeat-containing G protein-coupled receptor 5 (LGR5), a well-established stem cell marker in the adult mammalian intestine (Barker et al., 2007; Sato et al., 2009), has been shown to be specifically expressed in the adult progenitor cells of the *X. laevis* intestine, concomitantly with their appearance at stage 60 (Sun et al., 2010). Thereafter, as the intestinal folds form, the expression of LGR5 becomes localized in the trough region of the folds where the stem cells reside. These similarities in gene expression of the stem cell markers between the amphibian and mammalian intestines strongly suggest the usefulness of the *X. laevis* intestine as a model for the study of the adult stem cells.

3. ORIGIN OF ADULT STEM CELLS

To investigate mechanisms underlying the amphibian intestinal remodeling, we have previously established an organ culture system using *X. laevis* tadpoles at stage 57, when the small intestine is the longest throughout pre- and prometamorphosis. In this culture system, TH can organ-autonomously induce adult epithelial development *in vitro* as *in vivo* (Ishizuya-Oka & Shimozawa, 1992). Adult progenitor/stem cells are identified as small islets after 5 days, actively proliferate, and then differentiate into the adult absorptive epithelium expressing IFABP after 7 days. This indicates that the adult stem cells are derived from the tadpole intestine itself but not from other organs at latest at stage 57. If so, there are two possible origins of the stem cells: (1) the larval epithelial cells before metamorphic climax, as proposed by previous chronological observations (Amano, Noro, Kawabata, Kobayashi, & Yoshizato, 1998; Marshall & Dixon, 1978b; Schreiber, Cai, & Brown, 2005), or (2) nonepithelial cells that migrate into the epithelium after stage 57, as bone marrow-derived cells migrate into the epithelium during mammalian intestinal regeneration (Krause et al., 2001; Okamoto et al., 2002). Our recent study provided the first experimental evidence supporting the former by using Tg *X. laevis* tadpoles that constitutively express green fluorescent protein (GFP) for tissue recombinant culture experiments (Ishizuya-Oka et al., 2009). The larval epithelium isolated from wild-type (Wt) or GFP Tg intestines at stage 57 was recombined with homologous or heterologous nonepithelial tissues (non-E), and four kinds of recombinant intestines were cultured in the medium containing TH (Fig. 11.2A). In all the recombinant intestines, as in intact ones, adult progenitor/stem cells became detectable after 5 days and then differentiated into absorptive epithelial cells expressing IFABP

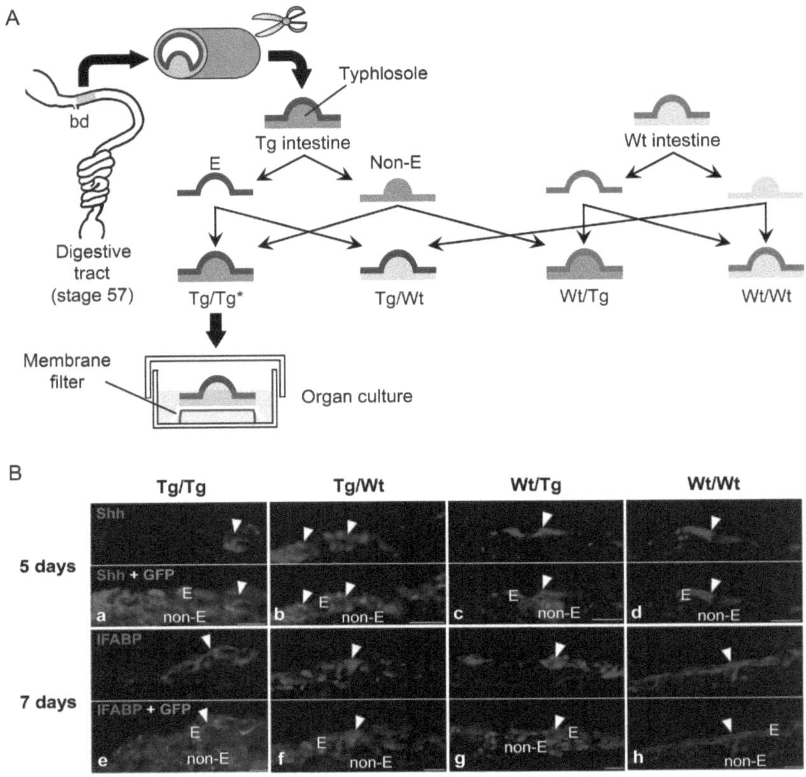

Figure 11.2 Adult stem cells originate from the larval epithelium in *X. laevis* intestines cultured *in vitro*. (A) Schematic diagram of tissue recombination and organ culture. Tubular fragments were isolated from the small intestine just behind the bile duct junction (bd) of stage 57-wild-type (Wt) or transgenic (Tg) tadpoles that express GFP ubiquitously under the CMV promoter. They were slit open lengthwise and separated into epithelial (E) or nonepithelial tissues (non-E). Each epithelium was then recombined with homologous or heterologous non-E. The four kinds of recombinant intestines were placed on membrane filters on grids and cultured in the medium containing thyroid hormone. * Type of E/type of non-E. (B) Recombinant intestines cultured for 5 (a–d) and 7 days (e–h). Sections were double-stained with anti-Shh (red) and anti-GFP (green) (a–d), or anti-IFABP (red) and anti-GFP (green) antibodies (e–h). Adult progenitor/stem cells positive for Shh (arrowheads) express GFP in Tg/Tg (a) and Tg/Wt intestines (b), but do not in Wt/Tg (c) and Wt/Wt intestines (d). Similarly, adult differentiated cells positive for IFABP (arrowheads) express GFP in Tg/Tg (e) and Tg/Wt intestines (f), but do not in Wt/Tg (g) and Wt/Wt intestines (h). Bars: 10 μm. (See Color Insert.)

after 7 days. Importantly, whenever the epithelium was derived from Tg intestine, both the adult progenitor/stem cells and their differentiated absorptive cells expressed GFP, whereas neither of them expressed GFP whenever the epithelium was derived from Wt intestine (Fig. 11.2B). These results indicate that the adult stem cells originate from the larval epithelium but not from the other non-E. However, previous light and electron microscopical studies reported that all of the epithelial cells in the tadpole intestine before stage 60 are essentially differentiated as larval type (Hourdry & Dauca, 1977; Marshall & Dixon, 1978b) and are immunohistochemically negative for any stem cell marker examined (Ishizuya-Oka et al., 2003, 2001). Taken together, it can be concluded that some of the larval epithelial cells dedifferentiate into the adult stem cells toward stage 60, similar to the mammalian epithelial "transit amplifying cells," which are partially differentiated but can dedifferentiate into adult stem cells during intestinal regeneration (Potten, 1998; Potten, Booth, & Pritchard, 1997). Supporting this conclusion, we recently found that the expression of nuclear lamins, which are intermediate filament proteins lining the inner nuclear membrane (Gruenbaum et al., 2003), changes in association with development of the adult stem cells in the *X. laevis* intestine (Hasebe, Kajita, Iwabuchi, Ohsumi, & Ishizuya-Oka, 2011). Before stage 60, almost all of the differentiated epithelial cells express lamin A (LA), whose expression is generally known to be specific to differentiated cells of mammalian organs (Rober, Weber, & Osborn, 1989) but not lamin LIII (LIII), an embryo-specific lamin (Doring & Stick, 1990). At the early metamorphic climax (stage 60), the expression of LA and LIII is down- and upregulated, respectively, only in the adult progenitor/stem cells concomitantly with their appearance. These expression profiles of LA and LIII are similar to those during mammalian somatic cell reprogramming induced by *X. laevis* egg extracts (Bru et al., 2008; Mitalipov et al., 2007; Miyamoto et al., 2007). Given that the larval epithelial cells dedifferentiate into the adult stem cells by the inductive action of TH, the next question arises whether TH acts on the larval epithelial cells directly or through interactions with the nonepithelial cells.

4. NICHE ESSENTIAL FOR ADULT STEM CELL DEVELOPMENT

It is noteworthy that adult stem cell development in the *X. laevis* intestine spatiotemporally correlates well with extensive changes in the surrounding connective tissue (Ishizuya-Oka & Shimozawa, 1987a; Marshall

& Dixon, 1978a). Throughout pre- and prometamorphosis, the connective tissue remains immature and very thin except in the typhlosole. When the adult stem cells appear at stage 60, the connective tissue suddenly increases in thickness, total cell number, and proliferative activity. In addition, remarkable ultrastructural changes occur in the connective tissue close to the epithelium. The thin and continuous basal lamina suddenly becomes thickened by vigorous folding and then amorphous (Fig. 11.3A and B; Ishizuya-Oka & Shimozawa, 1987b). Through such modified basal lamina, subepithelial fibroblasts that possess well-developed rough endoplasmic reticulum often make contacts with the adult progenitor/stem cells. Thereafter, as the adult epithelial cells differentiate, their basal lamina becomes thin again, and the cell contacts between the two tissues become rare. This modification of the basal lamina following the fibroblast–epithelial cell contacts can be reproduced in the tadpole intestine cultured *in vitro*, whenever the adult epithelium successfully develops (Ishizuya-Oka & Shimozawa, 1992). These observations

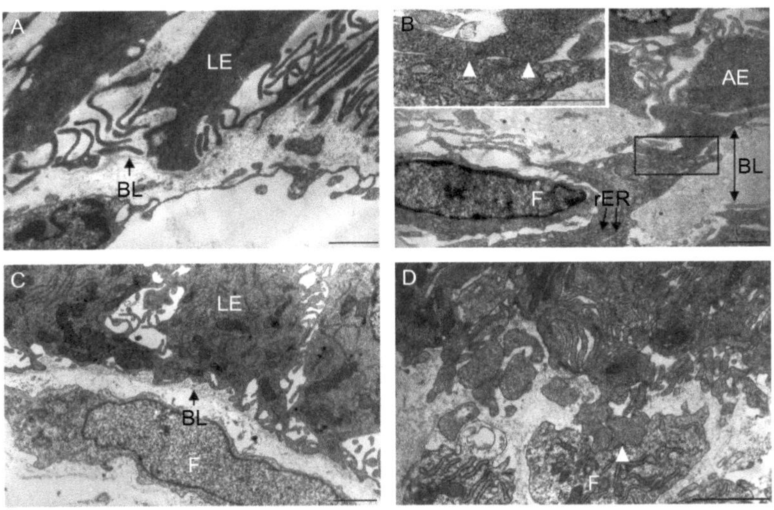

Figure 11.3 Epithelial–connective tissue interfaces in normal *X. laevis* intestines (A and B) and heat shock-treated intestines of wild-type (Wt) (C) or ST3 transgenic (Tg) tadpoles (D). Throughout pre- and prometamorphosis, the basal lamina (BL) underneath the larval epithelium (LE) is thin and continuous (A). At the start of metamorphic climax, the basal lamina becomes thickened through vigorous folding (B). Subepithelial fibroblasts (F) possessing well-developed rough endoplasmic reticulum (rER) often make contact with the adult progenitor/stem cells (AE) (inset; arrowheads). In premetamorphic (stage 54) Wt intestine, the basal lamina remains thin after heat shock (C). However, it becomes amorphous or absent in Tg intestine overexpressing ST3 (D), where subepithelial fibroblasts often make contact with the epithelium (arrowhead). Bars: 1 μm.

strongly suggest important roles of the epithelial–connective tissue interactions in adult epithelial development during amphibian intestinal remodeling just as epithelial–mesenchymal interactions during gut organogenesis of higher vertebrates (Yasugi & Mizuno, 2008) and during amphibian skin conversion (Yoshizato, 2007).

To demonstrate experimentally that the actions of TH on the connective tissue are essential for adult stem cell development during intestinal remodeling, we used Tg *X. laevis* tadpoles that express a dominant positive thyroid hormone receptor (dpTR) under the control of a heat shock-inducible promoter (Hasebe, Buchholz, Shi, & Ishizuya-Oka, 2011). In these Tg tadpoles, dpTR specifically binds to TH response elements within promoter regions of TH target genes and causes metamorphic changes *in vivo* without addition of TH (Buchholz, Tomita, Fu, Paul, & Shi, 2004). By making use of Wt and dpTR Tg intestines at stage 57 for tissue recombinant culture experiments, TH target genes can be specifically induced in either the epithelium or non-E or both in the absence of TH at any time by heat shock. As a result, whenever the epithelium was derived from Wt intestine, no stem cells were detected in any recombinant intestine. In contrast, whenever the epithelium was derived from Tg intestine, undifferentiated cells expressing Shh, which is the only stem cell marker known to be directly upregulated by TH (Stolow & Shi, 1995), became detectable after 5 days of cultivation with heat shock. More importantly, these cells expressed the other stem cell markers such as Msi1 when Tg epithelium was recombined with Tg non-E (Tg/Tg) but did not when it was recombined with Wt non-E (Tg/Wt). Then, after 7 days, differentiated absorptive epithelial cells expressing IFABP were detected in Tg/Tg intestine but not in Tg/Wt intestine. These results provide experimental evidence that, although some of the larval epithelial cells begin to express Shh by a direct action of TH, they require TH signaling in non-E to fully dedifferentiate into adult stem cells that generate the adult absorptive epithelium (Fig. 11.4). Thus, TH response genes expressed in non-E, other than the epithelium itself, are essential for stem cell development as the niche during *X. laevis* intestinal remodeling.

5. SIGNALING PATHWAYS INVOLVED IN ESTABLISHMENT OF STEM CELL NICHE

Previously, a large number of TH response genes have been identified in the *X. laevis* intestine by various PCR-based subtractive differential screens and cDNA microarrays (Amano & Yoshizato, 1998; Buchholz

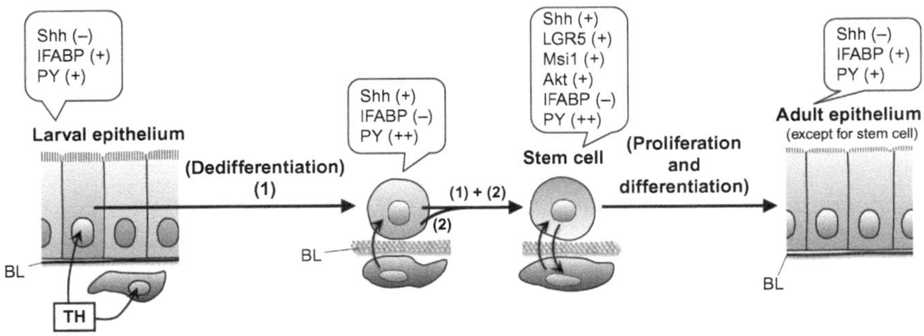

Figure 11.4 Schematic model showing adult stem cell development during amphibian intestinal remodeling. Cells that have a potency to dedifferentiate into adult stem cells reside in the larval epithelium before metamorphosis. By the direct action of thyroid hormone (TH), these cells partly dedifferentiate and begin to express Shh (1; intracellular pathway). However, they require TH signaling in nonepithelial tissues (2; extracellular pathway) to fully dedifferentiate into adult stem cells that express all of the stem cell markers including LGR5 and Msi1 and can generate the adult epithelium expressing IFABP. Thus, the cross talk with nonepithelial tissues is essential for adult stem cell development. BL, basal lamina; PY, pyronin Y. (For color version of this figure, the reader is referred to the online version of this chapter.)

et al., 2007; Heimeier et al., 2010; Shi & Brown, 1993). As mentioned earlier, expression and functional analyses of TH response genes should be informative for characterizing the niche factors. Although such analyses are now in progress, there is accumulating evidence that some key signaling pathways regulate the adult stem cells through interactions between the epithelium and the connective tissue.

5.1. MMPs and juxtacrine signaling

Matrix metalloproteinases (MMPs), a superfamily of Zn-dependent proteases, are generally known to degrade various extracellular matrix (ECM) components (Birkedal-Hansen et al., 1993; Matrisian, 1992; McCawley & Matrisian, 2001; Woessner, 1991). So far, a number of MMPs including collagenase-3, 4 (Stolow et al., 1996), gelatinase A, B (Fujimoto, Nakajima, & Yaoita, 2006), stromelysin-3 (ST3; MMP11) (Patterton, Hayes, & Shi, 1995), and membrane type-1-MMP (Hasebe, Hartman, Matsuda, & Shi, 2006) have been identified as TH response genes in the *X. laevis* intestine (Shi & Brown, 1993) and are noteworthy in connection with modification of the basal lamina consisting of ECM components. Among such MMPs, ST3 is the only one whose function has been demonstrated to modify the basal lamina structure both *in vivo* (Fu et al., 2005) and *in vitro* (Ishizuya-Oka et al., 2000).

TH directly upregulates the fibroblast-specific expression of ST3 (Fu, Tomita, Wang, Buchholz, & Shi, 2006; Patterton et al., 1995), which cleaves the 67-kDa receptor for laminin, a major ECM component of the basal lamina (Amano, Fu, Marshak, Kwak, & Shi, 2005), leading to degradation of the basal lamina and massive apoptosis of the larval epithelium (Mathew et al., 2009; Shi, Fu, Hasebe, & Ishizuya-Oka, 2007). Moreover, by using Tg tadpoles that express ST3 after heat shock, previous study has shown that ST3 expression alone causes modification of the basal lamina following frequent fibroblast–epithelial cell contacts, even in the absence of TH (Fig. 11.3C and D; Fu et al., 2005). This implies an essential role of ST3 for the cell contacts, although their biological significance is unclear.

In the *X. laevis* intestine during natural metamorphosis, the fibroblasts frequently make contact with only adult progenitor/stem cells but not with the other larval cells, while the basal lamina becomes amorphous underneath the entire epithelium. This suggests that a juxtacrine signaling caused by the cell contacts is involved in adult epithelial development but not in larval epithelial apoptosis. Similarly, in the mammalian intestine, previous study reported frequent epithelial–mesenchymal cell contacts only during intestinal maturation around birth (Mathan, Hermos, & Trier, 1972), when the adult stem cells develop (Harper, Mould, Andrews, Bikoff, & Robertson, 2011; Muncan et al., 2011). However, it still remains unknown to date what happens at the cell-to-cell contact sites in any vertebrate intestine. Given that neither the gap junction nor fusion of cell membranes are recognized at the contact sites of both amphibian (Ishizuya-Oka & Shimozawa, 1987b) and mammalian intestines (Mathan et al., 1972), it is possible that TH response genes encoding transmembrane proteins such as Notch (Buchholz et al., 2007) or transient receptor potential channels (our unpublished data) may be ones of the juxtacrine signaling components. If the genes that are specifically expressed at the contact sites will be identified in the near future, their functional analysis will pave a way to unravel molecular mechanisms underlying the cell contacts between the two tissues during postembryonic intestinal development.

5.2. Shh/BMP4 signaling pathway

As shown by tissue recombinant experiments using dpTR tadpoles mentioned earlier (Fig. 11.4), TH-upregulated expression of Shh is an early step toward epithelial dedifferentiation into the adult stem cells, suggesting its key role in this process. In fact, during natural metamorphosis, Shh is highly expressed in the adult epithelial progenitor/stem cells concomitantly with their appearance (Hasebe et al., 2008; Ishizuya-Oka et al., 2001). Shh is generally known to bind to a 12-transmembrane receptor, Patched (Ptc) (Fuse et al., 1999; Marigo, Davey, Zuo, Cunningham, & Tabin, 1996) and relieve Ptc-mediated inhibition of the activity of Smoothened (Smo), a second multipass membrane protein. This leads to the activation of the transcription factors Glis, which finally regulate Shh target genes (Ruiz i Altaba, 1999; Sasaki, Nishizaki, Hui, Nakafuku, & Kondoh, 1999; Villavicencio, Walterhouse, & Iannaccone, 2000). In the *X. laevis* intestine, the expression of all of the Shh pathway components examined, namely, Ptc1, Smo, and Gli1-3, has been shown to be upregulated by TH, although only Gli2, such as Shh, is considered to be a TH direct response gene (Hasebe, Kajita, Fu, Shi, & Ishizuya-Oka, 2012). In contrast to the epithelium-specific expression of Shh, their expression is specific for mesenchymal tissues as in the mammalian intestine (Kolterud et al., 2009). In addition, our organ culture study has shown that overexpression of Shh upregulates the expression of Ptc1, Smo, and all Glis even in the absence of TH. These expression profiles indicate that Shh by itself upregulates the expression of its pathway components and other target genes in a paracrine manner.

To investigate possible roles of Shh in larval-to-adult intestinal remodeling, we previously performed culture experiments using the *X. laevis* tadpole intestine (Ishizuya-Oka, Hasebe, Shimizu, Suzuki, & Ueda, 2006; Ishizuya-Oka et al., 2001). Exogenous Shh protein added to the TH-containing culture medium promoted cell proliferation in the connective tissue and muscles. In addition, Shh upregulated the expression of bone morphogenetic protein (BMP) 4 only in the connective tissue just as Shh upregulates the mesenchyme-specific expression of BMP4 in the avian embryonic gut (Roberts, Smith, Goff, & Tabin, 1998; Sukegawa et al., 2000). Further, prolonged addition of exogenous Shh to the medium until day 7, when the level of endogenous Shh expression becomes much lower, caused anomalies of the adult epithelial structure. The epithelium often failed to form a lumen as reported in the Shh-treated *X. laevis* embryonic gut (Zhang, Rosenthal, de Sauvage, & Shivdasani, 2001). This suggests that downregulation of the Shh expression is necessary for normal differentiation of the intestinal epithelium, although the action of Shh on the epithelium is indirect.

In contrast to Shh, exogenous BMP4 protein added to the culture medium suppressed cell proliferation in the connective tissue and caused precocious differentiation of the adult epithelium expressing IFABP in the *X. laevis* intestine (Ishizuya-Oka et al., 2006). It should be noted that addition of excessive

Chordin, an antagonist ofBMP4 (Piccolo, Sasai, Lu, & De Robertis, 1996), that is, lack of endogenous BMP4, leads to the reduction in proliferation and the total number of adult epithelial cells. This implies that a certain amount of BMP4 is required to maintain stem cells in the small intestine as in the case of mammalian embryonic stem cells and primordial germ cells (Fujiwara, Dunn, & Hogan, 2001; Qi et al., 2004; Ying, Nichols, Chambers, & Smith, 2003). Since a major type I receptor for BMP4, BMPR-1A is expressed in both the connective tissue and adult progenitor/stem cells (Ishizuya-Oka et al., 2006), it seems likely that the action of BMP4 on the adult epithelium is direct, differently from the indirect action of Shh. Further, the *Xenopus* homolog of *Drosophila* Tolloid closely related to BMP1 (Tolloid/BMP1), which regulates the activity of BMP4 by degrading Chordin in the early *X. laevis* embryos (Balemans & Van Hul, 2002; Blitz, Shimmi, Wunnenberg-Stapleton, O'Connor, & Cho, 2000; Wardle, Welch, & Dale, 1999), has also been identified as a TH response gene related BMP4 (Shimizu, Ishizuya-Oka, Amano, Yoshizato, & Ueda, 2002). Its expression is specific for the connective tissue and is considered to be directly upregulated by TH during intestinal remodeling. It is thus highly possible that the activity of BMP4 is spatiotemporally regulated by Tolloid/BMP1 through Chordin. These *in vitro* findings implicate Shh/BMP4 pathway as a key signaling essential for establishment of the stem cell niche. To clarify roles of Shh/BMP4 signaling in adult stem cell development more precisely, next step should be directed toward gain- or loss-of-function studies of this pathway *in vivo*, by using recent frog Tg technology (Rankin, Hasebe, Zorn, & Buchholz, 2009; Rankin, Zorn, & Buchholz, 2011).

5.3. Other signaling pathways

In the adult mammalian intestine, there is a growing body of evidence that canonical Wnt/β-catenin signaling pathway plays a central role in maintenance or proliferation of adult stem cells and their descendants (Medema & Vermeulen, 2011; Sato et al., 2009). Although roles of Wnt signaling pathways in the amphibian intestinal remodeling have not yet been examined, their components including β-catenin, Frizzled2, Wnt5a, and Ror2 have been identified as TH response genes in the *X. laevis* intestine (Buchholz et al., 2007). In addition, among TH response genes, there are many Wnt targets such as CD44 and Wnt regulators including R-spondin1, which amplifies canonical Wnt responses in the presence of LGR5 (Ootani et al., 2009), and secreted Frizzled-related protein 2, which is a so-called Wnt inhibitor (Bovolenta, Esteve, Ruiz, Cisneros, & Lopez-Rios, 2008) but has been shown to activate canonical Wnt pathway in the mammalian intestine (Kress, Rezza, Nadjar, Samarut, & Plateroti, 2009). Further, as mentioned, the adult progenitor/stem cells of the *X. laevis* intestine specifically express LGR5 (Sun et al., 2010), a Wnt target (Moore & Lemischka, 2006; Soubeyran et al., 2001; van de Wetering et al., 2002), and P-PTEN and Akt (Ishizuya-Oka & Shi, 2007), both of which are involved in the interplay between Wnt and BMP signals (He et al., 2007). These results strongly suggest the involvement of Wnt signaling pathways in adult stem cell development through interactions with other signals including BMP (He et al., 2004).

Notch signaling pathway is also known to be activated in the adult stem cells of the adult mammalian intestine (van Den Brink et al., 2001) and play a fundamental role in their maintenance (Fre et al., 2005; Pellegrinet et al., 2011). Considering that Msi1, a positive regulator of Notch pathway by suppressing translational expression of Numb (Okano et al., 2005), is specifically expressed in the adult progenitor/stem cells of the *X. laevis* intestine (Ishizuya-Oka et al., 2003), it is highly possible that Notch signaling pathway is required to develop and/or maintain the adult stem cells in the amphibian intestine. While Paneth cells express Notch ligand Dll4 in the adult mammalian intestine (Sato et al., 2011; Vooijs, Liu, & Kopan, 2011), they are absent in the amphibian intestine. It raises an intriguing question of what cells express Notch ligands in this amphibian model. Expression and functional analyses of these Notch- and Wnt-related TH response genes during amphibian intestinal remodeling await further investigation.

Taken together, one possible scenario at present is that TH upregulates Shh expression in some larval epithelial cells (precursors of adult stem cells) and ST3 expression in fibroblasts. Shh then directly acts only on mesenchymal tissues in a paracrine fashion and upregulates the expression of Shh target genes including Ptc, Smo, Glis, and BMP4. In turn, BMP4 proteins signal back to the epithelium under the local control of Tolloid/BMP1/Chordin signaling not only to maintain the adult progenitor/stem cells in an undifferentiated state but also to promote differentiation of their descendants. However, ST3 causes the basal lamina modification, which in turn allows contacts between the adult progenitor/stem cells and subepithelial fibroblasts, leading to the juxtacrine signaling involved in adult epithelial development. These TH-induced Shh/BMP4 and ST3 signaling pathways interact with Wnt and/or Notch ones and are integrated to establish the stem cell niche, which enables some of larval epithelial cells to completely dedifferentiate into the adult stem cells that newly generate the intestinal absorptive epithelium (Fig. 11.5).

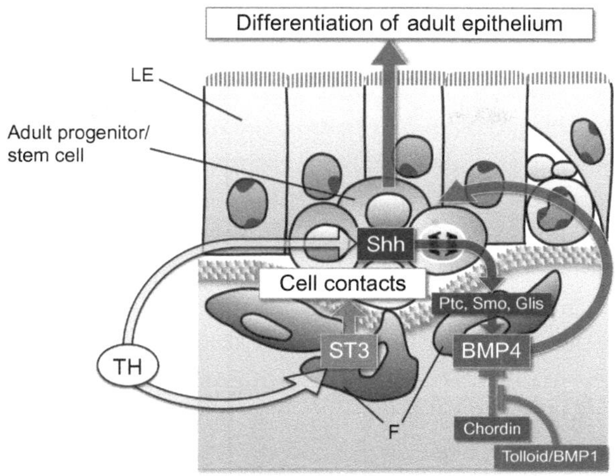

Figure 11.5 Thyroid hormone-induced signaling pathways leading to establishment of the stem cell niche in the amphibian intestine during metamorphosis. Thyroid hormone (TH) directly upregulates Shh expression specifically in adult progenitor/stem cells and ST3 expression in fibroblasts (F). Shh acts only on nonepithelial tissues in a paracrine manner and, together with TH, enhances the expression of other Shh pathway components including Ptc, Smo, and Glis. Shh not only activates proliferation of the fibroblasts but also upregulates their expression of BMP4, which in turn promotes differentiation of the adult epithelium and/or maintains the adult stem cells under the control of TH-upregulated expression of Tolloid/BMP1 through degrading Chordin. However, ST3 modifies the basal lamina and allows the fibroblast to make contact with the epithelial cells, leading to activation of juxtacrine signaling involved in adult stem cell development. In addition, these Shh/BMP4 and ST3 signaling pathways interact with other pathways such as Wnt one and are integrated to establish the stem cell niche. Arrows and T-shaped bars indicate activation and repression, respectively. (See Color Insert.)

6. EVOLUTIONARY CONSIDERATION AND PROSPECT

Amphibian metamorphosis shares common characteristics with mammalian birth, since the both phenomena are closely associated with adaptation from aquatic to terrestrial life (Shi, 1999; Tata, 1993). Indeed, as the amphibian intestine remodels during metamorphosis, the mammalian intestine matures for the dietary transition during the peak of endogenous TH levels around birth (Plateroti, Kress, Mori, & Samarut, 2006; Sirakov & Plateroti, 2011). The expression of both ST3 (Heimeier et al., 2010) and Shh (Kolterud et al., 2009) is also significantly upregulated in the mammalian intestine, concomitantly with the increase of the TH levels. It should be noted that TH-upregulated expression of Shh is specific for intervillus pockets where the adult stem cells develop in the mammalian intestine (Kolterud et al., 2009). This supports our proposal that Shh may function as an early key player to establish the stem cell niche common to the amphibian and mammalian intestines. More importantly, adult stem cells develop from fetal epithelial cells during mammalian intestinal maturation (Harper et al., 2011; Muncan et al., 2011) similar to those from the larval epithelial cells during amphibian intestinal remodeling (Ishizuya-Oka et al., 2009). Subsequently, the descendants of the adult stem cells in both mammalian and amphibian intestines replace the preexisting suckling-type or the larval epithelial cell by active proliferation and differentiation to generate the adult epithelium possessing a cell renewal system. These similarities strongly suggest the existence of evolutionarily conserved mechanisms underlying development of the adult stem cells and their niche (Ishizuya-Oka & Shi, 2011). What is unique to this amphibian model is that the stem cell niche can be experimentally induced by TH at any time both *in vitro* and *in vivo* and be studied through the use of numerous TH response genes. By taking advantages of these facts, it is an interesting and challenging issue to clarify how the stem cell niche is formed through cross talks between Shh and other paracrine and/or juxtacrine signaling pathways.

TH is generally known to be essential not only for amphibian metamorphosis but also homeostasis of frog organs as well as adult mammalian ones (Yoshizato, 2007). Although the identified TH response genes are still limited in the adult mammalian intestine (Kress, Samarut, & Plateroti, 2009), there is a growing body of evidence that signaling molecules induced by TH in the amphibian intestine are also expressed in the mammalian intestine to maintain its epithelial cell renewal. For example, Shh is specifically expressed in the crypt where adult stem cells reside throughout adulthood (Crosnier, Stamataki, & Lewis, 2006; de Santa Barbara et al., 2003; Varnat, Zacchetti, & Ruiz i Altaba, 2010), and its expression is upregulated when the stem cells actively proliferate during mammalian intestinal regeneration (Berman et al., 2003; Nielsen, Williams, van den Brink, Lauwers, & Roberts, 2004). In addition, many studies reported intestinal diseases caused by mutations in components of Shh/BMP4 signaling pathway (Batts, Polk, Dubois, & Kulessa, 2006; Beachy, Karhadkar, & Berman, 2004; Berman et al., 2003; Haramis et al., 2004; He et al., 2007, 2004; Litingtung, Lei, Westphal, & Chiang, 1998), although their functions on the adult stem cells have not yet been clarified enough. Further, β-catenin gene, one of TH response genes identified in the *X. laevis* intestine (Buchholz et al., 2007), has been shown to contain TH response elements in the mammalian intestine, suggesting the direct action of TH on canonical Wnt/β-catenin pathway (Plateroti et al., 2006). It is thus interesting from the standpoints of both evolution and stem cell biology to know how far molecular mechanisms regulating the adult stem cells are depend on TH and are conserved among amphibian intestinal remodeling, mammalian intestinal maturation, and homeostasis of the both intestines during adulthood. Their comparative studies should help us to fully elucidate the molecular nature of the intestinal stem cell niche, which can serve as a base for regenerative and cancer therapies.

ACKNOWLEDGMENTS

We thank Dr. Yun-Bo Shi for our longtime research collaboration and Emeritus Professor Takeo Mizuno for his continuous encouragement. This work was supported in part by the JSPS Grants-in-Aid for Scientific Research (C) (Grant number 24570078 to A. I.-O.).

REFERENCES

Amano, T., Fu, L., Marshak, A., Kwak, O., & Shi, Y.-B. (2005). Spatio-temporal regulation and cleavage by matrix metalloproteinase stromelysin-3 implicate a role for laminin receptor in intestinal remodeling during *Xenopus laevis* metamorphosis. *Developmental Dynamics, 234*, 190–200.

Amano, T., Noro, N., Kawabata, H., Kobayashi, Y., & Yoshizato, K. (1998). Metamorphosis-associated and region-specific expression of calbindin gene in the posterior intestinal epithelium of *Xenopus laevis* larva. *Development, Growth & Differentiation, 40*, 177–188.

Amano, T., & Yoshizato, K. (1998). Isolation of genes involved in intestinal remodeling during anuran metamorphosis. *Wound Repair and Regeneration, 6*, 302–313.

Balemans, W., & Van Hul, W. (2002). Extracellular regulation of BMP signaling in vertebrates: A cocktail of modulators. *Developmental Biology, 250*, 231–250.

Barker, N., van Es, J. H., Kuipers, J., Kujala, P., van den Born, M., Cozijnsen, M., et al. (2007). Identification of stem cells in small intestine and colon by marker gene Lgr5. *Nature, 449*, 1003–1007.

Batts, L. E., Polk, D. B., Dubois, R. N., & Kulessa, H. (2006). Bmp signaling is required for intestinal growth and morphogenesis. *Developmental Dynamics, 235*, 1563–1570.

Beachy, P. A., Karhadkar, S. S., & Berman, D. M. (2004). Tissue repair and stem cell renewal in carcinogenesis. *Nature, 432*, 324–331.

Berman, D. M., Karhadkar, S. S., Maitra, A., Montes De Oca, R., Gerstenblith, M. R., Briggs, K., et al. (2003). Widespread requirement for Hedgehog ligand stimulation in growth of digestive tract tumours. *Nature, 425*, 846–851.

Birkedal-Hansen, H., Moore, W. G., Bodden, M. K., Windsor, L. J., Birkedal-Hansen, B., DeCarlo, A., et al. (1993). Matrix metalloproteinases: A review. *Critical Reviews in Oral Biology and Medicine, 4*, 197–250.

Bjerknes, M., & Cheng, H. (1981). The stem-cell zone of the small intestinal epithelium. IV. Effects of resecting 30% of the small intestine. *The American Journal of Anatomy, 160*, 93–103.

Blitz, I. L., Shimmi, O., Wunnenberg-Stapleton, K., O'Connor, M. B., & Cho, K. W. (2000). Is chordin a long-range- or short-range-acting factor? Roles for BMP1-related metalloproteases in chordin and BMP4 autofeedback loop regulation. *Developmental Biology, 223*, 120–138.

Bonneville, M. A. (1963). Fine structural changes in the intestinal epithelium of the bullfrog during metamorphosis. *The Journal of Cell Biology, 18*, 579–597.
Bovolenta, P., Esteve, P., Ruiz, J. M., Cisneros, E., & Lopez-Rios, J. (2008). Beyond Wnt inhibition: New functions of secreted Frizzled-related proteins in development and disease. *Journal of Cell Science, 121*, 737–746.
Bru, T., Clarke, C., McGrew, M. J., Sang, H. M., Wilmut, I., & Blow, J. J. (2008). Rapid induction of pluripotency genes after exposure of human somatic cells to mouse ES cell extracts. *Experimental Cell Research, 314*, 2634–2642.
Buchholz, D. R., Heimeier, R. A., Das, B., Washington, T., & Shi, Y.-B. (2007). Pairing morphology with gene expression in thyroid hormone-induced intestinal remodeling and identification of a core set of TH-induced genes across tadpole tissues. *Developmental Biology, 303*, 576–590.
Buchholz, D. R., Tomita, A., Fu, L., Paul, B. D., & Shi, Y.-B. (2004). Transgenic analysis reveals that thyroid hormone receptor is sufficient to mediate the thyroid hormone signal in frog metamorphosis. *Molecular and Cellular Biology, 24*, 9026–9037.
Cheng, H., & Bjerknes, M. (1985). Whole population cell kinetics and postnatal development of the mouse intestinal epithelium. *Anatomical Record, 211*, 420–426.
Crosnier, C., Stamataki, D., & Lewis, J. (2006). Organizing cell renewal in the intestine: Stem cells, signals and combinatorial control. *Nature Reviews. Genetics*, 7, 349–359.
de Santa Barbara, P., van den Brink, G. R., & Roberts, D. J. (2003). Development and differentiation of the intestinal epithelium. *Cellular and Molecular Life Sciences, 60*, 1322–1332.
Dodd, M. H. I., & Dodd, J. M. (1976). The biology of metamorphosis. In B. Lofts (Ed.), *Physiology of amphibia* (pp. 467–599). New York: Academic Press.
Doring, V., & Stick, R. (1990). Gene structure of nuclear lamin LIII of *Xenopus laevis*; a model for the evolution of IF proteins from a lamin-like ancestor. *The EMBO Journal, 9*, 4073–4081.
Fox, H., Bailey, E., & Mahoney, R. (1972). Aspects of the ultrastructure of the alimentary canal and respiratory ducts in *Xenopus laevis* larvae. *Journal of Morphology, 138*, 387–405.
Fre, S., Huyghe, M., Mourikis, P., Robine, S., Louvard, D., & Artavanis-Tsakonas, S. (2005). Notch signals control the fate of immature progenitor cells in the intestine. *Nature, 435*, 964–968.
Fu, L., Ishizuya-Oka, A., Buchholz, D. R., Amano, T., Matsuda, H., & Shi, Y.-B. (2005). A causative role of stromelysin-3 in extracellular matrix remodeling and epithelial apoptosis during intestinal metamorphosis in *Xenopus laevis*. *The Journal of Biological Chemistry, 280*, 27856–27865.
Fu, L., Tomita, A., Wang, H., Buchholz, D. R., & Shi, Y.-B. (2006). Transcriptional regulation of the *Xenopus laevis* Stromelysin-3 gene by thyroid hormone is mediated by a DNA element in the first intron. *The Journal of Biological Chemistry, 281*, 16870–16878.
Fujimoto, K., Nakajima, K., & Yaoita, Y. (2006). One of the duplicated matrix metalloproteinase-9 genes is expressed in regressing tail during anuran metamorphosis. *Development, Growth & Differentiation, 48*, 223–241.
Fujiwara, T., Dunn, N. R., & Hogan, B. L. (2001). Bone morphogenetic protein 4 in the extraembryonic mesoderm is required for allantois development and the localization and survival of primordial germ cells in the mouse. *Proceedings of the National Academy of Sciences of the United States of America, 98*, 13739–13744.
Fuse, N., Maiti, T., Wang, B., Porter, J. A., Hall, T. M., Leahy, D. J., et al. (1999). Sonic hedgehog protein signals not as a hydrolytic enzyme but as an apparent ligand for patched. *Proceedings of the National Academy of Sciences of the United States of America, 96*, 10992–10999.
Gruenbaum, Y., Goldman, R. D., Meyuhas, R., Mills, E., Margalit, A., Fridkin, A., et al. (2003). The nuclear lamina and its functions in the nucleus. *International Review of Cytology, 226*, 1–62.

Haramis, A. P., Begthel, H., van den Born, M., van Es, J., Jonkheer, S., Offerhaus, G. J., et al. (2004). De novo crypt formation and juvenile polyposis on BMP inhibition in mouse intestine. *Science*, *303*, 1684–1686.

Harper, J., Mould, A., Andrews, R. M., Bikoff, E. K., & Robertson, E. J. (2011). The transcriptional repressor Blimp1/Prdm1 regulates postnatal reprogramming of intestinal enterocytes. *Proceedings of the National Academy of Sciences of the United States of America*, *108*, 10585–10590.

Hasebe, T., Buchholz, D. R., Shi, Y.-B., & Ishizuya-Oka, A. (2011). Epithelial-connective tissue interactions induced by thyroid hormone receptor are essential for adult stem cell development in the *Xenopus laevis* intestine. *Stem Cells*, *29*, 154–161.

Hasebe, T., Hartman, R., Matsuda, H., & Shi, Y.-B. (2006). Spatial and temporal expression profiles suggest the involvement of gelatinase A and membrane type 1 matrix metalloproteinase in amphibian metamorphosis. *Cell and Tissue Research*, *324*, 105–116.

Hasebe, T., Kajita, M., Fu, L., Shi, Y. B., & Ishizuya-Oka, A. (2012). Thyroid hormone-induced sonic hedgehog signal up-regulates its own pathway in a paracrine manner in the *Xenopus laevis* intestine during metamorphosis. *Developmental Dynamics*, *241*, 403–414.

Hasebe, T., Kajita, M., Iwabuchi, M., Ohsumi, K., & Ishizuya-Oka, A. (2011). Thyroid hormone-regulated expression of nuclear lamins correlates with dedifferentiation of intestinal epithelial cells during *Xenopus laevis* metamorphosis. *Development Genes and Evolution*, *221*, 199–208.

Hasebe, T., Kajita, M., Shi, Y.-B., & Ishizuya-Oka, A. (2008). Thyroid hormone-up-regulated hedgehog interacting protein is involved in larval-to-adult intestinal remodeling by regulating sonic hedgehog signaling pathway in *Xenopus laevis*. *Developmental Dynamics*, *237*, 3006–3015.

He, X. C., Yin, T., Grindley, J. C., Tian, Q., Sato, T., Tao, W. A., et al. (2007). PTEN-deficient intestinal stem cells initiate intestinal polyposis. *Nature Genetics*, *39*, 189–198.

He, X. C., Zhang, J., Tong, W. G., Tawfik, O., Ross, J., Scoville, D. H., et al. (2004). BMP signaling inhibits intestinal stem cell self-renewal through suppression of Wnt-β-catenin signaling. *Nature Genetics*, *36*, 1117–1121.

Heimeier, R. A., Das, B., Buchholz, D. R., Fiorentino, M., & Shi, Y.-B. (2010). Studies on *Xenopus laevis* intestine reveal biological pathways underlying vertebrate gut adaptation from embryo to adult. *Genome Biology*, *11*, R55.

Hourdry, J., & Dauca, M. (1977). Cytological and cytochemical changes in the intestinal epithelium during anuran metamorphosis. *International Review of Cytology. Supplement*, *5*, 337–385.

Ishizuya-Oka, A., Hasebe, T., Buchholz, D. R., Kajita, M., Fu, L., & Shi, Y.-B. (2009). Origin of the adult intestinal stem cells induced by thyroid hormone in *Xenopus laevis*. *The FASEB Journal*, *23*, 2568–2575.

Ishizuya-Oka, A., Hasebe, T., Shimizu, K., Suzuki, K., & Ueda, S. (2006). Shh/BMP-4 signaling pathway is essential for intestinal epithelial development during *Xenopus* larval-to-adult remodeling. *Developmental Dynamics*, *235*, 3240–3249.

Ishizuya-Oka, A., Li, Q., Amano, T., Damjanovski, S., Ueda, S., & Shi, Y.-B. (2000). Requirement for matrix metalloproteinase stromelysin-3 in cell migration and apoptosis during tissue remodeling in *Xenopus laevis*. *The Journal of Cell Biology*, *150*, 1177–1188.

Ishizuya-Oka, A., & Shi, Y.-B. (2007). Regulation of adult intestinal epithelial stem cell development by thyroid hormone during *Xenopus laevis* metamorphosis. *Developmental Dynamics*, *236*, 3358–3368.

Ishizuya-Oka, A., & Shi, Y. B. (2011). Evolutionary insights into postembryonic development of adult intestinal stem cells. *Cell & Bioscience*, *1*, 37.

Ishizuya-Oka, A., Shimizu, K., Sakakibara, S., Okano, H., & Ueda, S. (2003). Thyroid hormone-upregulated expression of Musashi-1 is specific for progenitor cells of the adult

epithelium during amphibian gastrointestinal remodeling. *Journal of Cell Science, 116*, 3157–3164.
Ishizuya-Oka, A., & Shimozawa, A. (1987a). Development of the connective tissue in the digestive tract of the larval and metamorphosing *Xenopus laevis*. *Anatomischer Anzeiger, 164*, 81–93.
Ishizuya-Oka, A., & Shimozawa, A. (1987b). Ultrastructural changes in the intestinal connective tissue of *Xenopus laevis* during metamorphosis. *Journal of Morphology, 193*, 13–22.
Ishizuya-Oka, A., & Shimozawa, A. (1992). Connective tissue is involved in adult epithelial development of the small intestine during anuran metamorphosis in vitro. *Roux's Archives of Developmental Biology, 201*, 322–329.
Ishizuya-Oka, A., & Ueda, S. (1996). Apoptosis and cell proliferation in the *Xenopus* small intestine during metamorphosis. *Cell and Tissue Research, 286*, 467–476.
Ishizuya-Oka, A., Ueda, S., Damjanovski, S., Li, Q., Liang, V. C., & Shi, Y.-B. (1997). Anteroposterior gradient of epithelial transformation during amphibian intestinal remodeling: Immunohistochemical detection of intestinal fatty acid-binding protein. *Developmental Biology, 192*, 149–161.
Ishizuya-Oka, A., Ueda, S., Inokuchi, T., Amano, T., Damjanovski, S., Stolow, M., et al. (2001). Thyroid hormone-induced expression of sonic hedgehog correlates with adult epithelial development during remodeling of the *Xenopus* stomach and intestine. *Differentiation, 69*, 27–37.
Kayahara, T., Sawada, M., Takaishi, S., Fukui, H., Seno, H., Fukuzawa, H., et al. (2003). Candidate markers for stem and early progenitor cells, Musashi-1 and Hes1, are expressed in crypt base columnar cells of mouse small intestine. *FEBS Letters, 535*, 131–135.
Kikuyama, S., Kawamura, K., Tanaka, S., & Yamamoto, K. (1993). Aspects of amphibian metamorphosis: Hormonal control. *International Review of Cytology, 145*, 105–148.
Kolterud, A., Grosse, A. S., Zacharias, W. J., Walton, K. D., Kretovich, K. E., Madison, B. B., et al. (2009). Paracrine Hedgehog signaling in stomach and intestine: New roles for hedgehog in gastrointestinal patterning. *Gastroenterology, 137*, 618–628.
Krause, D. S., Theise, N. D., Collector, M. I., Henegariu, O., Hwang, S., Gardner, R., et al. (2001). Multi-organ, multi-lineage engraftment by a single bone marrow-derived stem cell. *Cell, 105*, 369–377.
Kress, E., Rezza, A., Nadjar, J., Samarut, J., & Plateroti, M. (2009). The frizzled-related sFRP2 gene is a target of thyroid hormone receptor alpha1 and activates β-catenin signaling in mouse intestine. *The Journal of Biological Chemistry, 284*, 1234–1241.
Kress, E., Samarut, J., & Plateroti, M. (2009). Thyroid hormones and the control of cell proliferation or cell differentiation: Paradox or duality? *Molecular and Cellular Endocrinology, 313*, 36–49.
Leloup, J., & Buscaglia, M. (1977). La triiodothryronine: Hormone de la metamorphose des amphibians. *Comptes Rendus de l'Académie des Sciences, 284*, 2261–2263.
Litingtung, Y., Lei, L., Westphal, H., & Chiang, C. (1998). Sonic hedgehog is essential to foregut development. *Nature Genetics, 20*, 58–61.
Madara, J. L., & Trier, J. S. (1994). Functional morphology of the mucosa of the small intestine. In L. R. Johnson (Ed.), *Physiology of the gastrointestinal tract* (pp. 1577–1622). (3rd ed). New York: Raven Press Ltd.
Marigo, V., Davey, R. A., Zuo, Y., Cunningham, J. M., & Tabin, C. J. (1996). Biochemical evidence that patched is the Hedgehog receptor. *Nature, 384*, 176–179.
Marshall, J. A., & Dixon, K. E. (1978a). Cell proliferation in the intestinal epithelium of *Xenopus laevis* tadpoles. *The Journal of Experimental Zoology, 203*, 31–40.
Marshall, J. A., & Dixon, K. E. (1978b). Cell specialization in the epithelium of the small intestine of feeding *Xenopus laevis* tadpoles. *Journal of Anatomy, 126*, 133–144.

Mathan, M., Hermos, J. A., & Trier, J. S. (1972). Structural features of the epithelio-mesenchymal interface of rat duodenal mucosa during development. *The Journal of Cell Biology, 52*, 577–588.

Mathew, S., Fu, L., Fiorentino, M., Matsuda, H., Das, B., & Shi, Y. B. (2009). Differential regulation of cell type-specific apoptosis by stromelysin-3: A potential mechanism via the cleavage of the laminin receptor during tail resorption in *Xenopus laevis*. *The Journal of Biological Chemistry, 284*, 18545–18556.

Matrisian, L. M. (1992). The matrix-degrading metalloproteinases. *Bioessays, 14*, 455–463.

Matsuda, H., & Shi, Y.-B. (2010). An essential and evolutionarily conserved role of protein arginine methyltransferase 1 for adult intestinal stem cells during postembryonic development. *Stem Cells, 28*, 2073–2083.

McAvoy, J. W., & Dixon, K. E. (1977). Cell proliferation and renewal in the small intestinal epithelium of metamorphosing and adult *Xenopus laevis*. *The Journal of Experimental Zoology, 202*, 129–138.

McAvoy, J. W., & Dixon, K. E. (1978). Cell specialization in the small intestinal epithelium of adult *Xenopus laevis*: Structural aspects. *Journal of Anatomy, 125*, 155–169.

McCawley, L. J., & Matrisian, L. M. (2001). Matrix metalloproteinases: They're not just for matrix anymore!. *Current Opinion in Cell Biology, 13*, 534–540.

Medema, J. P., & Vermeulen, L. (2011). Microenvironmental regulation of stem cells in intestinal homeostasis and cancer. *Nature, 474*, 318–326.

Mitalipov, S. M., Zhou, Q., Byrne, J. A., Ji, W. Z., Norgren, R. B., & Wolf, D. P. (2007). Reprogramming following somatic cell nuclear transfer in primates is dependent upon nuclear remodeling. *Human Reproduction, 22*, 2232–2242.

Miyamoto, K., Furusawa, T., Ohnuki, M., Goel, S., Tokunaga, T., Minami, N., et al. (2007). Reprogramming events of mammalian somatic cells induced by *Xenopus laevis* egg extracts. *Molecular Reproduction and Development, 74*, 1268–1277.

Moore, K. A., & Lemischka, I. R. (2006). Stem cells and their niches. *Science, 311*, 1880–1885.

Muncan, V., Heijmans, J., Krasinski, S. D., Buller, N. V., Wildenberg, M. E., Meisner, S., et al. (2011). Blimp1 regulates the transition of neonatal to adult intestinal epithelium. *Nature Communications, 2*, 452.

Nielsen, C. M., Williams, J., van den Brink, G. R., Lauwers, G. Y., & Roberts, D. J. (2004). Hh pathway expression in human gut tissues and in inflammatory gut diseases. *Laboratory Investigation, 84*, 1631–1642.

Nieuwkoop, P. D., & Faber, J. (1967). *Normal table of* Xenopus laevis *(Daudin)*. Amsterdam: North-Holland.

Okamoto, R., Yajima, T., Yamazaki, M., Kanai, T., Mukai, M., Okamoto, S., et al. (2002). Damaged epithelia regenerated by bone marrow-derived cells in the human gastrointestinal tract. *Nature Medicine, 8*, 1011–1017.

Okano, H., Kawahara, H., Toriya, M., Nakao, K., Shibata, S., & Imai, T. (2005). Function of RNA-binding protein Musashi-1 in stem cells. *Experimental Cell Research, 306*, 349–356.

Ootani, A., Li, X., Sangiorgi, E., Ho, Q. T., Ueno, H., Toda, S., et al. (2009). Sustained in vitro intestinal epithelial culture within a Wnt-dependent stem cell niche. *Nature Medicine, 15*, 701–706.

Patterton, D., Hayes, W. P., & Shi, Y.-B. (1995). Transcriptional activation of the matrix metalloproteinase gene stromelysin-3 coincides with thyroid hormone-induced cell death during frog metamorphosis. *Developmental Biology, 167*, 252–262.

Pellegrinet, L., Rodilla, V., Liu, Z., Chen, S., Koch, U., Espinosa, L., et al. (2011). Dll1- and dll4-mediated notch signaling are required for homeostasis of intestinal stem cells. *Gastroenterology, 140*, 1230–1240.

Piccolo, S., Sasai, Y., Lu, B., & De Robertis, E. M. (1996). Dorsoventral patterning in *Xenopus*: Inhibition of ventral signals by direct binding of chordin to BMP-4. *Cell, 86*, 589–598.

Plateroti, M., Kress, E., Mori, J. I., & Samarut, J. (2006). Thyroid hormone receptor α1 directly controls transcription of the β-catenin gene in intestinal epithelial cells. *Molecular and Cellular Biology*, *26*, 3204–3214.

Potten, C. S. (1998). Stem cells in gastrointestinal epithelium: Numbers, characteristics and death. *Philosophical Transactions of the Royal Society of London. Series B, Biological Sciences*, *353*, 821–830.

Potten, C. S., Booth, C., & Pritchard, D. M. (1997). The intestinal epithelial stem cell: The mucosal governor. *International Journal of Experimental Pathology*, *78*, 219–243.

Potten, C. S., Booth, C., Tudor, G. L., Booth, D., Brady, G., Hurley, P., et al. (2003). Identification of a putative intestinal stem cell and early lineage marker; musashi-1. *Differentiation*, *71*, 28–41.

Qi, X., Li, T. G., Hao, J., Hu, J., Wang, J., Simmons, H., et al. (2004). BMP4 supports self-renewal of embryonic stem cells by inhibiting mitogen-activated protein kinase pathways. *Proceedings of the National Academy of Sciences of the United States of America*, *101*, 6027–6032.

Rankin, S. A., Hasebe, T., Zorn, A. M., & Buchholz, D. R. (2009). Improved cre reporter transgenic *Xenopus*. *Developmental Dynamics*, *238*, 2401–2408.

Rankin, S. A., Zorn, A. M., & Buchholz, D. R. (2011). New doxycycline-inducible transgenic lines in *Xenopus*. *Developmental Dynamics*, *240*, 1467–1474.

Rober, R. A., Weber, K., & Osborn, M. (1989). Differential timing of nuclear lamin A/C expression in the various organs of the mouse embryo and the young animal: A developmental study. *Development*, *105*, 365–378.

Roberts, D. J., Smith, D. M., Goff, D. J., & Tabin, C. J. (1998). Epithelial-mesenchymal signaling during the regionalization of the chick gut. *Development*, *125*, 2791–2801.

Ruiz i Altaba, A. (1999). Gli proteins encode context-dependent positive and negative functions: Implications for development and disease. *Development*, *126*, 3205–3216.

Sasaki, H., Nishizaki, Y., Hui, C., Nakafuku, M., & Kondoh, H. (1999). Regulation of Gli2 and Gli3 activities by an amino-terminal repression domain: Implication of Gli2 and Gli3 as primary mediators of Shh signaling. *Development*, *126*, 3915–3924.

Sato, T., van Es, J. H., Snippert, H. J., Stange, D. E., Vries, R. G., van den Born, M., et al. (2011). Paneth cells constitute the niche for Lgr5 stem cells in intestinal crypts. *Nature*, *469*, 415–418.

Sato, T., Vries, R. G., Snippert, H. J., van de Wetering, M., Barker, N., Stange, D. E., et al. (2009). Single Lgr5 stem cells build crypt-villus structures in vitro without a mesenchymal niche. *Nature*, *459*, 262–265.

Schreiber, A. M., Cai, L., & Brown, D. D. (2005). Remodeling of the intestine during metamorphosis of *Xenopus laevis*. *Proceedings of the National Academy of Sciences of the United States of America*, *102*, 3720–3725.

Shi, Y.-B. (1999). *Amphibian metamorphosis: From morphology to molecular biology*. New York: John Wiley & Sons, Inc.

Shi, Y.-B., & Brown, D. D. (1993). The earliest changes in gene expression in tadpole intestine induced by thyroid hormone. *The Journal of Biological Chemistry*, *268*, 20312–20317.

Shi, Y.-B., Fu, L., Hasebe, T., & Ishizuya-Oka, A. (2007). Regulation of extracellular matrix remodeling and cell fate determination by matrix metalloproteinase stromelysin-3 during thyroid hormone-dependent post-embryonic development. *Pharmacology & Therapeutics*, *116*, 391–400.

Shi, Y. B., Hasebe, T., Fu, L., Fujimoto, K., & Ishizuya-Oka, A. (2011). The development of the adult intestinal stem cells: Insights from studies on thyroid hormone-dependent amphibian metamorphosis. *Cell & Bioscience*, *1*, 30.

Shi, Y.-B., & Ishizuya-Oka, A. (1996). Biphasic intestinal development in amphibians: Embryogenesis and remodeling during metamorphosis. *Current Topics in Developmental Biology*, *32*, 205–235.

Shimizu, K., Ishizuya-Oka, A., Amano, T., Yoshizato, K., & Ueda, S. (2002). Isolation of connective-tissue-specific genes involved in *Xenopus* intestinal remodeling: Thyroid

hormone up-regulates Tolloid/BMP-1 expression. *Development Genes and Evolution, 212*, 357–364.
Sirakov, M., & Plateroti, M. (2011). The thyroid hormones and their nuclear receptors in the gut: From developmental biology to cancer. *Biochimica et Biophysica Acta, 1812*, 938–946.
Soubeyran, P., Haglund, K., Garcia, S., Barth, B. U., Iovanna, J., & Dikic, I. (2001). Homeobox gene Cdx1 regulates Ras, Rho and PI3 kinase pathways leading to transformation and tumorigenesis of intestinal epithelial cells. *Oncogene, 20*, 4180–4187.
Stolow, M. A., Bauzon, D. D., Li, J., Sedgwick, T., Liang, V. C., Sang, Q. A., et al. (1996). Identification and characterization of a novel collagenase in *Xenopus laevis*: Possible roles during frog development. *Molecular Biology of the Cell, 7*, 1471–1483.
Stolow, M. A., & Shi, Y.-B. (1995). *Xenopus* sonic hedgehog as a potential morphogen during embryogenesis and thyroid hormone-dependent metamorphosis. *Nucleic Acids Research, 23*, 2555–2562.
Sukegawa, A., Narita, T., Kameda, T., Saitoh, K., Nohno, T., Iba, H., et al. (2000). The concentric structure of the developing gut is regulated by Sonic hedgehog derived from endodermal epithelium. *Development, 127*, 1971–1980.
Sun, G., Hasebe, T., Fujimoto, K., Lu, R., Fu, L., Matsuda, H., et al. (2010). Spatio-temporal expression profile of stem cell-associated gene LGR5 in the intestine during thyroid hormone-dependent metamorphosis in *Xenopus laevis*. *PLoS One, 5*, e13605.
Tata, J. R. (1993). Gene expression during metamorphosis: An ideal model for post-embryonic development. *Bioessays, 15*, 239–248.
van de Wetering, M., Sancho, E., Verweij, C., de Lau, W., Oving, I., Hurlstone, A., et al. (2002). The beta-catenin/TCF-4 complex imposes a crypt progenitor phenotype on colorectal cancer cells. *Cell, 111*, 241–250.
van Den Brink, G. R., de Santa Barbara, P., & Roberts, D. J. (2001). Epithelial cell differentiation—A matter of choice. *Science, 294*, 2115–2116.
Varnat, F., Zacchetti, G., & Ruiz i Altaba, A. (2010). Hedgehog pathway activity is required for the lethality and intestinal phenotypes of mice with hyperactive Wnt signaling. *Mechanisms of Development, 127*, 73–81.
Villavicencio, E. H., Walterhouse, D. O., & Iannaccone, P. M. (2000). The sonic hedgehog-patched-gli pathway in human development and disease. *American Journal of Human Genetics, 67*, 1047–1054.
Vooijs, M., Liu, Z., & Kopan, R. (2011). Notch: Architect, landscaper, and guardian of the intestine. *Gastroenterology, 141*, 448–459.
Wardle, F. C., Welch, J. V., & Dale, L. (1999). Bone morphogenetic protein 1 regulates dorsal-ventral patterning in early *Xenopus* embryos by degrading chordin, a BMP4 antagonist. *Mechanisms of Development, 86*, 75–85.
Woessner, J. F., Jr. (1991). Matrix metalloproteinases and their inhibitors in connective tissue remodeling. *The FASEB Journal, 5*, 2145–2154.
Yasugi, S., & Mizuno, T. (2008). Molecular analysis of endoderm regionalization. *Development, Growth & Differentiation, 50*, S79–S96.
Ying, Q. L., Nichols, J., Chambers, I., & Smith, A. (2003). BMP induction of Id proteins suppresses differentiation and sustains embryonic stem cell self-renewal in collaboration with STAT3. *Cell, 115*, 281–292.
Yoshizato, K. (2007). Molecular mechanism and evolutional significance of epithelial-mesenchymal interactions in the body- and tail-dependent metamorphic transformation of anuran larval skin. *International Review of Cytology, 260*, 213–260.
Zhang, J., Rosenthal, A., de Sauvage, F. J., & Shivdasani, R. A. (2001). Downregulation of Hedgehog signaling is required for organogenesis of the small intestine in *Xenopus*. *Developmental Biology, 229*, 188–202.

CHAPTER TWELVE

Developmental Programs and Endocrine Disruption in Frog Metamorphosis: The Perspective from Microarray Analysis

Saurabh S. Kulkarni, Daniel R. Buchholz[1]
Department of Biological Sciences, University of Cincinnati, Cincinnati, Ohio, USA
[1]Corresponding author: e-mail address: buchhodr@ucmail.uc.edu

Contents

Abstract

A major goal for understanding the role of thyroid hormone (TH) in development has been to identify genes regulated by TH in different tissues during frog metamorphosis. The exquisite dependence of metamorphosis on TH also provides a model to study TH endocrine disruption. To identify such TH-regulated genes and select biomarkers for TH endocrine disruption, global gene expression analyses in tadpoles using microarrays have been done in 21 studies, involving five frog species, seven organs, and four endocrine disrupting chemicals. As expected, each organ has a unique set of genes

Current Topics in Developmental Biology, Volume 103
ISSN 0070-2153
http://dx.doi.org/10.1016/B978-0-12-385979-2.00012-5

associated with its tissue-specific metamorphic outcome, and functions ascribed to many of these genes correspond to histological changes induced by TH. Also, the large number of transcription factors identified in microarrays is consistent with the molecular mechanisms of TH action. On the other hand, microarray analysis has also revealed interesting findings not predicted from previous morphological or molecular studies. Furthermore, endocrine disruption studies identified candidate biomarkers for TH disruption, and the mechanisms of action of several endocrine disrupting chemicals have been examined. The microarray studies described here have produced a wealth of data on gene expression that requires further functional studies to elucidate the roles of these genes in development and endocrine disruption.

1. INTRODUCTION

All vertebrates require thyroid hormone (TH) for normal development, especially for the central nervous system (Howdeshell, 2002). The actions of hormones on development are particularly striking in frogs, where metamorphosis represents one of the most remarkable sets of hormone-dependent changes known among chordates (Just, Kraus-Just, & Check, 1981). The role of hormones in development is more easily studied in free-living tadpoles compared to uterus-enclosed mammalian embryos/fetuses because interpretations of hormonal manipulations in developing mammals are confounded by effects on the mother (Tata, 1993). Thus, the exquisite dependence of frog metamorphosis on TH and ease of analysis compared to mammals lends itself not only to studies of normal hormone-dependent development but also to studies of disturbed development due to endocrine disrupting chemicals.

Metamorphosis transforms the tadpole from one morphotype to another, involving extensive changes in morphology, biochemistry, and physiology (Dodd & Dodd, 1976). For example, tadpoles possess tails for swimming, their intestine is adapted for an herbivorous diet, and brain and skin are larval in character. During metamorphosis, the tail is completely resorbed, intestine, brain, and skin undergo extensive remodeling, and limbs grow out. In addition, the liver changes from ammonotelic to ureotelic metabolism of nitrogenous waste, and blood cells transform from larval to adult versions. TH is responsible for initiating all of these events of metamorphosis. If tadpoles are treated via surgical or chemical thyroidectomy such that they cannot make TH, they grow larger in size as a tadpole but do not undergo metamorphic changes. Conversely, TH added to the rearing water of premetamorphic tadpoles initiates all of the metamorphic changes that would have occurred later.

TH acts to accomplish these metamorphic changes predominantly if not exclusively via the nuclear receptors TRα and TRβ (Das et al., 2010). TRs are ligand-activated transcription factors that alter expression levels of genes responsible for metamorphosis (Buchholz, Paul, Fu, & Shi, 2006). Using transgenic animals, overexpression of a dominant negative form of TR that cannot bind TH inhibits metamorphosis (Buchholz, Hsia, Fu, & Shi, 2003). On the other hand, overexpression of a constitutively active TR (that does not require TH to induce T3-response genes) initiates metamorphic transformation in the absence of TH (Buchholz, Tomita, Fu, Paul, & Shi, 2004). TRs regulate gene expression by binding thyroid response elements (TREs) in promoter regions of TH-response genes. TRs have at least two modes to regulate genes, in that TH-response genes can have positive or negative TREs. On positive TREs, the presence of TH induces gene expression, whereas TH represses expression of genes containing negative TREs. Some TH-response genes with TREs have been identified as transcription factors that have gene targets of their own. Thus, TH induces TH-direct response genes, some of which subsequently regulate other genes, all of which comprises a gene regulation cascade, ultimately leading to metamorphosis.

Since the identification of TRs as nuclear receptors, more than 75 years after the discovery that TH induces frog metamorphosis (Gudernatsch, 1912; Yaoita, Shi, & Brown, 1990), a major research direction has been to identify the genes and pathways underlying TH-dependent metamorphosis. A unifying question for studies on frog metamorphosis is how a single molecule, TH, can trigger such varied developmental outcomes among tissues of the tadpole. The modern paradigm for answering this question centers around the molecular mechanisms underlying how different tissues can have TH-induced tissue-specific gene regulation cascades. A critical component to answering this question is to identify genes comprising these gene regulation cascades in different tissues.

At first, genes induced by TH were identified one by one based on their biochemical activities. For example, tail regression involved protease digestion of intra- and extracellular components, such as collagen, and thus collagenases and cathepsins were found (Yoshizato, 1996). The biochemistry of the liver undergoes metamorphic changes in nitrogen metabolism, such that urea cycle enzymes were found to be TH-regulated (Atkinson, Helbing, & Chen, 1996). Furthermore, distinct hemoglobins were found between tadpoles and frogs, and the adult hemoglobin genes were induced by TH (Weber, 1996). In a radically new approach, subtractive hybridization was employed to identify differentially regulated genes where *a priori* knowledge of protein activity was not required (Wang & Brown, 1991). These screens found between 14 and 34 genes in tail, hind limb, intestine, and brain (Buckbinder & Brown, 1992; Denver, Pavgi, & Shi, 1997; Shi & Brown, 1993; Wang & Brown, 1993). Massive sequencing efforts on ESTs, cDNA libraries, and genomic DNA in frogs led to increasing numbers of genes known, ultimately making virtually all genes known. Microarray technology allows the expression of many or all genes to be examined simultaneously in a single experiment (Fig. 12.1). The first microarray for frog

Array platforms

	cDNA/PCR product array	Agilent oligoarray	Affymetrix GeneChip
Probe source	PCR products	Oligosynthesis	Oligosynthesis
Probe length	200–600 base pairs	60 base pairs	25 base pairs
Chip production	Custom-built, spotted on membrane or slide	Commercial, *in situ* synthesis on slide	Commercial, *in situ* synthesis on slide
No. of transcripts	~400–1000 selected	~21,500, *X. laevis*	~14,400, *X. laevis* ~29,900, *X. laevis* 2.0 ~51,000, *X. tropicalis*

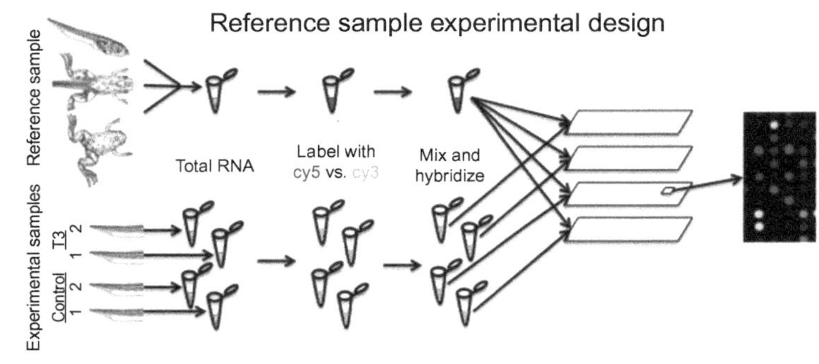

Figure 12.1 Types of microarray platforms and example microarray experimental design. Top panel indicates characteristics of the microarray platforms used in studies of frog metamorphosis. Lower panel depicts an experimental design using a reference sample, commonly used with the Agilent platform. The reference sample is a single preparation of RNA from a mixture of tissues and stages and labeled with one color (e.g., the red fluorescent dye, cy5). The experimental samples, including treatment and control samples, are labeled with a different color (e.g., the green fluorescent dye, cy3). Each experimental sample is mixed with a portion of the reference sample and hybridized to the microarray slide, resulting in each spot on the array with levels of red and green fluorescence corresponding to the level of expression of the RNA in the reference versus experimental sample. Expression differences among experimental samples are determined based on expression levels relative to the reference sample. The other platforms commonly use a one-color design and compare intensity values directly among samples. (For interpretation of the references to color in this figure legend, the reader is referred to the online version of this chapter.)

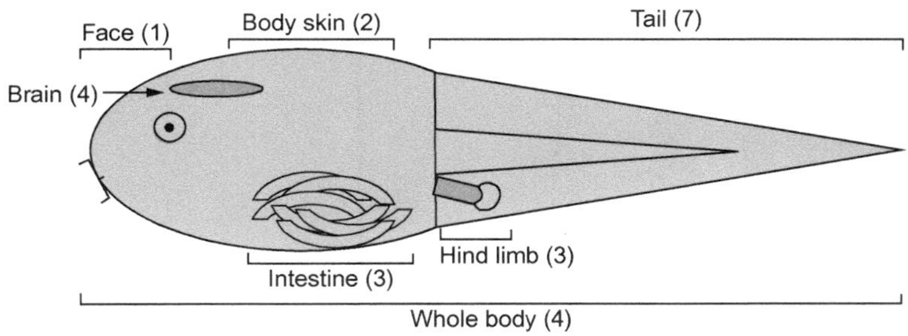

Species
Xenopus laevis (16)
Xenopus tropicalis (1)
Rana catesbiana (1)
Rana pirica (2)
Bufo marinus (1)

Endocrine disruptors
Bisphenol A (1)
Atrazine (2)
Polychlory biphenols (1)
Acetochlor (1)

Figure 12.2 Summary diagram of tadpole microarray studies. A total of 21 studies have been carried out on tadpoles using the five species and seven body regions indicated. The numbers in parentheses indicate the number of studies using the associated body region, species, or endocrine disrupting chemical.

metamorphosis was composed of 420 cDNAs, chosen from available sequences in GenBank in 2001 whose products were believed to be important in frog embryogenesis and metamorphosis (Crump, Werry, Veldhoen, Van Aggelen, & Helbing, 2002). With advances in microarray technology, genome-wide analysis of gene expression for more than 20,000 genes during metamorphosis became possible (Das et al., 2006).

The use of microarray technology in studies of frog metamorphosis is the subject of this review (see Fig. 12.2 and Table 12.1 for summaries of these studies). Two main goals have motivated microarray analysis in frog metamorphosis, namely, developmental genetic analysis of the TH gene regulation cascade and biomarker discovery and mechanism of action for endocrine disruption studies. In the first case, researchers want to identify all gene expression changes associated with a particular developmental stage or experimental treatment as a means toward understanding underlying pathways and mechanisms of tissue-specific TH-dependent metamorphosis. We discuss progress made using microarrays to understand developmental programs of tail, brain, hind limb, intestine, and skin. In the second case, researchers wish to identify a gene or genes robustly associated with TH signaling to serve as a reliable gauge of TH endocrine disruption. Also, several studies have used microarrays to examine the mechanism of action of TH disrupting chemicals, including bisphenol A, atrazine, polychlorinated biphenols, and acetochlor. Additional

Table 12.1 Summary of tadpole microarray studies (21 publications)

Tissue/species	Stages[a], conditions	No. of genes	Genes/pathways (upregulated, *downregulated*)	Platform	References
Whole body/ *X. laevis*	NF46, 14 h, ±100 nM T3, ±cycloheximide	188 up, 249 down in T3 + CHX versus control	Cell communication, transcription, protein degradation-dependent signaling, cell cycle, *cell differentiation*, *cytoskeletal biogenesis*, *cell division*	Agilent oligoarray, 21,495 Unigene clusters	Das et al. (2009)
Whole body/ *X. laevis*	400 ppb atrazine from NF43–NF62	44 in males, 44 in females	Ketone body metabolism, urea cycle, amino acid metabolism, *glycolysis/gluconeogenesis*, *fatty acid elongation*	Affymetrix, 14,400 elements	Langerveld et al. (2009), Zaya et al. (2011)
Whole body/ *Bufo marinus*	G29/30, 33/34, 38/40, 40–41, 42/43, 46	871 total; 20 up, 17 down in G46 versus G42	Induced during metamorphosis: hemoglobin α, β, harderin-like, apolipoprotein A1, γ-crystallin, serum albumin, gastrokine, trefoil factor, olfactomedin	cDNA array—5170 random cDNAs from toadlet library	Halliday et al. (2008)
Tail/*X. laevis*	NF47, 52, 56, 58, 60, 62	77 genes, 13 clusters	Transcription, metabolism, apoptosis, signal transduction, transport/binding	MAGEX cDNA[b], 420 cDNAs	Veldhoen et al. (2002)
Tail/*Rana catesbiana*	G31–33 tail tips, 10 nM T3, 0, 1, 2 days	92 genes	*PCNA*, K-ras, caspase 6, Na/dicarboxylate transporter, transcription, intracellular signaling, cell fate determination	MAGEX cDNA[b], 401 cDNAs	Veldhoen et al. (2006)

Tail/*X. laevis*	NF54, 100 nM T3, 0, 1, 2 days	1500 up, 700 down	Intracellular signaling, hydrolase activity, peptidase activity, cell death, *energy pathways*, *glycolysis*, *TCA cycle*, *mitochondrion*, *electron transport*	Agilent oligoarray, 21,654 Unigene clusters	Das et al. (2006)
Tail/*X. laevis*	NF54, 0, 2 days, T3[d], T4, MMI, PTU, PER	1–11 across treatments, time points	Metallothionein as biomarker for MMI	MAGEX cDNA[b], 434 cDNAs	Helbing et al. (2007)
Tail/*X. laevis*	NF54, 20 nM T3, 6, 48 h	6 h–177 up, 125 down; 48 h—432 up, 421 down	Genes at 6- and 48-h time points mostly not overlapping, enriched GO categories in metabolic processes; TRβ, deiodinase II, CRH-binding protein, fibroblast-activating protein α as biomarkers for TH	Affymetrix, 14,400 elements	Searcy et al. (2012)
Tail/*X. tropicalis*	NF54, 10 nM T3, 100 nM CORT, CORT + T3, 18 h	TH only—1636, cort only—573, CORTTH only—1208	TH only: proteases, apoptosis, transport, *amino acid metabolism*, CORT only: mitochondrion, intracellular localization and transport, CORTTH only: protein localization and transport	Affymetrix, 59,000 elements	Kulkarni and Buchholz (2012)
Tail/*X. laevis*	NF52/54, 100 nM T3, 10 nM acetochlor, T3 + AC, 48 h	26 genes in T3 versus T3 + AC	T3-induced genes further increased by acetochlor (e.g., gene 12, TH/bZIP, collagenase 3, KLF9, deiodinase III), non-T3-response genes induced by acetochlor only in the presence	MAGEX cDNA, 420 cDNAs	Crump et al. (2002)

Continued

Table 12.1 Summary of tadpole microarray studies (21 publications)—cont'd

Tissue/species	Stages, conditions	No. of genes	Genes/pathways (upregulated, *downregulated*)	Platform	References
			of T3 (e.g., Mad2, frizzled 7, caspase-9, nuclear factor I-C1, vitellogenin receptor, natriuretic peptide receptor type C), genes whose T3 repression was inhibited by acetochlor (e.g., 51-kDa cytokeratin type I, Genes 17, 18, 19, glutamine synthetase, Gli2)		
Intestine/*X. laevis*	NF53, 10 nM T3, 0, 1, 3, 6 days	Up—255, 710, 1040; down—280, 552, 585	Early genes—proteolysis, *cell cycle*; late genes—transcription, development, ECM, *glucose metabolism*, *transport*, *digestion*, *mitochondrion*	Agilent oligoarray, 21,495 Unigene clusters	Buchholz et al. (2007)
Intestine/*X. laevis*	NF53, 58, 61, 66	Up—463, 2613, 981; down—280, 552, 585	In chronological order: immune response, *catabolism*; transcription, signal transduction, cell–cell signaling, *metabolic processes*, *mitochondrion*; organismal process, development, *catalytic activity*, *RNA processing*	Agilent oligoarray, 21,495 Unigene clusters	Heimeier et al., (2010)
Intestine/*X. laevis*	NF54, 2 nM T3, 10 μM BPA, T3 + BPA, 4 days	T3—1051 up, 728 down; BPA—507 up, 290 down; T + B—1032 up, 738 down	Attenuated regulation of 58% of genes regulated by T3 occurs in the presence of BPA and 57% of the top 100 T3-regulated genes	Agilent oligoarray, 21,495 Unigene clusters	Heimeier et al. (2009)

Hind limb/ *X. laevis*	NF54, 100 nM T3, 0, 1, 2 days	2400 up, 1200 down	Cell cycle, MCM genes, RNA/DNA metabolism, electron transport	Agilent oligoarray, 21,654 elements	Das et al. (2006)
Hind limb/ *X. laevis*	Dox-inducible pCAR:TRDN[c]	2 weeks—42 up, 67 down; 6 weeks—1401 up, 1915 down	Nonmuscle genes unchanged, calsequestrin involved in transgenic muscle-wasting phenotype	Agilent oligoarray, 21,654 Unigene clusters	Cai et al. (2007)
Hind limb/ *X. laevis*	NF54, 0, 1, 4 days, T3[d], T4, MMI, PTU, PER	10–14 across treatments, time points	*RNA helicase II/Gu* as biomarkers for T4, metallothionein as biomarker for MMI, importin as biomarker for MMI and PER	MAGEX cDNA[b], 434 cDNAs	Helbing et al. (2007)
Brain/ *X. laevis*	NF54, 100 nM T3, 0, 1, 2 days	2000 up, 1000 down	Cell cycle, MCM genes, RNA/DNA metabolism, electron transport, Notch, Otx2	Agilent oligoarray, 21,654 Unigene clusters	Das et al. (2006)
Brain/*X. laevis*	NF54, 0, 1, 2, 4 days, T3[d], T4, MMI, PTU, PER	9–66 across treatments, time points	T3 versus T4 shows different expression profiles; MMI, PTU, PER-altered gene regulation by 24 h; cyclin H as biomarker for 24 h T3, *ferrochelatase* as biomarker for 48 h T4	MAGEX cDNA[b], 434 cDNAs	Helbing et al. (2007)

Continued

Table 12.1 Summary of tadpole microarray studies (21 publications)—cont'd

Tissue/species	Stages, conditions	No. of genes	Genes/pathways (upregulated, *downregulated*)	Platform	References
Brain/*X. laevis*	NF53/4, ±1 nM T3, 500 nM PCB, 4 days	526 T3 genes affected by PCB	Enriched GO terms: growth, brain development, cell proliferation, cell adhesion, neuron differentiation, apoptosis	Agilent oligoarray, 43,803 elements	Ishihara et al. (2011)
Hind brain/ *X. laevis*	NF53, recover from cut spine, ±1 mM MMI, 3, 7, 21 days	3 days—202, 7 days—331, 21 days—275, 50% up and down	72 annotated induced genes associated with recovery: transcription factors, membrane proteins, intracellular signaling, coregulators. Top 4 genes: SOCS2, XER81, neurotensin, MORC3	Affymetrix, 29,900 elements	Gibbs et al. (2011)
Body skin/ *X. laevis*	NF54/55, 5 nM T3; 0, 1, 3, 5, 7 days	240 up, 161 down	Genes: stromelysin-3, collagenase, TH/bZIP, adult keratin, *larval keratin*, *Genes 17–20*. Functional groups: transcription, proteolysis at 1 day, cell cycle at 3 days, defense response at 7 days	Affymetrix, 14,400 elements	Suzuki et al. (2009)

Body skin/*Rana pririca*	10 days ± predator from G25	8 up, 11 down	Bullous pemphigoid antigen, uromodulin, aldehyde dehydrogenase, *carboxypeptidase B*, *trypsinogen elastase I and II*, *fibrinogen*	297 PCR products from subtractive hybridization	Mori et al. (2005)
Face and brain/ *Rana pirica*	0, 6 h, 4 days, 8 days ± predator from G25	Not reported	Pirica (uromodulin-like gene) associated with predator-induced bulgy morphology, *type I and II keratins*	1020 PCR products from subtractive hybridization	Mori et al. (2009)

[a]NF is Nieuwkoop and Faber staging for *Xenopus*, and G is Gosner staging for other tadpoles.
[b]MAGEX is multispecies array for gene expression, with cDNA sequences representing approximately 90% *X. laevis* and 10% *R. catesbiana* selected for involvement in embryogenesis and metamorphosis.
[c]pCAR:TRDN is dominant negative TR expressed only in muscle cells only in the presence of doxycycline. NF53 tadpoles were treated for 2 or 6 weeks on Dox and harvested at NF56 and 61, respectively. Dox-treated samples were compared to nontransgenic NF56 and NF66 limbs.
[d]0.5–8 nM T3, 10–40 nM T4, 100 mg/L methimazole (MMI), 20 mg/L 6-propylthiouracil (PTU), and 4 mg/L sodium perchlorate (PER).

uses of microarray during frog metamorphosis have included identifying genes underlying phenotypic plasticity of predator-induced polyphenism (*Rana pirica*), finding genes associated with spinal cord regeneration in tadpoles, and looking for possible targets for pest control in *Bufo marinus*.

2. DEVELOPMENTAL PROGRAMS

2.1. Tail

The tail, being large and easily accessible, was the first organ to which microarray techniques were applied (Crump et al., 2002; Veldhoen, Crump, Werry, & Helbing, 2002). Much was already known about histological and biochemical effects of TH, where apoptosis and proteolysis play a predominant role in tail resorption (Yoshizato, 1996). The notochord and tail fin have collagen degraded by proteinases from fibroblasts. The muscle cells and tail fin epithelium die by apoptosis. Both *Xenopus laevis* and *Rana catesbiana* have been analyzed using a cDNA array with over 400 cDNAs. Several global expression analyses have also been carried out with oligoarrays representing many thousands of transcripts (Table 12.1).

Two tail gene expression programs were identified (Das et al., 2006). The tail hydrolytic program (tail fibroblasts switch from synthetic to hydrolytic genes) and muscle program (upregulation of proteases different from those in fibroblasts). As expected, genes involved in protein degradation were upregulated, such as matrix metalloproteases, lysosomal cathepsins, hyaluronidase, and therefore match histological and biochemical events of tail resorption. Also, cell death genes were upregulated, for example, caspase-6, and cell proliferation genes, for example, PCNA, were downregulated. Also as expected, a major program downregulated in tail fibroblasts was the synthesis of several kinds of collagen and extra cellular matrix proteins. Unexpectedly, some structural proteins were upregulated in the face of tissue resorption, including fibronectin, tubulin, actin, myosin heavy chain 3, and a keratin. The explanation for increases in these structural proteins awaits further investigation. Perhaps less obvious but predicted because TH acts through nuclear receptors is that genes involved in transcriptional regulation experience altered TH-induced expression levels, such as TRβ, TH/bZIP, KLF9, and SOX4. One study compared gene expression changes between 48 h of TH treatment and tails at metamorphic climax (Das et al., 2006). One-third of the genes on the array that were involved in muscle contraction were also downregulated at climax. Interestingly, genes most upregulated at climax and 48 h were largely overlapping, but genes most downregulated at climax and 48 h were very different from each other. The basis of this observation is not understood but may relate to different hormone environments.

Transcriptional repression of most genes encoding the steps leading to glycolysis, electron transport chain (including most of the genes that encode the ATP synthase complex), and tricarboxylic acid cycle genes are downregulated in tadpole tail muscle, which is consistent with cell death and tail resorption. These major metabolic pathways are shut off in all of the tail muscle at climax when the muscle fibers appear to be healthy and before the upregulation of the proteolytic enzymes and the morphological evidence of dying fibers. This global downregulation is followed by the more localized upregulation pattern that coincides with the activation of caspase-3. The upregulation of proteins involved in transport (e.g., Na^+/dicarboxylate transporter) and proteins of the ATP Na/K pump are perhaps required to support altered metabolism and apoptosis, but elucidating their actual role requires functional studies.

Many regulated genes with various biological functions did not have predicted roles in metamorphosis, because their biological activities did not match known physiological or morphological or biochemical phenomena. One of the largest groups of upregulated tail genes is signal transduction, including MAPK phosphatase, TGF-β pathway, GTPase-mediated signaling, and Rho-mediated signaling. The signaling genes that were tested by *in situ* hybridization were expressed in fibroblasts (Das et al., 2006). A role for most of these signaling pathways in metamorphosis is not known, but cyclin C/cdk8 phosphorylation was examined (Skirrow, Veldhoen, Domanski, & Helbing, 2008). One of the most highly upregulated genes in the tail is corticotropin releasing hormone binding protein, whose function was elegantly tracked down as contributing to regulation of timing of tail resorption (Boorse, Kholdani, Seasholtz, & Denver, 2006). The roles of many other genes await similar analysis.

TH action during metamorphosis is modulated by other hormones, principally corticosterone (CORT), the stress hormone in frogs. CORT synergizes with TH via many mechanisms, including at the level of gene regulation, to accelerate metamorphic processes induced by TH (Bonett, Hoopfer, & Denver, 2010). Two genes have been identified as induced by TH and CORT and TRβ and KLF9. Both of these genes are induced in synergy and/or by either hormone alone, depending on tissue. A microarray analysis was performed to identify more regulated genes and more gene regulation interaction patterns by these two hormones (Kulkarni & Buchholz, 2012). Premetamorphic *X. tropicalis* tadpole tails were harvested 18 h after treatment *in vivo* with 100 nM CORT, 10 nM T3, or CORT + T3. Using Venn diagrams, genes were segregated into clusters based on up- and downregulation by one or both TH, CORT, and CORTTH in all combinations. Of the thousands of genes regulated, the predominant patterns were genes regulated by TH only and in synergy with TH and CORT. Some genes were regulated by CORT uninfluenced by TH. Most known tail TH-response genes were recovered in this analysis as TH-only genes, but some were also regulated in the CORT and CORTTH treatments. A set of genes was regulated by TH in the absence of CORT but in the presence of CORT, genes expression returned back to control. Similarly, a set of genes regulated by CORT had this regulation blocked by TH. The roles of these genes in metamorphosis are not well understood, but for detailed analysis and discussion please refer to Kulkarni and Buchholz 2012.

2.2. Brain

The first histological effect of TH on the tadpole brain is to initiate cell proliferation within 1–2 days after TH treatment, mostly in the cells lining the ventricles (Cai & Brown, 2004; Denver, Hu, Scanlan, & Furlow, 2009). Subsequent remodeling events involve elimination of neurons innervating degenerating larval structures (e.g. Malthner neurons that innervate the tail) and rewiring of neurons in remodeled organs (e.g., development of binocular vision) (Kollros, 1981). Details about the fate of the newly replicated cells in the brain are not well established.

To identify genes comprising the brain metamorphic program, whole brains from premetamorphic tadpoles of *X. laevis* treated with 100 nM TH for 24 and 48 h were used in a microarray analysis (Das et al., 2006). The tadpoles were also pretreated in 1 mM methimazole (MMI) for 1 week to reduce the effect of endogenous levels of TH. The brain program includes many expected categories corresponding to known histological events, such as cell cycle, protein folding, RNA and DNA metabolism, and translation. The most dramatically TH-induced genes are the members of the mini-chromosome maintenance (MCM) complex. Five of the six MCM genes on the *X. laevis* array are upregulated after 48 h of TH induction. Induction of these newly identified TH-responsive cell cycle-related genes in the brain is limited to the cells that line the ventricles, the same cells that are stimulated by TH to replicate. Most of these genes that were upregulated within 48 h in the brain were also upregulated in the limb as well (see Section 2.3 and Fig. 12.3).

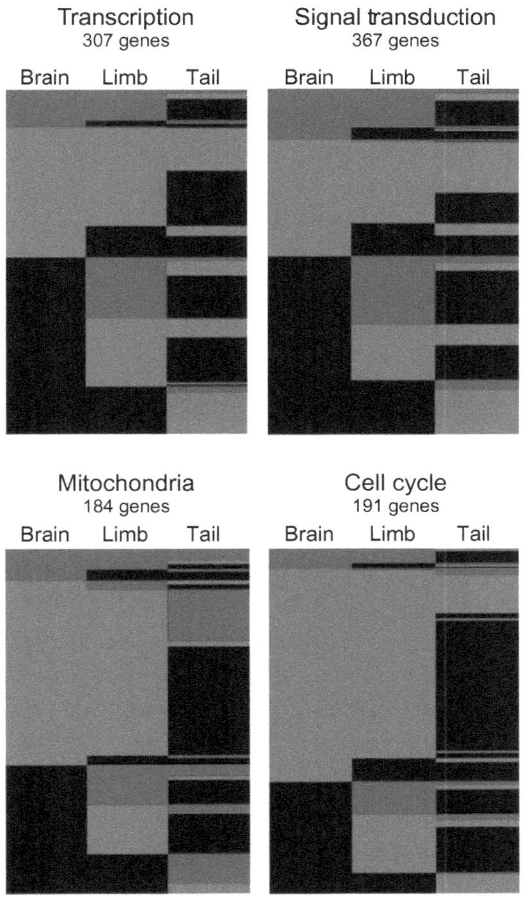

Figure 12.3 Comparison of altered gene expression among brain, limb, and tail. From Das et al. (2006), significantly regulated genes in brain, limb, or tail after 48 h of 100 nM T3 were grouped by gene ontology categories. Here, we chose to display genes from transcription, signal transduction, mitochondria, and cell cycle. Genes up- or down-regulated or no change were assigned green, red, or black, respectively, then sorted by color starting with brain, then limb, and then tail. Among tissues, the gene regulation patterns for each GO category indicate that brain and limb are more similar to each other than to tail, matching their response to TH after 2 days, that is, proliferation versus apoptosis. Among GO categories, gene regulation patterns were similar between transcription and signal transduction and between mitochondria and cell cycle. For example, gene expression patterns among tissues are much more similar to each other in the transcription and signal transduction categories compared to the mitochondria and cell cycle categories. (See Color Insert.)

Two genes were identified as possible candidates for neural-specific development, Notch and Otx2 (Das et al., 2006). Notch activation by TH occurs in the same replicating cells that line the brain and spinal cord ventricle, and the homeobox protein Otx2 has been implicated in mouse brain development. Because brain remodeling requires 1–2 weeks, 48 h of TH treatmentmay not be the appropriate time point to capture altered expression of genes regulating adult brain differentiation and thus more studies are needed to identify such genes.

2.3. Hind limb

The first response to TH in the limb is cell proliferation, which begins within 1–2 days after TH treatment (Brown & Cai, 2007). Subsequently, limb differentiation also occurs where muscle, nerves, and skeletal elements form coordinately. The predominant TH-induced morphological change in the limb is elongation of the elements, as limb patterning can occur in the absence of TH. TH-induced gene expression profiles were carried out on hind limbs of premetamorphic tadpoles treated with 100 nM TH for 14, 24, 48 h (Das et al., 2006). Microarray studies targeting the beginning of the gene regulation cascade in the hind limb were compared to initial steps of the cascade in tail and brain to gain insight into how different tissues can respond differently to TH.

As in the brain, the hind limb program during the first 48 h of TH treatment reflects the intense proliferation that occurs then. The TH-induced limb program shares many genes and virtually all functional categories with the early brain program (Fig. 12.3). Specifically, more than 92% of 955 genes that are upregulated in the brain are also upregulated in the limb after 48 h of TH treatment. Similarly, more than 88% of 199 genes that are downregulated in brain are also downregulated in the hind limb. Genes involved in every step of the cell cycle are TH-regulated and the two programs are remarkably similar containing many cell cycle-related genes. Also, there is high representation of enriched gene products that reside in the nucleolus, which are related to this increase in translation-related proteins. Half of the genes in the array that encode protein-folding proteins are upregulated in the hind limb including chaperone proteins. Many of the components that are involved in the ubiquitination and proteosome pathways are also upregulated, especially in the limb program.

In every functional category, there are more genes upregulated in the hind limb than in the brain and the extent that they are differentially regulated is usually greater. This difference between limb and brain may be attributable to the higher proportion of cells initially stimulated to begin proliferation. Indeed, as examined by *in situ* hybridization in the limb, TH induces all cell types to upregulate the cell cycle genes, compared to only the ventricle boundaries in the brain. Comparing tail, brain, and hind limb, many transcription factors are upregulated in all three tissues, such as KLF9, c-myc, and NFI-X2. Similarly, chromatin remodelers, such as BAF53, BAZ1B, CHD4, DNMT1, HDAC1, 2, despite their different fates, are also commonly upregulated among tissues. In addition, corepressors and coactivators such as EZH2, HES1, NCoR1, NCoR2, and TRIP3 are also commonly upregulated among tissues. Nevertheless, although there are many genes similarly regulated in all three programs (tail, limb, and brain), the tail (death) program is very different from the limb and brain (growth) programs. In order to identify candidate genes that may determine tissue-specific T3-responses, Das et al. (2006) identified 36 differentially regulated genes from the hind limb that regulate transcription at the earliest time point studied, that is, 14 h of TH treatment. Several of these genes are upregulated in the limb and brain but not in the tail and might control the opposite expression patterns in these two programs, including CEBPD, HBP1, HDAC9, HES1, MYOD1, TRMM55, and ZFP36. In addition to differentially expressed transcriptional regulators and chromatin modifiers, different preexisting chromatin states and presence of tissue-specific coactivators and/or corepressors likely interact with TH-response genes in a tissue-specific manner to explain different gene regulation cascades among tissues.

Downstream of the TH-induced gene regulation cascade, genes in several key functional categories are regulated in the opposite manner in tail versus limb programs. Many of the genes that encode energy pathway proteins that are localized in the mitochondria (e.g., the genes in the mitochondrial electron transport chain) are downregulated in tail muscle but upregulated during limb growth. Also, 8 out of 17 genes encoding tricarboxylic acid cycle enzymes are upregulated in the limb and 6 of these are downregulated in the tail. Additionally, half of the 32 genes in the cytoplasm-localized glycolytic pathway are downregulated in tails at metamorphic climax, and four of those are upregulated in the limb.

In a clever use of microarray analysis, Brown and colleagues investigated the developmental basis of a limb muscle-wasting phenotype (Cai, Das, & Brown, 2007). When a dominant negative version of the TH receptor (TRDN) was inducibly expressed in limb muscle cells in transgenic frogs undergoing metamorphosis, limb muscle fibers disintegrated but the muscle cells did not die. Microarray was used to identify candidate genes responsible for this TRDN-induced, rhabdomyolysis-like phenotype. They compared nontransgenic prometamorphic (NF55) limbs and frog limbs after 2 and 6 weeks of transgene induction from NF55. Because hind limb is very sensitive to TH, NF55 limbs have experienced significant TH-induced developmental changes. Candidate genes were those induced or repressed in TRDN-expressing transgenic animals relative to NF55 controls and opposite to expression changes in the nontransgenic frog limb. They found 24 candidate genes with altered regulation after 2 weeks of induction, most of which had known muscle expression. Because the muscle lysis phenotype became visible within 2 days, highly regulated genes were analyzed by *in situ* hybridization, and calsequestrin 1 was found to have downregulation within 1 day after induction of the TRDN transgene. Roles for the five other candidate genes tested, which were downregulated by 4 days, are also possible. The other 18 candidate genes await similar analysis. Another pattern observed from this analysis was that a number of genes downregulated in tail just before tail resorption were repressed in limb muscle expressing the TRDN transgene. Thus, tail degeneration requires T3 induction, and limb degeneration requires the blockade of T3 induction.

2.4. Intestine

The intestine remodels extensively during metamorphosis going from a long, coiled herbivorous gut with a single infolding to a short, carnivorous gut with a villus-trough structure in cross section (Shi & Ishizuya-Oka, 2001). The three main tissue types, epithelium, connective tissue (mostly fibroblasts), and muscle, have distinct responses to TH. During transformation, most cells of the larval epithelium die by apoptosis, and later adult epithelial cells appear and proliferate to form the adult absorptive surface. Fibroblasts and muscle cells also proliferate and their layers become thicker. The same remodeling events are believed to occur whether induced by TH or allowed to progress naturally.

To understand gene programs involved in remodeling the intestine, two microarray analyses were carried out, one on natural metamorphosis and one on TH-induced metamorphosis (Buchholz, Heimeier, Das, Washington, & Shi, 2007; Heimeier, Das, Buchholz, Fiorentino, & Shi, 2010). For natural metamorphosis, intestines were collected at different stages, namely premetamorphosis (NF53, prior to endogenous TH in circulation), prometamorphosis (NF58, beginning of larval epithelial apoptosis), climax

(NF61, beginning of adult epithelial proliferation), and end of metamorphosis (NF66, tail resorption). For induced metamorphosis, the guts from premetamorphic tadpoles treated with 10 nM TH for 0, 1, 3, and 6 days were analyzed. The intestine of tadpoles treated with TH for 3–6 days had extensive cell death and adult cell proliferation but no adult epithelial differentiation, and thus resembled stages at climax of natural metamorphosis (NF61). Despite this morphological similarity, only 20.9% and 33% of the genes upregulated at prometamorphosis (stage 58) overlapped with genes upregulated after 3 and 6 days of T3 treatment, respectively. A similar percentage overlap was observed for downregulated genes. Thus, even though similar histological changes occur during induced and spontaneous metamorphosis, the majority of gene expression changes between natural and induced intestine remodeling are not congruent, likely due to different levels of TH and/or presence of additional hormones during natural metamorphosis.

The functional activities of many genes altered during early metamorphic stages or early TH treatment time points are consistent with the histological epithelial tissue degeneration. Genes transiently up- and downregulated have roles consistent with larval epithelial degeneration, the predominant early event in TH-induced remodeling. The proteolysis gene ontology (GO) category had significantly regulated genes, such as stromelysin-3 in larval cell apoptosis and collagenase 3 in tissue breakdown. Also, genes associated with the cell cycle were significantly downregulated. After the initial wave of apoptosis, adult cells proliferate and differentiate, accompanied by the proliferation of connective tissue and muscle cells, which would presumably include extensive changes in gene regulation. Indeed, a significant number of transcription factors have altered expression at later time points or developmental stages, as well as development genes and extracellular matrix genes. In late stages of remodeling, genes for cell proliferation, signal transduction, and cell–cell signaling are significantly upregulated. Therefore, the two waves of morphological remodeling, namely cell death then cell proliferation and differentiation, are reflected by corresponding waves of gene expression. One of the pathways strongly upregulated during this period is TGF-β signaling including TGF-β ligands (e.g., BMP-1, -2, -3, -4, -7), type I and II receptors, and SMADs. As the tadpole progressed through prometamorphosis, four genes in this pathway were upregulated. By climax, 15 genes were upregulated, and by stage 66, the number of genes upregulated compared to premetamorphosis was only 5, and one gene was downregulated. TGF-β pathway genes were also identified after 6 days of TH treatment. Many of these genes may be indirect response genes because another induced signaling molecule, sonic hedgehog, is required for BMP-4 induction. Functional studies have supported these results suggesting that upregulation of the TGF-β pathway is important for the remodeling taking place at the climax of metamorphosis (Ishizuya-Oka, Hasebe, Shimizu, Suzuki, & Ueda, 2006).

As with other tissues, downregulated genes made up approximately half of the genes regulated by TH. Strongly downregulated GO categories of genes were glucose and fatty acid metabolism and the electron transport chain. Metabolic pathways such as glycolysis, digestion, and the complexes that transfer electrons and synthesize ATP in the mitochondrial inner membrane all appear to shut down at metamorphic climax. Such downregulation of genes, nearly 30, involved with energy consumption was temporary during the remodeling period associated with cell death and lack of proliferation. Later developmental stages saw the expression levels of these genes return to those found in the tadpole to achieve a functional absorptive organ, including intestinal fatty acid binding protein allowing nutrient absorption.

An additional analysis from the natural metamorphosis data set was the identification of larval-specific and adult-specific genes. In all, 17 larval- and 52 adult-specific genes were identified, 11 and 30 of which could be associated with human homologs. The larval-specific genes varied in their potential functions and included GO categories associated with catalytic activity and RNA processing. The larval-specific genes such as UDP-glucose dehydrogenase and solute carrier 22A6 may play a role in signal transduction and cell migration, while mucin, cytochrome P450, thioredoxin reductase, keratin 8, and lactotransferrin are described as potential molecular markers for colon, breast, and other cancers, suggesting that inappropriate expression of larval-specific genes in adult intestine may cause or be indicative of cancer formation. Many adult-specific genes are associated with digestion. For example, dietary enzymes, including the serine proteinases PRSS2 and PRSS3, are significantly increased in the adult intestine. Trypsin is essential for food digestion, but is also involved in other physiological and pathological processes, such as inflammation and tumor invasion. Two other adult-specific genes evaluated, ELA3 and NFI-X2, were predominantly expressed in the epithelium. These genes may provide good adult-specific markers to evaluate adult epithelial regeneration.

Similar processes are occurring in the tail and intestine at the beginning of remodeling, where 100% of the cells in the tail and the vast majority of cells in the intestine, the larval epithelial cells, are destined to die. The early response to TH in the limb is proliferation and the same is true for the brain, even though, like the intestine, the brain undergoes extensive remodeling. The difference between early responses between brain and intestine is that tissue degeneration is not as extensive in the brain. Comparison of gene expression data across these tissues indicates that tissues with similar cellular responses to TH have the most overlap in TH-induced genes. Thus, the initial proliferative responses of the limb and brain to TH explain their similar sets of regulated genes, which share little overlap with genes similarly induced in tail and intestine associated with cell death occurring in the initial response to TH. Despite the association between morphology and altered gene expression among tissues, a core set of 59 genes similarly expressed among all these tissues was identified. How these similarly expressed genes are involved in the mechanisms of gene regulation underlying tissue-specific remodeling has yet to be determined.

2.5. Skin

The larval skin is composed of three main cell types, apical cells, skein cell, and basal cells, organized in two layers (Suzuki, Utoh, Kotani, Obara, & Yoshizato, 2002). During metamorphosis or precocious exposure to TH, basal cells differentiate into adult-type skin stem cells and replace the apoptotic apical and skein cells with cornified, multilayered adult skin with mucus, and serous glands, the latter producing antimicrobial peptides among other proteins.

Global gene expression during TH-induced metamorphosis in body skin was analyzed after 0 versus 1, 3, 5, and 7 days of 5 nM T3 (Suzuki et al., 2009). They found 240 upregulated genes and 161 downregulated genes organized into nine functional groups. Within the first 24 h, the T3-response genes, TRβ, TH/bZIP, KLF9, and ST3, were induced in the skin, as observed in all other organs from previous studies. Transcription and proteolysis were also functional categories recognized by 24 h, followed by cell cycle genes by 3 days, and then defense genes (including antimicrobial peptides) by 7 days. Adult keratin genes and antimicrobial peptide genes were among the most upregulated genes, and genes 17–20 (apical cell-specific genes identified by subtractive hybridization; Wang & Brown, 1991) were among the most downregulated. A major change in gene expression unanticipated by histology was the gradual down regulation of immune response genes (mostly MHC genes). Explanation for this change awaits further work.

2.6. Direct response genes

Because TH induces all metamorphic changes and its effects are mediated through transcriptional regulation by TRs, the identification and characterization of the TH early, direct target genes are of critical importance in elucidating mechanisms underlying the tissue-specific responses to TH.

To identify these genes, two chemicals, cycloheximide and anisomycin, that inhibit protein translation were included in an experiment directed toward identifying early, direct target genes (Das, Heimeier, Buchholz, & Shi, 2009). Specifically, stage 46 *X. laevis* tadpoles were treated with vehicle, 100 nM TH, inhibitors, and 100 nM TH + inhibitors for 14 h, and RNA from whole tadpoles was used in the microarray. This analysis identified 188 and 249 inhibitor-resistant up- and downregulated genes, respectively (i.e., TH + inhibitor vs. vehicle). For TH only, 311 and 162 genes were up- and downregulated. Because TH alone treatment is expected to have a combination of direct response genes and late-response genes, one might expect the TH + inhibitor to be a subset of the TH-only genes. Surprisingly, less than half of the genes were in common between these two treatments. Although most inhibitor-resistant TH-response genes were not in the TH only set of genes, they are believed to be direct TH-response genes. The inhibitors stabilized many mRNA transcripts and likely increased the ability of these genes to be detected by the microarray compared to TH treatment alone.

GO analysis showed that categories related to metabolism/catabolism, cell proliferation (cell cycle/DNA replication), and metallopeptidase activity were found to be common between the TH and TH + inhibitor groups, suggesting that the genes in these categories are involved in the early events of metamorphosis. In particular, increased metallopeptidase activity suggests that the remodeling of the extracellular matrix and other protein degradation events are important early signaling events induced by TH during metamorphosis. In fact, one of the first TH-direct response genes found during metamorphosis was stromelysin-3 (MMP-11), which is important for TH-induced cell death during metamorphosis (Fu et al., 2005). In addition, GO categories like cell communication, steroid hormone receptor activity, and transcription factor activity were found to be significant only in the TH + inhibitor group, indicating that these activities are also involved in early events of metamorphosis. Some of these transcription factors have been reported previously as direct targets of TH, such as TRβ and TH/bZIP. This may not be surprising as it is expected that transcriptional regulation should be an important initial step of metamorphic events.

Interestingly, DNA replication genes, apoptosis, and ribosomal biogenesis and assembly genes were highly enriched in TH upregulated genes but not significantly enriched in the inhibitor-resistant upregulated genes. These results suggest that, although cell cycle progression was induced by both TH and TH + inhibitors, some late-response genes, sensitive to inhibitors, are required for the cells to progress to DNA replication. These data suggest that, in the absence of protein synthesis, TH may induce cell cycle changes.

The direct target genes by themselves, however, are not sufficient for DNA replication but rather prepare the cells to enter the S-phase. This is strongly supported by the enrichment of G1/S-phase transition genes but the absence of DNA replication genes in the microarray analysis. Likewise, the absence of genes in GO categories related to ribosomal biogenesis and assembly in the inhibitor-resistant TH-response genes but their presence in the genes induced by TH alone suggest the possibility that downstream genes are required to sustain cell proliferation. In addition, genes involved in apoptosis are also not significant in TH + inhibitor-treated animals, indicating that these genes are upregulated downstream of direct response transcription factors.

In contrast to the number of upregulated GO categories shared by TH and TH + inhibitors versus control, little overlap was found among the significantly enriched GO categories between the genes downregulated by TH and the genes downregulated by TH + inhibitors. Interestingly, a number of categories involved in cellular processes, such as cell differentiation and cytoskeletal changes, were significantly enriched in genes downregulated by TH in the presence of inhibitors, whereas a number of categories at the organismal level, such as organ development and multicellular organismal processes, were significantly enriched in genes downregulated by TH alone. Such changes suggest that downregulation of cellular processes occurs prior to the changes at the organ levels and that the latter requires late/indirect TH-response genes.

Interaction of the direct TH-induced transcription factors and signaling pathways with the premetamorphic set of transcription factors undoubtedly contribute to tissue specificity of the TH response. Despite the presence of TREs in their promoters, some direct TH-response genes are not expressed in all tissues, where ST3 is restricted to fibroblasts and sonic hedgehog to intestinal epithelium. Thus, the complement of transcription factors as well as epigenetically modified DNA and histones determine how a tissue responds to TH. Use of microarrays to identify TH-regulated genes is a critical step to understanding causes and consequences or tissue-specific responses to TH, but further studies involving these genes are required to determine how different tissues respond to the same hormone differently.

3. ENDOCRINE DISRUPTION

3.1. Biomarkers for TH disruption

A major goal of endocrine disruption research is to identify molecular biomarkers for detection of factors that may inhibit, mimic, or accentuate hormone action. Because TH-induced morphological changes in tadpoles are not observed before 3 days, molecular changes may serve as rapid and sensitive endpoints. Thus, microarray analysis is well suited to enable selection of one or a few genes from the entire set of TH-response genes that are particularly appropriate for detecting TH disruption. Additionally, microarray analysis can be used to identify biomarker genes for detection of the presence of a particular endocrine disruptor by identifying genes uniquely induced by it.

One of the first microarray analyses for biomarker development used the indigenous species, *R. catesbiana* (Veldhoen et al., 2006). Using the MAGEX cDNA array with 401 cDNAs, they identified K-ras (Kirsten rat sarcoma) and PCNA (proliferating cell nuclear antigen) as being strongly up- and downregulated by TH, respectively. These genes were similarly TH regulated in *X. laevis* and mammals, as well. The microarray results for these two genes, as well as for TRα and TRβ, were further analyzed in *R. catesbiana*, where robust changes in gene expression of TRβ, K-ras, and PCNA were identified in 6, 24, and 48 h, respectively. This study showcased the utility of using heterologous cDNA arrays compared to oligonucleotide arrays for analysis of nonmodel species. Ninety percent of the cDNAs on the MAGEX array are from *X. laevis* sequences, and the other 10% are from *R. catesbiana*. Because cDNA arrays have extended sequence with which the sample cDNAs can hybridize, it is robust to cross-species sequence divergence (Helbing, Maher, Han, Gunderson, & Borheras, 2010). In contrast, short oligoarrays (25–60 nucleotides) will not as readily hybridize with divergent sequences. Furthermore, sequences from EST libraries used to make global gene expression microarray assays often have 3′UTRs, which are typically highly divergent even among closely related species, thereby making these sequences unsuitable for nonmodel taxa. cDNA arrays have the further advantage that individual cDNAs can be selected for actual or possible relevance to endocrine disruption.

Robustly induced TH-response genes, such as TRα, TRβ, and KLF9, were identified prior to microarray studies that were sufficient to detect the action of TH agonists. However, their transcripts were not significantly changed in the presence to TH synthesis inhibitors, such as MMI, propylthiouracil (PTU), and perchlorate (PER). To identify useful genes to detect disruption caused by these TH synthesis inhibitors, Helbing et al. (2007) compared results from a 14-day morphological exposure assay to short-term (0–4 day) gene expression profiles of brain transcripts after T4, T3, MMI, PTU, and PER treatments (Helbing et al., 2007) (T4 and T3 are the precursor and active versions of TH). As expected, genes involved in transcription, signal transduction, and hormone regulation were significantly regulated by TH. Surprisingly, there was little overlap in genes regulated by T3 and T4, even though they have similar morphological outcomes. This finding cannot be accounted for by differences in timing of T4 to T3 conversion by deiodinases and awaits an explanation. For the inhibitors, similar numbers of genes in similar categories were identified, and association analysis of genes regulated by MMI, PTU, and PER suggested they have similar gene regulation changes. Thus, the inhibitors may have similar modes of action on brain gene regulation. However, the rapid responses, that is, within 1–2 days, suggests that the mode of action may be a direct effect on brain tissue, rather than blockade of TH synthesis with attendant delay in running out of stores of hormone in the thyroid gland.

A complimentary analysis to the previous MAGEX brain study using the same animals was carried out using hind limb and tail (Helbing et al., 2007). As with the brain, the strategy was to find candidate genes using microarray that could then be used in quantitative comparisons using qPCR. Fewer genes overall were identified in these tissues compared to the brain. As in the brain, T4 and T3 had nonoverlapping sets of transcripts. Also, the sets of transcripts significantly regulated by MMI, PTU, and PER were very different from each other. Some genes were identified as potential biomarkers, such as RNA helicase II/Gu or DAD1 (defender against death 1). However, their use as biomarkers is dependent on the tissue and time point after exposure. These studies revealed the important issue that tissue and exposure duration need to be carefully determined to identify biomarkers with high and robust predictive value.

In a recent microarray analysis, a meta-analysis approach was used to identify robust TH-responsive genes as determined by detection across several array platforms (Searcy et al., 2012). They used an Affymetrix array with tail tissue treated for 6 and 48 h, with 20 nM TH. Many of the regulated genes included those previously identified, such as transcription factors and proteases. However, when comparing across platforms, they found 30% or less overlap of TH-regulated genes, which may be explained by genes not represented on one of the arrays and the different TH doses used. On the other hand, there was a high proportion of overlap with TH-induced tail genes identified by subtractive hybridization. They reasoned that overlap among platforms would represent reliable and robust genes for use as biomarkers for TH disruption. As a proof of principle, they examined the effect of wastewater effluent (WWE) on four TH-response genes, TRβ, D2, CRHBP, FAPα. WWE augmented the induction of these transcripts in the presence of TH, and transcripts of CRHBP and FAPα were increased with WWE even without TH.

3.2. Effects/mechanism of EDC action

3.2.1 BPA

BPA has well-known estrogenic effects, but BPA also affects brain differentiation, a process not known to be strongly influenced by estrogen (vom Saal & Hughes, 2005). TH, on the other hand, is critical for proper brain development, opening the possibility that BPA may be a TH endocrine disruptor as well. Indeed, BPA inhibited TH-induced external morphological changes as well as intestinal remodeling (Heimeier, Das, Buchholz, & Shi, 2009). The effect of BPA on established TH-response genes was to partially block induced changes. To examine effects of BPA on transcription across all genes, tadpoles were treated with 2 nM TH, 10 μM BPA, or both together for 4 days and then a microarray analysis was carried out on the intestine. Genes regulated by TH and TH + BPA compared to controls were relatively similar, despite the general inhibition of BPA on metamorphic transformation. However, because BPA only partially inhibited TH-induced changes in gene regulation, it is possible that BPA attenuates but does not abolish TH-response gene induction. Thus, when comparing gene sets from TH versus TH + BPA, BPA inhibited regulation either up or down in 501 genes, more than 50% of which were TH-response genes. Because BPA reduced most TH-induced gene regulation, BPA likely blocks TH-dependent metamorphosis via inhibition of most if not all molecular and developmental pathways required for metamorphosis.

3.2.2 Atrazine

Atrazine has negative effects on growth, metabolism, immune function, and sexual differentiation in fish and frogs (Hayes, 2005). However, the molecular mechanisms underlying these organismal phenotypes are not well understood. To better understand how atrazine interacts with the organism at the level of gene expression, Ide and colleagues carried out a microarray analysis of tadpoles chronically exposed to 400 ppb atrazine from hatching to climax of metamorphosis (Langerveld, Celestine, Zaya, Mihalko, & Ide, 2009). This atrazine treatment throughout the larval period affected growth, development, and survival. Fat bodies were also reduced in both male and female tadpoles. Microarray analysis from whole bodies identified changes in males (44 genes) and females (77 genes) at metamorphic climax, with less than 50% overlap among sexes. The genes identified were involved in growth and metabolism genes, proteolysis, fibrinogen complex formation, and immune system function. Reproductive function genes were not observed, but gonads constitute a small proportion of whole-body RNA. A reanalysis of these microarray data looking for up- or downregulated metabolic pathways found genes involved in increased fat usage and protein degradation and decreased storage of fat, protein synthesis, and glucose conversion to energy (Zaya, Amini, Whitaker, & Ide, 2011). These changes correlate with morphological effects of atrazine, and could be due to a direct effect of atrazine in these genes at the transcriptional level. Alternatively, the authors suggested that atrazine may reduce food consumption thus indirectly leading to altered gene regulation of metabolism genes. Detoxification of atrazine is an energy intensive process and may place high demand for energy, perhaps an explanation for the etiology of atrazine's effects on slowed growth and development. Reduced survivorship may be due to deleterious effects on immune function perhaps leading to increased susceptibility to infection. However, the cause of death was not determined. The role of altered blood cell function in atrazine-treated tadpoles is not clear and represents another example of microarray analysis revealing cryptic changes not obvious from gross morphological or histological analysis. Different effects of atrazine observed between the sexes are not well understood.

3.2.3 PCBs

PCBs affect neural development, thereby implicating a potential role in TH endocrine disruption based on TH's role in development of the nervous system (Jacobson & Jacobson, 1997). Tadpole metamorphosis is used as a model system to understand the role of PCBs on development to avoid the difficulties of studying endocrine disruption in uterus-enclosed fetuses in mammals. In tadpoles, hydroxylated polychlorinated biphenyls (OH-PCBs) significantly delayed TH-induced metamorphosis (Ishihara, Makita, & Yamauchi, 2011). Microarray on brain tissue showed that 526 TH-response genes (25% of the 2139 T3-regulated genes) had altered TH responses in the presence of 4-OH-PCB-106 and/or -159. GO categories significantly associated with these genes were growth, brain development, cell proliferation, cell adhesion, neuron differentiation, and apoptosis, providing molecular underpinning of the presumed effects of PCBs on the tadpole brain. The fact that most TH-response gene regulation was unaffected suggests that PCBs may not act directly through TRs or that PCBs influence TRs only in a subset of TH-response genes. The authors suggested the possibility that the 526 genes could be a source of biomarkers for PCB disruption of T3-dependent brain development but did not characterize any.

3.2.4 Acetochlor

The pre-emergent herbicide acetochlor accelerated TH-induced metamorphosis in ranid species and *X. laevis* (Crump et al., 2002). In the absence of TH, acetochlor had no significant effect on body area (reduced body area is a measure of metamorphosis due to gill resorption). Gene expression induced by TH was altered in the presence of acetochlor in a manner corresponding to acetochlor's effect on morphological transformation. Acetochlor increased expression of genes that were (1) induced by TH, (2) not affected by TH, and (3) repressed by TH. Another group of genes induced by TH were unaffected by the addition of acetochlor. Given these patterns of gene expression that not all TH-response genes were affected, a direct agonist or antagonist action of acetochlor on TR is unlikely, though promoter-specific or tissue-specific affects allowing direct contact with TR is not ruled out. However, it is more likely that acetochlor, like BPA, affects a subset of TR-containing gene regulatory complexes explaining the complex effects of the endocrine disruptor on TH-response gene expression.

4. SPECIAL TOPICS

4.1. Phenotypic plasticity

Microarray analysis was used to examine the molecular basis of *R. pirica* plasticity to salamander larvae predators (Mori et al., 2005, 2009). Upon exposure to caged salamander larvae, tadpoles get a bulgy fluid-filled body with a deeper tail, and this phenotype goes away with predator removal. To examine the endpoint of this phenotypic plasticity, namely, the altered skin morphology, subtractive hybridization between control and body skin was performed, where 297 differentially expressed transcripts were identified. In an unusual use of microarray, these transcripts were then used as targets on a cDNA array to confirm the subtractive hybridization results. Downregulated genes included carboxypeptidase B, trypsinogen, elastase, fibrinogen, which may have the effect of increasing rate of fibrinolysis allowing skin to stretch and become bulgy. Upregulated genes included bullous pemphigoid antigen, uromodulin, aldehyde dehydrogenase most of which have unclear relation to the bulgy phenotype. The researchers then repeated the experiment and obtained larger number of subtracted clones, which were again used in a microarray validation analysis (Mori et al., 2009). This expanded set of genes identified downregulation of some keratins, and upregulation of a novel uromodulin-like protein in patterns expected for genes involved in the bulgy phenotype. Keratins are downregulated in the bulgy phenotype whose expression comes back upon removal of the predator. Uromodulin has the opposite expression pattern. *In situ* hybridization showed these genes are expressed in epidermis. Brain genes were identified by subtractive hybridization, but the proportion of brain tissue in the RNA sample was too low for positive signals in the microarray analysis. Uromodulin is an abundant protein in mammalian and frog kidneys, but its role is unclear, although alteration of water permeability of tadpole skin may be a possibility. Keratins function to retain cell shape, and their downregulation may allow cell deformation upon water retention in body skin. These molecular associations of uromodulin and keratins with the bulgy phenotype provide potential molecular explanations of the bulgy phenotype. An interesting area of future research is to examine the control upstream of these genes, for example, the steps from predator presence to regulation of these genes remains unanswered.

4.2. Biocontrol

Another goal for a microarray study was to identify genes upregulated during metamorphosis that could be used in a strategy to control the invasive cane toad, *B. marinus* (Halliday et al., 2008). The authors produced a 5170-spot cDNA array of PCR products from randomly selected clones from a toadlet cDNA library. Whole-body microarray comparisons were made among 9, 18, 28, 30, and 53 days postfertilization (9, 18, 28 dpf are premetamorphic stages, and 30 and 53 are postmetamorphic stages). The authors concentrated on genes upregulated during metamorphosis, and identified nine strongly upregulated genes, namely, serum albumin, hemoglobin α and β, olfactomedin, trefoil factor, gastrokine, harderin-like protein, gamma crystallin, apolipoprotein. The potential utility of these highly upregulated genes for biocontrol requires further research.

4.3. Regeneration

The promise of recovery from spinal cord injury in humans spurs research in animals that have this ability. The ability of tadpoles to regenerate axons across a spinal cord transection is lost during metamorphosis (Gibbs, Chittur, & Szaro, 2011). Thus, microarray analysis lends itself well to finding genes permitting regeneration in tadpoles and losing this ability during metamorphosis. Recovery from spinal cord transection in premetamorphosis was slowed by treatment with MMI and blocked by TH, suggesting that low levels of TH may promote regeneration but are not required and that

TH-induced development goes toward a developmental state not capable of regeneration. To look for genes potentially involved in this process, microarray was done using hindbrain, the source of regenerating axons. The authors focused on genes upregulated by spinal cord injury in the regeneration permissive state (MMI treated) and found 72 annotated genes, including transcription factors, membrane proteins, cell–cell and intracellular signaling, and transcriptional coregulators. The five most regulated genes were SOCS2, XER81, neurotensin, MORC3, and a protein associated with RNA polymerase. All but the last of this list have known functions in promoting neural development, survival, and neurite outgrowth. The larval state appears to be important for regeneration, and characterizing precisely the larval state and which specific genes involved in regeneration will require functional studies on candidate genes.

5. SUMMARY AND CONCLUSIONS

Knowledge of how TH can induce wildly divergent responses among tissues during frog metamorphosis will include identification of the component genes of TH-induced tissue-specific gene regulation cascades. Microarray analyses have contributed most significantly to this effort and have also revealed aspects of TH-induced metamorphosis unanticipated by previous morphological and physiological studies. As expected, each organ has unique TH-regulated genes concordant with its distinct role in the organism, for example, larval and adult keratins in the skin, proteins associated with nutrient absorption in the intestine. On the other hand, some tissues, such as brain, limb, and intestine, have a large set of commonly regulated genes despite having distinct morphological outcomes. These common genes are comprised of cell cycle genes and consistent with cell proliferation as a component of their metamorphic remodeling. In this respect, the correlation between expected functions of genes in cell biology and histological observation were consistent, thereby matching known morphological events with molecular signatures. Similarly, the tail and intestine had overlapping sets of genes, including decreased metabolism and electron transport chain proteins and increased extracellular proteases, consistent with tail regression and apoptosis of larval intestinal epithelium. Visualization of these comparisons among tissues reveals that cellular responses to TH induction, such as mitochondrial and cell cycle regulation, are much more divergent among tissues than changes in transcription and signal transduction (Fig. 12.3).

Not predicted from premicroarray studies was a set of genes upregulated by TH in every tissue examined, such as TRβ, TH/bZIP, KLF9, NFI-X2, all of which are transcription factors. This common set of development genes in organs with different responses to TH, from cell death to cell proliferation, suggests that it is quite likely that a combination of organ-specific TH-regulated early genes, especially downregulated ones, and organ-specific responses to the changes in common TH-regulated genes dictates organ-specific metamorphic changes. Thus, an important area of interest will be to determine the downstream targets of the common early regulated genes.

Additional unexpected findings from the microarray analyses are the total number of TH-response genes involved, well above 1000 genes as components of the TH gene regulation cascade and well over 200 direct response genes. Our bias in understanding the molecular mechanisms of direct induction versus repression of gene expression by TH did not prepare us for the identification of equivalent numbers of up- and downregulated genes. Finally, a newly uncovered complexity to the gene regulation cascade, namely, the web of intra- and extracellular signaling pathways induced by TH, was not initially predicted but undoubtedly contributes to the outcome of tissue-specific responses to TH. These novel and important findings about the number and identity of TH-induced genes and their spatial and temporal changes during tadpole development secures the value of microarray studies on frog metamorphosis.

Not only developmental studies but also studies on endocrine disruption have benefitted from microarray analyses. Based on morphological or physiological endpoints, many chemicals appear to interfere with endocrine physiology. However, as hormonal control of development and homeostasis is complex and as hormone systems can interact with each other, the mechanisms of action of a given chemical are not often obvious. Identifying the global pattern of altered gene regulation of an endocrine disruptor gives insight into how that chemical is acting and interacting with the endocrine system. Another use of microarray in endocrine disruption studies lies in the search for biomarkers of TH endocrine disruption. A useful biomarker requires particular properties of its response to TH, namely it must be a sensitive and robust marker of TH signaling. Global analysis of potential genes allows one to select the most appropriate gene or small set of genes from the large group of TH-response genes that may not all have desirable fold-change levels or expression kinetics. This strategy for identifying suitable genes has also been employed in development studies where larval- versus adult-specific genes have been found for intestinal epithelium and skin and genes responding to CORT or TH only without influence from the other hormone. Thus, microarray studies are well suited to the goal of identifying biomarkers for developmental as well as endocrine disruption studies.

The wealth of knowledge obtained using microarray experiments has come despite some technical issues associated with the technique. The high variability among workers in terms of genes identified as TH-regulated stems in part from the microarray methods and also in part from different rearing and TH treatment conditions among studies. These interstudy discrepancies indicate that microarray data are not definitive but rather indicative of important genes and pathways. Another issue is that currently many elements on the microarrays have no ascribed function because they are ESTs representing nonconserved 3′UTRs or hypothetical proteins with low or no homology to a mammalian sequence, giving little information about gene function. Improvements in gene annotation and genomics should refine the complement of known response genes.

While much has been learned about gene regulation cascades, much is still left to understand. Noncoding RNAs are now believed to have a major impact on gene regulation and undoubtedly contribute to tissue-specific TH responses. Differential expression of noncoding RNAs due to TH can be analyzed by microarray technology but thus far have not been examined. Another gap in our knowledge is distinguishing cause from effect in the gene regulation cascade. Microarray analysis provides molecular correlations with cellular and histological changes. In many cases cause and effect relationships can be inferred, such as upregulation of proteases with ECM degradation or upregulation of cell cycle genes with cell proliferation. However, it is often difficult to sort out causes versus effect underlying control of the tissue-specific gene regulation cascades. Determining the web of interactions starting with direct response genes to how they interact with already present tissue-specific transcription factors and TH-indirect response genes may benefit from additional microarray analyses using transgenic-inducible overexpression of TH-direct response transcription factors but likely other research approaches as well. Another important consideration is that the relationship between mRNA levels, as detected by microarray analysis, and the cognate protein levels for most genes is not currently known. The relationship between protein levels and their activity levels altered by posttranslational modifications is also not known. Thus, identifying TH-induced genes is a critical first step, but other types of research, including proteomics and functional studies, are required to elucidate protein levels and the downstream effects of protein activity.

The tissues that comprise organs typically undergo their own TH-induced changes independent of other associated tissues, and they have their own tissue-specific TH-response genes, such as sonic hedgehog in the intestinal epithelium and stromelysin-3 in fibroblasts. Also, the same tissue can undergo distinct responses depending on their location in the body, for example, fibroblasts and epidermis in tail versus body. However, tissues within an organ may be of low abundance and critical cells within tissues (such as stem cells) will be rare. In these cases, mRNA from rare cells may not be well represented in the large pool of mRNA from abundant cells and thus gene expression changes may not be detectable from the rare cells using microarray. Special techniques, such as dissecting out particular tissues from an organ or isolating particular cells using laser capture microdissection may allow detection of rare but critical TH-induced gene expression changes. Also, due to the nature of how the microarray works, low abundance, but high fold-change genes may not be detected. For example, TH/bZIP was identified using subtractive hybridization and is a very high fold-change gene, but it is often not detected using microarrays due to its apparent low abundance even when induced. In addition to TH/bZIP, there may be other such potentially critical genes, which may yet be detected by other means, such as transcriptomics or RNA-seq.

ACKNOWLEDGMENTS

This publication was made possible by University of Cincinnati Dissertation Completion Fellowship to S. S. K. and NSF IOS 0950538 to D. R. B.

REFERENCES

Atkinson, B. G., Helbing, C. C., & Chen, Y. (1996). Chapter 15: Reprogramming of genes expressed in amphibian liver during metamorphosis. In L. I. Gilbert, J. R. Tata & B. G. Atkinson (Eds.), *Metamorphosis: Postembryonic reprogramming of gene expression in amphibian and insect cells* (pp. 539–565). San Diego: Academic Press.

Bonett, R. M., Hoopfer, E. D., & Denver, R. J. (2010). Molecular mechanisms of corticosteroid synergy with thyroid hormone during tadpole metamorphosis. *General and Comparative Endocrinology*, *168*, 209–219.

Boorse, G. C., Kholdani, C. A., Seasholtz, A. F., & Denver, R. J. (2006). Corticotropin-releasing factor is cytoprotective in Xenopus tadpole tail: Coordination of ligand, receptor, and binding protein in tail muscle cell survival. *Endocrinology*, *147*, 1498–1507.

Brown, D. D., & Cai, L. (2007). Amphibian metamorphosis. *Developmental Biology*, *306*, 20–33.

Buchholz, D. R., Heimeier, R. A., Das, B., Washington, T., & Shi, Y. B. (2007). Pairing morphology with gene expression in thyroid hormone-induced intestinal remodeling and identification of a core set of TH-induced genes across tadpole tissues. *Developmental Biology*, *303*, 576–590.

Buchholz, D. R., Hsia, S. C., Fu, L., & Shi, Y. B. (2003). A dominant-negative thyroid hormone receptor blocks amphibian metamorphosis by retaining corepressors at target genes. *Molecular and Cellular Biology, 23*, 6750–6758.

Buchholz, D. R., Paul, B. D., Fu, L., & Shi, Y. B. (2006). Molecular and developmental analyses of thyroid hormone receptor function in Xenopus laevis, the African clawed frog. *General and Comparative Endocrinology, 145*, 1–19.

Buchholz, D. R., Tomita, A., Fu, L., Paul, B. D., & Shi, Y. B. (2004). Transgenic analysis reveals that thyroid hormone receptor is sufficient to mediate the thyroid hormone signal in frog metamorphosis. *Molecular and Cellular Biology, 24*, 9026–9037.

Buckbinder, L., & Brown, D. D. (1992). Thyroid hormone-induced gene expression changes in the developing frog limb. *The Journal of Biological Chemistry, 267*, 25786–25791.

Cai, L., & Brown, D. D. (2004). Expression of type II iodothyronine deiodinase marks the time that a tissue responds to thyroid hormone-induced metamorphosis in Xenopus laevis. *Developmental Biology, 266*, 87–95.

Cai, L., Das, B., & Brown, D. D. (2007). Changing a limb muscle growth program into a resorption program. *Developmental Biology, 304*, 260–271.

Crump, D., Werry, K., Veldhoen, N., Van Aggelen, G., & Helbing, C. C. (2002). Exposure to the herbicide acetochlor alters thyroid hormone-dependent gene expression and metamorphosis in *Xenopus laevis. Environmental Health Perspectives, 110*, 1199–1205.

Das, B., Cai, L., Carter, M. G., Piao, Y. L., Sharov, A. A., Ko, M. S., et al. (2006). Gene expression changes at metamorphosis induced by thyroid hormone in Xenopus laevis tadpoles. *Developmental Biology, 291*, 342–355.

Das, B., Heimeier, R. A., Buchholz, D. R., & Shi, Y. B. (2009). Identification of direct thyroid hormone response genes reveals the earliest gene regulation programs during frog metamorphosis. *The Journal of Biological Chemistry, 284*, 34167–34178.

Das, B., Matsuda, H., Fujimoto, K., Sun, G., Matsuura, K., & Shi, Y. B. (2010). Molecular and genetic studies suggest that thyroid hormone receptor is both necessary and sufficient to mediate the developmental effects of thyroid hormone. *General and Comparative Endocrinology, 168*, 174–180.

Denver, R. J., Hu, F., Scanlan, T. S., & Furlow, J. D. (2009). Thyroid hormone receptor subtype specificity for hormone-dependent neurogenesis in Xenopus laevis. *Developmental Biology, 326*, 155–168.

Denver, R. J., Pavgi, S., & Shi, Y. (1997). Thyroid hormone-dependent gene expression program for *Xenopus* neural development. *The Journal of Biological Chemistry, 272*, 8179–8188.

Dodd, M. H. I., & Dodd, J. M. (1976). The biology of metamorphosis. In B. Lofts (Ed.), *Physiology of the amphibia* (pp. 467–599). New York: Academic Press.

Fu, L., Ishizuya-Oka, A., Buchholz, D. R., Amano, T., Matsuda, H., & Shi, Y. B. (2005). A causative role of stromelysin-3 in extracellular matrix remodeling and epithelial apoptosis during intestinal metamorphosis in Xenopus laevis. *The Journal of Biological Chemistry, 280*, 27856–27865.

Gibbs, K. M., Chittur, S. V., & Szaro, B. G. (2011). Metamorphosis and the regenerative capacity of spinal cord axons in *Xenopus laevis. European Journal of Neuroscience, 33*, 9–25.

Gudernatsch, J. F. (1912). Feeding experiments on tadpoles. I. Tile influence of specific organs given as food on growth and differentiation. *Archiv für Entwicklungsmechanik, D Organ, 35*, 457–484.

Halliday, D. C. T., Kennedy, G. C., Hamilton, N. H. R., Tarmo, S., Alderman, J., Siddon, N. A., et al. (2008). Genes induced during the early developmental stages of the Cane Toad, Bufo (Chaunus) marinus. *Gene Expression Patterns, 8*, 424–432.

Hayes, T. B. (2005). Welcome to the revolution: Integrative biology and assessing the impact of endocrine disruptors on environmental and public health. *Integrative and Comparative Biology*, *45*, 321–329.

Heimeier, R. A., Das, B., Buchholz, D. R., Fiorentino, M., & Shi, Y. B. (2010). Studies on Xenopus laevis intestine reveal biological pathways underlying vertebrate gut adaptation from embryo to adult. *Genome Biology*, *11*, R55.

Heimeier, R. A., Das, B., Buchholz, D. R., & Shi, Y. B. (2009). The xenoestrogen bisphenol A inhibits postembryonic vertebrate development by antagonizing gene regulation by thyroid hormone. *Endocrinology*, *150*, 2964–2973.

Helbing, C. C., Bailey, C. M., Ji, L., Gunderson, M. P., Zhang, F., Veldhoen, N., et al. (2007). Identification of gene expression indicators for thyroid axis disruption in a Xenopus laevis metamorphosis screening assay. Part 1. Effects on the brain. *Aquatic Toxicology*, *82*, 227–241.

Helbing, C. C., Maher, S. K., Han, J., Gunderson, M. P., & Borheras, C. (2010). Peering into molecular mechanisms of action with frog SCOPE. *General and Comparative Endocrinology*, *168*, 190–198.

Howdeshell, K. L. (2002). A model of the development of the brain as a construct of the thyroid system. *Environmental Health Perspectives*, *110*(Suppl. 3), 337–348.

Ishihara, A., Makita, Y., & Yamauchi, K. (2011). Gene expression profiling to examine the thyroid hormone-disrupting activity of hydroxylated polychlorinated biphenyls in metamorphosing amphibian tadpole. *Journal of Biochemical and Molecular Toxicology*, *25*, 303–311.

Ishizuya-Oka, A., Hasebe, T., Shimizu, K., Suzuki, K., & Ueda, S. (2006). Shh/BMP-4 signaling pathway is essential for intestinal epithelial development during Xenopus larval-to-adult remodeling. *Developmental Dynamics*, *235*, 3240–3249.

Jacobson, J. L., & Jacobson, S. W. (1997). Evidence for PCBs as neurodevelopmental toxicants in humans. *Neurotoxicology*, *18*, 415–424.

Just, J. J., Kraus-Just, J., & Check, D. A. (1981). Chapter 9: Survey of chordate metamorphosis. In L. I. Gilbert & E. Frieden (Eds.), *Metamorphosis: A problem in developmental biology* (pp. 265–326). (2nd ed). New York: Plenum Press.

Kollros, J. J. (1981). Chapter 13: Transitions in the nervous system during amphibian metamorphosis. In L. I. Gilbert & E. Frieden (Eds.), *Metamorphosis: A problem in developmental biology* (pp. 445–460). (2nd ed). New York: Plenum Press.

Kulkarni, S. S. & Buchholz, D. R. (in press). Beyond synergy: Corticosterone and thyroid hormone have numerous interaction effects on gene regulation in *Xenopus tropicalis* tadpoles. *Endocrinology*, *153*(11), 5309–5324.

Langerveld, A. J., Celestine, R., Zaya, R., Mihalko, D., & Ide, C. F. (2009). Chronic exposure to high levels of atrazine alters expression of genes that regulate immune and growth-related functions in developing *Xenopus laevis* tadpoles. *Environmental Research*, *109*, 379–389.

Mori, T., Hiraka, I., Kurata, Y., Kawachi, H., Kishida, O., & Nishimura, K. (2005). Genetic basis of phenotypic plasticity for predator-induced morphological defenses in anuran tadpole, Rana pirica, using cDNA subtraction and microarray analysis. *Biochemical and Biophysical Research Communication*, *330*, 1138–1145.

Mori, T., Kawachi, H., Imai, C., Sugiyama, M., Kurata, Y., Kishida, O., et al. (2009). Identification of a novel uromodulin-like gene related to predator-induced bulgy morph in anuran tadpoles by functional microarray analysis. *PLoS One*, *4*, e5936.

Searcy, B. T., Beckstrom-Sternberg, S. M., Beckstrom-Sternberg, J. S., Stafford, P., Schwendiman, A. L., Soto-Pena, J., et al. (2012). Thyroid hormone-dependent development in Xenopus laevis: A sensitive screen of thyroid hormone signaling disruption by municipal wastewater treatment plant effluent. *General and Comparative Endocrinology*, *176*, 481–492.

Shi, Y. B., & Brown, D. D. (1993). The earliest changes in gene expression in tadpole intestine induced by thyroid hormone. *The Journal of Biological Chemistry, 268*, 20312–20317.

Shi, Y. B., & Ishizuya-Oka, A. (2001). Thyroid hormone regulation of apoptotic tissue remodeling: Implications from molecular analysis of amphibian metamorphosis. *Progress in Nucleic Acid Research and Molecular Biology, 65*, 53–100.

Skirrow, R., Veldhoen, N., Domanski, D., & Helbing, C. C. (2008). Roscovitine inhibits thyroid hormone-induced tail regression of the frog tadpole and reveals a role for cyclin C/Cdk8 in the establishment of the metamorphic gene expression program. *Developmental Dynamics, 237*, 3787–3797.

Suzuki, K., Machiyama, F., Nishino, S., Watanabe, Y., Kashiwagi, K., Kashiwagi, A., et al. (2009). Molecular features of thyroid hormone-regulated skin remodeling in *Xenopus laevis* during metamorphosis. *Development, Growth and Differentiation, 51*, 411–427.

Suzuki, K., Utoh, R., Kotani, K., Obara, M., & Yoshizato, K. (2002). Lineage of anuran epidermal basal cells and their differentiation potential in relation to metamorphic skin remodeling. *Development, Growth and Differentiation, 44*, 225–238.

Tata, J. R. (1993). Gene expression during metamorphosis: An ideal model for post-embryonic. *Development, 15*, 239–248.

Veldhoen, N., Crump, D., Werry, K., & Helbing, C. C. (2002). Distinctive gene profiles occur at key points during natural metamorphosis in the Xenopus laevis tadpole tail. *Developmental Dynamics, 225*, 457–468.

Veldhoen, N., Skirrow, R. C., Ji, L., Domanski, D., Bonfield, E. R., Bailey, C. M., et al. (2006). Use of heterologous cDNA arrays and organ culture in the detection of thyroid hormone-dependent responses in a sentinel frog, Rana catesbiana. *Comparative Biochemistry and Physiology Part D, 1*, 187–199.

vom Saal, F., & Hughes, C. (2005). An extensive new literature concerning low dose effects of bisphenol A shows the need for a new risk assessment. *Environmental Health Perspectives, 113*, 926–933.

Wang, Z., & Brown, D. D. (1991). A gene expression screen. *Proceedings of the National Academy of Sciences of the United States of America, 88*, 11505–11509.

Wang, Z., & Brown, D. D. (1993). Thyroid hormone-induced gene expression program for amphibian tail resorption. *The Journal of Biological Chemistry, 268*, 16270–16278.

Weber, R. (1996). Chapter 16: Switching of globin genes during anuran metamorphosis. In L. I. Gilbert, J. R. Tata & B. G. Atkinson (Eds.), *Metamorphosis: Postembryonic reprogramming of gene expression in amphibian and insect cells* (pp. 567–598). San Diego: Academic Press.

Yaoita, Y., Shi, Y., & Brown, D. (1990). Xenopus laevis alpha and beta thyroid hormone receptors. *Proceedings of the National Academy of Sciences of the United States of America, 87*, 8684.

Yoshizato, K. (1996). Cell death and histolysis in amphibian tail during metamorphosis. In L. I. Gilbert, J. R. Tata & B. G. Atkinson (Eds.), *Metamorphosis: Postembryonic reprogramming of gene expression in amphibian and insect cells. 1996* San Diego: Academic Press.

Zaya, R. M., Amini, Z., Whitaker, A. S., & Ide, C. F. (2011). Exposure to atrazine affects the expression of key genes in metabolic pathways integral to energy homeostasis in *Xenopus laevis* tadpoles. *Aquatic Toxicology, 104*, 254–262.

CHAPTER THIRTEEN

Is Thyroid Hormone Signaling Relevant for Vertebrate Embryogenesis?

G. Morvan-Dubois, J.B. Fini, B.A. Demeneix[1]
Laboratoire d'Evolution des Régulations Endocriniennes, Muséum National d'Histoire Naturelle, UMR 7221 CNRS, Department RDDM, Paris, France
[1]Corresponding author: e-mail address: demeneix@mnhn.fr

Contents

Abstract

Classically, thyroid hormones (THs) have been primarily associated with postembryonic development (Tata, 1968), notably metamorphosis in anuran amphibians and flat fish. This period is parallel to the perinatal period in man and many marked developmental transitions in other species. As amply described in other chapters, metamorphosis is characterized by a peak of thyroxine (T_4) and triiodothyronine (T_3) that is synchronous with the metamorphic climax.

In contrast, the developmental period that characterizes embryonic development prior to the significant production of TH by the endogenous thyroid gland has received

Current Topics in Developmental Biology, Volume 103
ISSN 0070-2153
http://dx.doi.org/10.1016/B978-0-12-385979-2.00013-7

little attention. Furthermore, the prevailing concepts of TH physiology during this period have been framed by two observations in amphibians and mammals: first, TRs are expressed, while circulating TH levels are much lower than those during metamorphosis and, second, extrapolating from the knowledge largely obtained from *in vitro* models, in the absence of TH, the aporeceptor represses target gene transcription during premetamorphic development. We propose to revisit both concepts in the light of accumulating data, first, on TH availability both in eggs and in embryos and, second, on the increasing knowledge of the complexity of TR and TH control of transcription.

1. INTRODUCTION

In this review, we revisit two principal concepts that have framed current thinking on thyroid hormone (TH) signaling during embryonic development over the past two decades: first, TH levels are too low prior to the induction of the thyroid gland to activate target gene transcription and, second, that as receptors, TRs are expressed during early development and their principal role is to act as aporeceptors to repress target gene transcription (Chassande, 2003). These ideas arise in some part from the clear distinction of premetamorphic states from the marked TH-dependent prometamorphic and metamorphic phases (see Chapter 10). Although metamorphosis in anuran amphibians, and to some extent in flat fish, has received the most attention, most vertebrate species display a peak of TH during a key phase around birth or hatching (see Chapter 14). As discussed by Holzer and Laudet, in amniotes with nonclassical metamorphosis, a TH peak is also associated with drastic morphological and functional changes (Paris, Brunet, Markov, Schubert, & Laudet, 2008).

In mammals, embryonic development can be defined as the process that covers development from the zygote until birth including organogenesis and early morphogenesis. Similarly, in birds and reptiles, hatching is regarded as the end of embryonic development. In the context of aquatic vertebrate development, the limits are less well defined. Hatching can be used to define the end of embryonic development, but in many species, embryogenesis is not finalized at this stage, and the embryo is still yolk-feeding. Consequently, in the European Union, the legal definition for aquatic species is to consider that the embryonic period terminates within dependent feeding (2010/63/UE).

In this chapter, we review current knowledge on TH signaling during early development and focus on the following ideas:

- During embryonic development in *Xenopus* and other species, not only are TH receptors (TRs) and all the deiodinases expressed but also TH

levels are within physiological ranges, despite being much lower than during metamorphosis.
- Given the increased knowledge of the complexity of TR and TH control of transcription, there is a need to go beyond the idea that in the absence of TH, the nonliganded receptor (aporeceptor) represses target gene transcription and take into account the increasing evidence for activation of gene transcription by the aporeceptor.
- Consideration of each of the above concepts leads us to suggest that TH signaling during early development is much more complex than previously envisaged, with a potential scenario of both activation and repression of target genes, notably in certain tissues, such as the developing nervous system.
- Finally, given the potentially complex role of TH during this developmental period, this time window could be a period of sensitivity to endocrine disruption.

2. BOTH TRs AND DEIODINASES ARE EXPRESSED DURING EMBRYOGENESIS

2.1. TRs are differentially expressed during embryonic development

In vertebrates, different TH-binding TR isoforms have evolved, leading to neofunctionalization for distinct TRs (Escriva, Holland, Gronemeyer, Laudet, & Holland, 2002). Expression data and *in situ* screens demonstrate that in most species harboring TR, they are expressed from the embryonic stage onward. The expression patterns of TR show that the various isoforms are differentially expressed during development in mammals as well as in birds and amphibians. Schematically, two genes *TRα* (*NR1A1*) and *TRβ* (*NR1A2*) encode the different isoforms of TRs, even if duplication or tetraploidization has occurred in some species. TRα1, TRβ1, and TRβ2 (Laudet, 2011; Yaoita & Brown, 1990) are the isoforms with both a DNA-binding domain and a ligand-binding domain. In mouse, TRs are expressed early in development, at least from stage E11.5 for TRα, and from stage E13.5 for TRβ (according to Allen Developing Mouse Brain Reference Atlas; Lein et al., 2007). TRα and TRβ have distinct roles in development, as exemplified by Lezoualc'h, Hassan, Abdel-Tawab, Puymirat, and Demeneix (1994), Lezoualc'h et al. (1995), and O'Shea et al. (2005).

The numerous mouse models of TR invalidation exhibit slighter developmental defects than those of hypothyroid pup. In the mouse models, as for cretinism in humans, lack of ligand causes striking neurological and bone formation defects that can be explained by the deleterious effect of inappropriate gene regulation by the unliganded aporeceptor (Bassett et al., 2007; Chassande, 2003; Gothe et al., 1999). Numerous studies analyzing the effects of hypothyroidism (largely in mammals) during embryogenesis have shown that liganded TRs have numerous, essential roles in developmental processes (see, for instance, Escobar-Morreale, del Rey, Obregon, & de Escobar, 1996).

In *Xenopus laevis*, TRα and TRβ are present at very low levels in the oocyte, and then decrease and rise from NF24 stage (Duarte-Guterman, Langlois, Pauli, & Trudeau, 2010; Fini, Le Mével, et al., 2012; Morvan Dubois et al., 2006); this stage of development corresponds to the maturation of the nervous system (Nieuwkoop & Faber, 1994). TRα mRNA levels then increase from hatching (NF35–37) to stage NF45, levels stabilizing at stage NF48, prior to prometamorphosis (Fini, Le Mével, et al., 2012; Havis et al., 2006; Morvan Dubois et al., 2006). Generally, TRβ was observed at lower levels than TRα and was considered to be insignificant. However, Havis et al. (2006) showed that both TRα and TRβ transcripts were present at comparable levels and in that both displayed dynamic patterns of expression. More recently, we have revisited this question using qPCR and showed that TRβ mRNA levels are in fact slightly higher than those of TRα mRNA in eggs but the two display similar expression profiles, with the most marked increases occurring between stages NF37 and NF45 (Fini, Le Mével, et al., 2012; Havis et al., 2006). Furthermore, the TR proteins can be deduced to be functionally active well prior to metamorphosis, as triiodothyronine (T_3) treatment of stage 45 embryos induces precocious metamorphosis (Shi, Wong, Puzianowska-Kuznicka, & Stolow, 1996; Tata, 1968; Yaoita & Brown, 1990). It is in fact this type of experiment, the addition of large amounts of T_3 to the stage 45 embryo, with the consequent, but incomplete and disorganized, metamorphic-like response, that has largely contributed to the perduring concept that apo-TRs have repressive roles during early development. Some experimental work has addressed the roles of unliganded receptor during *Xenopus* embryonic development. Notably, using germinal transgenesis to overexpress a mutated TR that cannot recruit corepressor complexes (nor coactivator complexes) allowed the demonstration of the importance of unliganded TRβ in craniofacial development and particularly eye formation (Havis et al., 2006). In the same set of experiments, the authors used a pharmacological approach to interfere with TH signaling, the deiodinase inhibitor iopanic acid and the mixed TH agonist/antagonist NH-3 (Lim, Nguyen, Yang, Scanlan, & Furlow, 2002), also observing effects on eye formation.

In chicken embryos, TRs are present during early embryogenesis, particularly in neuroectoderm, and TH-activated reporter gene expression was found in several embryonic tissues, including Hensen's node and the primitive streak (Flamant & Samarut, 1998). TRs are expressed from early development, with TRα exhibiting a quite ubiquitous pattern but variable expression levels, whereas TRβ expression is restricted to some organs such as the brain, kidney, eye, yolk sac, and liver (Forrest, Sjoberg, & Vennstrom, 1990). The same group also demonstrated that both receptors are expressed (but with distinct patterns) in the cerebellum at E9, and questioned the presence of liganded receptor at this stage (Forrest, Hallbook, Persson, & Vennstrom, 1991).

In teleost fishes, TRs are also detected in eggs and are expressed during early development (see Power et al., 2001, reviewing T_3/thyroxine (T_4) contents of many teleosts species). The role of unliganded receptor was demonstrated soon after the cloning of the TRs in zebrafish (Essner, Breuer, Essner, Fahrenkrug, & Hackett, 1997), but the question of the importance of liganded TRs during early development remains unsolved. Nevertheless, it was demonstrated, as it was in *Xenopus*, that all the actors of TH metabolism and signaling are present and functional during zebrafish embryonic development (Walpita, Van der Geyten, Rurangwa, & Darras, 2007). However, here again, there is no demonstration of a developmental role for an "endogenous-liganded" receptor.

In lamprey, two TR isoforms are present but they are resulted from a duplication that is specific to agnathes (Laudet, 2011). In contrast to tetrapods and teleost fishes, where a peak of TH is mandatory for metamorphosis to occur, lamprey metamorphosis is dependent on a decrease in TH (Manzon & Youson, 1997; Youson, Manzon, Peck, & Holmes, 1997). TRs and TH are present in most lamprey embryonic tissues such as the liver, kidney, and gill, with particularly high levels in intestine (reviewed in Youson & Manzon, 2012). The lamprey could therefore provide a valuable model for studying the role of a liganded TR during embryonic development. However, the molecular pathways activated by TH in lamprey embryo are still to be analyzed.

The increasing number of sequenced genomes shows that TRs are present in most bilateria (Laudet, 2011). TH can have a biological action in Aplysia and sea urchin, and TR homologs have been described in both groups, but their role during embryogenesis has not been investigated

(Heyland, Price, Bodnarova-Buganova, & Moroz, 2006). TR function has been more extensively investigated in the prochordate, Amphioxus. It has a poor affinity for T_3 and its endogenous ligand is likely to be TRIAC (Paris et al., 2008). Here, again it is not clear whether the liganded receptor plays any developmental role. In the same study, Ciona TR was also shown not to bind TH, even if it was demonstrated that TH inhibitors are able to block Ciona metamorphosis (Patricolo, Cammarata, & D'Agati, 2001). These invertebrate and prochordate models can represent alternatives to the use of vertebrates in experimental research, but their physiology appears to be quite distinct from chordate/vertebrate systems.

2.2. Both activating and inactivating deiodinases are present in embryos

The cloning of the genes corresponding to the activating (*Dio2* and *Dio1*) and inactivating (*Dio3 and Dio1*) deiodinases (the enzymes are referred to as D1, D2, and D3) is more recent than that of the TRs (*Dio1*: Berry & Larsen, 1993; *Dio2*: Becker, Schneider, Davey, & Galton, 1995; Croteau, Whittemore, Schneider, & St Germain, 1995; *Dio3*: Croteau, Davey, Galton, & St Germain, 1996). However, deiodinases have recently moved to the fore as major actors in TH signaling with their roles in tight regulation of TH availability (reviewed in Dentice & Salvatore, 2011).

Measurable levels of mRNA encoding the three deiodinases were found in *X. laevis* embryos from neurulation stages onward (Fini, Le Mével, et al., 2012; Morvan Dubois et al., 2006). Expression is also found in the egg where *Dio1* mRNA is expressed at high levels, whereas *dio2* and *dio3* mRNAs are expressed only at much lower levels (Fini, Le Mével, et al., 2012; Morvan Dubois et al., 2006). Similarly, *dio2* and *dio3* mRNAs were also found at the two-cell stage in *Xenopus tropicalis*, indicating that these mRNAs are present in *X. tropicalis* egg (Duarte-Guterman et al., 2010). During embryological development in *X. laevis*, *dio1* expression decreases significantly between egg and gastrula stages and then increases dramatically from stage NF24 onward, stabilizing between NF45 and 48 (Fini, Le Mével, et al., 2012). Similar to *dio2*, during embryogenesis two major waves of expression are discerned in *X. laevis* (Morvan Dubois et al., 2006): the first around stages NF10–14 (neurula) and the second at stages NF31–NF37 (late embryogenesis), levels then increase significantly from stage NF41 onward, but not increasing to the same extent as the deactivating deiodinases mRNA of D1 and D3 (Fini, Le Mével, et al., 2012). D2 activity accounts for most of the outer ring (activating) deiodinase activity in the developing *Xenopus* embryo.

Table 13.1 References describing specific localization of TRβ (TRα is apparently ubiquitious) and deiodinase expression in embryonic *X. laevis* and *X. tropicalis*(*)

	Notochord	Spinal chord	Brain	Retina	Otic vesicle
Trβ		Havis et al. (2006)		Cossette and Drysdale (2004), Havis et al. (2006); Oofusa et al. (2001)	Havis et al. (2006)
Dio2	Tindall et al. (2007)*	Morvan Dubois et al. (2006)	Morvan Dubois et al. (2006), Tindall et al. (2007)*	Morvan Dubois et al. (2006), Tindall et al. (2007)*	Morvan Dubois et al. (2006), Tindall et al. (2007)*
Dio3	Tindall et al. (2007)*	Morvan Dubois et al. (2006)	Morvan Dubois et al. (2006), Tindall et al. (2007)*	Morvan Dubois et al. (2006), Tindall et al. (2007)*	

Interestingly, specifically in the brain of early stage *X. laevis*, we also saw higher expression of *dio2* as compared to the other deiodinases transcripts (Fini, Le Mével, et al., 2012). *Dio3* mRNA levels in embryos are fairly constant, but low, throughout early embryogenesis, only increasing at stage NF37, then continuing to rise through stages NF41–48 (Fini, Le Mével, et al., 2012). Interestingly, similar results to those found for *X. laevis* were obtained for *Dio2* mRNA expression in *X. tropicalis* (Tindall, Morris, Pownall, & Isaacs, 2007), suggesting conservation in the mechanisms regulating transcription and potential functions (Table 13.1). *Dio1* mRNA expression has not been investigated in *X. tropicalis*.

In chick, the three deiodinases are expressed early on during development (Van Herck, Geysens, Delbaere, Tylzanowski, & Darras, 2012). These authors measured both mRNA levels and deiodinase activity during the first half of embryogenesis. Although *Dio1* mRNA expression was found, no activity could be detected, whereas both *Dio2* expression and D2 activity increased steadily from E4 to E12 in all brain areas examined, that is, telencephalon, diencephalon, mesencephalon, and rhombencephalon. In contrast, in the same brain areas, *Dio3* mRNA levels showed decreasing levels as development progressed from stage E4 to E12, but the corresponding D3 activity, though highly variable in the same areas over the same developmental time frame, showed a distinct peak in the mesencephalon at E8. *Dio2* is in particular expressed in the developing hypothalamus, devoid of *Dio3* mRNA during chicken hypothalamic ontogenesis. *Dio3* is expressed in developing structures related to the visual system from E3. *Dio3* is in particular expressed in the developing retina from E4 to E10 (Geysens et al., 2012).

Given their role in regulating ligand availability, deiodinases-like enzymes are also of interest in prochordates and their presence has been characterized in numerous species, including Ciona (recently reviewed in Darras & Van Herck, 2012). Despite the apparently ubiquitous expression of deiodinases during vertebrate development, the only cases where a developmental role has been formally demonstrated are the mouse models describing *Dio3* and *Dio2* invalidation. If the individual global phenotypes show surprisingly little alteration, these models demonstrate that certain developmental processes are accelerated in the *Dio3* mutant, whereas they are delayed in *Dio2* mutant (reviewed in Galton, 2005; Williams & Bassett, 2011).

2.3. Transporters are expressed in embryonic tissues

TH transporters were cloned and characterized even later than deiodinases. It had been assumed for a very long time that TH entered cells by passing through membranes. However, in the 1970s, it was demonstrated that TH uptake by cells is ATP-dependent and is therefore an active process (Rao, Eckel, Rao, & Breuer, 1976). The demonstration that transporters are present in the cytoplasmic membrane was not until two decades later (Docter et al., 1997). Similarly for deiodinases, the properties of TH transport were analyzed before the actual cloning of a transporter (see Hennemann et al., 2001 for review of transporter characteristics identified before cloning). The cloning of the first transporter (i.e., MCT8) (Friesema et al., 2003) immediately opened up a new field of research, a field that is increasingly active. Examination of patients bearing a mutation in the MCT8 transporter led to the hypothesis that the neurological damage was due to impaired TH availability in the developing brain, therefore suggesting the importance of TH transporters in embryogenesis.

Recent experiments have described the expression of *MCT8* and *OATP1C1* in the developing chicken brain, suggesting that different brain areas do not have the same TH requirements, in accordance with differences in *Dio2* and *Dio3* expression (Geysens et al., 2012; Van Herck et al., 2012). Nevertheless, no functional role has been demonstrated for these transporters in chick brain up to now. The expression of mRNA encoding MCT8 was also characterized in the zebrafish embryo, peaking at 48 h postfertilization, which is consistent with its potential role in the control of local TH availability during organogenesis (Arjona, de Vrieze, Visser, Flik, & Klaren, 2011).

It has to be mentioned that despite the molecular characterization of transporters during *X. tropicalis* metamorphosis (Connors, Korte, Anderson, & Degitz, 2010), no information is available yet on the potential role of TH transporters in the developing *Xenopus* embryo or tadpole of either species. Nevertheless, an efficient TH transporter, 4F2hc-IU12, was characterized by overexpression in *X. laevis* oocytes (Ritchie, Peter, Shi, & Taylor, 1999). The transcript encoding the IU12 subunit was previously demonstrated to be an early T_3 responding gene during intestinal development (Liang, Sedgwick, & Shi, 1997). The expression of IU12 is also upregulated in the head and limb during metamorphosis (Liang et al., 1997).

3. BOTH T_3 AND T_4 ARE PRESENT DURING EMBRYONIC STAGES

Taken together, the data reviewed above show that TRs, deiodinases, and transporters are present and functional in all vertebrate embryos examined to date. In some species, including humans, changed phenotypes can be linked to impaired TH signaling during early development, raising the central question of ligand availability during this period. A related question is how TH delivered to the embryonic tissues that require TH signaling and degraded in those that do not? There were several attempts to visualize TH availability and action using reporter systems (Flamant & Samarut, 1998; Quignodon et al., 2004; Wallis et al., 2010). For example, Flamant and Samarut used a cocultivation assay in which different parts of the early chick embryo (from day 1 of incubation) were cocultured with a T_3 sensitive reporter cell line. Region-specific and developmental stage-specific production of T_3 was found. T_3 was produced successively by the blastoderm (*area opaca*), Hensen's node (the equivalent of Spemann's organiser in *Xenopus*), and then by the caudal part of the embryo. However, the methodological bias of this approach is that TH can be detected only in tissue where the TH signaling activates transcription and where cofactors required for the specific construct used are present. The most straightforward method to date to test for TH presence in tissue is RIA. The main drawback of this method is that the precise spatial distribution of TH within tissues is lost. Another means of addressing the question is the use of autoradiography using Iodine isotope in order to visualize TH distribution. This technique is the only one providing any clue on spatial distribution of TH within tissues and has been applied to mature tissues such as the brain (Greenberg et al., 2006), but not to the authors' knowledge to the developing embryo.

Human thyroid gland development starts to develop at the day 24 of gestation. At that time, the embryonic thyroid is visible as a bulge in the pharyngeal floor. Then, the thyroid gland tissues merge and migrate to their final location. By gestational day 74, the thyroid gland starts to concentrate iodide and produce MIT, DIT, T_4, and T_3 (Shepard, 1967).

In the free-living *X. laevis* embryo, the thyroid gland forms between stages NF40 and NF45 (Jayatilaka, 1978) with visible follicles at NF46 (about 1 week of age in standard rearing conditions). In zebrafish, the thyroid gland forms at 40 hpf (Rohr & Concha, 2000), but the use of radioactive iodine uptake shows that endogenous production of THs starts at 3 dpf in zebrafish (Brown, 1997). The origin of TH present in embryo has therefore to be investigated, as thyroid gland only become active in late development.

3.1. T_3 and T_4 are maternally provided in mammals

THs, T_4 and T_3, are essential for normal physiology and development in all vertebrates, with particularly marked effects on metamorphosis and brain development in mammals (Bernal, 2007; Hamburgh, 1969; Lezoualc'h et al., 1995). Traditionally, research in human and rodents has focused on what appears to the most sensitive period, usually the period of most rapid brain growth, which in humans is that early postnatal period.

However, during pregnancy many changes occur in maternal TH metabolism. Peripheral metabolism of TH and serum TH distributor protein production increase during this time. Notably, there is a huge increase in free thyroxin during the first trimester followed a decrease to reach levels of a nonpregnant woman (Burrow, Fisher, & Larsen, 1994). In parallel, total T_4 and protein distributors (thyroid binding globulin TBG and transthyretin TTR) also increase significantly during the first trimester and remain high until delivery (Burrow et al., 1994; Glinoer, 2000). Interestingly, in hypothyroid mothers treated before becoming pregnant, the dose of T_4 given to fit the demand has to be increased by 25–50% to maintain normal serum thyrotropin concentrations during pregnancy (Kaplan, 1992b; Mandel, Hanna, & LaFranchi, 1990). Thus, the total (free and bound-to-serum protein) levels of maternal T_4 and T_3 increase during pregnancy, whereas the maternal level of free T_4 remains constant (Contempre et al., 1993; Kaplan, 1992a) due to the increase in distributor proteins.

It had long been thought that maternal TH could not reach the fetus as the placenta expresses high levels of the inactivating deiodinase, D3. However, Vulsma, Gons, and de Vijlder (1989) demonstrated that T_4 did reach the fetus by observing infants with a total organification defect. Total T_4 concentrations in maternal serum and amniotic fluid at 6–12 weeks of development are, respectively, 11.3 μg and 1.6 ng/dL, (Burrow et al., 1994), but the differences in total hormone have to take account the high amounts of distributor proteins in the maternal serum. The fetal thyroid gland and pituitary–thyroid axis become functional by 10 weeks after conception (Thorpe-Beeston, Nicolaides, & McGregor, 1992).

Surprisingly, in a study of homogenized human brain tissues at 10 and 12 weeks of gestation, T_3 was detected, but not T_4 (Bernal & Pekonen, 1984). This corroborates the study of Obregon, Calvo, Del Rey, and de Escobar (2007) on the presence of T_3 and T_4 in rat brain tissue before thyroid gland formation. In addition, a number of clinical studies show that low T_4 maternal concentrations during the first trimester of pregnancy are associated with impaired neurological and visuospatial development of fetus (Haddow et al., 1999; Pop et al., 1999) and that the consequences of low maternal TH levels in the first trimester are worser than lower maternal TH levels in the second and third trimesters. These results in humans reflect data from studies on rodents showing that underactive maternal thyroid function in early gestation (before the fetal thyroid is formed) is associated with impaired fetal brain growth (Obregon et al., 2007).

3.2. T_3 and T_4 are present in the egg and early embryo of vertebrates

The first reports of TH presence in vertebrate eggs date from over 50 years ago. T_3 and T_4 were reported in the vitellus of different amphibian species as early as the 1950s (Volpert, Michel, & Roche, 1958) results that were confirmed later (Gancedo, Alonso-Gomez, de Pedro, Delgado, & Alonso-Bedate, 1997). THs of maternal origin are present in the egg yolk of birds (McNabb, 2006; Power et al., 2001; Prati, Calvo, Morreale, & Morreale de Escobar, 1992), reptiles (Shepherdley, Richardson, Evans, Kuhn, & Darras, 2002), and fish eggs where TH is present in variable amount (Power et al., 2001; Szisch, Papandroulakis, Fanouraki, & Pavlidis, 2005, Walpita, Crawford, Janssens, Van der Geyten, & Darras, 2009). In the latter study, the authors demonstrated that TH content decreased during embryonic development. If the presence of T_4 and T_3 in teleost fish embryo is well documented, the role(s) of TH signaling during embryogenesis remains to be defined (Power et al., 2001; Raine & Leatherland, 2000). TRs are present in teleost fish embryos and it has been demonstrated that increasing rainbow trout oocyte TH content can affect later embryo TR gene expression, which suggests that both maternal THs and *in ovo* TH manipulation can affect subsequent fish development (Power et al., 2001; Raine & Leatherland, 2000).

In *X. laevis* embryo, early work on THs only investigated TH signaling pathways from NF50 (premetamorphosis) onward, plasma TH concentrations at stages earlier than NF52 being assumed to be too low to measure (Leloup & Buscaglia, 1977; Regard, Taurog, & Nakashima, 1978). The fact that activating deiodinases are functional at early larval stages (NF35–37), prior to thyroid organogenesis and free feeding raised the question of the ligand/substrate availability during *Xenopus* development. We have measured the quantities of T_3 and T_4 both in eggs and embryos at progressive stages. Eggs from different females show quite variable TH content (Fini, Le Mével, et al., 2012; Morvan Dubois et al., 2006). However, the most important finding is that all NF35–37 embryos from different females exhibit measurable, physiological relevant amounts in the femtomole range, with both hormones being found in eggs and embryos. These amounts are compatible with the affinity levels of deiodinases. Krain and Denver (2004) have reported levels of T_4 and T_3 in NF50 *Xenopus* tadpoles. These authors found very low levels of T_4, in the order of 0.2 fmol/tadpole and 1.0 fmol T_3/tadpole at NF50. Measuring total TH levels in whole tadpoles does not take into account tissue-specific changes. Indeed, the fact that overexpressing mutant TRs, unable to recruit corepressors, produces an eye-specific phenotype (Havis et al., 2006), underscores tissue specificity of TR and TH signaling in early development. There were no significant variations in T_4 levels between eggs and embryos. In contrast, a significant difference ($P < 0.05$) was observed between the mean of T_3 measured in eggs and that found in NF35–37 embryos. T_3 is increased (threefold) at stage 35–37 compared to eggs (Fini, Le Mével, et al., 2012; Morvan Dubois et al., 2006), suggesting an endogenous conversion of T_4 into T_3 by the embryonic-activating deiodinases. As this pathway is energy consuming, it should therefore be of interest to examine if this aspect of thyroid signaling during embryonic development is conserved through evolution.

In oviparous species, TH is maternally supplied. For instance, it is well documented that TH accumulates in the avian yolk (Roche, Michel, Volpert, & Sanz, 1957), and T_3 response signals were found in early transgenic chick embryos (Flamant & Samarut, 1998). We also reviewed data on T_4 and T_3 measurements in reptiles, fishes, and amphibians. The most likely interpretation would be that T_4 in the avian and *Xenopus* embryos is exclusively of maternal origin. However, Tindall et al. (2007) have shown that thyroid peroxidase (TPO) is also expressed during embryonic *Xenopus* development, raising the possibility that some synthesis occurs, but for this to be the case thyroglobulin would also most probably have to be present. However, up to now no functional study has been carried out to show whether this enzyme is active during *Xenopus* development. Nevertheless, expression patterns of TPO during *Xenopus* embryogenesis are similar to those observed for deiodinases. Tindall et al. (2007) show that TPO is expressed well before thyroid gland anlage formation. They show that blocking TPO with methimazole during embryogenesis affects both morphogenesis and organogenesis.

4. WHAT ROLES DOES TH SIGNALING HAVE DURING EARLY DEVELOPMENT?

As described above, many components of TH signaling pathways, including transporters, receptors, ligands, and deiodinases, have been shown to be present in the early stages of embryonic development in chicken, mouse, and human. Given their presence in these species and the presence of at least the ligands in the eggs of other vertebrates, it is feasible that TH signaling is also likely to be functional during early development of many vertebrates. Despite the demonstrations of the three components mentioned, other elements of the pathway are not characterized yet, such as transporter and distributor protein expression during early development. Furthermore, the actual functions and mechanisms of TH signaling are however very difficult to investigate, especially because of the difficulty of controlling TH availability with precision. Valuable insight came from the analysis of neurological disorders due to impaired TH availability during early pregnancy and from the careful analysis of endocrine disruptors effect on embryonic development. Using the *Xenopus* model, we recently showed that *Xenopus* are competent to respond to T_3 prior to NF40, and that TH may have a role in neuronal differentiation.

4.1. Reduced maternal TH levels affect neuromotor development in the child

The first demonstrations that treating young patients with thyroid extract could improve the mental retardation characteristic dated from the end of the nineteenth century (Railton, 1894; Smith, 1894). However, even at that time, it soon became evident that to be effective thyroid extract had to be given to young children, as the effects of treating older cretins were deleterious. In the past few decades, the roles of TH in brain maturation have been documented in detail. Most research concentrates on the most rapid period of brain growth, that is, in mammals, the perinatal period (Bernal, 2007; Nunez, Celi, Ng, & Forrest, 2008). However, data from numerous epidemiological and clinical studies point to the possibility that earlier stages of neuronal development might also be TH dependent (de Escobar, Obregon, & del Rey, 2004; Gyamfi, Wapner, & D'Alton, 2009; Pharoah, Ellis, Ekins, & Williams, 1976; Pop et al., 1999). For instance, both radial glial and microglial cells are negatively affected by hypothyroid conditions, indicating that TH may influence very early stages of brain development (Howdeshell, 2002). Measuring maternal TH levels in humans and assessing the neurodevelopmental scores of infants show that the most marked neurological deficits are associated with low maternal TH levels in the first trimester of pregnancy as opposed to the later stages (Kooistra, Crawford, van Baar, Brouwers, & Pop, 2006; Pop et al., 1999). One of the most important ideas to emerge is that the mother's TH levels prior to the development of the fetal thyroid (Obregon et al., 2007) are critical in determining her child's postnatal neuromotor development. Most significantly, data from different groups examining maternal TH levels during pregnancy show a child's intellectual quotient to better correlate with its mother's thyroid status during the first trimester of pregnancy as opposed to maternal levels in the second or third trimester (de Escobar et al., 2004; Henrichs et al., 2010; Li et al., 2011; Pop et al., 1999; Zimmermann, 2011). Henrichs et al. (2010) reported that even slight maternal hypothyroxinemia at 13 weeks of gestation was associated with an increased risk of certain forms of language delay, with greater risk for severe hypothyroxinemia.

After birth, TH is equally (if not more) important, hence the current testing for TH levels in newborns and, in many countries, the use of iodine-supplemented salt. However, it should be noted that despite such governmental actions, epidemiological studies show that in many countries pregnant women and young children suffer from lack of iodine with potentially deleterious effects on their TH levels (Kutlu & Kara, 2012; Szybinski et al., 2010).

Experimentally addressing the question of the significance of TH signaling in early mammalian embryos is experimentally challenging, implicating surgery of the dam. In contrast, using aquatic embryos such as fish and amphibian provides easy access to and treatment of any developmental stage, making such vertebrate models precious tools to analyze the molecular mechanism implied in TH signaling during embryogenesis. As we have already stated, in amphibians, plasma TH concentrations at stages earlier than NF52 were assumed to be too low to measure (Leloup & Buscaglia, 1977; Regard et al., 1978). Analyzing morphological and biochemical data as well as profiles of mRNA encoding the *Xenopus* TH receptors (*thr*), led to the proposition that *X. laevis* embryos start to show competence to respond to T_3 treatment between NF stages 40–44 (Shi et al., 1996; Tata, 1968; Yaoita & Brown, 1990). However, such studies were carried out before the dynamics of different actors of thyroid signaling in early embryogenesis was known and hence no studies at that time addressed whether TH target genes were sensitive to antagonism or endocrine disruption during this period, hence our recent reanalysis of this possibility.

5. DOES TH SIGNALING DURING EARLY DEVELOPMENT REPRESENT A NEW SCENARIO FOR ENDOCRINE DISRUPTION?

5.1. Epidemiological evidence for the presence of endocrine-disrupting chemicals during the early development

The above arguments in favor of active thyroid signaling during early embryogenesis lead to the question of what could be the consequences on embryogenesis of an intermittent, or even constant, exposure to xenobiotics with the potential to interfere with TH signaling (Boas, Feldt-Rasmussen, & Main, 2012; Zoeller et al., 2002). This question has relevance for both human health and biodiversity. Not only the water of rivers and lakes is increasingly polluted, threatening developing aquatic organisms but also human amniotic fluid can contain significant amounts of environmental contaminants that have TH-disrupting effects (Cariou et al., 2008; Kawashiro et al., 2008; Kim, Lee, Kim, & Oh, 2011; Miller, Chernyak, Domino, Batterman, & Loch-Caruso, 2012). Of most concern are brominated flame retardants (BFRs) that include polybrominated diphenyl ethers (PBDE) and tetra-bromo-bisphenol A (TBBPA). Significant levels of these BFRs have been found in numerous human tissues and fluids, notably in maternal milk, serum, umbilical cord tissue, and cord blood (Cariou et al., 2008; Kawashiro et al., 2008). Both of these substances are known to interfere with TH signaling.

5.2. PBDE effects on TH signaling

As mentioned, PBDE are ubiquitous flame retardants and known endocrine disruptors. These chemicals have been used in many household products since 1970s (Birnbaum & Staskal, 2004). The more bromine is present on the ether, the more efficient against the fire propagation the BFR is. However, given their adverse effects on human health, most PBDEs have been banned from use in Europe and the United States.

Despite these bans, numerous reports have indicated that PBDEs are present in significant amounts in the fetal (and postnatal) environment. A recent report shows mean total PBDE concentrations of 3795 pg/ml (337–21.842 pg/ml) in amniotic fluid sampled in women from Michigan (USA). The same authors found measurable BDE-47 and BDE-99 in all samples tested (Miller et al., 2012). Similarly, recent data on South Korean populations (Kim et al., 2011) found measurable levels of seven PDBE congeners in maternal blood, umbilical cord blood, and breast milk. Again, BDE-47 was the predominant form identified. Zhang et al. also measured PBDEs levels in breast milk of 60 primiparous women in Shenzhen (China) to estimate the daily exposure to PBDEs in breast-fed infants. Levels of PBDE body burden in these mothers were higher than those previously reported for other Chinese regions. BDE-47 and BDE-153 were the major congeners of PBDEs in milk samples (Zhang, Sun, & Ai, 2012).

As to the TH-disrupting effects of PBDEs, Blanco, Mulero, Lopez, Domingo, and Sanchez (2011) studied the neurotoxicity of BDE-99 in primary cultures of cerebellar neurons of rats. This compound alters the expression of TRα and TRβ and brain proteins, including the T_3-dependent brain-derived neurotrophic factor and the antiapoptotic protein Bcl-2 (Blanco et al., 2011). Regarding enzymes involved in TH metabolism, *in vivo* studies by Szabo et al. (2009) showed that PBDEs could inhibit deiodinase activity. Moreover, experimental exposure to PBDE resulted in decreased serum T_4 *via* induction of uridine-diphospho-glucuronosyltransferase (Hallgren & Darnerud, 2002; Zhou, Ross, DeVito, & Crofton, 2001). Similarly, BDE-47 and BDE-99 displayed metamorphic-disrupting effects in *X. tropicalis* (Carlsson, Kulkarni, Larsson, & Norrgren, 2007), whereas BDE-47 96-h exposure induced thyroglobulin and sodium iodide symporter gene expressions in zebrafish embryos (Yu et al., 2010). Among the effects reported, metabolites formed by degradation of compounds *in vivo* could also affect TH signaling. Hydroxylated metabolites of PBDE were found to bind to TRs (Kojima et al., 2009) and to human transthyretin (Meerts et al., 2000). More recently, Ren and Guo (2012) showed that OH-BDE-47 and OH-BDE-154 bound, respectively, to transthyretin and thyroglobulin with greater affinities than thyroxin.

A pilot study was conducted to explore relationships between concentrations of PBDE low-brominated (BDE-17-154), highly brominated PBDEs (BDE-183-209), and hydroxylated metabolites of PBDEs (OH-PBDEs) and other measures of thyroid function in pregnant women in California. The results indicate that PBDE levels are elevated in pregnant women and may be correlated with elevated serum TSH (Zota et al., 2011). However, this remains controversial (Eggesbo et al., 2011).

5.3. The case of TBBPA and its effects on TH signaling

TBBPA was first employed to replace PBDEs, the use of which caused concern because of their high environmental persistency (Eriksson, Jakobsson, & Fredriksson, 2001). TBBPA was supposed to be inert thanks to its relatively short half-life in mammals (Kuester, Solyom, Rodriguez, & Sipes, 2007) and its extensive metabolism in mammals including humans (Hakk & Letcher, 2003; Schauer, Volkel, & Dekant, 2006). However, TBBPA has been found repeatedly in environmental and human samples (Darnerud, 2008; Sjodin, Patterson, & Bergman, 2003). In a French study, Cariou and colleagues measured levels of TBBPA in cord serum and found 0.1 μg TBBPA/g lipid showing first that TBBPA crosses placenta barrier and second that TBBPA can be present in significant amounts during human neurogenesis. Leijs et al. (2012) showed that both TBBPA and BDE-99 were associated with TH metabolism modifications.

Studies of thyroid disruption caused by TBBPA are mostly more recent than for PBDEs. *In vitro* studies showed competition with TH receptors (Kitamura, Jinno, Ohta, Kuroki, & Fujimoto, 2002, Kitamura et al., 2005) when placed with T_3. Meerts et al. (2000) have shown that TBBPA, like certain PBDEs, could bind transthyretin, with a 10-fold greater affinity than that of T_4. These data were confirmed by Marchesini et al. (2008). More recently, transient transfection studies in CV-1 cells described antagonistic effects of TBBPA on TH signaling (Sun et al., 2009).

In vivo studies on aquatic species have reported inhibitory effects of TBBPA on *Rana pipiens* (Iwamuro, Yamada, Kato, & Kikuyama, 2006) and on *X. laevis* TH signaling (Fini et al., 2007; Jagnytsch, Opitz, Lutz, & Kloas, 2006; Fig. 13.2B). Until recently, data about the metabolic fate of TBBPA in animal models were limited to rodents. These studies and one study with human volunteers (Schauer et al., 2006) showed that TBBPA is metabolized in both glucuronide–TBBPA and sulfate–TBBPA revealing conjugation via phase II enzyme pathway. We recently showed that same conjugation pathways are used in *Xenopus*, a finding that emphasizes that effects on TH signaling obtained using amphibians make drawing parallels with human entirely plausible (Fini, Riu, et al., 2012).

Besides these BFRs, over 100 environmental contaminants have been identified as potential thyroid-disrupting chemicals (Boas et al., 2012; Brucker-Davis, 1998), with certain authors suggesting that TH signaling with its complex, multicomponent requirements for synthesis, activation, and metabolism is the endocrine system that could be most prone to endocrine disruption. Compounds such as perchlorate have long been known to interfere with TH synthesis by blocking iodine accumulation in thyreocytes. Emerging compounds such as perfluoro compounds are prime candidates for TH disruptors (Lopez-Espinosa, Mondal, Armstrong, Bloom, & Fletcher, 2012). Historically, two groups of chemicals that have received considerable attention for their endocrine-disrupting effects are the polychlorinated bisphenols (PCBs) and the bisphenols, particularly bisphenol A (BPA). The PCBs are well-characterized disruptors of TH signaling, whereas BPA, despite its TH-disrupting actions, has more marked effects on signaling through the estrogen receptor (ERs) (for review, Welshons, Nagel, & vom Saal, 2006) and the estrogen-related receptors (ERRs) (Gibert et al., 2011; Matsushima et al., 2007).

5.3.1 PCBs in the fetal environment and their effects on TH signaling

PCBs were developed and used early on in the past century. PCBs represent a large group of compounds, with a theoretical number of 209 congeners. These polymers found many uses in paints, plastics, rubbers, adhesives, inks, insecticides, or as a flame retardant (Lincer & Peakall, 1970). In 1928, their toxic effects were discovered after poisoning of workers. However, they were widely used between 1930 and 1970 until they were banned in the 1970s. Unfortunately, PCBs and organochlorine insecticides belong to the class of persistant organic pollutants and persist in the environment due to their lipophilic properties. They are still present in the environment nowadays. Many effects on endocrine systems have been reported, among them male reproduction disruption (Sager, 1983) and thyroid system disruption (Gauger et al., 2007; Zoeller, Dowling, & Vas, 2000).

In human, exposure to PCBs causes decreases in circulating levels of TH (Hallgren & Darnerud, 2002; Hallgren, Sinjari, Hakansson, & Darnerud, 2001; Salay & Garabrant, 2009). PCBs, such as PCB180, PCB52, have repeatedly been found in amniotic fluid (Bergonzi et al., 2009; Cao et al., 2011; Lopez-Espinosa et al., 2007). Given the crucial role of TH in brain development, the question of the impact of *in utero* exposure to PCBs on intellectual faculties was raised. The results are controversial: a correlation was found between the levels of PCBs *in utero* and loss of intellectual capacity in 9-year-old children (Jacobson & Jacobson, 1996; Stewart et al., 2008), whereas other studies show no effects. These controversial effects are discussed in a review (Stewart, Reihman, Lonky, & Pagano, 2011). Lopez-Espinosa et al. (2011) showed that newborns exposed *in utero* to organochlorine exhibit size, birth weight, and head circumference below normal.

In studies using rodents as models, *in utero* exposure of rats to PCBs results in lower circulating levels of TH and diminishes Purkinje cell number (Roegge et al., 2006). The injection of PCBs in chicken embryos at the early gestation alters the peak of TH which normally occurs before hatching and considerably delays hatching (Beck, Roelens, & Darras, 2006, Beck, Roelens, Maervoet, Schepens, & Darras, 2005; Powell, Aulerich, Stromborg, & Bursian, 1996; Roelens et al., 2005).

As PBDEs, PCBs are metabolized into hydroxylated PCBs (OH-PCBs) that accumulate in all organs and particularly in fetal tissues in gravid rats. Interestingly, the deiodination of T_4 to T_3 is increased in the forebrain of newborn rodent (Meerts et al., 2002) after PCB exposure. These perturbations on TH signaling were also described at the receptor level (Miyazaki, Iwasaki, Takeshita, Kuroda, & Koibuchi, 2004).

As to the well-known plasticizer bisphenol A (BPA), its effects on the thyroid axis are less well-studied than its effects on reproduction. However, BPA is found in fetal tissues and in the placenta in humans (Cao et al., 2012) and could therefore interfere with normal TH signaling. Rats exposed to BPA have a larger thyroid without obvious histopathological changes (Tan, Couvineau, Van Rampelbergh, & Laburthe, 2003). A positive correlation between exposure to BPA and activity of UDP-glucuronyl transferases was also demonstrated. This conjugation pathway is also used for the metabolism of TH (Wu, Green, Huang, Hays, & Chopra, 2005) and could modify TH metabolism. BPA is probably a promiscuous endocrine-disrupting chemical affecting multiple pathways, having been shown to bind with different affinities to ER, ERR, and PPAR receptors, as well as TRs (Moriyama et al., 2002). More recently, other teams showed antithyroidian effect of BPA on aquatic species (Fini et al., 2007; Heimeier & Shi, 2010).

6. CONCLUSION

In conclusion, numerous studies converge to show that all the components of an active TH signaling are present and functional in early vertebrate embryos (Figs. 13.1 and 13.2). Whereas the role, in early development, of the unliganded receptor is documented in some species, the function of

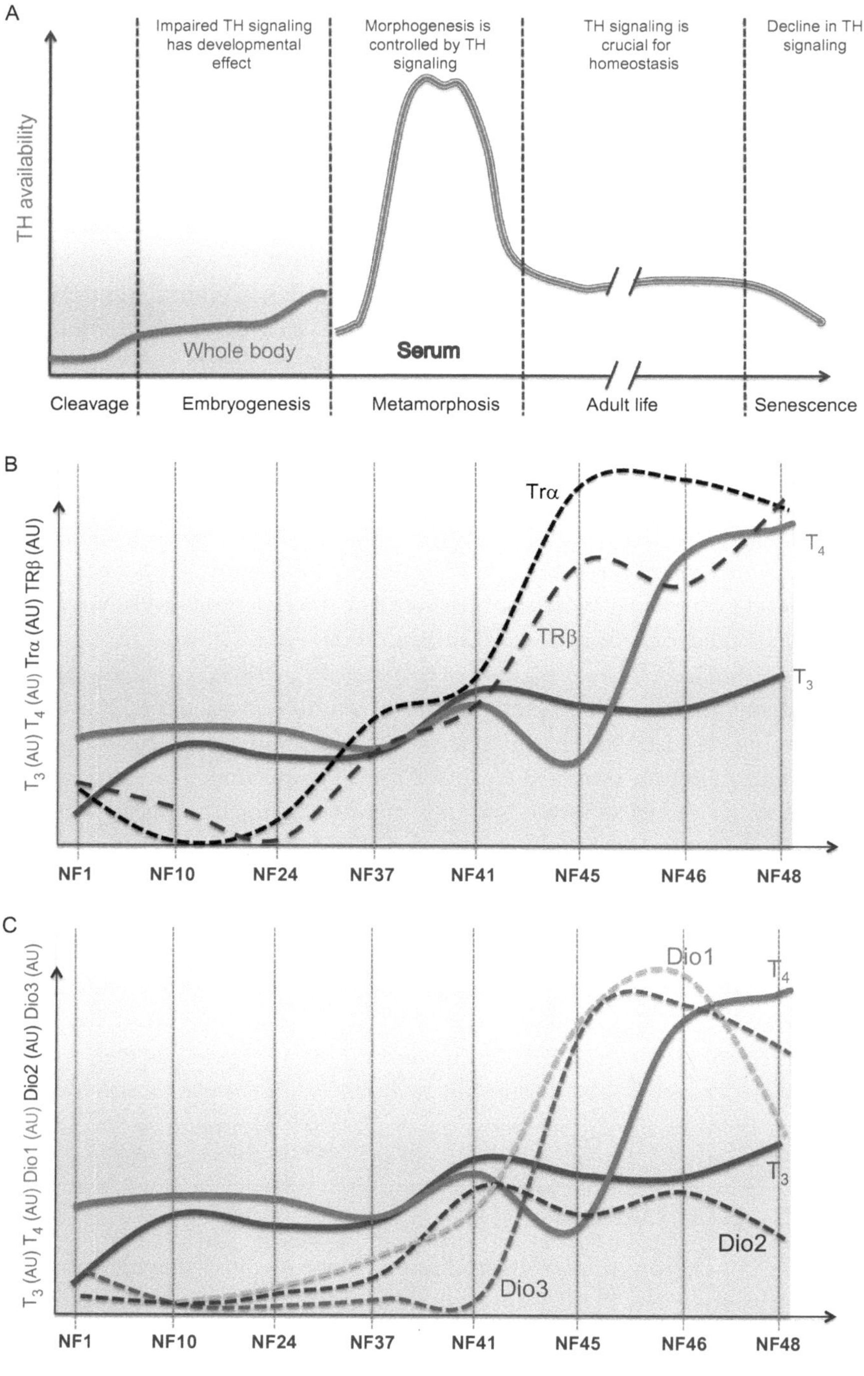
A
Impaired TH signaling has developmental effect
Morphogenesis is controlled by TH signaling
TH signaling is crucial for homeostasis
Decline in TH signaling
TH availability
Whole body
Serum
Cleavage
Embryogenesis
Metamorphosis
Adult life
Senescence
B
Trα
TRβ
T4
T3
T3 (AU) T4 (AU) Trα (AU) TRβ (AU)
NF1
NF10
NF24
NF37
NF41
NF45
NF46
NF48
C
Dio1
T4
T3
Dio2
Dio3
T3 (AU) T4 (AU) Dio1 (AU) Dio2 (AU) Dio3 (AU)
NF1
NF10
NF24
NF37
NF41
NF45
NF46
NF48

the liganded receptor during this period remains to be determined. However, the expression patterns of deiodinases can be used to derive hypothesis as to which tissues may be sites of TH metabolism and action in the embryo. For instance in neurogenic areas of stage NF 35–37 *X. laevis* tadpoles, not only are deiodinases expressed but also one or several isoforms of TRs are expressed (see Table 13.1). Taking such data into account, we hypothesize that in *Xenopus*, TH signaling plays a role in neurogenesis and in the early formation of sensory organs (eye, ear, and olfactory system) as well as cartilage. In these organs, the expression patterns of the different components are highly dynamic, suggesting very precise windows of action for TH signaling, as described for retinoic acid.

The functionality of liganded TR is however difficult to analyze as TH levels in embryos are hard to manipulate. The most straightforward treatment, that of adding exogenous T_3, can be directly neutralized and counteracted by local D3-mediated deiodination, even in the external developing models, such as zebrafish and *Xenopus*. Despite this limitation, these small model organisms remain more flexible models than mammalian systems, as besides the fact that there is no need for surgery to access them, there is, moreover, no maternal influence in TH content during embryogenesis and the amount of TH present in the egg can be evaluated. Moreover, the analysis of TH signaling function during embryogenesis will provide new insight into understanding how endocrine disruptors affect embryonic development (Fig. 13.2).

Finally, several major endocrine disruptors have been shown not only to interfere with TH signaling but also to be present at significant levels in the general environment and specifically in the fetal environment. It is, however, important to remember that a given endocrine disruptor may affect more than one endocrine pathway thus affecting multiple developmental

Figure 13.1 The different components of TH signaling pathway are present during embryogenesis. (A) The general consensus on TH signaling is that T_3 is the key trigger and orchestrator of amphibian metamorphosis. Nevertheless, it is also crucial for adult physiology. TH role during embryogenesis is less defined. Whereas seric concentrations are well known in metamorphic and postmetamorphic tadpoles, they are irrelevant in embryos. We therefore reported whole body data. (B) The analysis of T_3 and T_4 availability in regard with TRs expression shows that when the receptors are expressed during embryogenesis, T_3 is present in (some) tissues. T_3 is therefore likely to bind TRs locally. (C) Deiodinases are expressed in some tissues from early development. They can be responsible for the slight T_3 increase and T_4 decrease during the first phases of embryogenesis (up to NF45). The increase in T_4 content between NF45 and NF46 indicates *de novo* TH synthesis. *Schematic view, adapted from Fini, Le Mével, et al. (2012).* (See Color Insert.)

A

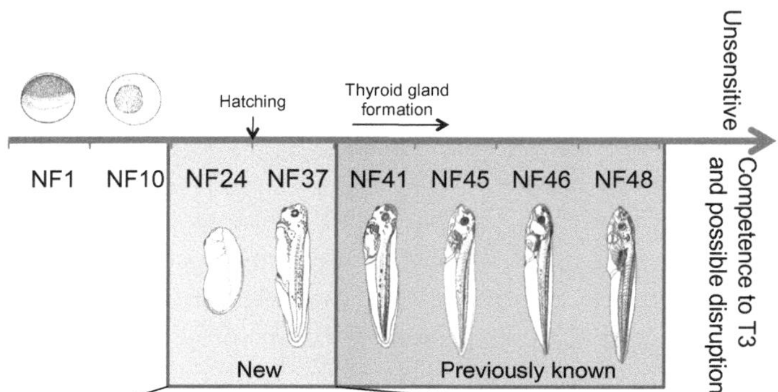

B

Treatment	Start (stage)	Stop (stage)	Exposure length	Downregulation	Upregulation
T3	24	37	24 h	*dio1, thra**	
NH3	24	37	24 h		*dio1, dio3, thra**
T3	37	41	24 h	*oct91, sox2, rest*, thra**	*thibz, thrb, klf9, tubb2b, rxrg**
NH3	37	41	24 h	*pcna*	*klf9, thrb, mbp*, rxrg**
TBBPA	37	41	24 h	*pcna, sox2*	*thrb, klf9, pparg, oct91°*

Figure 13.2 (A) Schematic representation of competence to TH signaling in developing *X. laevis*. Competence to TH is present in early embryos prior to formation of the endogenous thyroid gland. These functional responses to TH appear to be highly tissue specific and will be governed by stocks of TH laid down in the egg. The presence, during early embryogenesis, of xenobiotics with the capacity to interfere with TH signaling might therefore exert deleterious effects on development during this period. (B) TH-regulated genes: *Xenopus* embryos were treated at different stages either from NF24 to NF37 or from NF37 to NF41 for 24 h with either DMSO (controls) or T_3 (5 nM) or NH-3 (1 μM) or TBBPA (1 μM). Heads of NF37 embryos or brains of NF41 were dissected and mRNA total extraction was done. qPCR was done at RT using specific primers couples for TRbeta (*thrb*), TRalpha (*thra*), THbZIP (*thibz*), BTEB (*klf9*), PPARg (*pparg*), RXRg (*rxrg*), proliferation marker PCNA (*pcna*), mature neuron marker tubulin 2 beta (*tubb2b*), stem cell markers (*sox2, oct91,* and *rest*), deiodinase type I (*dio1*), type III (*dio3*). Relative fold changes presented were calculated using geometric mean of *ef1alpha* and *odc* as normalizers and DMSO-treated animals (CTRL) values for the 1.0 reference. Significant downregulation or upregulation ($P < 0.05$) using Mann-Whitney nonparametric test are presented. Results are presented in figures in Fini, Le Mével, et al. (2012) and except those with "*" (unpublished data) and "°" means that this result was obtained on total heads for NF41 and not from brains.

systems. Such is the case of BPA (see above) and of TBBPA that binds to both PPAR and RXR receptors, and in some species to TRs. Such scenarios make the specter of the deleterious effects of endocrine disruption of early development even more disquieting and underline yet again the urgent need for rapid, physiologically screening methods and regulations.

ACKNOWLEDGMENTS

The work in the authors' laboratory is supported by the European Commission (SwitchBox) and by the French Government (ANRs Kismet, Thrast).

REFERENCES

Arjona, F. J., de Vrieze, E., Visser, T. J., Flik, G., & Klaren, P. H. (2011). Identification and functional characterization of zebrafish solute carrier Slc16a2 (Mct8) as a thyroid hormone membrane transporter. *Endocrinology, 152*, 5065–5073.

Bassett, J. H., Nordstrom, K., Boyde, A., Howell, P. G., Kelly, S., Vennstrom, B., et al. (2007). Thyroid status during skeletal development determines adult bone structure and mineralization. *Molecular Endocrinology, 21*, 1893–1904.

Beck, V., Roelens, S. A., & Darras, V. M. (2006). Exposure to PCB 77 induces tissue-dependent changes in iodothyronine deiodinase activity patterns in the embryonic chicken. *General and Comparative Endocrinology, 148*, 327–335.

Beck, V., Roelens, S. A., Maervoet, J., Schepens, P., & Darras, V. M. (2005). Interaction of PCBs with thyroid hormone levels and time of hatching in chicken embryos. *Annals of the New York Academy of Sciences, 1040*, 224–226.

Becker, K. B., Schneider, M. J., Davey, J. C., & Galton, V. A. (1995). The type III 5-deiodinase in *Rana catesbeiana* tadpoles is encoded by a thyroid hormone-responsive gene. *Endocrinology, 136*, 4424–4431.

Bergonzi, R., Specchia, C., Dinolfo, M., Tomasi, C., De Palma, G., Frusca, T., et al. (2009). Distribution of persistent organochlorine pollutants in maternal and foetal tissues: Data from an Italian polluted urban area. *Chemosphere, 76*, 747–754.

Bernal, J. (2007). Thyroid hormone receptors in brain development and function. *Nature Clinical Practice. Endocrinology & Metabolism, 3*, 249–259.

Bernal, J., & Pekonen, F. (1984). Ontogenesis of the nuclear 3,5,3'-triiodothyronine receptor in the human foetal brain. *Endocrinology, 114*, 677–679.

Berry, M. J., & Larsen, P. R. (1993). Molecular cloning of the selenocysteine-containing enzyme type I iodothyronine deiodinase. *The American Journal of Clinical Nutrition, 57*, 249S–255S.

Birnbaum, L. S., & Staskal, D. F. (2004). Brominated flame retardants: Cause for concern? *Environmental Health Perspectives, 112*, 9–17.

Blanco, J., Mulero, M., Lopez, M., Domingo, J. L., & Sanchez, D. J. (2011). BDE-99 deregulates BDNF, Bcl-2 and the mRNA expression of thyroid receptor isoforms in rat cerebellar granular neurons. *Toxicology, 290*, 305–311.

Boas, M., Feldt-Rasmussen, U., & Main, K. M. (2012). Thyroid effects of endocrine disrupting chemicals. *Molecular and Cellular Endocrinology, 355*, 240–248.

Brown, D. D. (1997). The role of thyroid hormone in zebrafish and axolotl development. *Proceedings of the National Academy of Sciences of the United States of America, 94*, 13011–13016.

Brucker-Davis, F. (1998). Effects of environmental synthetic chemicals on thyroid function. *Thyroid, 8*, 827–856.

Burrow, G. N., Fisher, D. A., & Larsen, P. R. (1994). Maternal and foetal thyroid function. *The New England Journal of Medicine, 331*, 1072–1078.

Cao, L. L., Yan, C. H., Yu, X. D., Tian, Y., Zhao, L., Liu, J. X., et al. (2011). Relationship between serum concentrations of polychlorinated biphenyls and organochlorine pesticides and dietary habits of pregnant women in Shanghai. *The Science of the Total Environment, 409*, 2997–3002.

Cao, X. L., Zhang, J., Goodyer, C. G., Hayward, S., Cooke, G. M., & Curran, I. H. (2012). Bisphenol A in human placental and foetal liver tissues collected from Greater Montreal area (Quebec) during 1998–2008. *Chemosphere, 89*, 505–511.

Cariou, R., Antignac, J. P., Zalko, D., Berrebi, A., Cravedi, J. P., Maume, D., et al. (2008). Exposure assessment of French women and their newborns to tetrabromobisphenol-A: Occurrence measurements in maternal adipose tissue, serum, breast milk and cord serum. *Chemosphere*, *73*, 1036–1041.

Carlsson, G., Kulkarni, P., Larsson, P., & Norrgren, L. (2007). Distribution of BDE-99 and effects on metamorphosis of BDE-99 and -47 after oral exposure in *Xenopus tropicalis*. *Aquatic Toxicology*, *84*, 71–79.

Chassande, O. (2003). Do unliganded thyroid hormone receptors have physiological functions? *Journal of Molecular Endocrinology*, *31*, 9–20.

Connors, K. A., Korte, J. J., Anderson, G. W., & Degitz, S. J. (2010). Characterization of thyroid hormone transporter expression during tissue-specific metamorphic events in *Xenopus tropicalis*. *General and Comparative Endocrinology*, *168*, 149–159.

Contempre, B., Jauniaux, E., Calvo, R., Jurkovic, D., Campbell, S., & de Escobar, G. M. (1993). Detection of thyroid hormones in human embryonic cavities during the first trimester of pregnancy. *The Journal of Clinical Endocrinology and Metabolism*, *77*, 1719–1722.

Cossette, S. M., & Drysdale, T. A. (2004). Early expression of thyroid hormone receptor beta and retinoid X receptor gamma in the Xenopus embryo. *Differentiation*, *72*(5), 239–249.

Croteau, W., Davey, J. C., Galton, V. A., & St Germain, D. L. (1996). Cloning of the mammalian type II iodothyronine deiodinase. A selenoprotein differentially expressed and regulated in human and rat brain and other tissues. *The Journal of Clinical Investigation*, *98*, 405–417.

Croteau, W., Whittemore, S. L., Schneider, M. J., & St Germain, D. L. (1995). Cloning and expression of a cDNA for a mammalian type III iodothyronine deiodinase. *The Journal of Biological Chemistry*, *270*, 16569–16575.

Darnerud, P. O. (2008). Brominated flame retardants as possible endocrine disrupters. *International Journal of Andrology*, *31*, 152–160.

Darras, V. M., & Van Herck, S. L. (2012). Iodothyronine deiodinase structure and function: From ascidians to humans. *The Journal of Endocrinology*, *215*(2), 189–206. http://dx.doi.org/10.1530/JOE-12-0204. Epub 2012 Jul 23.

de Escobar, G. M., Obregon, M. J., & del Rey, F. E. (2004). Maternal thyroid hormones early in pregnancy and foetal brain development. *Best Practice & Research. Clinical Endocrinology & Metabolism*, *18*, 225–248.

Dentice, M., & Salvatore, D. (2011). Deiodinases: The balance of thyroid hormone: Local impact of thyroid hormone inactivation. *The Journal of Endocrinology*, *209*(2), 273–282.

Docter, R., Friesema, E. C., van Stralen, P. G., Krenning, E. P., Everts, M. E., Visser, T. J., et al. (1997). Expression of rat liver cell membrane transporters for thyroid hormone in *Xenopus laevis* oocytes. *Endocrinology*, *138*, 1841–1846.

Duarte-Guterman, P., Langlois, V. S., Pauli, B. D., & Trudeau, V. L. (2010). Expression and T3 regulation of thyroid hormone- and sex steroid-related genes during *Silurana* (*Xenopus*) *tropicalis* early development. *General and Comparative Endocrinology*, *166*, 428–435.

Eggesbo, M., Thomsen, C., Jorgensen, J. V., Becher, G., Odland, J. O., & Longnecker, M. P. (2011). Associations between brominated flame retardants in human milk and thyroid-stimulating hormone (TSH) in neonates. *Environmental Research*, *111*, 737–743.

Eriksson, P., Jakobsson, E., & Fredriksson, A. (2001). Brominated flame retardants: A novel class of developmental neurotoxicants in our environment? *Environmental Health Perspectives*, *109*, 903–908.

Escobar-Morreale, H. F., del Rey, F. E., Obregon, M. J., & de Escobar, G. M. (1996). Only the combined treatment with thyroxin and triiodothyronine ensures euthyroidism in all tissues of the thyroidectomized rat. *Endocrinology*, *137*, 2490–2502.

Escriva, H., Holland, N. D., Gronemeyer, H., Laudet, V., & Holland, L. Z. (2002). The retinoic acid signaling pathway regulates anterior/posterior patterning in the nerve cord

and pharynx of amphioxus, a chordate lacking neural crest. *Development, 129,* 2905–2916.
Essner, J. J., Breuer, J. J., Essner, R. D., Fahrenkrug, S. C., & Hackett, P. B., Jr. (1997). The zebrafish thyroid hormone receptor alpha 1 is expressed during early embryogenesis and can function in transcriptional repression. *Differentiation, 62,* 107–117.
Fini, J. B., Le Mével, S., Palmier, K., Darras, V. M., Punzon, I., Richardson, S. J., et al. (2012). Thyroid hormone signalling in the *Xenopus laevis* embryo is functional and susceptible to endocrine disruption. *Endocrinology, 153,* 5068–5081.
Fini, J. B., Le Mevel, S., Turque, N., Palmier, K., Zalko, D., Cravedi, J. P., et al. (2007). An *in vivo* multiwell-based fluorescent screen for monitoring vertebrate thyroid hormone disruption. *Environmental Science and Technology, 41,* 5908–5914.
Fini, J. B., Riu, A., Debrauwer, L., Hillenweck, A., Le Mevel, S., Chevolleau, S., et al. (2012). Parallel biotransformation of tetrabromobisphenol A in *Xenopus laevis* and mammals: *Xenopus* as a model for endocrine perturbation studies. *Toxicological Sciences, 125,* 359–367.
Flamant, F., & Samarut, J. (1998). Involvement of thyroid hormone and its alpha receptor in avian neurulation. *Developments in Biologicals, 197,* 1–11.
Forrest, D., Hallbook, F., Persson, H., & Vennstrom, B. (1991). Distinct functions for thyroid hormone receptors alpha and beta in brain development indicated by differential expression of receptor genes. *The EMBO Journal, 10,* 269–275.
Forrest, D., Sjoberg, M., & Vennstrom, B. (1990). Contrasting developmental and tissue-specific expression of alpha and beta thyroid hormone receptor genes. *The EMBO Journal, 9,* 1519–1528.
Friesema, E. C., Ganguly, S., Abdalla, A., Manning Fox, J. E., Halestrap, A. P., & Visser, T. J. (2003). Identification of monocarboxylate transporter 8 as a specific thyroid hormone transporter. *The Journal of Biological Chemistry, 278,* 40128–40135.
Galton, V. A. (2005). The roles of the iodothyronine deiodinases in mammalian development. *Thyroid, 15,* 823–834.
Gancedo, B., Alonso-Gomez, A. L., de Pedro, N., Delgado, M. J., & Alonso-Bedate, M. (1997). Changes in thyroid hormone concentrations and total contents through ontogeny in three anuran species: Evidence for daily cycles. *General and Comparative Endocrinology, 107,* 240–250.
Gauger, K. J., Giera, S., Sharlin, D. S., Bansal, R., Iannacone, E., & Zoeller, R. T. (2007). Polychlorinated biphenyls 105 and 118 form thyroid hormone receptor agonists after cytochrome P4501A1 activation in rat pituitary GH3 cells. *Environmental Health Perspectives, 115,* 1623–1630.
Geysens, S., Ferran, J. L., Van Herck, S. L., Tylzanowski, P., Puelles, L., & Darras, V. M. (2012). Dynamic mRNA distribution pattern of thyroid hormone transporters and deiodinases during early embryonic chicken brain development. *Neuroscience, 221,* 69–85.
Gibert, Y., Sassi-Messai, S., Fini, J. B., Bernard, L., Zalko, D., Cravedi, J. P., et al. (2011). Bisphenol A induces otolith malformations during vertebrate embryogenesis. *BMC Developmental Biology, 11,* 4.
Glinoer, D. (2000). Potential repercussions for the progeny of maternal hypothyroxinmia during pregnancy. *Thyroid, 10,* 59–62.
Gothe, S., Wang, Z., Ng, L., Kindblom, J. M., Barros, A. C., Ohlsson, C., et al. (1999). Mice devoid of all known thyroid hormone receptors are viable but exhibit disorders of the pituitary-thyroid axis, growth, and bone maturation. *Genes & Development, 13,* 1329–1341.
Greenberg, J. H., Reivich, M., Gordon, J. T., Schoenhoff, M. B., Patlak, C. S., & Dratman, M. B. (2006). Imaging triiodothyronine binding kinetics in rat brain: A model for studies in human subjects. *Synapse, 60,* 212–222.

Gyamfi, C., Wapner, R. J., & D'Alton, M. E. (2009). Thyroid dysfunction in pregnancy: The basic science and clinical evidence surrounding the controversy in management. *Obstetrics and Gynecology*, *113*, 702–707.

Haddow, J. E., Palomaki, G. E., Allan, W. C., Williams, J. R., Knight, G. J., Gagnon, J., et al. (1999). Maternal thyroid deficiency during pregnancy and subsequent neuropsychological development of the child. *The New England Journal of Medicine*, *341*, 549–555.

Hakk, H., & Letcher, R. J. (2003). Metabolism in the toxicokinetics and fate of brominated flame retardants—A review. *Environmental International*, *29*, 801–828.

Hallgren, S., & Darnerud, P. O. (2002). Polybrominated diphenyl ethers (PBDEs), polychlorinated biphenyls (PCBs) and chlorinated paraffins (CPs) in rats-testing interactions and mechanisms for thyroid hormone effects. *Toxicology*, *177*, 227–243.

Hallgren, S., Sinjari, T., Hakansson, H., & Darnerud, P. O. (2001). Effects of polybrominated diphenyl ethers (PBDEs) and polychlorinated biphenyls (PCBs) on thyroid hormone and vitamin A levels in rats and mice. *Archives of Toxicology*, *75*, 200–208.

Hamburgh, M. (1969). The role of thyroid and growth hormones in neurogenesis. *Current Topics in Developmental Biology*, *4*, 109–148.

Havis, E., Le Mevel, S., Morvan Dubois, G., Shi, D. L., Scanlan, T. S., Demeneix, B. A., et al. (2006). Unliganded thyroid hormone receptor is essential for *Xenopus laevis* eye development. *The EMBO Journal*, *25*, 4943–4951.

Heimeier, R. A., & Shi, Y. B. (2010). Amphibian metamorphosis as a model for studying endocrine disruption on vertebrate development: Effect of bisphenol A on thyroid hormone action. *General and Comparative Endocrinology*, *168*, 181–189.

Hennemann, G., Docter, R., Friesema, E. C., de Jong, M., Krenning, E. P., & Visser, T. J. (2001). Plasma membrane transport of thyroid hormones and its role in thyroid hormone metabolism and bioavailability. *Endocrine Reviews*, *22*, 451–476.

Henrichs, J., Bongers-Schokking, J. J., Schenk, J. J., Ghassabian, A., Schmidt, H. G., Visser, T. J., et al. (2010). Maternal thyroid function during early pregnancy and cognitive functioning in early childhood: The generation R study. *The Journal of Clinical Endocrinology and Metabolism*, *95*, 4227–4234.

Heyland, A., Price, D. A., Bodnarova-Buganova, M., & Moroz, L. L. (2006). Thyroid hormone metabolism and peroxidase function in two non-chordate animals. *Journal of Experimental Zoology. Part B, Molecular and Developmental Evolution*, *306*, 551–566.

Howdeshell, K. L. (2002). A model of the development of the brain as a construct of the thyroid system. *Environmental Health Perspectives*, *110*(Suppl. 3), 337–348.

Iwamuro, S., Yamada, M., Kato, M., & Kikuyama, S. (2006). Effects of bisphenol A on thyroid hormone-dependent up-regulation of thyroid hormone receptor alpha and beta and down-regulation of retinoid X receptor gamma in *Xenopus* tail culture. *Life Sciences*, *79*, 2165–2171.

Jacobson, J. L., & Jacobson, S. W. (1996). Dose-response in perinatal exposure to polychlorinated biphenyls (PCBs): The Michigan and North Carolina cohort studies. *Toxicology and Industrial Health*, *12*, 435–445.

Jagnytsch, O., Opitz, R., Lutz, I., & Kloas, W. (2006). Effects of tetrabromobisphenol A on larval development and thyroid hormone-regulated biomarkers of the amphibian *Xenopus laevis*. *Environmental Research*, *101*, 340–348.

Jayatilaka, A. D. (1978). An ultrastructural study of the thyroid gland in the pre-metamorphic *Xenopus laevis* (Daudin) tadpole. *Journal of Anatomy*, *125*, 579–591.

Kaplan, M. M. (1992a). Assessment of thyroid function during pregnancy. *Thyroid*, *2*, 57–61.

Kaplan, M. M. (1992b). Monitoring thyroxin treatment during pregnancy. *Thyroid*, *2*, 147–152.

Kawashiro, Y., Fukata, H., Omori-Inoue, M., Kubonoya, K., Jotaki, T., Takigami, H., et al. (2008). Perinatal exposure to brominated flame retardants and polychlorinated biphenyls in Japan. *Endocrine Journal*, *55*, 1071–1084.

Kim, U. J., Lee, I. S., Kim, H. S., & Oh, J. E. (2011). Monitoring of PBDEs concentration in umbilical cord blood and breast milk from Korean population and estimating the effects of various parameters on accumulation in humans. *Chemosphere, 85*, 487–493.

Kitamura, S., Jinno, N., Ohta, S., Kuroki, H., & Fujimoto, N. (2002). Thyroid hormonal activity of the flame retardants tetrabromobisphenol A and tetrachlorobisphenol A. *Biochemical and Biophysical Research Communications, 293*, 554–559.

Kitamura, S., Kato, T., Iida, M., Jinno, N., Suzuki, T., Ohta, S., et al. (2005). Anti-thyroid hormonal activity of tetrabromobisphenol A, a flame retardant, and related compounds: Affinity to the mammalian thyroid hormone receptor, and effect on tadpole metamorphosis. *Life Sciences, 76*, 1589–1601.

Kojima, H., Takeuchi, S., Uramaru, N., Sugihara, K., Yoshida, T., & Kitamura, S. (2009). Nuclear hormone receptor activity of polybrominated diphenyl ethers and their hydroxylated and methoxylated metabolites in transactivation assays using Chinese hamster ovary cells. *Environmental Health Perspectives, 117*, 1210–1218.

Kooistra, L., Crawford, S., van Baar, A. L., Brouwers, E. P., & Pop, V. J. (2006). Neonatal effects of maternal hypothyroxinmia during early pregnancy. *Pediatrics, 117*, 161–167.

Krain, L. P., & Denver, R. J. (2004). Developmental expression and hormonal regulation of glucocorticoid and thyroid hormone receptors during metamorphosis in *Xenopus laevis*. *The Journal of Endocrinology, 181*, 91–104.

Kuester, R. K., Solyom, A. M., Rodriguez, V. P., & Sipes, I. G. (2007). The effects of dose, route, and repeated dosing on the disposition and kinetics of tetrabromobisphenol A in male F-344 rats. *Toxicological Sciences, 96*, 237–245.

Kutlu, A. O., & Kara, C. (2012). Iodine deficiency in pregnant women in the apparently iodine-sufficient capital city of Turkey. *Clinical Endocrinology, 77*, 615–620.

Laudet, V. (2011). The origins and evolution of vertebrate metamorphosis. *Current Biology, 21*, R726–737.

Leijs, M. M., Ten Tusscher, G. W., Olie, K., van Teunenbroek, T., van Aalderen, W. M., de Voogt, P., et al. (2012). Thyroid hormone metabolism and environmental chemical exposure. *Environmental Health, 11*(Suppl. 1), S10.

Lein, E. S., Hawrylycz, M. J., Ao, N., Ayres, M., Bensinger, A., Bernard, A., et al. (2007). Genome-wide atlas of gene expression in the adult mouse brain. *Nature, 445*, 168–176.

Leloup, J., & Buscaglia, M. (1977). La triiodothyronine, hormone de la métamorphose des amphibiens. *Comptes rendus hebdomadaires des scéances de l'académie des sciences [In french] Serie D, 284*, 2261–2263.

Lezoualc'h, F., Hassan, A., Abdel-Tawab, H., Puymirat, J., & Demeneix, B. A. (1994). Precocious auto-induction of thyroid hormone receptors in embryonic chick hypothalamic neurons. *Neuroscience Letters, 180*, 197–202.

Lezoualc'h, F., Seugnet, I., Monnier, A. L., Ghysdael, J., Behr, J. P., & Demeneix, B. A. (1995). Inhibition of neurogenic precursor proliferation by antisense alpha thyroid hormone receptor oligonucleotides. *The Journal of Biological Chemistry, 270*, 12100–12108.

Li, Y., Shan, Z., Teng, W., Yu, X., Li, Y., Fan, C., et al. (2011). Abnormalities of maternal thyroid function during pregnancy affect neuropsychological development of their children at 25-30 months. *Clinical Endocrinology, 72*, 825–829.

Liang, V. C., Sedgwick, T., & Shi, Y. B. (1997). Characterization of the *Xenopus* homolog of an immediate early gene associated with cell activation: Sequence analysis and regulation of its expression by thyroid hormone during amphibian metamorphosis. *Cell Research, 7*, 179–193.

Lim, W., Nguyen, N. H., Yang, H. Y., Scanlan, T. S., & Furlow, J. D. (2002). A thyroid hormone antagonist that inhibits thyroid hormone action *in vivo*. *The Journal of Biological Chemistry, 277*, 35664–35670.

Lincer, J. L., & Peakall, D. B. (1970). Metabolic effects of polychlorinated biphenyls in the American kestrel. *Nature, 228*, 783–784.

Lopez-Espinosa, M. J., Granada, A., Carreno, J., Salvatierra, M., Olea-Serrano, F., & Olea, N. (2007). Organochlorine pesticides in placentas from Southern Spain and some related factors. *Placenta, 28*, 631–638.
Lopez-Espinosa, M. J., Mondal, D., Armstrong, B., Bloom, M. S., & Fletcher, T. (2012). Thyroid function and perfluoroalkyl acids in children living near a chemical plant. *Environmental Health Perspectives, 120*, 1036–1041.
Lopez-Espinosa, M. J., Murcia, M., Iniguez, C., Vizcaino, E., Llop, S., Vioque, J., et al. (2011). Prenatal exposure to organochlorine compounds and birth size. *Pediatrics, 128*, e127–134.
Mandel, S., Hanna, C., & LaFranchi, S. (1990). Thyroid function of infants born to mothers with Graves disease. *The Journal of Pediatrics, 117*, 169–170.
Manzon, R. G., & Youson, J. H. (1997). The effects of exogenous thyroxin (T4) or triiodothyronine (T3), in the presence and absence of potassium perchlorate, on the incidence of metamorphosis and on serum T4 and T3 concentrations in larval sea lampreys (Petromyzon marinus L.). *General and Comparative Endocrinology, 106*, 211–220.
Marchesini, G. R., Meimaridou, A., Haasnoot, W., Meulenberg, E., Albertus, F., Mizuguchi, M., et al. (2008). Biosensor discovery of thyroxin transport disrupting chemicals. *Toxicology and Applied Pharmacology, 232*, 150–160.
Matsushima, A., Kakuta, Y., Teramoto, T., Koshiba, T., Liu, X., Okada, H., et al. (2007). Structural evidence for endocrine disruptor bisphenol A binding to human nuclear receptor ERR gamma. *Journal of Biochemistry, 142*, 517–524.
McNabb, F. M. (2006). Avian thyroid development and adaptive plasticity. *General and Comparative Endocrinology, 147*, 93–101.
Meerts, I. A., Assink, Y., Cenijn, P. H., Van Den Berg, J. H., Weijers, B. M., Bergman, A., et al. (2002). Placental transfer of a hydroxylated polychlorinated biphenyl and effects on foetal and maternal thyroid hormone homeostasis in the rat. *Toxicological Sciences, 68*, 361–371.
Meerts, I. A., van Zanden, J. J., Luijks, E. A., van Leeuwen-Bol, I., Marsh, G., Jakobsson, E., et al. (2000). Potent competitive interactions of some brominated flame retardants and related compounds with human transthyretin *in vitro*. *Toxicological Sciences, 56*, 95–104.
Miller, M. F., Chernyak, S. M., Domino, S. E., Batterman, S. A., & Loch-Caruso, R. (2012). Concentrations and speciation of polybrominated diphenyl ethers in human amniotic fluid. *The Science of the Total Environment, 417–418*, 294–298.
Miyazaki, W., Iwasaki, T., Takeshita, A., Kuroda, Y., & Koibuchi, N. (2004). Polychlorinated biphenyls suppress thyroid hormone receptor-mediated transcription through a novel mechanism. *The Journal of Biological Chemistry, 279*, 18195–18202.
Moriyama, K., Tagami, T., Akamizu, T., Usui, T., Saijo, M., Kanamoto, N., et al. (2002). Thyroid hormone action is disrupted by bisphenol A as an antagonist. *The Journal of Clinical Endocrinology and Metabolism, 87*, 5185–5190.
Morvan Dubois, G., Sebillot, A., Kuiper, G. G., Verhoelst, C. H., Darras, V. M., Visser, T. J., et al. (2006). Deiodinase activity is present in *Xenopus laevis* during early embryogenesis. *Endocrinology, 147*, 4941–4949.
Nieuwkoop, P. D., & Faber, J. (Eds.), (1994). *Normal table of* Xenopus laevis *(Daudin)*. New York: Garland Publishing Inc.
Nunez, J., Celi, F. S., Ng, L., & Forrest, D. (2008). Multigenic control of thyroid hormone functions in the nervous system. *Molecular and Cellular Endocrinology, 287*, 1–12.
Obregon, M. J., Calvo, R. M., Del Rey, F. E., & de Escobar, G. M. (2007). Ontogenesis of thyroid function and interactions with maternal function. *Endocrine Development, 10*, 86–98.
Oofusa, K., Tooi, O., Kashiwagi, A., Kashiwagi, K., Kondo, Y., Watanabe, Y., et al. (2001). Expression of thyroid hormone receptor betaA gene assayed by transgenic Xenopus laevis carrying its promoter sequences. *Mol Cell Endocrinol, 181*(1–2), 97–110.

O'Shea, P. J., Bassett, J. H., Sriskantharajah, S., Ying, H., Cheng, S. Y., & Williams, G. R. (2005). Contrasting skeletal phenotypes in mice with an identical mutation targeted to thyroid hormone receptor alpha1 or beta. *Molecular Endocrinology, 19*, 3045–3059.

Paris, M., Brunet, F., Markov, G. V., Schubert, M., & Laudet, V. (2008). The amphioxus genome enlightens the evolution of the thyroid hormone signaling pathway. *Development Genes and Evolution, 218*, 667–680.

Paris, M., Escriva, H., Schubert, M., Brunet, F., Brtko, J., Ciesielski, F., et al. (2008). Amphioxus postembryonic development reveals the homology of chordate metamorphosis. *Current Biology, 18*, 825–830.

Patricolo, E., Cammarata, M., & D'Agati, P. (2001). Presence of thyroid hormones in ascidian larvae and their involvement in metamorphosis. *The Journal of Experimental Zoology, 290*, 426–430.

Pharoah, P. O., Ellis, S. M., Ekins, R. P., & Williams, E. S. (1976). Maternal thyroid function, iodine deficiency and foetal development. *Clinical Endocrinology, 5*, 159–166.

Pop, V. J., Kuijpens, J. L., van Baar, A. L., Verkerk, G., van Son, M. M., de Vijlder, J. J., et al. (1999). Low maternal free thyroxin concentrations during early pregnancy are associated with impaired psychomotor development in infancy. *Clinical Endocrinology, 50*, 149–155.

Powell, D. C., Aulerich, R. J., Stromborg, K. L., & Bursian, S. J. (1996). Effects of 3,3',4,4'-tetrachlorobiphenyl, 2,3,3',4,4'-pentachlorobiphenyl, and 3,3',4,4',5-pentachlorobiphenyl on the developing chicken embryo when injected prior to incubation. *Journal of Toxicology and Environmental Health, 49*, 319–338.

Power, D. M., Llewellyn, L., Faustino, M., Nowell, M. A., Bjornsson, B. T., Einarsdottir, I. E., et al. (2001). Thyroid hormones in growth and development of fish. *Comparative Biochemistry and Physiology. Toxicology & Pharmacology: CBP, 130*, 447–459.

Prati, M., Calvo, R., Morreale, G., & Morreale de Escobar, G. (1992). L-Thyroxin and 3,5,3'-triiodothyronine concentrations in the chicken egg and in the embryo before and after the onset of thyroid function. *Endocrinology, 130*, 2651–2659.

Quignodon, L., Legrand, C., Allioli, N., Guadano-Ferraz, A., Bernal, J., Samarut, J., et al. (2004). Thyroid hormone signaling is highly heterogeneous during pre- and postnatal brain development. *Journal of Molecular Endocrinology, 33*, 467–476.

Railton, T. C. (1894). Sporadic cretinism treated by administration of the thyroid gland. *British Medical Journal, 1*, 1180–1181.

Raine, J. C., & Leatherland, J. F. (2000). Morphological and functional development of the thyroid tissue in rainbow trout (*Oncorhynchus mykiss*) embryos. *Cell and Tissue Research, 301*, 235–244.

Rao, G. S., Eckel, J., Rao, M. L., & Breuer, H. (1976). Uptake of thyroid hormone by isolated rat liver cells. *Biochemical and Biophysical Research Communications, 73*, 98–104.

Regard, E., Taurog, A., & Nakashima, T. (1978). Plasma thyroxin and triiodothyronine levels in spontaneously metamorphosing *Rana catesbeiana* tadpoles and in adult anuran amphibia. *Endocrinology, 102*, 674–684.

Ren, X. M., & Guo, L. H. (2012). Assessment of the binding of hydroxylated polybrominated diphenyl ethers to thyroid hormone transport proteins using a site-specific fluorescence probe. *Environmental Science and Technology, 46*, 4633–4640.

Ritchie, J. W., Peter, G. J., Shi, Y. B., & Taylor, P. M. (1999). Thyroid hormone transport by 4F2hc-IU12 heterodimers expressed in *Xenopus* oocytes. *The Journal of Endocrinology, 163*, R5–9.

Roche, J., Michel, R., Volpert, E., & Sanz, B. (1957). Fixation of radioactive rhenium and manganese by the thyroid and ovocytes in egg-laying chickens. *Comptes Rendus des Seances de la Societe de Biologie et de ses Filiales, 151*, 1098–1100.

Roegge, C. S., Morris, J. R., Villareal, S., Wang, V. C., Powers, B. E., Klintsova, A. Y., et al. (2006). Purkinje cell and cerebellar effects following developmental exposure to PCBs and/or MeHg. *Neurotoxicology and Teratology, 28*, 74–85.

Roelens, S. A., Beck, V., Aerts, G., Clerens, S., Vanden Bergh, G., Arckens, L., et al. (2005). Neurotoxicity of polychlorinated biphenyls (PCBs) by disturbance of thyroid hormone-regulated genes. *Annals of the New York Academy of Sciences, 1040*, 454–456.

Rohr, K. B., & Concha, M. L. (2000). Expression of nk2.1a during early development of the thyroid gland in zebrafish. *Mechanisms of Development, 95*, 267–270.

Sager, D. B. (1983). Effect of postnatal exposure to polychlorinated biphenyls on adult male reproductive function. *Environmental Research, 31*, 76–94.

Salay, E., & Garabrant, D. (2009). Polychlorinated biphenyls and thyroid hormones in adults: A systematic review appraisal of epidemiological studies. *Chemosphere, 74*, 1413–1419.

Schauer, U. M., Volkel, W., & Dekant, W. (2006). Toxicokinetics of tetrabromobisphenol a in humans and rats after oral administration. *Toxicological Sciences, 91*, 49–58.

Shepard, T. H. (1967). Onset of function in the human foetal thyroid: Biochemical and radioautographic studies from organ culture. *The Journal of Clinical Endocrinology and Metabolism, 27*, 945–958.

Shepherdley, C. A., Richardson, S. J., Evans, B. K., Kuhn, E. R., & Darras, V. M. (2002). Thyroid hormone deiodinases during embryonic development of the saltwater crocodile (*Crocodylus porosus*). *General and Comparative Endocrinology, 126*, 153–164.

Shi, Y. B., Wong, J., Puzianowska-Kuznicka, M., & Stolow, M. A. (1996). Tadpole competence and tissue-specific temporal regulation of amphibian metamorphosis: Roles of thyroid hormone and its receptors. *Bioessays, 18*, 391–399.

Sjodin, A., Patterson, D. G., Jr., & Bergman, A. (2003). A review on human exposure to brominated flame retardants—Particularly polybrominated diphenyl ethers. *Environmental International, 29*, 829–839.

Smith, T. (1894). Case of sporadic cretinism treated with thyroid gland. *British Medical Journal, 1*, 1178–1180.

Stewart, P. W., Lonky, E., Reihman, J., Pagano, J., Gump, B. B., & Darvill, T. (2008). The relationship between prenatal PCB exposure and intelligence (IQ) in 9-year-old children. *Environmental Health Perspectives, 116*, 1416–1422.

Stewart, P. W., Reihman, J., Lonky, E., & Pagano, J. (2011). Issues in the interpretation of associations of PCBs and IQ. *Neurotoxicology and Teratology, 34*, 96–107.

Sun, H., Shen, O. X., Wang, X. R., Zhou, L., Zhen, S. Q., & Chen, X. D. (2009). Anti-thyroid hormone activity of bisphenol A, tetrabromobisphenol A and tetrachlorobisphenol A in an improved reporter gene assay. *Toxicology In Vitro, 23*, 950–954.

Szabo, D. T., Richardson, V. M., Ross, D. G., Diliberto, J. J., Kodavanti, P. R., & Birnbaum, L. S. (2009). Effects of perinatal PBDE exposure on hepatic phase I, phase II, phase III, and deiodinase 1 gene expression involved in thyroid hormone metabolism in male rat pups. *Toxicological Sciences, 107*, 27–39.

Szisch, V., Papandroulakis, N., Fanouraki, E., & Pavlidis, M. (2005). Ontogeny of the thyroid hormones and cortisol in the gilthead sea bream, *Sparus aurata*. *General and Comparative Endocrinology, 142*, 186–192.

Szybinski, Z., Walas, S., Zagrodzki, P., Sokolowski, G., Golkowski, F., & Mrowiec, H. (2010). Iodine, selenium, and other trace elements in urine of pregnant women. *Biological Trace Element Research, 138*, 28–41.

Tan, Y. V., Couvineau, A., Van Rampelbergh, J., & Laburthe, M. (2003). Photoaffinity labeling demonstrates physical contact between vasoactive intestinal peptide and the N-terminal ectodomain of the human VPAC1 receptor. *The Journal of Biological Chemistry, 278*, 36531–36536.

Tata, J. R. (1968). Early metamorphic competence of *Xenopus* larvae. *Developments in Biologicals, 18*, 415–440.

Thorpe-Beeston, J. G., Nicolaides, K. H., & McGregor, A. M. (1992). Foetal thyroid function. *Thyroid, 2*, 207–217.

Tindall, A. J., Morris, I. D., Pownall, M. E., & Isaacs, H. V. (2007). Expression of enzymes involved in thyroid hormone metabolism during the early development of *Xenopus tropicalis*. *Biology of the Cell*, *99*, 151–163.

Van Herck, S. L., Geysens, S., Delbaere, J., Tylzanowski, P., & Darras, V. M. (2012). Expression profile and thyroid hormone responsiveness of transporters and deiodinases in early embryonic chicken brain development. *Molecular and Cellular Endocrinology*, *349*, 289–297.

Volpert, E., Michel, R., & Roche, J. (1958). Concentration of iodides and thyroid hormones by the ovary of the frog (*Rana temporaria* L.). *Comptes Rendus des Seances de la Societe de Biologie et de ses Filiales*, *152*, 406–408.

Vulsma, T., Gons, M. H., & de Vijlder, J. J. (1989). Maternal-foetal transfer of thyroxin in congenital hypothyroidism due to a total organification defect or thyroid agenesis. *The New England Journal of Medicine*, *321*, 13–16.

Wallis, K., Dudazy, S., van Hogerlinden, M., Nordstrom, K., Mittag, J., & Vennstrom, B. (2010). The thyroid hormone receptor alpha1 protein is expressed in embryonic postmitotic neurons and persists in most adult neurons. *Molecular Endocrinology*, *24*, 1904–1916.

Walpita, C. N., Crawford, A. D., Janssens, E. D., Van der Geyten, S., & Darras, V. M. (2009). Type 2 iodothyronine deiodinase is essential for thyroid hormone-dependent embryonic development and pigmentation in zebrafish. *Endocrinology*, *150*, 530–539.

Walpita, C. N., Van der Geyten, S., Rurangwa, E., & Darras, V. M. (2007). The effect of 3,5,3'-triiodothyronine supplementation on zebrafish (*Danio rerio*) embryonic development and expression of iodothyronine deiodinases and thyroid hormone receptors. *General and Comparative Endocrinology*, *152*, 206–214.

Welshons, W. V., Nagel, S. C., & vom Saal, F. S. (2006). Large effects from small exposures. III. Endocrine mechanisms mediating effects of bisphenol A at levels of human exposure. *Endocrinology*, *147*, S56–69.

Williams, G. R., & Bassett, J. H. (2011). Deiodinases: The balance of thyroid hormone: Local control of thyroid hormone action: Role of type 2 deiodinase. *The Journal of Endocrinology*, *209*, 261–272.

Wu, S. Y., Green, W. L., Huang, W. S., Hays, M. T., & Chopra, I. J. (2005). Alternate pathways of thyroid hormone metabolism. *Thyroid*, *15*, 943–958.

Yaoita, Y., & Brown, D. D. (1990). A correlation of thyroid hormone receptor gene expression with amphibian metamorphosis. *Genes & Development*, *4*, 1917–1924.

Youson, J. H., & Manzon, R. G. (2012). Lamprey metamorphosis. In S. Doufour, K. Rousseau & B. G. Kapoor (Eds.), *Metamorphosis in fish*. Enfield, New Hampshire: Science Publishers.

Youson, J. H., Manzon, R. G., Peck, B. J., & Holmes, J. A. (1997). Effects of exogenous thyroxin (T4) and triiodothyronine (T3) on spontaneous metamorphosis and serum T4 and T3 levels in immediately premetamorphic sea lampreys, *Petromyzon marinus*. *The Journal of Experimental Zoology*, *279*, 145–155.

Yu, L., Deng, J., Shi, X., Liu, C., Yu, K., & Zhou, B. (2010). Exposure to DE-71 alters thyroid hormone levels and gene transcription in the hypothalamic-pituitary-thyroid axis of zebrafish larvae. *Aquatic Toxicology*, *97*, 226–233.

Zhang, J. G., Sun, X. W., & Ai, H. (2012). Levels and congener profiles of polybrominated diphenyl ethers (PBDEs) in primipara breast milk from Shenzhen and exposure risk for breast-fed infants. *Journal of Environmental Monitoring*, *14*, 893–900.

Zhou, T., Ross, D. G., DeVito, M. J., & Crofton, K. M. (2001). Effects of short-term *in vivo* exposure to polybrominated diphenyl ethers on thyroid hormones and hepatic enzyme activities in weanling rats. *Toxicological Sciences*, *61*, 76–82.

Zimmermann, M. B. (2011). The role of iodine in human growth and development. *Seminars in Cell & Developmental Biology*, *22*, 645–652.

Zoeller, T. R., Dowling, A. L., Herzig, C. T., Iannacone, E. A., Gauger, K. J., & Bansal, R. (2002). Thyroid hormone, brain development, and the environment. *Environmental Health Perspectives, 110*(Suppl. 3), 355–361.

Zoeller, R. T., Dowling, A. L., & Vas, A. A. (2000). Developmental exposure to polychlorinated biphenyls exerts thyroid hormone-like effects on the expression of RC3/neurogranin and myelin basic protein messenger ribonucleic acids in the developing rat brain. *Endocrinology, 141*, 181–189.

Zota, A. R., Park, J. S., Wang, Y., Petreas, M., Zoeller, R. T., & Woodruff, T. J. (2011). Polybrominated diphenyl ethers, hydroxylated polybrominated diphenyl ethers, and measures of thyroid function in second trimester pregnant women in California. *Environmental Science and Technology, 45*, 7896–7905.

CHAPTER FOURTEEN

Thyroid Hormones and Postembryonic Development in Amniotes

Guillaume Holzer, Vincent Laudet[1]
Molecular Zoology Team, Institut de Génomique Fonctionnelle de Lyon, École Normale Supérieure de Lyon, UMR5242 CNRS, Université Lyon, France
[1]Corresponding author: e-mail address: vincent.laudet@ens-lyon.fr

Contents

Abstract

In chordates, metamorphosis is a developmental event well described in amphibians in which thyroid hormone triggers this event. Interestingly, among amphibians, several variations upon the eggs/tadpole/frog developmental sequence are observed such as direct development or neoteny. The fact that TH-regulated metamorphosis is conserved in invertebrate chordates such as amphioxus implies that this event is an ancient

Current Topics in Developmental Biology, Volume 103
ISSN 0070-2153
http://dx.doi.org/10.1016/B978-0-12-385979-2.00014-9

feature of all vertebrates. This allows us to propose that TH may play an important role in coordinating the postembryonic development of apparently nonmetamorphosing vertebrates such as mammals or sauropsids. Indeed, the observations of thyroid hormone levels in mammals and sauropsids draw interesting parallels with what is observed during amphibian metamorphosis. At the physiological level, the increase of thyroid hormone signaling is required for the normal development particularly for the intestine and the brain. At the behavioral level, a peak of TH often precedes the autonomy of the young from parental care. At the ecological level, offspring with a TH peak close to birth/hatching tends to be precocial young whereas offspring with a TH peak long after birth/hatching tends to be altricial young. Taken together, these observations in amniotes, which are not considered as undergoing metamorphosis during their development, are consistent with the idea of a late developmental step controlled by TH and allowing the accession to the adult ecological niche. Thus, according to this view, at the molecular level all vertebrates undergo a period of remodeling controlled by TH that is reminiscent of metamorphosis.

1. INTRODUCTION

Metamorphosis is classically defined as a spectacular and abrupt postembryonic transformation of a larva into a juvenile (Shi, 2000). Metamorphosis has received most attention in vertebrate model species such as the toad *Xenopus laevis* but also some teleost fishes like the flatfishes, where thyroid hormones (TH) produced by the thyroid gland, trigger the changes observed at the morphological and physiological levels. However, despite its main association with amphibians and teleost fish, it is a really widespread life history transition, with most animals undergoing some form of metamorphosis. Our recent analyses of metamorphosis in invertebrate chordates, specifically in amphioxus, have revealed that metamorphosis is triggered by a TH derivative and its receptor (Paris et al., 2008). This has led to the suggestion that TH-regulated metamorphosis is a shared ancestral character common to all chordates. According to this view, metamorphosis in frogs is an extreme example of a widespread life history transition and all vertebrates, including mammals, do pass during their postembryonic development through a remodeling period governed by TH that is extremely similar, at the molecular and endocrine levels, to metamorphosis (see Laudet, 2011 for a review).

In this review, we examine the role of TH during postembryonic development by comparing, across vertebrate species, what we know about TH levels and their roles in the newborn and the young. If THs are well studied during amphibians and teleost fishes postembryonic development because of their role in controlling metamorphosis, much less is known in other vertebrates. Indeed in mammals, THs are mostly studied for their endocrine role in adult, especially for medical purposes such as heart rate control, thermogenesis, or regulation of metabolism. Here, we focus mainly on amniotes, the group of vertebrates clustering mammals and sauropsids (birds, crocodiles, turtles, lizard, and snakes), in order to provide a general view on the role played by TH in the postembryonic development of these taxa. This is needed in order to compare on firm ground how metamorphosing and nonmetamorphosing vertebrates use TH for controlling their postembryonic development.

2. TH SIGNALING: THE BASICS

THs are quite unusual signaling molecules, as they are iodinated: the hormone is formed by the condensation of two aromatic rings of tyrosine residues coupled with iodine atoms. The cycle linked to the amino acid radical is called the inner ring and the other one, the outer ring. Each cycle can carry two iodine atoms in positions 3 and 5 (Hulbert, 2000). Two forms of the hormone are considered as biologically relevant: the thyroxine or 3,5,3′,5′ tetraiodothyronine also known as T4 that carries four iodine atoms and the 3,3′,5 triiodothyronine or T3 that is deiodinated at position 5′.

In most species, the thyroid gland mainly produces T4, which is transformed into T3 in peripheral organs by the action of outer-ring deiodinase (Fig. 14.1). T4 and T3 do not have the same affinity for the TRs; indeed, T3 has a 10-fold higher affinity than T4 and is thus considered as the biologically active compound (Hulbert, 2000). However, it is possible that T4 plays a specific role in some organs in which outer-ring deiodinases are not active. Similarly, several data suggest that a T3 metabolite 3,5 diiodothyronine or 3,5-T2 may be an active compound in specific conditions (Ball, Sokolov, & Chin, 1997). It is becoming increasingly clear that deiodination, that is the removal of an iodine atom in the outer or inner ring, is an important control mechanism of the TH activity. Deiodination is performed by three paralogous enzymes in vertebrates, D1, D2, and D3 deiodinases, that are selenoproteins (Bianco & Kim, 2006). The outer-ring deiodination transforms the native form T4 into the active form T3 and is performed by D1 and D2. It is thus an activating deiodination. By acting on T3, it can produce 3,5-T2, as mentioned previously (Ball et al., 1997). The inner-ring deiodinase transforms T4 into the inactive reverse T3 (rT3; 3,3′,5′

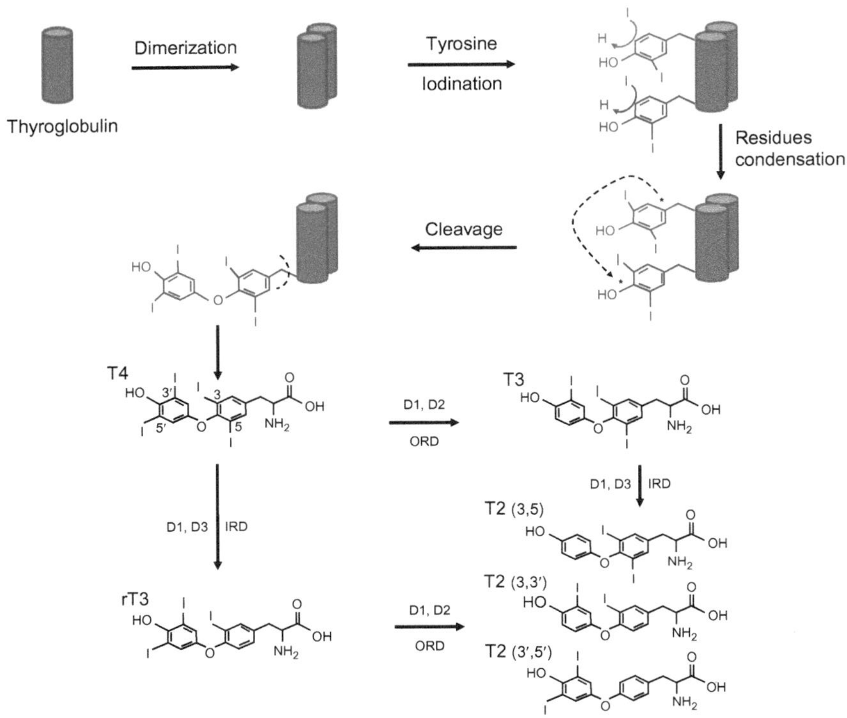

Figure 14.1 Synthesis of thyroid hormones. ORD: outer-ring deiodinase; IRD: inner-ring deiodinase; T4: thyroxine, native form, the numbers correspond to the iodine atoms; T3: 3,3′,5-triiodothyronine, most active form; rT3: 3,3′,5′-triiodothyronine, inactive form; T2: 3,3′-diiodothyronine, inactive form; T2: 3′,5′-diiodothyronine, inactive form; T2: 3,5-diiodothyronine, active in certain condition. *Adapted from Bianco and Kim (2006) and Gavaret et al. (1981).* (For color version of this figure, the reader is referred to the online version of this chapter.)

triiodothyronine) or the active T3 into the inactive 3,3′-T2. This inactivating deiodination is catalyzed by D1 and D3 (Kuiper, Kester, Peeters, & Visser, 2005). Thus, deiodinases can have opposite roles, D2 being associated with active hormone production, whereas D3 is considered as a degrading enzyme. The balance between the two enzymes is important in the homeostatic regulation of T3 levels in peripheral organs. The function of D1, however, is not yet fully understood. It is the main deiodinase expressed in the liver but the part of its contribution to the circulating T3 is not clearly established (for review, see Darras & Van Herck, 2012). This enzyme might be used as a mechanism to recycle the iodine atom (for review, see Orozco, Valverde, Olvera, & García, 2012).

The synthesis of THs is a complex pathway that occurs in the thyroid gland (Fig. 14.1). During this process, a large protein, thyroglobulin, is used as a reservoir of tyrosine residues on which iodine atoms are coupled by thyroperoxidase (TPO). These mono- or di-iodotyrosines are then coupled to form either T4 (mainly) or T3 (Dunn & Dunn, 1999; Gavaret, Cahnmann, & Nunezet, 1981). The newly synthesized molecules are then separated from the matrix protein, secreted and transported in the blood by various carrier, or transporter proteins that are more or less specific for THs (for review, see Visser, Friesema, & Visser, 2011). Several molecules, known as goitrogens, are widely used to specifically block THs synthesis pathway. These chemicals mainly act by inhibiting the iodine uptake in the thyroid gland or by the binding of iodine to the tyrosine residues. For instance, methimazole or propylthiouracil inhibits TPO and specifically its action in oxidizing the iodine anion before its incorporation. In contrast, sodium perchlorate inhibits the sodium-dependant iodide transporter known as NIS that is required for iodine uptake in the thyrocyte (Opitz et al., 2006).

TH binds to thyroid hormone receptors (TRs) that are members of the nuclear receptor family. There are two TRs genes TRα and TRβ in most vertebrates that originate from an ancestral vertebrate duplication and are thus considered as paralogous genes (Escriva, Manzon, Yousson, & Laudet, 2002; Paris & Laudet, 2008). These receptors act as ligand-dependant transcription factors regulating negatively the transcription of their target genes in the absence of TH and positively in their presence (Buchholz, Paul, Fu, & Shi, 2006). In the past two decades, the establishment of several TR mutant mice line provided powerful tools to investigate the role of TH particularly during development and homeostasis in mammals (Flamant, Gauthier, & Samarut, 2007; Flamant & Samarut, 2003).

It is important to note that in all vertebrates the amount of TH produced by the thyroid gland is controlled by the hypothalamic–pituitary–thyroid (HPT) axis (Chiamolera & Wondisford, 2009). The thyrotropin-releasing hormone (TRH) production by the hypothalamus is controlled by environmental cues and regulates the levels of thyroid-stimulating hormone (TSH) production by the pituitary, which in turn stimulates TH production by the thyroid. In turn, T3 has long been known to control its own production via negative feedback loops acting at the levels of TRH and TSH production in the hypothalamus and pituitary, respectively. The molecular control of these feedback loops is far from being well understood and is currently deciphered in mammals using genetically modified mouse models. Although it is out of the scope of this review, it is important to mention that in amphibians it is

CRF and not TRH that controls TSH level during metamorphosis, and as CRF is controlling the corticoid axis, this explains the coupling between stress and metamorphosis (see Laudet, 2011 and references therein).

3. TH CONTROLS METAMORPHOSIS IN AMPHIBIANS

3.1. Metamorphosis in anurans

Anurans (frogs and toads) provided the historical models in which the link between metamorphosis and TH was established. The first hints were given when tadpole fed with horse thyroid extract underwent a precocious metamorphosis compared to others fed with nonthyroid tissue (liver, muscle, etc.) (Gudernatsch, 1912). Since this experiment, a lot of publications investigating metamorphosis used the toad *Xenopus tropicalis* or *X. laevis* as model organism, and TH levels are today well known along the various stages in these models. Several authors have shown that a peak of T4 and T3 occurs at the climax of metamorphosis, its most anatomically spectacular phase when the animal exhibits both regressing larval and developing adult features (Leloup & Buscaglia, 1977). In a similar way, there is a peak of *TRβ* mRNA, but not *TRα* mRNA, at the climax of metamorphosis (Fig. 14.2). Interestingly, the comparison of *TRα* expression levels between different toad species have shown that a higher expression correlates with an earlier and faster metamorphosis suggesting that the level of this receptor is important for the control of the onset of metamorphosis (Hollar, Choi, Grimm, & Buchholz,

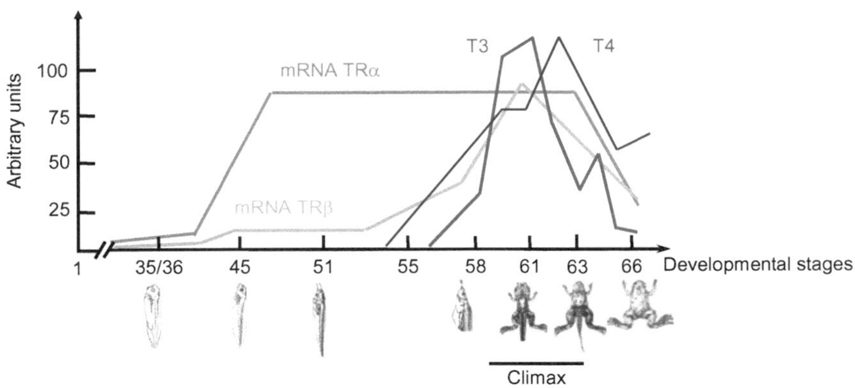

Figure 14.2 THs and TRs mRNA level during *Xenopus laevis* development. *Adapted from Leloup and Buscaglia (1977). Stage illustrations are adapted from Nieuwkoop and Faber (1994).* (For color version of this figure, the reader is referred to the online version of this chapter.)

2011). This difference of profile indicates that, even if both are implicated, *TRα* and *TRβ* do not play the same role during metamorphosis. The peak of *TRβ* is a key event in triggering the whole process and could be used as a signature of metamorphosis in Anurans.

Treatment with exogenous TH induces a precocious metamorphosis. On the contrary, tadpole treated by goitrogen becomes much bigger than the control and fails to undergo metamorphosis, indicating that the tadpoles still grow but are not able to achieve their postembryonic development (see Buchholz et al., 2006; Laudet, 2011; and references therein). THs are produced only by the thyroid gland and are distributed by the blood stream in the peripheral organs of the animal. As mentioned above, locally, the amount of THs and particularly the amount of T3 are controlled by deiodinase activity that has been shown as important to regulate the remodeling of tissues in amphibian metamorphosis such as the tail resorption or limb growth (Huang, Cai, Remo, & Brown, 2001). The action of deiodinases explains at least in part the difference in the timing of metamorphosis between various tadpole organs, such as the limbs that appear early and the tail that shrinks at the final stages.

During amphibian metamorphosis, most organs undergo complex and coordinated changes and this event is an excellent model to study the effects of THs at different levels. The intestine undergoes spectacular changes including both apoptosis and cell proliferation (for review, see Ishizuya-Oka, Hasebe, & Shi, 2010). In *X. laevis*, the adult intestine exhibits villosities and is 90% shorter than the long and simple larval intestine which consists of a monolayer of larval epithelial cells, with thin layers of surrounding connective tissue and muscles. When T3 levels surge the vast majority, if not all, of the larval epithelial cells undergo apoptosis, and concurrently, adult epithelial stem/progenitor cells appear *de novo* and rapidly proliferate (Ishizuya-Oka & Ueda, 1996; Shi, Hasebe, Fu, Fujimoto, & Ishizuya-Oka, 2011). Thus, the intestine undergoes opposite changes at the morphological and cellular level, triggered by T3. The brain also exhibits dramatic changes, with a massive apoptosis of certain central nervous system neurons (Coen, Le Blay, Rowe, & Demeneix, 2007) as well as motor and sensory neurons of the resorbing tail. Other organs, such as gills that disappear or skin that acquire features necessary for a terrestrial life are drastically remodeled during metamorphosis (for review, see Ishizuya-Oka et al., 2010). We will not review in detail the action of TH at the cellular and gene regulation levels during *Xenopus* metamorphosis as several excellent reviews are available on this topic (see Chapters 2, 4 and 9 in this issue).

3.2. Metamorphosis variation in amphibians

Within amphibians, all species do not undergo such marked metamorphosis as seen in Anurans such as *Xenopus*. For most Urodeles (salamanders and newts), we have as yet no clear description of how THs and their receptors trigger metamorphosis at the cellular and molecular levels. Even in classical biphasic species such as *Pleurodeles waltl* in which metamorphosis is well known at the morphological level (Galien & Durocher, 1957), it is not known if and how the *Xenopus* model can be applied. This is a clear limitation for understanding the frequent variations of metamorphosis that are observed in several species of Urodeles.

The axolotl (*Ambystoma mexicanum*) is a facultatively neotenic Urodele that belongs to a large complex of species (Voss & Shaffer, 1997). Under natural condition, some species undergo metamorphosis and develop the classical adult features of a salamander, but many species of Ambystoma, and in particular *A. mexicanum*, acquire the ability to reproduce while keeping their larval features. In axolotl, TRs are functional; they respond to TH, bind DNA, are expressed at high levels in several tissues (Safi et al., 2004), and exogenous T4 treatment on axolotl induces metamorphosis (Brown, 1997). The facultative neoteny of axolotl has been attributed to low levels of circulating THs due to weak activation of the thyroid axis by the pituitary gland, when the rest of TH signaling pathway function normally (Boorse & Denver, 2002; Safi et al., 2004).

The mudpuppy (*Necturus maculosus*) is an example of an obligate paedomorphic amphibian. Obligate means that the adult of this species exhibits the same morphological traits as the larvae of closely related species. Morphologically, the mudpuppy resembles a typical Urodele larva with external gills and an entirely aquatic life history. It was even described as an animal resistant to the action of THs (Yaoita, Shi, & Brown, 1990). Nevertheless, it has been shown that, as for *Xenopus*, mudpuppy TRs can bind DNA and fully respond to THs (Safi et al., 2006). Moreover, *TRβ* is expressed in the gills (Vlaeminck-Guillem et al., 2006) and is upregulated by TH, as during *Xenopus* metamorphosis, indicating that this animal is not resistant to TH. In *Necturus*, the visible absence of metamorphosis is more likely due to changes in the genes regulated by TH rather than by a global arrest of metamorphosis. Indeed, the intestine of *Necturus* resembles that of an adult *Xenopus*, suggesting that this organ undergoes a postembryonic remodeling similar to the one occurring in *Xenopus*. We thus suggest that *Necturus* undergoes a weak, or subtle, metamorphosis in which the pathway controlling TH levels and TH tissue response is indeed present with no obvious morphological changes. This example provides an excellent illustration of similar situations that can be found in other species of vertebrates, in particular, mammals and other amniotes.

3.3. Direct development in amphibians

One of the most striking examples of heterochrony of metamorphosis comes from the terrestrial frog *Eleutherodactylus coqui* (see Chapter 5 in this issue). In this species, a small froglet hatches directly from the egg without any free-living tadpole stage. This situation is called direct development. In this species, some of the larval features, as lateral line organs or some cranial features do not appear at all during development but others, such as the tail, are maintained within the egg and are transformed before hatching (Callery & Elinson, 2000). *E. coqui* and *Xenopus* exhibit similar TRs expression profile. TRα expression is detectable at early developmental stages and remains fairly constant during development. There is a peak of *TRβ* at the same time as the organ remodeling occurs (Callery & Elinson, 2000). T3 treatment increases tail resorption in the developing *E. coqui* (Kulkarni, Singamsetty, & Buchholz, 2010), while goitrogen inhibits *E. coqui* development and their effect can be rescued by T3 treatment (Callery & Elinson, 2000), as for classical amphibian metamorphosis in *Xenopus*. One of the features of the *E. coqui* egg is the large size of the yolk compared to other amphibian species. For instance, *E. coqui* eggs are 20 times the volume of *X. laevis* egg, mainly due to the yolk size (Buchholz et al., 2007), whereas the adult *E. coqui* is smaller than *X. laevis*, indicating that important energy reserves seem necessary for direct development. Thus, in this species, a true metamorphosis occurs in the embryonic period, within the egg. This can also be used as a proxy to explain the situation in other vertebrate species in which no metamorphosis is visible.

4. TH-REGULATED METAMORPHOSIS IS AN ANCESTRAL CHORDATE CHARACTER

The molecular characterization of metamorphosis in model species (*Xenopus* but also some teleost fishes like the flatfishes) has shown that this process is governed in all these species by TH that triggers the changes observed at the morphological and physiological levels via their receptors TRα and TRβ. Recent analyses of metamorphosis conducted in invertebrate chordates, specifically in the cephalochordate amphioxus (*Branchiostoma*), have revealed that in this basal chordate, metamorphosis is also triggered by TH and its receptor (Paris et al., 2008, 2010). Amphioxus undergoes metamorphosis with a spectacular morphological change: the asymmetric larval mouth and gill slits becoming symmetric in the adult. This is accompanied by an ecological change, as the pelagic larva becomes a benthic filter-feeding adult living in sand. Amphioxus exhibits a peak of its unique TR expression just before metamorphosis, although the canonical forms of TH (i.e., T3 and T4) do not activate the receptor. We demonstrated that TRIAC, a T3 derivative, plays a T3-like role in this species: binding the amphioxus TR and inducing metamorphosis (Paris et al., 2008, 2010). Taken together, these data support the idea that the TH pathway is involved in metamorphosis in the cephalochordate amphioxus. Given both the phylogenetic position of amphioxus and the strong evidence for TH-controlled metamorphosis in urochordates (Patricolo, Cammarata, & D'Agati, 2001), TH-regulated metamorphosis appears thus as an ancient process. This, together with the observations of variations on the way metamorphosis occurs in development of some species as the neotenic *Necturus* or the direct developer *E. coqui* (Laudet, 2011), leads to the suggestion that TH-regulated metamorphosis is a shared ancestral character common to all chordates (Laudet, 2011; Paris et al., 2008; Paris & Laudet, 2008).

According to this model, vertebrate species that do not exhibit a clear metamorphosis during postembryonic development would either be direct developers, with intra-egg metamorphosis and a juvenile hatching from the egg, or alternatively, undergo a morphologically cryptic posthatching metamorphosis such as observed in *Necturus*. These ideas challenge the classical view of vertebrate postembryonic development, a rather neglected developmental period, especially in species that apparently do not undergo metamorphosis. In the remaining part of this chapter, we will thus concentrate on amniotes (Fig. 14.3), their postembryonic development, and the role of TH.

5. MAMMALS

If indeed there is a role of TH in the control of postembryonic development in mammals, similar to the one observed in biphasic species such as *Xenopus* or flatfishes, we should observe two phenomena: (i) a peak of TH during this period and (ii) a role of TH in the remodeling of organs and tissues during this peak. We will thus analyze the situation in mammals, taken into account that the actual data available are extremely biased toward mouse and human.

5.1. TH levels in eutherian mammals

In mice (*Mus musculus*), TH levels are relatively low at birth and a peak of circulating TH is observed at 15 days postpartum. T4 levels increase about 10-fold and T3 levels increase even more (Hadj-Sahraoui, Seugnet, Ghorbel, & Demeneix, 2000). In rat, the situation is very similar to the mouse (Babu et al., 2011). In human (*Homo sapiens*), embryonic blood TH levels increase during the gestation in contrast with maternal levels that remain stable (Thorpe-Beeston, Nicolaides, Felton, Butler, & McGregor,

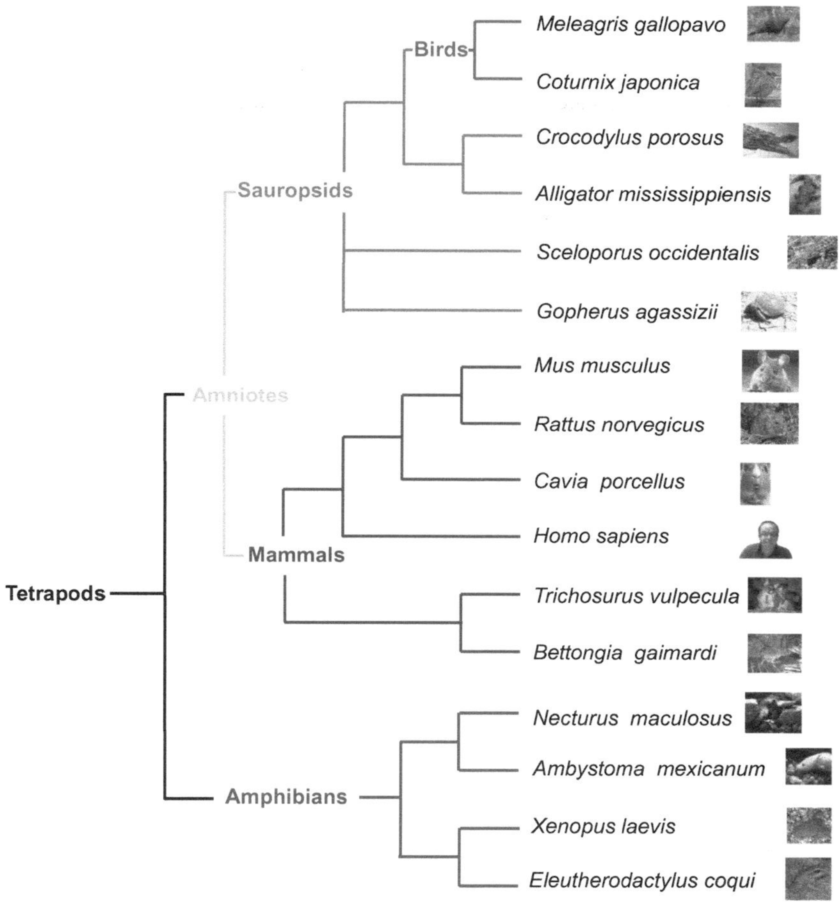

Figure 14.3 Phylogenetic tree of the species described in this review. The several relevant phylum are highlighted in different colors. *Adapted from Werneburg and Sánchez-Villagra (2009) and Shen, Liang, Wen, and Zhang (2011).* (See Color Insert.)

1991). Peaks of free plasma T4 and T3 are observed between 24 and 48 h following birth; smoother increases are also observed for total plasmatic T3 and T4 at the same period (Erenberg, Phelps, Lam, & Fisher, 1974).

There is only sparse data regarding other developmental TH levels in mammals but from, often quite early, published data one can conclude that in sheep (Breall, Rudolp, & Heymann, 1984), cow (Paulíková, Seidel, Nagy, Tóthová, & Ková, 2011), pig (Carrol, Veum, & Matteri, 1998), and horse (Irvine & Evans, 1975), there is, as in human, a peak of TH that corresponds to birth. In other species such as seals, several data indicate that a

TH peak occurs after birth in a period of molt that precedes the weaning period (Atkinson, Amould, & Mashburn, 2011). This suggests that the precise timing of the TH peak can vary from one species to another and that it would be very interesting to study more systematically these levels in several mammalian orders to see if a clear pattern emerges and whether it is related to specific developmental strategies.

Another rodent species, the guinea pig (*Cavia porcellus*), seems particularly interesting in this respect. This species has a very long gestation period as compared to the mouse, the level of free T4 in the serum doubles between 45 and 62 days of gestation, reaching a maximum near term (at 63–68 days of gestation). However, the levels of total or free T3 in serum do not follow the same path, whereas the levels of reverse T3 increase as those of T4 (Castro, Alex, Young, Braveman, & Emerson, 1986). This suggests that a significant proportion of the T4 is degraded into rT3 in the serum. Whether or not this corresponds to an increase of T3 concentration within tissues remains to be studied. What is clear from these data, however, is that a high level of T4 occurs during postembryonic development in the mouse whereas in guinea pig, such high levels are reached at the end of gestation. It is not clear, however, how the level of T4 and T3 change during postembryonic life in guinea pig. However, the authors of this study conclude that, in contrast to mouse and rat, maturation of the thyroid system occurs *in utero* in guinea pig (Castro et al., 1986). Taken together, the data coming from mammals show that there is a TH peak during late gestation or early postembryonic life, but that the precise timing of this peak, relative to birth, may change from one species to another.

5.2. Role of TH in late gestation and early postembryonic life: data from mice

In mammals, during weaning, the pup becomes independent from the mother, ceasing to suckle. This transition requires physiological modification and remodeling of several organs among which the intestine develops significantly during this period. Changes in other organs are also observed: bone ossification acceleration, appearance of autonomous thermoregulation, and brain remodeling (for review, see Forrest & Vennström, 2000). These events are dependent on a functional TH signaling pathway. Indeed, analysis of Pax8 KO mice, that are severely hypothyroid due to the failure of the thyroid gland to develop, confirms this notion: these mice display growth arrest and fail to properly remodel their organs before dying around weaning at 3 weeks (Mansouri, Chowdhury, & Gruss, 1998). The TRα-deficient mice exhibit similar defects: after 2 weeks, these mice (either the $TR\alpha^{-/-}$ that lack only the TRα1 isoform or the TRα0/0 that lack all the isoforms encoded by the TRα locus; see Flamant et al., 2007) become progressively hypothyroid, displaying growth arrest (Fraichard et al., 1997; Gauthier et al., 1999, 2001). During this period, many other physiological functions are modified, correlating with organ maturation. The increase in intestine size is characterized by an extensive development of the villi and muscular layers (Sirakov & Plateroti, 2011). Metabolic controls also mature and specific genes involved in *de novo* lipogenesis are turned on preparing the soon-to-be independent pup to obtain lipids from food. All these data therefore suggest that TRα gene is essential for the transformation of a maternally dependent pup to a juvenile-autonomous mouse. Thus, the pup diet switches for the adult diet at weaning. Then, this switch is related to the digestive system remodeling and nervous system development, both of these events being dependent of the TH pathway. Overall, this mechanism is essential for the transformation of a maternally dependent pup to a juvenile-autonomous mouse and this is highly reminiscent of TH-controlled metamorphosis in amphibians.

It is well known in human that THs are critical for brain development with a particularly sensitive window of maturation at birth. In fact, TH levels are one of the first blood parameters measured in human newborns. The absence of TH during the very first weeks of life of a human baby can have profound and definitive consequences on its brain maturation (Biswas et al., 2002; Williams, 2008). The effects of hypothyroidism effect on human brain development have long been studied and described (see for an early review Legrand, 1979). Severe and untreated hypothyroidism from birth, usually due to congenital hypothyroidism results in the phenotype known as cretinism characterized by delayed mental and physical development, hearing and speech defects, and hypotonia (for review, see Bernal, 2007).

Using mouse models, it has been shown that THs and their receptors are involved in normal brain development, particularly in the cerebellum (Billon, Tokumoto, Forrest, & Raff, 2001; Flamant & Samarut, 2003; Morte, Manzano, Scanlan, Vennstörm, & Bernal, 2002). The rodent cerebellum is composed of a small number of neuronal types (Purkinje cells, granule cells, and GABAergic interneurons) that are morphologically and molecularly well characterized (for review, see Dusart & Flamant, 2012). Neurogenesis in the cerebellum extends over a protracted period of time between embryonic day 10.5 and postnatal day 15 in mice. The post-embryonic period is particularly important as there is a maturation event that is especially spectacular for the Purkinje neurons. Indeed, the development of mouse Purkinje cells occurs largely during the first three postnatal weeks of life, during which Purkinje cells undergo a two-phase remodeling process of their dendrites and establish synaptic connections (Sotelo & Dusart, 2009). This also coincides with granule cell proliferation, maturation of GABAergic interneuron precursors in neighboring cerebellar layers, as well as with the terminal differentiation of oligodendrocytes. This postnatal maturation program is clearly driven by THs (Dusart & Flamant, 2012). In rodents, T3 deficiency results in a number of histological alterations visible in cerebellum. Purkinje cell alignment is affected, their dendritic arborisations are drastically reduced, and they have fewer synapses (Fauquier et al., 2011). This cellular phenotype can be rescued only if the TH level is restored at an early stage. Oligodendrocyte proliferation arrest that occurred during their postnatal differentiation process is controlled in a cell autonomous fashion (Picou, Fauquier, Chatonnet, & Flamant, 2012). Recent data have suggested that Purkinje neurons have a central role in the coordination of cellular interactions that are instrumental for this maturation process and that TH may play an important role in setting the timing of Purkinje cell differentiation (Fauquier et al., 2011; Picou et al., 2012). However, the maturation event is controlled by TH, the precise cellular target(s) of the hormone, the role played by TRα and TRβ (that are both expressed in cerebellum), and the target genes controlled by the receptors are far from fully understood. Nevertheless, it is clear from the situation in mice that TH plays a critical role in nervous system maturation in mice.

5.3. What about other mammals?

The question is whether these data largely derived from mouse models, and to a lesser extent human cases, can be generalized to other mammalian species. Again, it is important to mention that the data available on other species are rare, and having a general view is still impossible. However, some isolated observations are revealing.

Returning to the guinea pig model, it has been shown that the only difference between weaning stage and adult intestine is the neuron density that is lower in weaning intestine which suggests that, except for a subpopulation of neurons in the ileum, the intestine during weaning is physiologically close to the adult intestine. This indicates that in the guinea pig, the main steps of the intestine development occur before weaning (Abalo, Vera, Rivera, Moro-Rodríguez, & Martín-Fontelles, 2009). These results may suggest that the main maturation of the intestine takes place before birth in this species, consistent with the fact that the peak of TH also occurs before birth. A study on brain development in guinea pig has shown high $TR\alpha$ and $TR\beta$ mRNA expression in the guinea pig brain, particularly in the hippocampus (Chan et al., 2005) at 52 days of gestation. At the same time, deiodinases D2 and D3 are found throughout the brain (Chan et al., 2005). Interestingly, detailed observations of brain development in this species indicate that the most important part of brain development occurs from 50 days postgestation until birth in guinea pig (Matthews, 1998). Taken together, these data suggest that brain maturation is already advanced at birth in guinea pig, on the contrary to mouse and rat. As for intestine maturation, this provides indirect evidence indicating that THs are involved in this prebirth maturation.

5.4. The young mammal as a larva?

As discussed above, mouse, rat, and, to a lesser extent, humans are the only mammals for which we have sufficient data to deduce that TH is controlling a postembryonic remodeling process that can be compared to metamorphosis, as described in the canonical amphibian model. But in fact, many mammalian species exhibit pronounced differences in their postembryonic life history strategies. Strong variations in the length of the gestation period among eutherian mammals can be observed. Gestation length may be linked to behavioral factors as long gestations give rise to precocial neonates that are immediately able to be active (e.g., horse), whereas species displaying a long period of maternal dependency (altricial, e.g., cat) have much shorter gestations. Such differences can occur even between quite closely related species, for example, between mouse and guinea pig. The length of the gestation is extremely different between the two species: 21 days for the mouse and more than 63–68 days for the guinea pig. The aspect of the neonate is also different: the guinea pig is a precocial species with newborn harboring fur and teeth, open eyes able to move, to fully regulate their temperature and eat solid food just 2 days after birth (Dawes & Mestyán, 1963). In contrast, mouse pups are altricial: they are totally maternally dependent, naked, and incapable of moving around. One cannot avoid concluding that mouse pups are "larvae," whereas guinea pigs neonates are already "juveniles."

It is interesting to correlate the thyroid metabolism of these species with their precocial or altricial character. Interestingly, such a correlation between altricial/precocial strategy and TH metabolism has already been developed by McNabb, based on studies on birds (McNabb, Lyons, & Hughes, 1984; Schew, McNabb, & Scanes, 1996). Mice, the less autonomous of our examples, complete the development of their brain and intestine during the postembryonic weaning period when a TH peak induces the maturation of these organs. In humans, the brain exhibits a burst of development at birth when a TH peak is observed (Erenberg et al., 1974). Weaning guinea pigs have their intestine (Abalo et al., 2009; Matthews, 1998) physiologically close to the adult intestine and the main step of brain maturation occurs before birth, at a time during which TH level is high. If, as suggested above, the mouse pups are considered as "larvae" and the guinea pig newborn as "juveniles," then we may propose that guinea pigs undergo their TH-controlled maturation at the end of the long gestation period within the mother. Thus, guinea pig may represent a case of direct development in mammals in which the "metamorphosis" occurred during development (Fig. 14.4)

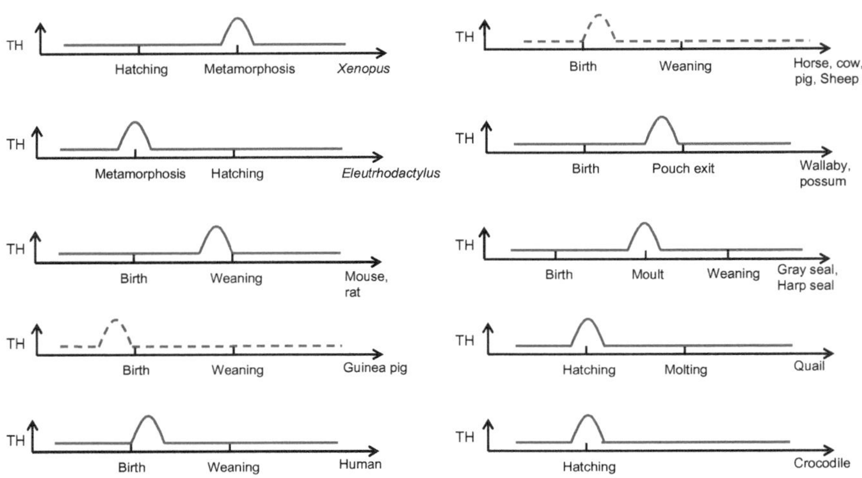

Figure 14.4 Schematic representation of the thyroid level during the development of the species discussed in this review. The *Y*-axis represents an arbitrary TH level, and the *X*-axis represents the time of development. The *X*-axis is at an arbitrary scale for the different species. Continuous line means that data are reliable enough to draw the TH levels during development. Interrupted line means that only partial data are available and the TH levels during development are still speculative to our knowledge. (For color version of this figure, the reader is referred to the online version of this chapter.)

5.5. The special case of marsupials

Metatherians (marsupials) have a very different development strategy from that of eutherian mammals. In this group, the newborn, called a joey, born blind, deaf, and furless. It reaches the ventral pouch of the mother crawling on the fur where it finds a mammary gland and feeds on it for several months (Murphy & Smith, 1970; Wilkes & Janssens, 1986). The joey terminates its development in the pouch for several months until it is fully able to feed and move independently and ensure its thermic autonomy (Shield, 1966). In metatherians, the gestation time in itself lasts several weeks and the postpartum development several months (with differences according to the size of the animal). If we consider that leaving the pouch is concomitant with the weaning of the juvenile, metatherians spend more time in postembryonic development than in gestation itself.

In the tammar wallaby (*Macropus eugenii*), a peak of T4 is observed ca. 180 days after birth when the joey's thyroid gland becomes functional. Maximum deiodinase outer-ring deiodination activity is observed near 205 days after birth and subsequently a peak of T3 occurs between 210 and 235 days after birth (Richardson et al., 2002). The T3 peak occurs between the young's first exit of the pouch near 200 days and the definitive leaving near 260 days (Setchell, 1974). The peaks of TH correspond to the beginning of homeothermic autonomy, pelage development, and digestive tract remodeling of the young (Janssens, Grigg, Doce, & Hulbert, 1990). THs are important for development, as thyroidectomized wallabies are unable to thermoregulate and exhibit growth and fur development defects (Janssens et al., 1990; Setchell, 1974), indicating that even if the plasma TH peak starts just after the onset of remodeling in wallaby, the hormone is indeed necessary to achieve these changes.

Very similar pictures emerge from the data coming from two other species. In the brushtail possum (*Trichosurus vulpecula*), a peak of T4 is observed at day 120 after birth (Buaboocha & Gemmell, 1995). The T4 peak occurs a little after the young possum is able to exit the pouch at day 118 (Buaboocha & Gemmell, 1995) and prior the definitive leaving around 180 days postpartum. In the tasmanian bettong (*Bettongia gaimardi*), a peak of T4 is observed at 12–14 weeks after birth which is concomitant with homeothermy and the fur development (Rose & Kuswanti, 2004). The peak of T4 corresponds to the time when the young exits the pouch for the first time, about 14 weeks, before the final leaving at 15 weeks.

Indirect clues highlighting the importance of TH level during wallaby development come from the level of TH carrier proteins. During postembryonic development, two proteins have been identified as involved in TH transport in the blood, the thyroxine-binding globin (TBG), and the transthyretin (TTR). TBG level is high during pouch life and disappears after pouch leaving, whereas TTR level is low during pouch life and rise in adulthood (Richardson et al., 2002, 2005). This switch of TH carrying protein is concomitant with the timing of the T3 peak. These data, gathered in a marsupial, show that a switch in TH transport protein expression is concomitant with young's exit of the pouch. This provides indirect evidence that THs are involved in the pouch exit event and that their levels must be finely regulated during development.

5.6. A partial conclusion on mammals

There are large disparities among mammals, between metatherians and eutherians but also within eutherians depending on their altricial or precocial development strategies. Nevertheless, in the late development/early postembryonic life of mammals, it seems clear that a peak of TH is mandatory for the normal development of organs such as intestine or brain. In eutherians, this peak is concomitant with brain and intestine development (Fraichard et al., 1997; Plateroti et al., 2001). Moreover, this peak is apparently delayed after birth in the species, such as mouse, requiring a long postbirth development until autonomy. In contrast, this peak can occur very early in species with long gestation time exhibiting precocial strategies: this is the case of the guinea pig. In human, the peak is concomitant to birth. In metatherians, it is apparently concomitant with the establishment of endothermy, fur, intestine development, and the first leaving of the pouch, that is to say, the beginning of autonomy (Table 14.1).

Table 14.1 Summary of the features observed at birth in the mammals described in this review. "+" indicates the presence of the feature when "−" indicate its absence

	Feature				
Species	**Eyes open**	**Self-feeding**	**Self-heating**	**Fur/feather**	**TH peak**
Mice	−	−	−	−	At weaning
Guinea pig	+	+	+	+	Evidence before birth
Human	+	−	+	+/−	Right after birth
Possum	−	−	−	−	Right before the first pouch exit

If we consider weaning in mammals as an ecological change during which the young is no longer maternally protected but exposed to the outside environment, it could be analogous to metamorphosis in biphasic vertebrates such as amphibians since, as mentioned earlier, metamorphosis always corresponds to an ecological transition. According to this view, the TH peak could trigger in amphibian and mammals the remodeling of intestine and brain and other organs necessary to reach the adult stage and to cope with all the changes induced by the living in a new ecological environment. In amphibian, this peak indeed correlates with metamorphosis. In most Anurans, it leads to the spectacular and complete metamorphosis of tadpole into frog. In mammals, as in some amphibians, we would observe some variations between the TH peak and birth (or hatching) as an example of heterochrony that is an evolutionary change in developmental timing. According to this view, both the guinea pig and *E. coqui* are direct developers that undergo their TH-controlled remodeling as embryos and hatch (or born) as juveniles. The TH system, that is connected to the environment via the HPT axis, has all the necessary flexibility to allow plastic response to environmental cues. Such a system allows for heterochronic changes on the respective timing of birth and "metamorphosis" in various species.

6. SAUROPSIDS

In this review, we took the phylogenetic point of view to assess a physiological and developmental question among amniotes. Thus, if the hypothesis stands and THs control organ remodeling in newborn amniotes, we should observe the same relation between TH level and organ remodeling in sauropsids, as in mammals. Only few studies specifically address TH levels during hatching (McNabb et al., 1984), and as most sauropsids lay eggs, as do the well-studied amphibians or teleost fishes, this simplifies the situation because in mammals, it is obvious that maternal change of TH levels during pregnancy can have a profound influence on the developing embryo.

6.1. TH levels during development in birds

In birds, there is neither clear larval to adult transition as in amphibian nor weaning as in mammal, but there is independent feeding (as in mammals) and changes of ecological niche: egg/non-egg or nest/non-nest. Nevertheless, TH has been identified as important in the late development of precocial birds particularly during stages close to hatching. A first piece of evidence comes from the Japanese quail (*Coturnix japonica*) in which TH levels in the laid egg yolk has been shown to be similar to the hen. However, importantly, the THs levels within the embryo have been found as not influenced by the hens and yolk THs level (Wilson & McNabb, 1997).

As in mammals, one can distinguish two sorts of birds according to their development processes. Precocial birds hatch with their eyes open and with mature feathers and, consequently, are able to become independent from parental care early. On the contrary, altricial birds hatch with their eyes closed, without feathers and stay a long period in the nest totally depending on parental feeding and warmth (Ehrlich, Dobkin, & Wheye, 1988). Of course, this is a schematic description and all the intermediaries between these two extremes are found and this is linked to the life history of each species. In the Japanese quail, a precocial bird, thyroid gland develops early in the embryonic life, and hormone production begins ca. 5 days before hatching and a peak of TH is observed in these precocial birds near or at hatching (for review, see McNabb, 2006). On the contrary, altricial birds do not exhibit such a TH peak at hatching and the peak arose much later, ca. 10–15 days after hatching, as, for example, in the dove (McNabb, 2006). In these species, the thyroid gland seems to become functional later in the development, a few days after hatching, and the synthesis of TH starts progressively during postembryonic development (McNabb, 2006). However, some data, in turkey (*Meleagris gallopavo*) shade some trouble on such a link between TH level and the precocial or altricial mode of development. In turkey, a precocial bird, a peak of TH, is observed near 12/16 weeks after hatching (unpublished data cited in Siopes, Millam, & Steinman, 2010). Unfortunately, these data on TH level during turkey development have not been published to our knowledge. Thus, we do not know if there is a peak of TH at hatching in turkey as in other precocial birds or if that peak is delayed until 12/16 weeks of development. However, despite these discrepancies, studies performed in bird seem consistent with the existence of a peak of TH after birth in precocial birds and a smoother increase of TH until adult level in altricial bird. Thus, as in mammals, TH may be implicated in the control of postembryonic life history strategies, although this is far from being proved, yet.

6.2. TH levels in heterotherm sauropsids

Among sauropsids, TH levels vary significantly between birds on the one hand and tortoise, lizard, and alligator (forming the ancient zoological group called "reptilia" that is no longer valid on a zoological point of view; see

Fig. 14.3) on the other hand. One important difference between these species concerns thermoregulation, as birds are homeotherm species, whereas the other sauropsids are heterotherms. This is important to mention given the well-known role of TH in thermoregulation. In the adult tortoise (*Gopherus agassizii*), lizard (*Sceloporus occidentalis*), and alligator (*Alligator mississippiensis*), TH levels in adult vary seasonally (Boggs, Hamlin, Lowers, & Guillette, 2011; John-Alder, 1984; Kohel, MacKenzie, Rostal, Grumbles, & Lance, 2001). These variations are problematic to evaluate the changes in TH level in these species during development and especially during the early postembryonic period. Indeed, the hierarchy of the parameters controlling TH level, among which metabolism, food intake, and temperature, makes it much more difficult to interpret the levels observed in these heterothermic sauropsids (Brasfield, Talent, & Janz, 2008; Kohel et al., 2001).

In addition, in nonavian sauropsids, some data on TH level or TH transporters exist. Investigation of TTR, a TH carrier protein, in the saltwater crocodile (*Crocodylus porosus*) has shown that the protein is expressed only before hatching but not after (Richardson et al., 2005) indicating a specificity of TH transport in the embryo. It has been shown in this species that free T3 and T4 rise at hatching compared to prehatching level (Shepherdley, Daniels, et al., 2002) and are higher than juvenile TH level (unpublished observation cited in Shepherdley, Richardson, Evans, Kühn, & Darras, 2002). Consistent with the T3 increase, the T4 outer-ring deiodinase activity rises in the gut and brain of the crocodile prior hatching. Interestingly, at the same period, the inner-ring deiodinase activity that controls TH degradation is stopped. Taken together, the embryonic specific expression of TTR and the rise of TH level at hatching indicate a fine regulation of TH levels during development, particularly around hatching, suggesting once again the existence of a peak during this critical period.

6.3. Effect of TH in bird

Japanese quail treated with T4 lay eggs with an enriched T4 yolk content compared with nontreated hens. Nevertheless, the embryonic level of T4 is not directly influenced by the T4 level of the yolk indicating a fine-tuned regulation of T4 level within the embryo (Wilson & McNabb, 1997). No morphological or developmental difference is observed between chick of adult Japanese quails treated or not with T4. But T4 treatment of the hens delays the hatching of the chick for several hours (Wilson & McNabb, 1997).

In Japanese quail, goitrogen treatment (methimazole) does not significantly decrease plasma T4 level nor the T4 and T3 content of yolk of the laid eggs highlighting a high inertia of goitrogen response in the hen probably because of the storage capacity of the egg. However, this treatment increases the weight of the thyroid gland and disrupts egg laying (Wilson & McNabb, 1997). Treatment of young chicks started at 3 days posthatching with methimazole decrease T4 thyroid content and increase thyroid weight, but T4 plasma content and body weight are affected only by heavy and long treatment (McNabb, Jang, & Larsen, 2004) as in adult. Similar results were found in bobwhite quail chick (*Colinus virginianus*); blood T4 content and body weight are little or not affected by goitrogen treatment (McNabb, Larsen, & Pooler, 2004) as in adult. Taken together, these data suggest that the quail chick has a TH system functional enough to bear a decrease in T4 synthesis, indicating that in the precocial quails, TH gland seems to mature early.

As goitrogen treatment in the Japanese quail (Wilson & McNabb, 1997) has shown that the hens are resistant to this form of TH disruption, it is difficult to infer the role of TH in the hatching quail. Goitrogen (ammonium perchlorate) treatment for 4 weeks in Bobwhite quail of 4 days-old hatchling has little effect or does not disturb the growth of the young (McNabb, Jang, & Larsen, 2004; McNabb, Larsen, & Pooler, 2004). Two nonexclusive explanations can be provided: (i) The quail hatchling has sufficient TH and iodide stock in the yolk derived from the hen to handle goitrogen treatment. (ii) TH is not required for a few days after hatching in the hatchling. Taking into account the TH peak at hatching in the closely related Japanese quail (Wilson & McNabb, 1997), maybe after hatching, the highly TH-dependent development period is over in quail.

6.4. A "metamorphosis" in sauropsids?

Most of the data on TH level around hatching come from studies performed in birds. Taken together, these data suggest a different TH regulation between precocial and altricial birds with a near-hatch peak for the former and a smoother late one during posthatch development for the later. This fits with the notion of a TH peak related to autonomy of the hatchling, an idea that was proposed by McNabb several years ago (McNabb et al., 1984; see also a more recent review in McNabb, 2006). Indeed, precocial birds tend to have thermoregulatory independence and feed themselves, whereas altricial birds rely on their parent for heating and feeding. Moreover, precocial hatchling such as quail appears to be resistant to thyroid manipulation with goitrogens, indicating that (i) no further important maturation controlled by TH seems to occur in these birds after hatching and (ii) TH metabolism is already mature in these birds as the hatchling respond to goitrogen in a manner indistinguishable to the adults. As far as other sauropsids are concerned, indirect evidence indicates that a peak of TH could exist at hatching in some species such as crocodiles, but few data are available. As we previously proposed for mammals, the TH peak at birth could be an indicator of the degree of independence of the hatchling in sauropsids with a peak of TH just before or at hatching correlating with an independent autonomous offspring. This acquisition of autonomy allowed by the TH trigger could be considered as being equivalent to the amphibian metamorphosis, although this is only a speculation given the data available.

7. CONCLUSION

In this review, we focused on the data available about TH level in late embryonic development and early postembryonic development of several amniote species (Fig. 14.4). TH peaks are observed across amniotes around birth/hatching or at other crucial developmental time points, as weaning, and these peaks are important for organ maturation, particularly in brain and digestive system. Moreover, the timing of the peak is variable across species and seems related to the degree of autonomy of the young at birth/hatching with a delayed peak in species with nonautonomous young. On the contrary, in species with autonomous young, this peak is close to birth/hatching, allowing a shorter period of organ maturation, particularly for the intestine (that renders self-feeding possible) and the brain (that allows an independent behavior and metabolic control).

Among direct developing amphibians, such as *E. coqui*, the young froglet that hatches is a fully autonomous miniature adult, as amphibians do not need a pond to develop. In this species, the tadpole-to-frog transformation occurs very early, within the egg. Thus, we can draw a parallel with amniote species in which the more the young is autonomous, the earlier the peak of TH seems to happen. In sauropsids, this would be exemplified by the Japanese quail with an endothermic autonomous young, and in mammals this would be exemplified by the guinea pigs with a young than can eat solid food as soon as 2 days after birth. In metatherian, this period occurs after birth, when the joey starts to leave its mother and may correspond to the onset of weaning.

Overall, what we observed in mammals and bird species for which data are available is a stage of organ maturation that corresponds to a period of autonomy acquisition by the young that is concomitant to a TH peak, and probably a period of high sensitivity to the hormone. We propose that this period should be considered as equivalent to Anuran metamorphosis, and defined as a TH-controlled functional remodeling of organs allowing the maturation of main physiological system preparing the entry into a new ecological niche. This clearly suggests that in most vertebrates there is a deployment of the TH signaling pathway that allows this transition. This process is tightly coupled to the environment via the nervous system and the connection with the internal milieu due to the roles of the HPT axis. This explains why variations of this period, linked to variation of life history strategy, occurred and this offers interesting prospects for future research in order to explore how in some species heterochronic shifts of the TH-sensitive period might have occurred.

To conclude, it is important to acknowledge that this new unifying vision of postembryonic development of vertebrates is very speculative, full of unsolved questions. It is clear that experimental data should challenge this vision but it has the advantage of being a framework on which experimental evidence could be confronted. We are convinced that shaking this model will allow to shed a new light on postembryonic development in amniotes.

ACKNOWLEDGMENTS

We are very grateful to Barbara Demeneix for critical reading and many fruitful suggestions. All errors and omissions are entirely ours. We apologize to our colleagues whose work was not cited adequately for lack of space. Work from our laboratory is supported by the CNRS, the Agence National de la Recherche, the French Ministry of Ecology (PNRPE), and the European Commission.

REFERENCES

Abalo, R., Vera, G., Rivera, A. J., Moro-Rodríguez, E., & Martín-Fontelles, M. I. (2009). Postnatal maturation of the gastrointestinal tract: A functional and immunohistochemical study in the guinea-pig ileum at weaning. *Neuroscience Letters*, *467*(2), 105–110.

Atkinson, S., Amould, J. P. Y., & Mashburn, K. L. (2011). Plasma cortisol and thyroid concentration in pre-weaning Australian fur seal pups. *General and Comparative Endocrinology*, *172*, 277–281.

Babu, S., Sinha, R. A., Mohan, V., Rao, G., Pal, A., Pathak, A., et al. (2011). Effect of hypothyroxinemia on thyroid hormone responsiveness and action during rat postnatal neocortical development. *Experimental Neurology*, *228*, 91–98.

Ball, S. G., Sokolov, J., & Chin, W. W. (1997). 3,5-Diiodo-L-thyronine (T2) has selective thyroliletic effects in vivo and in vitro. *Journal of Molecular Endocrinology*, *19*, 137–147.

Bernal, J. (2007). Thyroid hormone receptors in brain development and function. *Nature Reviews. Endocrinology*, *3*, 249–259.

Bianco, A. C., & Kim, B. W. (2006). Deiodinases: Implications of the local control of thyroid hormone action. *Journal of Clinical Investigation, 116*(10), 2571–2579.

Billon, N., Tokumoto, Y., Forrest, D., & Raff, M. (2001). Role of thyroid hormone receptors in timing oligodendrocyte differentiation. *Developmental Biology, 235*, 110–120.

Biswas, S., Buffery, J., Enich, H., Bland, J. M., Walters, D., & Markiewicz, M. (2002). A longitudinal assessment of thyroid hormone concentration in preterm infants younger than 30 weeks' gestation during the first 2 weeks of life and their relationship to outcome. *Pediatrics, 109*(2), 222–227.

Boggs, A. S., Hamlin, H. J., Lowers, R. H., & Guillette, L. J., Jr. (2011). Seasonal variation in plasma thyroid hormone concentrations in coastal versus inland populations of juvenile American alligators (Alligator mississippiensis): Influence of plasma iodide concentrations. *General and Comparative Endocrinology, 174*(3), 362–369.

Boorse, G. C., & Denver, R. J. (2002). Acceleration of Ambystoma tigrinum metamorphosis by corticopin-releasing hormone. *The Journal of Experimental Zoology, 293*(1), 94–98.

Brasfield, S. M., Talent, L. G., & Janz, D. M. (2008). Reproductive and thyroid hormone profiles in captive Western fence lizards (Sceloporus occidentalis) after a period of brumation. *Zoo Biology, 27*(1), 36–48.

Breall, J. A., Rudolp, A. M., & Heymann, M. A. (1984). Role of thyroid hormone in postnatal circulatory and metabolic adjustments. *Journal of Clinical Investigation, 73*, 1418–1424.

Brown, D. D. (1997). The role of thyroid hormone in zebrafish and axolotl development. *Proceedings of the National Academy of Sciences of the United States of America, 94*(24), 13011–13016.

Buaboocha, W., & Gemmell, R. T. (1995). Thyroid gland development in the brushtail possum, *Trichosurus vulpecula. Anatomical Record, 243*(2), 254–260.

Buchholz, D. R., Paul, B. D., Fu, L., & Shi, Y.-B. (2006). Molecular and developmental analyses of thyroid hormones receptor function in *Xenopus laevis*, the African clawed frog. *General and Comparative Endocrinology, 145*, 1–9.

Buchholz, D. R., Singamsetty, S., Karadge, U., Wiliamson, S., Langer, C. E., & Elinson, R. P. (2007). Nutritional endoderm in a direct developing frog: A potential parallel to the evolution of the amniote egg. *Developmental Dynamics, 236*, 1259–1272.

Callery, E. M., & Elinson, R. P. (2000). Opercular development and ontogenetic reorganization in a direct-developing frog. *Development Genes and Evolution, 210*(7), 377–381.

Carrol, J. A., Veum, T. L., & Matteri, R. L. (1998). Endocrine response to weaning and changes in post-weaning diet in the young pig. *Domestic Animal Endocrinology, 15*(3), 183–194.

Castro, M. I., Alex, S., Young, R. A., Braveman, L. E., & Emerson, C. H. (1986). Total and free serum thyroid hormone concentrations in fetal and adult pregnant and nonpregnant guinea pigs. *Endocrinology, 118*(2), 533–537.

Chan, S. Y., Andrews, M. H., Lingas, R., McCabe, C. J., Franklyn, J. A., Kilby, M. D., et al. (2005). Maternal nutrient deprivation induces sex-specific changes in thyroid hormone receptor and deiodinase expression in the fetal guinea pig brain. *The Journal of Physiology, 566*(2), 467–480.

Chiamolera, M. I., & Wondisford, F. E. (2009). Minireview: Thyrotropin-releasing hormone and the thyroid hormone feedback mechanism. *Endocrinology, 150*(3), 1091–1096.

Coen, L., Le Blay, K., Rowe, I., & Demeneix, B. A. (2007). Caspase-9 regulates apoptosis/proliferation balance during metamorphic brain remodeling in *Xenopus. Proceedings of the National Academy of Sciences of the United States of America, 104*(20), 8502–8507.

Darras, V. M., & Van Herck, L. J. (2012). Iodothyronine deiodinase structure and function: From ascidians to humans. *Journal of Endocrinology, 216*, 189–206.

Dawes, G. S., & Mestyán, G. (1963). Changes in the oxygen consumption of new-born guinea-pigs and rabbits on exposure to cold. *The Journal of Physiology*, *168*, 22–42.
Dunn, J. T., & Dunn, A. D. (1999). The importance of thyroglobulin structure for thyroid hormone biosynthesis. *Biochomie*, *81*, 506–509.
Dusart, I., & Flamant, F. (2012). Profound morphological and functional changes of rodents Purkinje cells between the first and the second postnatal week: A metamorphosis? *Frontiers in Neuroanatomy*, *6*(11).
Ehrlich, P. R., Dobkin, D. S., & Wheye, D. (1988). *The birder's handbook. A field guide to the natural history of North American birds.* New York: Simon & Schuster.
Erenberg, A., Phelps, D. L., Lam, R., & Fisher, D. A. (1974). Total and free thyroid hormone concentrations in the neonatal period. *Pediatrics*, *53*(2), 211–216.
Escriva, H., Manzon, L., Yousson, J., & Laudet, V. (2002). Analysis of Lamprey and Hagfish genes reveals a complex history of gene duplications during early vertebrate evolution. *Molecular Biology and Evolution*, *19*(9), 1440–1450.
Fauquier, T., Romero, E., Picou, F., Chatonnet, F., Nguyen, X. N., Quignodon, L., et al. (2011). Severe impairment of cerebellum development in mice expressing a dominant-negative mutation inactivating thyroid hormone receptor alpha1 isoform. *Developmental Biology*, *356*(2), 350–358.
Flamant, F., Gauthier, K., & Samarut, J. (2007). Thyroid hormones signalling is getting more complex: STORMs are coming. *Molecular Endocrinology*, *21*(2), 321–333.
Flamant, F., & Samarut, J. (2003). Thyroid hormone receptors: Lessons from knockout and knock-in mutant mice. *Trends in Endocrinology and Metabolism: TEM*, *14*(2), 85–90.
Forrest, D., & Vennström, B. (2000). Functions of thyroid hormone receptors in mice. *Thyroid*, *10*(1), 41–52.
Fraichard, A., Chassande, O., Plateroti, M., Roux, J. P., Trouillas, J., Dehay, C., et al. (1997). The T3R alpha gene encoding a thyroid hormone receptor is essential for post-natal development and thyroid hormone production. *EMBO Journal*, *16*(14), 4412–4420.
Galien, L., & Durocher, M. (1957). Tables de développement chez *Pleurodeles waltl. Bulletin Biologique de la France et de la Belgique*, *91*, 97–117.
Gauthier, K., Chassande, O., Plateroti, M., Roux, J. P., Legrand, C., Pain, B., et al. (1999). Different functions for the thyroid hormone receptors TRα and TRβ in the control of thyroid hormone production and post-natal development. *EMBO Journal*, *18*(3), 623–631.
Gauthier, K., Plateroti, M., Harvey, C. B., Williams, G. R., Weiss, R. E., Refetoff, S., et al. (2001). Genetic analysis reveals different functions for the products of the thyroid hormone receptor alpha locus. *Molecular and Cellular Biology*, *21*(14), 4748–4760.
Gavaret, J. M., Cahnmann, H. J., & Nunez, J. (1981). Thyroid hormone synthesis in thyroglobulin. *Journal of Biological Chemistry*, *256*(17), 9167–9173.
Gudernatsch, J. F. (1912). Feeding experiments on tadpoles: I. The influence of specific organs given as food on growth and differentiation. A contribution to the knowledge of organs with internal secretion. *Wilhelm Roux Arch. Entwicklungsmech Organismen*, *35*, 457–483.
Hadj-Sahraoui, N., Seugnet, I., Ghorbel, M. T., & Demeneix, B. (2000). Hypothyroidism prolongs mitotic activity in the post-natal mouse brain. *Neuroscience Letters*, *280*(2), 79–82.
Hollar, A. R., Choi, J., Grimm, A. T., & Buchholz, D. (2011). Higher thyroid hormone receptor expression correlates with short larval periods in spadefoot toads and increases metamorphic rate. *General and Comparative Endocrinology*, *173*(1), 190–198.
Huang, H., Cai, L., Remo, B. F., & Brown, D. D. (2001). Timing of metamorphosis and the onset of the negative feedback loop between the thyroid gland and the pituitary is controlled by type II iodothyronine deiodinase in *Xenopus* laevis. *Proceedings of the National Academy of Sciences of the United States of America*, *98*(13), 7348–7353.

Hulbert, A. J. (2000). Thyroid hormones and their effects: A new perspective. *Biological Reviews*, *75*, 519–631.
Irvine, C. H., & Evans, M. J. (1975). Postnatal changes in total and free thyroxine and triiodothyronine in foal serum. *Journal of Reproduction and Fertility. Supplement*, *23*, 709–715.
Ishizuya-Oka, A., Hasebe, T., & Shi, Y. B. (2010). Apoptosis in amphibian organs during metamorphosis. *Apoptosis*, *15*(3), 350–364.
Ishizuya-Oka, A., & Ueda, S. (1996). Apoptosis and cell proliferation in the *Xenopus* small intestine during metamorphosis. *Cell and Tissue Research*, *286*(3), 467–476.
Janssens, P. A., Grigg, J. A., Doce, H., & Hulbert, A. J. (1990). Thyroid hormones during development of a marsupial, the tammar wallaby, *Macropus eugenii*. *Journal of Endocrinology*, *127*(3), 427–436.
John-Alder, H. B. (1984). Reduced aerobic capacity and locomotory endurance in thyroid-deficient lizards. *Journal of Experimental Biology*, *109*, 175–189.
Kohel, K. A., MacKenzie, D. S., Rostal, D. C., Grumbles, J. S., & Lance, V. A. (2001). Seasonality in plasma thyroxine in the desert tortoise, *Gopherus agassizii*. *General and Comparative Endocrinology*, *121*(2), 214–222.
Kuiper, G. G. J. M., Kester, M. H. A., Peeters, R. P., & Visser, T. J. (2005). Biochemical mechanisms of thyroid hormone deiodination. *Thyroid*, *15*(8), 787–798.
Kulkarni, S. S., Singamsetty, S., & Buchholz, D. R. (2010). Corticotropin-releasing factor regulates the development in the direct developing frog, *Eleutherodactylus coqui*. *General and Comparative Endocrinology*, *169*(3), 225–230.
Laudet, V. (2011). The origin and evolution of Vertebrate metamorphosis, a life history transition. *Current Biology*, *21*(18), 726–737.
Legrand, J. (1979). Morphogenetic actions of thyroid hormones. *Trends in Neurological Sciences*, *2*, 236–243.
Leloup, J., & Buscaglia, M. (1977). La triiodothyronine, hormone de la métamorphose des amphibiens. *Comptes rendus de l'Académie des sciences*, *284*, 2261–2263.
Mansouri, A., Chowdhury, K., & Gruss, P. (1998). Follicular cells of the thyroid gland require Pax8 gene function. *Nature Genetics*, *19*(1), 87–90.
Matthews, S. G. (1998). Dynamic changes in glucocorticoid and mineralocorticoid receptor mRNA in the developing guinea pig brain. *Brain Research. Developmental Brain Research*, *107*(1), 123–132.
McNabb, F. M. A. (2006). Avian thyroid development and adaptive plasticity. *General and Comparative Endocrinology*, *147*(2), 93–101.
McNabb, F. M. A., Jang, D. A., & Larsen, C. T. (2004). Does thyroid function in developing birds adapt to sustained ammonium perchlorate exposure? *Toxicological Sciences*, *82*(1), 106–113.
McNabb, F. M. A., Larsen, C. T., & Pooler, P. S. (2004). Ammonium perchlorate effects on thyroid function and growth in bobwhite quail chicks. *Environmental Toxicology and Chemistry*, *23*(4), 997–1003.
McNabb, F. M. A., Lyons, L. J., & Hughes, T. E. (1984). Free thyroid hormones in altricial (ring doves) versus precocial (Japanese quail) development. *Endocrinology*, *115*(6), 2133–2136.
Morte, B., Manzano, J., Scanlan, T., Vennstörm, B., & Bernal, J. (2002). Deletion of the thyroid hormone receptor alpha 1 prevents the structural alterations of the cerebellum induced by hypothyroidism. *Proceedings of the National Academy of Sciences of the United States of America*, *99*(6), 3985–3989.
Murphy, C. R., & Smith, J. R. (1970). Age determination of pouch young and juvenile Kangaroo Island wallabies. *Transactions of the Royal Society of South Australia*, *94*, 15–20.
Nieuwkoop, P. D., & Faber, J. (1994). *Normal table of* Xenopus laevis *(Daudin)*. New York: Garland.
Opitz, R., Trubiroha, A., Lorenz, C., Lutz, I., Hartmann, S., Blank, T., et al. (2006). Expression of sodium-iodide symporter mRNA in the thyroid gland of *Xenopus laevis*

tadpole: Developmental expression effects of antithyroidal compounds, and regulation by TSH. *Journal of Endocrinology, 190*, 157–170.

Orozco, A., Valverde, R. C., Olvera, A., & García, G. (2012). Iodothyronine deiodinase. A functional and evolutionary perspective. *Journal of Endocrinology, 215*(2), 207–219.

Paris, M., Escriva, H., Schubert, M., Brunet, F., Brtko, J., Ciesielski, F., et al. (2008). Amphioxus postembryonic development reveals the homology of chordate metamorphosis. *Current Biology, 18*(11), 825–830.

Paris, M., Hillenweck, A., Bertrand, S., Delous, G., Escriva, H., Zalko, D., et al. (2010). Active metabolism of thyroid hormone during metamorphosis of amphioxus. *Integrative and Comparative Biology, 50*(1), 63–74.

Paris, M., & Laudet, V. (2008). The history of a developmental stage: Metamorphosis in chordates. *Genesis, 46*(11), 657–672.

Patricolo, E., Cammarata, M., & D'Agati, P. (2001). Presence of thyroid hormones in ascidian larvea and their involvement in metamorphosis. *Journal of Experimental Biology, 290*, 426–430.

Pauliková, I., Seidel, H., Nagy, O., Tóthová, C., & Ková, G. (2011). Concentrations of thyroid hormones in various age categories. *Acta Veterinaria, 61*(5–6), 489–503.

Picou, F., Fauquier, T., Chatonnet, F., & Flamant, F. (2012). A bimodal influence of thyroid hormone on cerebellum oligodentrocyte differentiation. *Molecular Endocrinology, 26*(4), 608–618.

Plateroti, M., Gauthier, K., Domon-Dell, C., Freund, J. N., Samarut, J., & Chassande, O. (2001). Functional interference between thyroid hormone receptor alpha (TRα) and natural truncated TRΔα isoforms in the control of intestine development. *Molecular and Cellular Biology, 21*(14), 4761–4772.

Richardson, S. J., Aldred, A. R., Leng, S. L., Renfree, M. B., Hulbert, A. J., & Schreiber, G. (2002). Developmental profile of thyroid hormone distributor proteins in a marsupial, the tammar wallaby *Macropus eugenii*. *General and Comparative Endocrinology, 125*(1), 92–103.

Richardson, S. J., Monk, J. A., Shepherdley, C. A., Ebbesson, L. O., Sin, F., Power, D. M., et al. (2005). Developmentally regulated thyroid hormone distributor proteins in marsupials, a reptile, and fish. *American Journal of Physiology. Regulatory, Integrative and Comparative Physiology, 288*(5), 1264–1272.

Rose, R. W., & Kuswanti, N. (2004). Thyroid function and the development of endothermy in a marsupial, the Tasmanian bettong, *Bettongia gaimardi* (Demarest 1822). *General and Comparative Endocrinology, 136*(1), 17–22.

Safi, R., Bertrand, S., Marchand, O., Duffraisse, M., de Luze, A., Vanacker, J.-M., et al. (2004). The axolotl (*Ambystoma mexicanum*), a neotenic amphibian, expresses functional thyroid hormone receptors. *Endocrinology, 145*(2), 760–772.

Safi, R., Vlaeminck-Guillem, V., Duffraisse, M., Seugnet, I., Plateroti, M., Margotat, A., et al. (2006). Pedomorphosis revisited: Thyroid hormone receptors are functional in *Necturus* maculosus. *Evolution & Development, 8*(3), 284–292.

Schew, W. A., McNabb, F. M. A., & Scanes, C. G. (1996). Comparison of the ontogenesis of thyroid hormones, growth hormone, and insulin-like growth factor-I in ad libitum and food-restricted (altricial) European starling and (precocial) Japanese quail. *General and Comparative Endocrinology, 101*, 304–316.

Setchell, P. J. (1974). The development of thermoregulation and thyroid function in the marsupial *Macropus eugenii* (Desmares). *Comparative Biochemistry and Physiology. Part A, 47*, 1115–1121.

Shen, X.-X., Liang, D., Wen, J.-Z., & Zhang, P. (2011). Multiple genome alignments facilitate development of NPCL markers: A case study of tetrapods phylogeny focusing on the position of turtles. *Molecular Biology and Evolution, 28*(12), 3237–3252.

Shepherdley, C. A., Daniels, C. B., Orgeig, S., Richardson, S. J., Evans, B. K., & Darras, V. M. (2002). Glucocorticoids, thyroid hormones, and iodothyronine deiodinases in embryonic saltwater crocodiles. *American Journal of Physiology. Regulatory, Integrative and Comparative Physiology, 283*(5), 1155–1163.
Shepherdley, C. A., Richardson, S. J., Evans, B. K., Kühn, E. R., & Darras, V. M. (2002). Thyroid hormone deiodinases during embryonic development of the saltwater crocodile (*Crocodylus porosus*). *General and Comparative Endocrinology, 126*(2), 153–164.
Shi, Y.-B. (2000). *Amphibian metamorphosis: From morphology to molecular biology*. New York: John Wiley & Sons, Inc.
Shi, Y.-B., Hasebe, T., Fu, L., Fujimoto, K., & Ishizuya-Oka, A. (2011). The development of the adult intestinal stem cells: Insights from studies on thyroid hormones-dependent amphibian metamorphosis. *Cell & Bioscience, 1*(1), 30.
Shield, J. (1966). Oxygen consumption during pouch development of the macropod marsupial *Setonix brachyurus. The Journal of Physiology, 187*, 257–270.
Siopes, T. D., Millam, J. R., & Steinman, M. Q. (2010). Initiating egg production in turkey breeder hens: Thyroid hormone involvement. *Poultry Science, 89*(10), 2265–2272.
Sirakov, M., & Plateroti, M. (2011). The thyroid hormones and their nuclear receptors in the gut: From developmental biology to cancer. *Biochimica et Biophysica Acta, 1812*(8), 938–946.
Sotelo, C., & Dusart, I. (2009). Intrinsic versus determinants during the development of Purkinje cell dendrites. *Neuroscience, 162*(3), 589–600.
Thorpe-Beeston, J. G., Nicolaides, K. H., Felton, C. V., Butler, J., & McGregor, A. M. (1991). Maturation of the secretion of thyroid hormone and thyroid-stimulating hormone in the fetus. *The New England Journal of Medicine, 324*(8), 532–536.
Visser, W. E., Friesema, E. C. H., & Visser, T. J. (2011). Minireview: Thyroid hormone transporters: The knowns and the unknowns. *Molecular Endocrinology, 25*(1), 1–14.
Vlaeminck-Guillem, V., Safi, R., Guillem, P., Leteurtre, E., Duterque-coquillard, M., & Laudet, V. (2006). Thyroid hormone receptor expression in the obligatory paedomorphis salamander *Necturus maculosus. International Journal of Developmental Biology, 50*, 553–560.
Voss, S. R., & Shaffer, H. B. (1997). Adaptive evolution via a major gene effect: Paedomorphosis in the Mexican axolotl. *Proceedings of the National Academy of Sciences of the United States of America, 94*, 14185–14189.
Werneburg, I., & Sánchez-Villagra, M. R. (2009). Timing of organogenesis support basal position of turtles in the Amniote tree of life. *BMC Evolutionary Biology, 9*(82).
Wilkes, G. E., & Janssens, P. A. (1986). Physiological changes associated with weaning in the tammar wallaby *Macropus eugenni. Journal of Comparative Physiology. Part B, 156*, 829–837.
Williams, G. R. (2008). Neurodevelopment and neurophysiological actions of thyroid hormone. *Journal of Neuroendocrinology, 20*(6), 784–794.
Wilson, C. M., & McNabb, F. M. (1997). Maternal thyroid hormones in Japanese quail eggs and their influence on embryonic development. *General and Comparative Endocrinology, 107*(2), 153–165.
Yaoita, Y., Shi, Y.-B., & Brown, D. D. (1990). *Xenopus* laevis α and β thyroid hormone receptors. *Proceedings of the National Academy of Sciences of the United States of America, 87*, 7090–7094.

INDEX

Note: Page numbers followed by "*f*" indicate figures, and "*t*" indicate tables.

E

J

L

M

N

Q

R

S

T

U

X

Y

Z

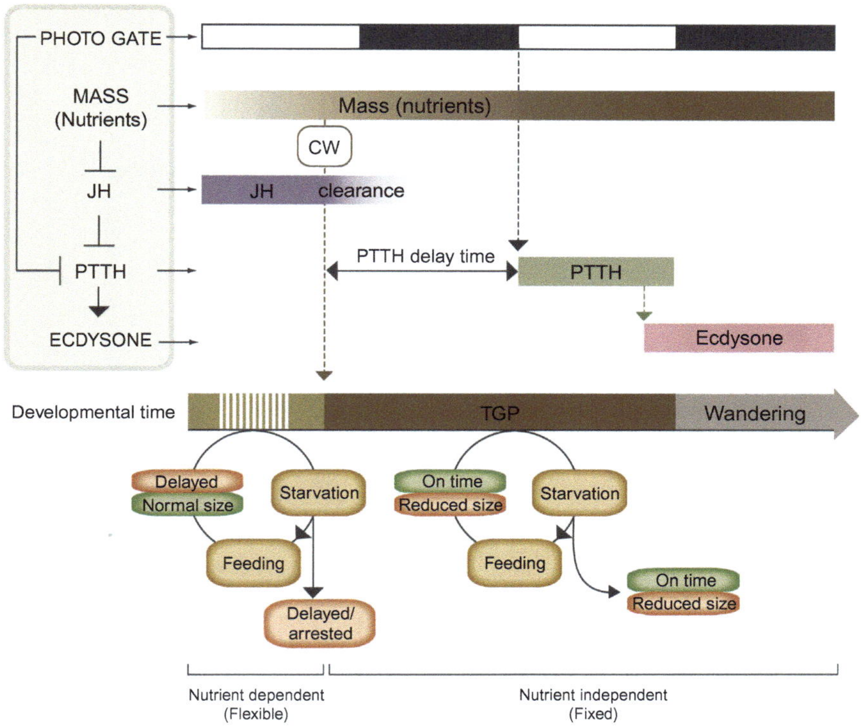

Kim F. Rewitz *et al.*, Figure 1.1 Hormonal regulation of body size and timing of metamorphosis. According to classical work the larval accumulation of mass corresponding to the critical weight results in the breakdown of JH that inhibits PTTH release. Above critical weight, PTTH is release after a delay period determined by the time of JH clearance and the photoperiod gating of PTTH. Starvation before critical weight results in a developmental delay of pupariation. If feeding is resumed critical weight is attained and normal body size achieved. In contrast, larvae starved above critical weight, when the metamorphic program is activated, pupariate on time but with a reduced size. As critical weight is independent of nutrition, final size is determined by the amount of growth in the interval between critical weight and cessation of feeding (wandering) called the terminal growth period (TGP). Photo gate, a PTTH gating mechanism imposed by the photoperiod; CW, critical weight.

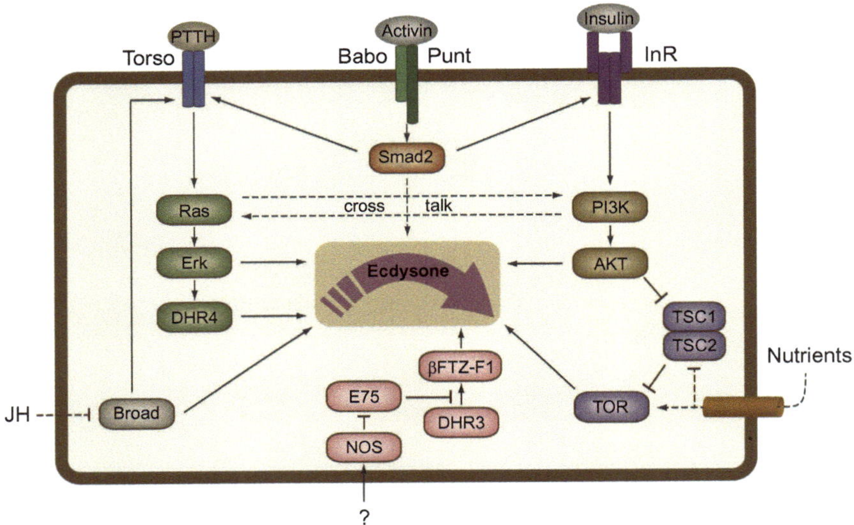

Kim F. Rewitz *et al.*, Figure 1.3 Mechanisms converging on the PG time ecdysone production. TGFβ/Activin is required for normal expression of the *insulin receptor* (*InR*) and *torso*, which provides glandular competence to PTTH (developmental) and insulin (nutrient) cues. Furthermore, the PG harbors a TOR-dependent nutrients sensor that presumably allows compensation for poor nutrient environments. Under such conditions, ecdysone release is delayed which prolongs the growth period. Broad is required for normal expression of the Halloween genes encoding the enzymes mediating ecdysone synthesis, although it is not clear if JH regulates *broad* in the PG. Nitric oxide (NO) generated by the nitric oxide synthase (NOS) regulates nuclear receptor signaling in the PG. In turn, NO inhibits E75, a repressor of DHR3 which then activates expression of *βFTZ-F1*, a nuclear receptor required for expression of at least two key ecdysone biosynthetic enzymes, Phantom and Disembodied.

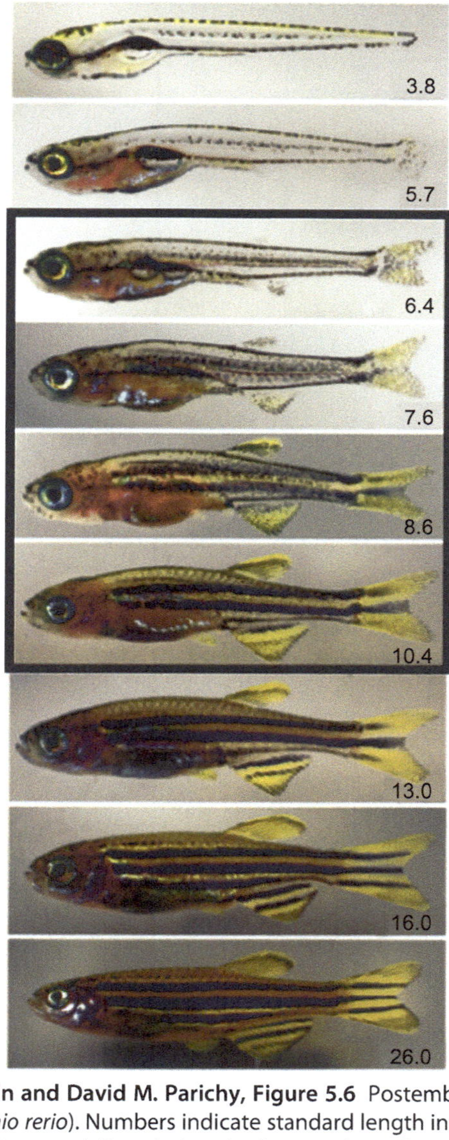

Sarah K. McMenamin and David M. Parichy, Figure 5.6 Postembryonic development in the zebrafish (*Danio rerio*). Numbers indicate standard length in mm, with the period of peak metamorphic remodeling during the larva-to-juvenile transition indicated by the black box. At standard temperature and typical rearing densities, this transition begins at about 10–14 days and takes about 2 weeks. As postembryonic fish can develop and grow at markedly different rates, days of development is an inadequate staging criterion; stages based on morphological criteria, and their size equivalents have been defined for this species and are preferable for developmental studies (Parichy et al., 2009).

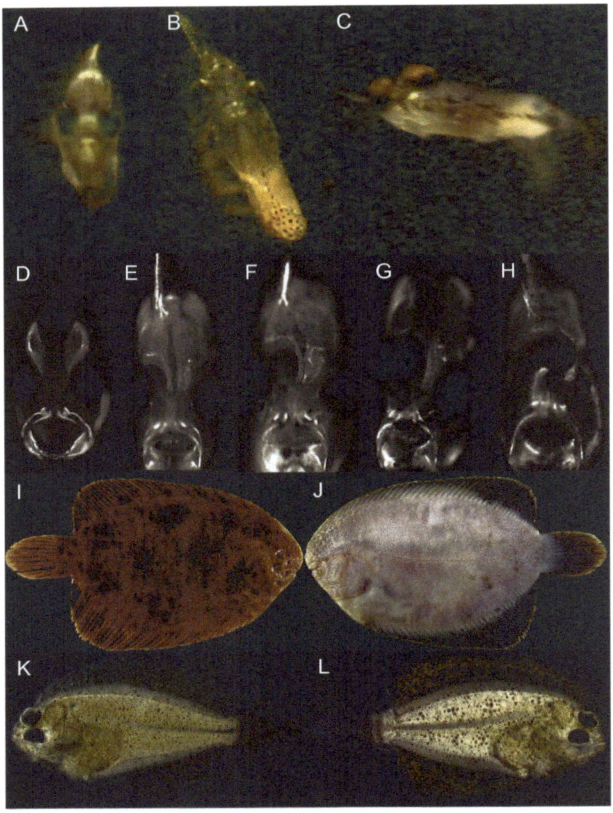

Alexander M. Schreiber, Figure 6.2 Development of lateralized swimming posture and morphological asymmetry during flatfish metamorphosis. Bilaterally symmetrical premetamorphic Southern flounder larvae (*Paralichthys lethostigma*) swim with an upright posture in the light (A). Migration of one eye to the opposite side of the head is accompanied by a transition to a tilted posture during metamorphic climax (B), and by the end of metamorphosis flounder swim on one side (C). Asymmetric remodeling of the skull (visualized *in vivo* by the fluorescent calcium-binding stain, calcein) accommodates eye translocation (D–H) in *P. lethostigma*. After metamorphosis, dextral flatfish, such as this hogchoker (*Trinectes maculatus*), have a pigmented "eyed" (right) side (I) and a nonpigmented "blind" (left) side (J). Some flatfish species, such as *P. lethostigma*, can metamorphose with either sinistral (K) or dextral (L) morphology and behavior. *Figure panels A–H are reproduced with permission from Schreiber (2006).*

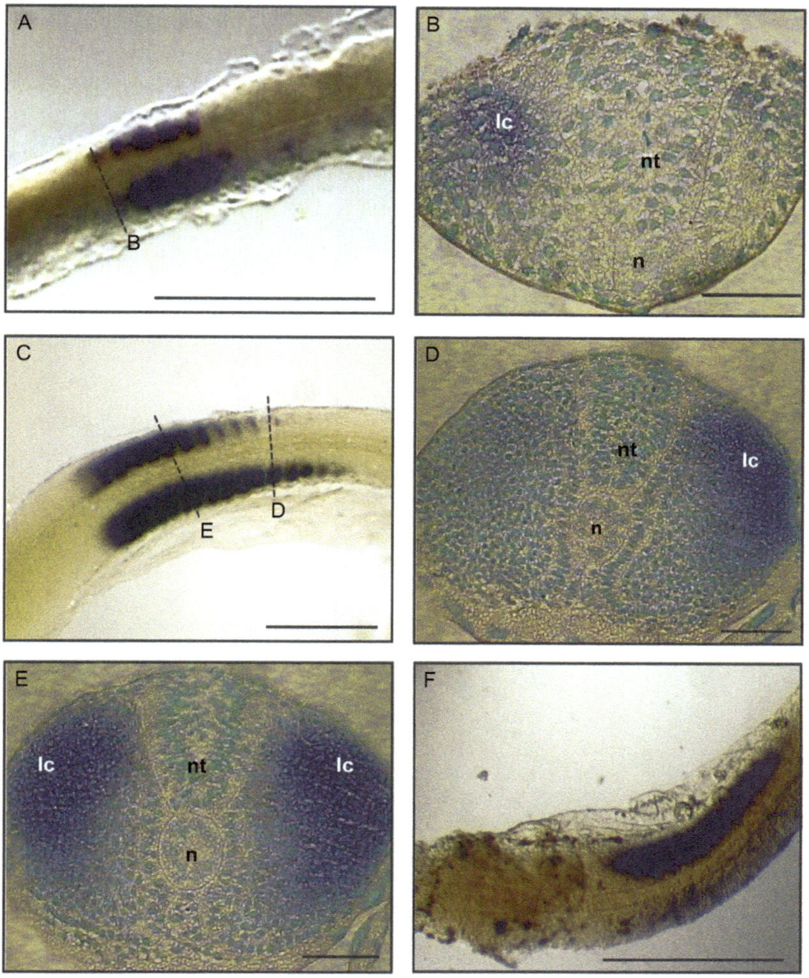

Alexander M. Schreiber, Figure 6.6 Left–right asymmetric expression of *myod2* in Atlantic halibut during somitogenesis analyzed by whole-mount in situ hybridization. (A) Asymmetric *myod2* signal shown as one additional positive somite at five somites (75% epiboly). Stippled line indicates site of transverse section. (B) Transverse section showing left-sided expression of *myod2* in lateral cells at five somites. (C) Right-sided asymmetric expression of *myod2* in embryo at 20 somites. Transverse sections shown in picture (D) and (E) are indicated by stippled lines. (D) Transverse section of the embryo in picture C demonstrating *myod2* expression in the lateral cells on the right side. (E) Transverse section of the embryo in picture C with stronger *myod2* signal on the right side. (F) Embryo at 45 somites with no *myod2* signal on the right side of the midline. All embryos are positioned with the head toward right and are seen dorsally. The sections are positioned with the dorsal side on the top of the picture. lc, lateral cells; n, notochord; nt, neural tube. Scale bars for the whole embryos and transverse sections equal 0.5 mm and 50 μm, respectively. *Reproduced with permission from Andersen et al. (2009).*

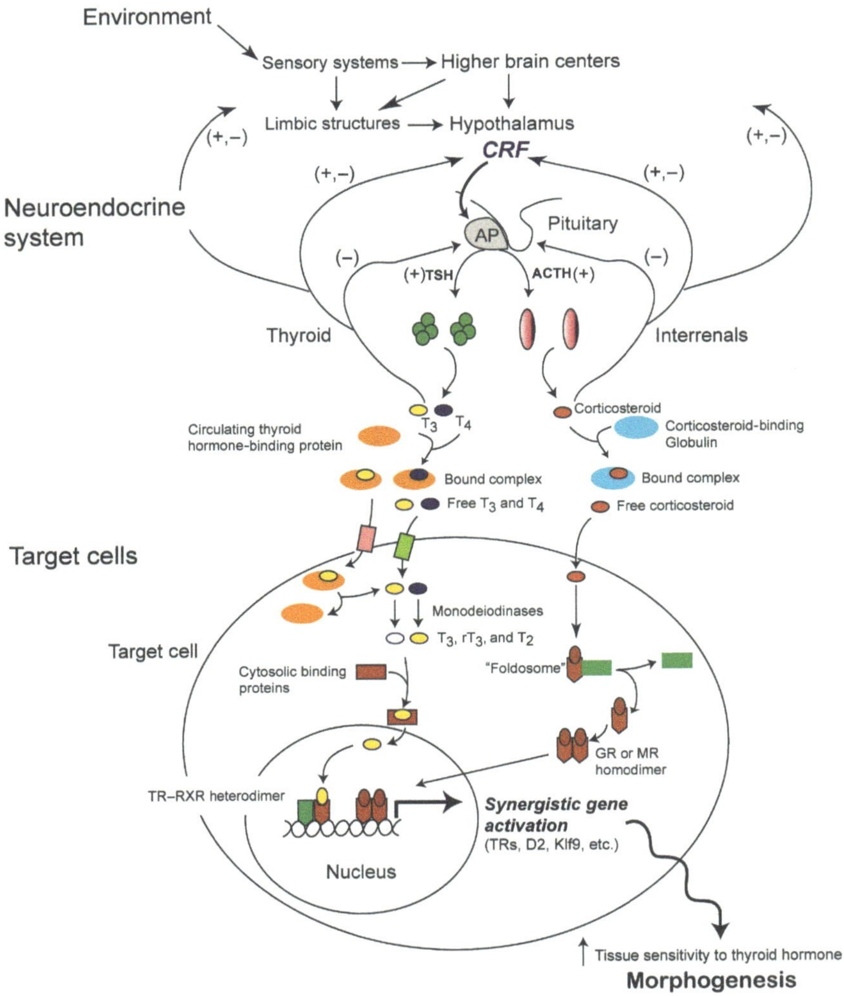

Robert J. Denver, Figure 7.1 Central and peripheral organization of the thyroid and stress endocrine axes controlling amphibian metamorphosis. A schematic representation of the hypothalamo–pituitary–thyroid (HPT) and hypothalamo–pituitary–adrenal (HPA; stress) axes in amphibian tadpoles, their regulation by input from the external environment, transduction of this input by neural and neuroendocrine pathways, and synergistic interactions among thyroid hormones and corticosteroids in target cells leading to the promotion of metamorphosis. The two endocrine axes are controlled centrally by corticotropin-releasing factor (CRF) which acts on the anterior pituitary gland (AP) to stimulate the release of thyrotropin (TSH) and corticotropin (ACTH). TSH acts on the thyroid gland to stimulate release of thyroxine (T_4) and 3,5,3′-triiodothyronine (T_3). Thyroid hormones are transported in the blood bound by serum binding proteins (transthyretin, TBG, and albumin). ACTH acts on adrenal cortical cells in the interrenal glands to stimulate

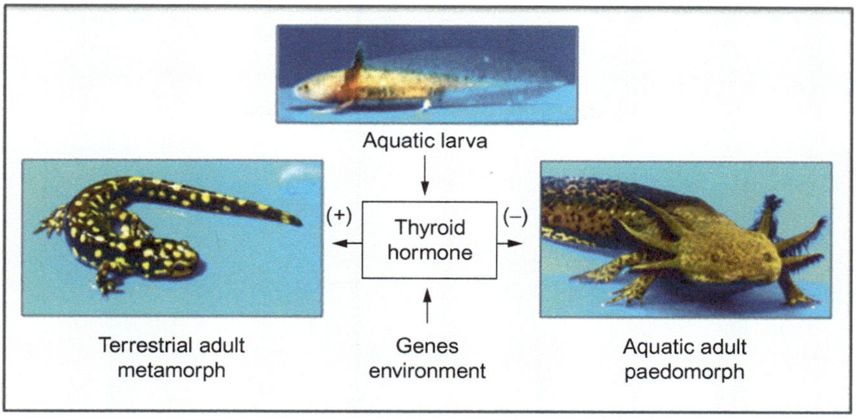

Carlena K. Johnson and S. Randal Voss, Figure 8.1 A representative salamander larva, adult metamorph (*Ambystoma tigrinum*), and adult paedomorph (*A. mexicanum*). Critical levels of thyroid hormone induce metamorphosis in salamanders. Environmental and genetic factors may cause thyroid hormone levels to be low, resulting in paedomorphic salamanders. *The pictures were taken by Jeramiah Smith.*

---

biosynthesis and release of glucocorticoids which are transported in the blood bound to corticosteroid-binding globulin. Cellular uptake of T_3 and T_4 is achieved by organic anion, monocarboxylate, and amino acid transporters; there is also evidence that thyroid hormones may enter cells bound to transthyretin via a receptor-mediated process. Glucocorticoids enter cells by passive diffusion across the plasma membrane. Upon entering the cell, thyroid hormone is bound by cytosolic-binding proteins, some of which (the monodeiodinases) convert the hormone to either active (T_3; deiodinases types I and II) or inactive forms (reverse T_3 [rT_3], diiodothyronine [T_2]; deiodinase type III). TRs form heterodimers with RXRs and are bound to DNA in the unliganded form where they actively repress gene transcription. Upon thyroid hormone binding to TR, gene transcription is derepressed and activated. Upon entering the cell, glucocorticoids bind to corticosteroid receptors (glucocorticoid receptor [GR] or mineralocorticoid receptor [MR]) that are located in the cytosol bound to heat shock proteins (the "foldosome"). Hormone binding causes a conformational change in the receptor, the release of heat shock proteins, and dimerization and translocation of receptors to the nucleus where they bind DNA to activate or repress target genes. When cells are exposed to low concentrations of thyroid hormone plus glucocorticoids, genes such as the TRs, deiodinase type 2, and the thyroid hormone-inducible transcription factor *Klf9* are activated in a synergistic manner. This leads to enhanced sensitivity of cells to the actions of thyroid hormone, which accelerates metamorphosis. "+" indicates an increase and "−" indicates a decrease in the regulated variable.

Carlena K. Johnson and S. Randal Voss, Figure 8.3 *Ambystoma dumerilii* is a paedomorph that has secondarily evolved webbed-feet and extra filaments that develop on the dorsal surfaces of each gill. The bottom panel is a magnified image of the dorsal gill surface. The arrow indicates dorsal filaments ($<$1.0 mm in length) that are sprouting on the gill surface. These extra fibers form late in the larval period, at the time metamorphosis normally occurs in related metamorphic species. *The* A. dumerilii *picture was taken by Brad Shaffer.*

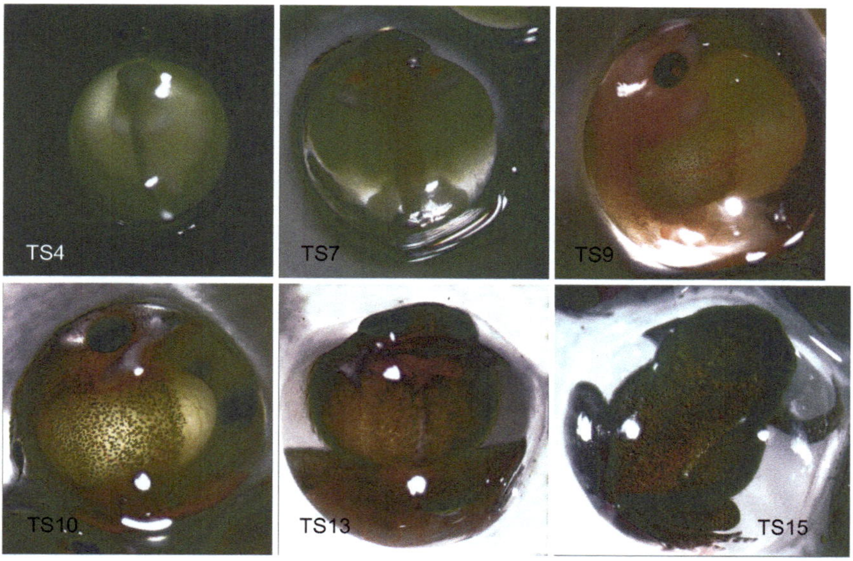

Richard P. Elinson, Figure 9.1 Direct development of *Eleutherodactylus coqui*. *E. coqui* develops on land in jelly capsules brooded by the father. These photographs, taken at the same magnification, show Townsend–Stewart (TS) stages of embryos within their capsules. Limb buds are apparent as early as TS4, about 4 days after fertilization. After 3 weeks, the TS15 froglet is ready to hatch.

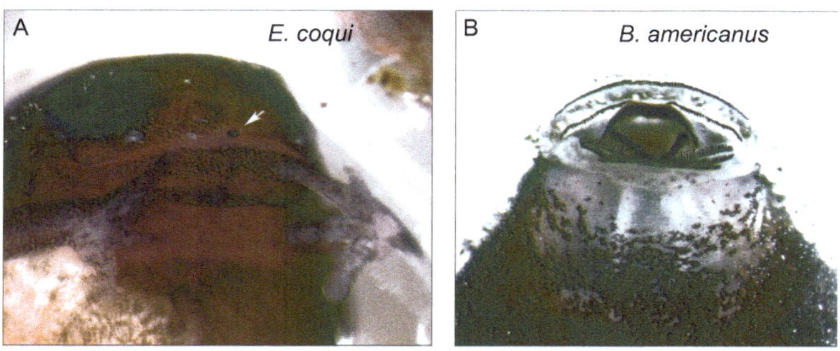

Richard P. Elinson, Figure 9.3 Comparison between jaws of an *E. coqui* TS13 embryo and a tadpole. (A) The *E. coqui* embryo has a froglike mouth, with a single keratinous egg tooth (arrow) in the middle of the upper jaw. The egg tooth is used to hatch from the jelly capsule. (B) A tadpole, here of the toad *Bufo americanus*, has rows of keratinous teeth and a keratinous beak. The teeth and beak are used for scraping surfaces to collect decayed plant material and other detritus. Note the very different shape of the tadpole jaw, compared to that of *E. coqui*. These photographs are at the same magnification.

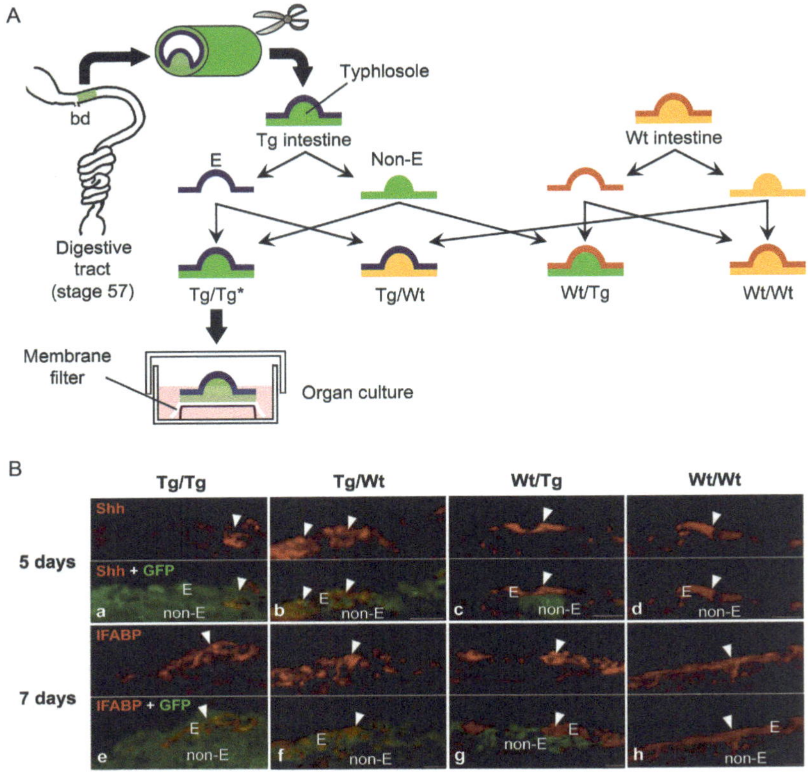

Atsuko Ishizuya-Oka and Takashi Hasebe, Figure 11.2 Adult stem cells originate from the larval epithelium in *X. laevis* intestines cultured *in vitro*. (A) Schematic diagram of tissue recombination and organ culture. Tubular fragments were isolated from the small intestine just behind the bile duct junction (bd) of stage 57-wild-type (Wt) or transgenic (Tg) tadpoles that express GFP ubiquitously under the CMV promoter. They were slit open lengthwise and separated into epithelial (E) or nonepithelial tissues (non-E). Each epithelium was then recombined with homologous or heterologous non-E. The four kinds of recombinant intestines were placed on membrane filters on grids and cultured in the medium containing thyroid hormone. * Type of E/type of non-E. (B) Recombinant intestines cultured for 5 (a–d) and 7 days (e–h). Sections were double-stained with anti-Shh (red) and anti-GFP (green) (a–d), or anti-IFABP (red) and anti-GFP (green) antibodies (e–h). Adult progenitor/stem cells positive for Shh (arrowheads) express GFP in Tg/Tg (a) and Tg/Wt intestines (b), but do not in Wt/Tg (c) and Wt/Wt intestines (d). Similarly, adult differentiated cells positive for IFABP (arrowheads) express GFP in Tg/Tg (e) and Tg/Wt intestines (f), but do not in Wt/Tg (g) and Wt/Wt intestines (h). Bars: 10 μm.

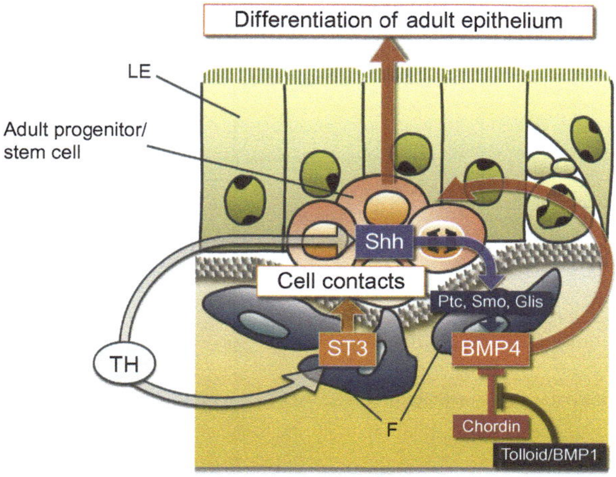

Atsuko Ishizuya-Oka and Takashi Hasebe, Figure 11.5 Thyroid hormone-induced signaling pathways leading to establishment of the stem cell niche in the amphibian intestine during metamorphosis. Thyroid hormone (TH) directly upregulates Shh expression specifically in adult progenitor/stem cells and ST3 expression in fibroblasts (F). Shh acts only on nonepithelial tissues in a paracrine manner and, together with TH, enhances the expression of other Shh pathway components including Ptc, Smo, and Glis. Shh not only activates proliferation of the fibroblasts but also upregulates their expression of BMP4, which in turn promotes differentiation of the adult epithelium and/or maintains the adult stem cells under the control of TH-upregulated expression of Tolloid/BMP1 through degrading Chordin. However, ST3 modifies the basal lamina and allows the fibroblast to make contact with the epithelial cells, leading to activation of juxtacrine signaling involved in adult stem cell development. In addition, these Shh/BMP4 and ST3 signaling pathways interact with other pathways such as Wnt one and are integrated to establish the stem cell niche. Arrows and T-shaped bars indicate activation and repression, respectively.

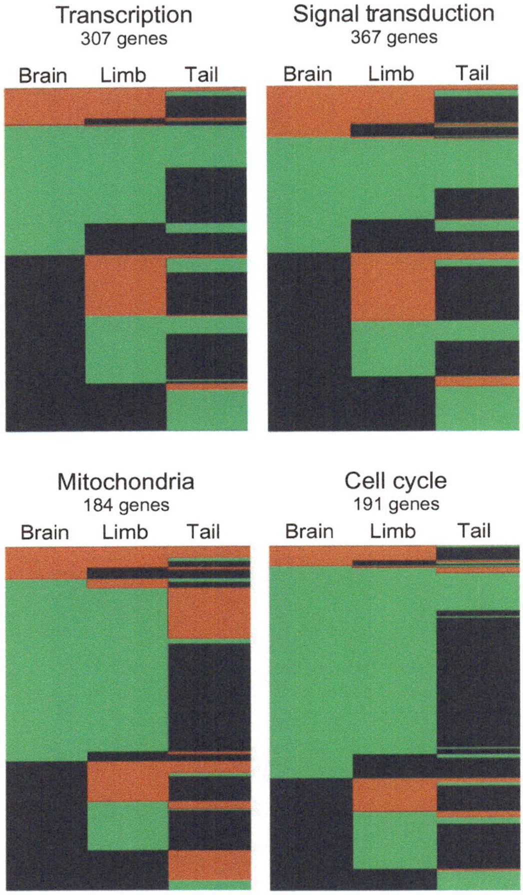

Saurabh S. Kulkarni and Daniel R. Buchholz, Figure 12.3 Comparison of altered gene expression among brain, limb, and tail. From Das et al. (2006), significantly regulated genes in brain, limb, or tail after 48 h of 100 nM T3 were grouped by gene ontology categories. Here, we chose to display genes from transcription, signal transduction, mitochondria, and cell cycle. Genes up- or downregulated or no change were assigned green, red, or black, respectively, then sorted by color starting with brain, then limb, and then tail. Among tissues, the gene regulation patterns for each GO category indicate that brain and limb are more similar to each other than to tail, matching their response to TH after 2 days, that is, proliferation versus apoptosis. Among GO categories, gene regulation patterns were similar between transcription and signal transduction and between mitochondria and cell cycle. For example, gene expression patterns among tissues are much more similar to each other in the transcription and signal transduction categories compared to the mitochondria and cell cycle categories.

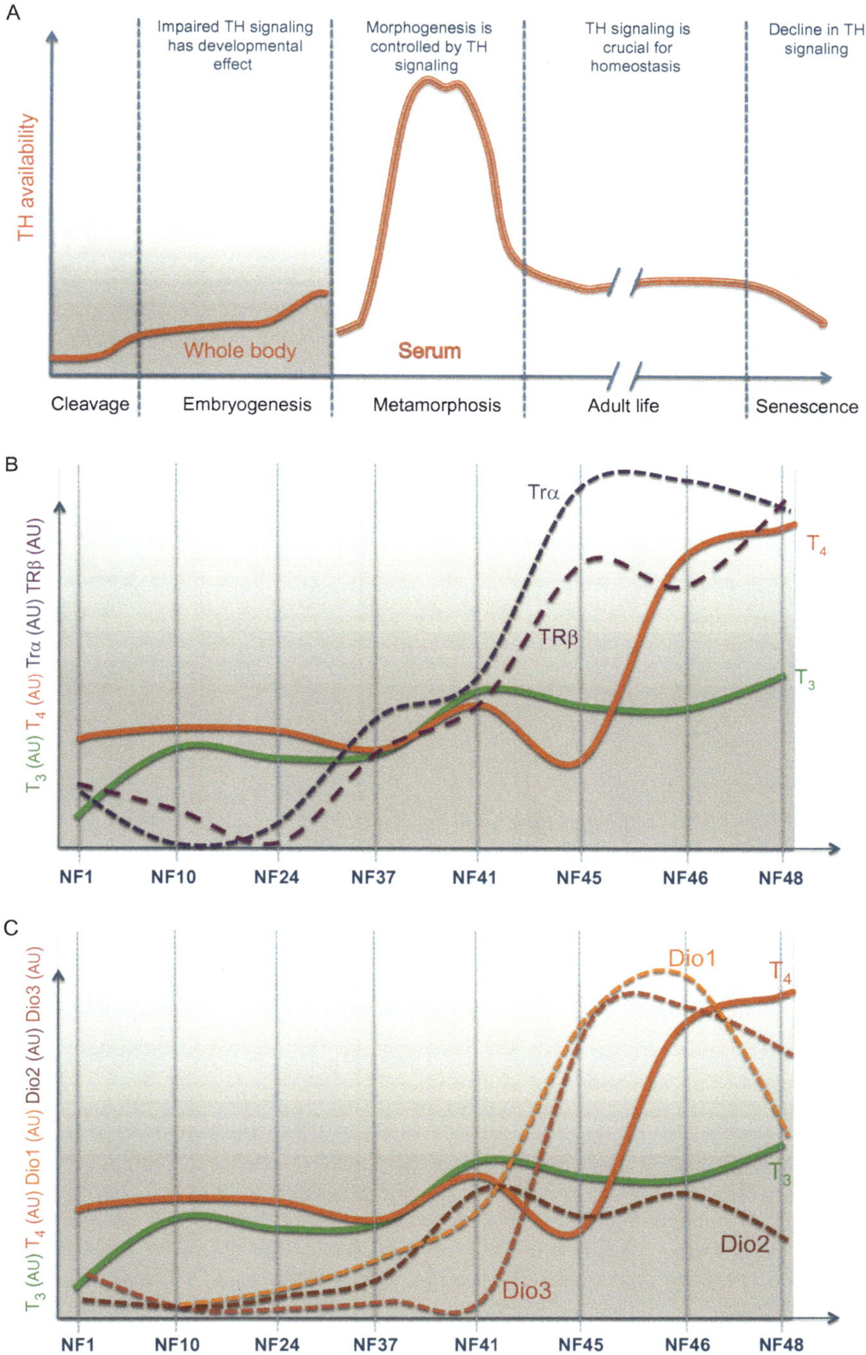

G. Morvan-Dubois *et al.*, Figure 13.1 The different components of TH signaling pathway are present during embryogenesis. (A) The general consensus on TH signaling is that T_3 is

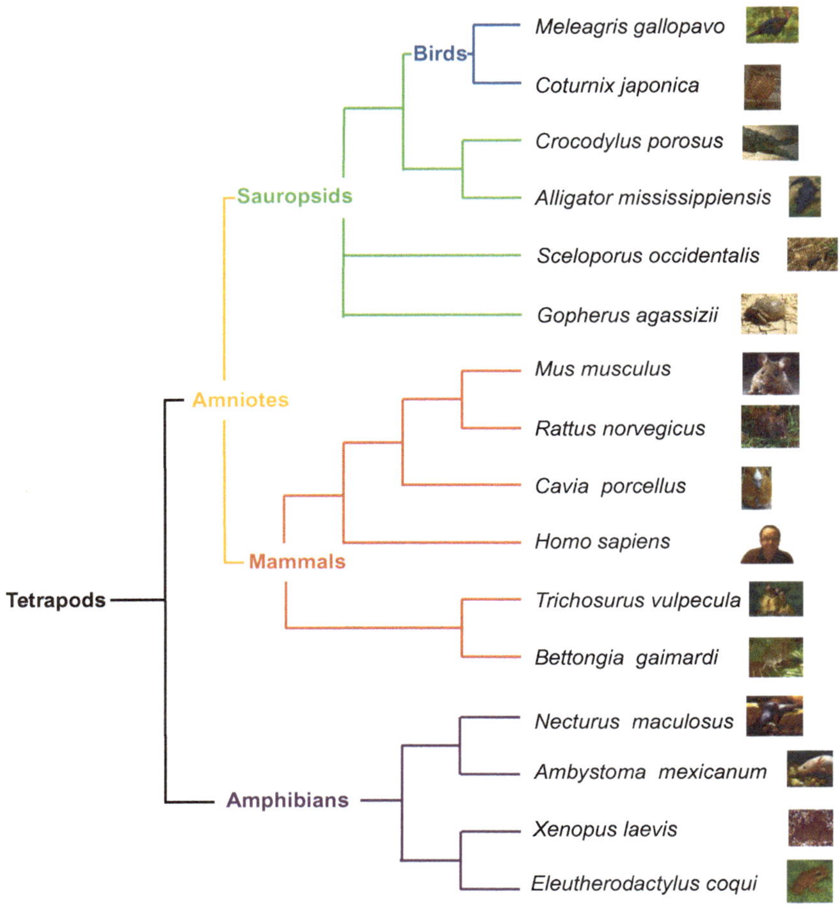

Guillaume Holzer and Vincent Laudet, Figure 14.3 Phylogenetic tree of the species described in this review. The several relevant phylum are highlighted in different colors. *Adapted from Werneburg and Sánchez-Villagra (2009) and Shen, Liang, Wen, and Zhang (2011).*

---

the key trigger and orchestrator of amphibian metamorphosis. Nevertheless, it is also crucial for adult physiology. TH role during embryogenesis is less defined. Whereas seric concentrations are well known in metamorphic and postmetamorphic tadpoles, they are irrelevant in embryos. We therefore reported whole body data. (B) The analysis of T_3 and T_4 availability in regard with TRs expression shows that when the receptors are expressed during embryogenesis, T_3 is present in (some) tissues. T_3 is therefore likely to bind TRs locally. (C) Deiodinases are expressed in some tissues from the early development. They can be responsible for the slight T_3 increase and T_4 decrease during the first phases of embryogenesis (up to NF45). The increase in T_4 content between NF45 and NF46 indicates *de novo* TH synthesis. *Schematic view, adapted from Fini, Le Mével, et al. (2012).*

Kolibri 360 GmbH
Ferdinand-Jühlke-Straße 7
99095 Erfurt, Deutschland
produktsicherheit@kolibri360.de